Handbook of diet, nutrition and the skin

Handbook of
diet, nutrition and the skin

edited by:
Victor R. Preedy

Human Health Handbooks no. 2

ISSN 2212-375X

Wageningen Academic
Publishers

ISBN: 978-90-8686-175-0
e-ISBN: 978-90-8686-729-5
DOI: 10.3920/978-90-8686-729-5

ISSN 2212-375X

First published, 2012

Wageningen Academic Publishers
P.O. Box 220
6700 AE Wageningen
The Netherlands
www.WageningenAcademic.com
copyright@WageningenAcademic.com

This book is dedicated to my daughter Hannah

Table of contents

Introduction

General aspects of skin, nutrition and diet

Micronutrients

Nutraceuticals and skin

Skin cancer, nutrition and diet

Specific skin conditions in relation to diet and nutrition

Introduction

Summary points

- The skin, the largest organ of the body, is a complex system that co-ordinates barrier function, cutaneous immunology and neurosensory interactions with the environment.
- The outer layer of the skin is the epidermis which creates a barrier through keratinocyte layers, organizes immune recognition using Langerhans cells and binds to the deeper skin via a complex basement membrane zone.
- The dermis contains fibroblasts which provide skin turgor through the production of the extracellular matrix of collagen and elastin, neurosensory and vascular structures as well as a variety of immune cells including antigen presenting dermal dendrocytes and circulating skin lymphocytes.
- The homeostasis of the skin is maintained through the functions of the skin layers as well as through temperature regulation and colonization by commensal organisms.

1. The skin: an introduction

N.B. Silverberg
Departments of Dermatology, St. Luke's-Roosevelt and Beth Israel Medical Centers, 1090 Amsterdam Avenue, Suite 11D, New York, NY 10025, USA; nsilverb@chpnet.org

1.1 Introduction

The skin is the largest organ of the body and represents a complex barrier structure composed of surface keratinocytes, inter-keratinocyte substances, dermis and sub-dermal structures. The cutaneous structures serve a variety functions that preserve homeostasis of the body, including cooling and electrolyte balance, formation and processing of vitamins and hormones, and physical and thermal protection of underlying muscle, bone and internal organs (Pincelli, 2010).

The skin has a complex superficial arrangement that creates a barrier through retention of anucleate corneocytes at the surface, production of intercellular substances, colonization with commensal organisms, intercellular communication and adhesion molecules.

Last, but not least, the skin represents a tremendous interface with the environment, processing both allergens, irritants and pathogens through a broad-based immunological system consisting of exclusively cutaneous immune cells, circulating immune cells and intrinsic production of anti-pathogenic substances. The following chapter represents a brief overview of the manifold cutaneous structures and their functions to provide a foundation for greater understanding of obligatory nutrients and how nutrition influences the function of the complex skin on a daily basis and in specific disease states (Pincelli, 2010).

1.2 Structure and ultrastructure of the skin

The skin is composed roughly of three layers: the epidermis, an exterior of layered keratinocytes, the orifices of appendages and intercellular substances, the dermis composed primarily of collagen and elastin and acting as the scaffold for the epidermis, the conduit for the middle portion of the appendages and the final destination of branched vascular structures, and the hypodermis a complex of fat overlying fascia and home to larger blood vessels, and the site of vascular interaction with the hair follicle and the sweat glands.

1.2.1 Epidermis and the skin barrier

The epidermis is a constantly renewing structure that regenerates from epidermal stem cells. Regeneration maintains the skin homeostasis through normal exfoliation and in response to a multitude of injuries (Pincelli, 2010). Keratinocytes are the main cellular component of the epidermis, followed by cells which have pod-like extensions or dendritic cells which intercalate between the keratinocytes, i.e. melanocytes and Langerhans' cells, which will be addressed later in this chapter.

The interfollicular keratinocyte stem cells begin at the basal layer (Pincelli, 2010), some of which are resting, others active progenitors of the upper layers of epidermis. Moving from the base upward to the skin surface the layers are termed the basal layer, the spinous cell layer (*stratum spinosum*), the granulocyte layer (*stratum granulosum*) and the corneocytes of the *stratum corneum*. The basal cell layer has epidermal stem cells most of which are resting. When in the active division phase, creating 2 cells, one that rises through the epidermis and a transit amplifying cell that continues to divide producing the basal cell layer (Pincelli, 2010). Clonogenic capacity correlates to β1 integrin levels in the cell. EGF receptor antagonist Lrig1 acts as a marker for resting epidermal stem cells (Pincelli, 2010). This allows for homeostasis of the skin by allowing stem cells to rest and allowing a limitation on epidermal growth. This also allows for cells with DNA damage to come out of resting phase and produce pre-cancerous and cancerous lesions periodically.

Keratins

Keratins, of which there are 54, are intermediate filaments that are present in the epithelia of the body, including the skin, hair, corneal epithelium, mucosa, gastrointestinal tract, breast and lungs. Keratins in the skin serve to maintain the cytoskeletal structure, providing mechanical stability to the cells and stability of intercellular adhesion and the basement membrane zone where they insert (Moll, 2008). There are 2 types of epithelial (non-hair) keratins, type I (Chromosome 17q21.2; 1-8, 71-80) and type II (Chromosome 12q13; 9-28), a filament of each one together complex into a heteropolymer with central α-helical domain (Omary, 2009). Specific keratin pairs are seen in the epidermis, e.g. K5/K14 pairs seen in the basal layer, while K1/K10 are seen in the suprabasal layers (3). Many genodermatoses have been reported associated with keratin anomalies including epidermolytic hyperkeratosis (K1/K10), Dowling- Meara type epidermolysis bullosa (K5/K14) and Ichthyosis Bullosa of Siemens (K2). Keratin production can be influenced by hormones and vitamins including glucocorticoids, retinoic acid, vitamin D and thyroid hormone (Ramot, 2009).

Desmosomes

As the keratinocytes differentiate (the process called keratinization), they rise through the layers forming desmosomes between the cells to maintain cellular adhesion (Sandjeu, 2009). The desmosomes are composed of proteins of the cadherin i.e. calcium dependent adhesion proteins (e.g. desmogleins, desmocollins), Armadillo proteins bind cadherins and keratins (e.g.

plakoglobin, plakophilin) and plakin- plaque complex and anchoring filaments (e.g. desmoplakin, peri-plakin and plectin) families (Sandjeu, 2009). A newer intercellular inhabitant that has been identified is desmosealin, a proteoglycan, which increases in expression from the basal to the corneal layers.

Of the cadherins, desmogleins 1 and 4 are expressed in the *stratum corneum*, allowing for premature loss of *stratum corneum* with anti-desmoglein 1 antibodies in pemphigus foliaceous, while desmoglein 3 is the target of antibodies in the statum spinosum in pemphigus, allowing for deeper blisters. Phenotypes of abnormalities in the Armadillo proteins include palmoplantar keratoderma and cardiomyopathy in Naxos disease (plakoglobin) (Protonotarios, 2006) vs. skin fragility and ectodermal dysplasia with plakophilin1 abnormalities.

The extracellular matrix of the *stratum granulosum* and *stratum corneum* contains lamellar layers of lipid released by keratinosomes (lamellar bodies) that constitute the skin barrier, i.e. ceramides, cholesterol and free fatty acids in equimolar portions which present a hydrophobic barrier. Within these layers of lipid are enzymes that eventually degrade the intercellular adhesion molecules allowing for desquamation (Sandjeu, 2009). In the movement from the spinous to the corneocyte stage/layer keratinocytes undergo apoptosis with gradual reduction of the intercellular adhesion, such that the surface can shed, thereby maintaining homeostasis in terms of thickness of the barrier. The cells lose their nuclei and keratins stop forming a broad scaffold, becoming more granular in nature and eventually shedding as surface keratinocytes. Reduction in shedding results in hyperkeratosis as seen due to intrinsic abnormal production of proteins (e.g. Ichthyosis Vulgaris) and due to inflammatory and immune causes (e.g. Netherton's Disease).

Basement membrane zone

Basal keratinocytes produce keratin, which attaches to the basal attachment plaques, providing enhanced adhesion, strength and scaffolding of the keratinocytes. The basal keratinocytes attach to the dermis continuously through the basement membrane, a complex series of intracellular and/or extracellular proteins that form a puzzle-like scaffolding. The basal cells are also nourished through a variety of extracellular matrix proteins and growth factors (Pincelli, 2010). One of the proteins binding the basement membrane zone is the α6β4 integrin. These bindings both maintain skin integrity and enhance nourishment of the keratinocytes.

The basement membrane consists of the lower border of the keratinocytes and the extracellular matrix at the base of the keratinocytes, i.e. the dermo-epidermal junction. This region consists of a complex three-dimensional structure composed of collagens IV and XVII, laminins, nidogen, fibronectin and proteoglycans.

An extracellular matrix can be found in the epidermis and dermis. In the epidermis, the extracellular matrix becomes distinctively important in the *stratum corneum*, because the cells are actually no longer living, they do not maintain the normal intercellular adhesion seen in the basal and spinous layers.

1.2.2 Dermis

The dermis consists of a fribroblast-produced extracellular matrix made up of fibers of collagen and elastin, glycoproteins such as fibronectin and laminin and the hydrated polysaccharides-glycosaminoglycans (GAG) and proteoglycans, which is GAG attached to a protein core. GAGS are involved in many activities in the skin, including hydration and ionic filtration. GAGs include hyaluronin (the most abundant), heparan sulfate, keratin sulfate, and chondroitin/ dermatan sulfate (Sandjeu, 2009). These substances provide a firm scaffold for nutrients, to absorb mechanical stress and to maintain vascular channels flowing through the skin. The dermis provides the turgor of the skin. Elastic tissue production is generally in an orderly pattern which becomes lost with breakdown from ultraviolet damage or the abnormal fibrillar production of elastin that is seen in some forms of cutis laxa (Sandjeu, 2009).

1.2.3 Hypodermis – fat as a bioactive substance

Fat, the portion of the hypodermis composed of adipocytes, is a collection of cells that surround the vascular structures and provide a conduit between the dermis and the systemic vasculature. Abdominal fat can produce TNF-α and other substances that feed (no pun intended) into the cycle of the metabolic syndrome. Unlike abdominal or visceral fat, subcutaneous fat, especially in the peripheral extremities, does not participate in the metabolic syndrome. Inflammation of the fat is seen in a variety of conditions and may relate to the larger size of the vessels seen in the fat (medium-sized) as opposed to the smaller vessels of the dermis (Hamdy, 2006; Polcari, 2010).

1.2.4 Appendages of the skin

Hair

The hair follicle is generated by the hair follicle stem cells (Pincelli, 2010), located in the bulge region of the follicle, which is along the shaft, just below the sebaceous gland attachment and near the attachment of the arrector pili muscle. These stem cells are characterized by CD 34 expression and K15 promoter expression (Jaks, 2010).

The hair follicle is nourished through a capillary tuft located within the fat at the base. This region is termed the dermal papilla (Pincelli, 2010). Stat-3 is integral to the maintenance of hair stem cells, preventing these lesions from undergoing apoptosis (Pincelli, 2010). Hairs undergo three phases of growth- anagen active phase, catagen intermediate phase and telogen resting phase.

In the hair, keratins constitute a major component of the core of the hair shaft. Hair keratins are distinct from skin keratins (Moll, 2008).The hair keratins are high in sulfur in the non-helical regions allowing for cross-linkage.

Sweat glands

Sebaceous glands

The sebaceous gland attaches to the hair follicles of the face, upper chest, back, cornea, areola and groin region. The sebaceous gland is generated by the sebaceous stem cells that lie above the bulge of the hair follicle and below the orifice of the gland into the hair shaft. These stem cells form sebocytes, which secrete sebum through holocrine degradation. Sebaceous glands not associated with the hair follicle have other functions including pheromone production. Sebum is under the influence of androgens and is overproduced in acne (Schneider, 2010).

Eccrine glands

Eccrine sweat glands are responsible for thermal regulation through the cooling effect of sweat production and via the control of electrolyte secretion through the skin. The presence of eccrine sweat glands helps thermoregulate the human body in normal states, stress states (e.g. exercise) and in disease states (e.g. fever) (Shibasaki, 2010).

Apocrine glands

Apocrine glands release sweat via apocrine production, i.e. release of part of the cytoplasm. This is noted in the axillae, palms, soles, groin and the inframammary locations and appears to be under autonomic control, allowing for anxiety based excess palmoplantar sweating in some individuals. Hidradenitis suppurativa is an inflammatory disorder of the apocrine sweat glands which is part of the follicular occlusion triad (Gesase, 2003).

1.2.5 Vascular components of the skin

The vasculature of the skin consists of small vessels and capillary loops in the dermal papillae, and in the distal locations e.g. fingertips, ears, and the larger vessels of the deeper dermis and hypodermis. The vasculature of the skin oxygenates the tissue and is under the control of three systems: adrenergic stimulation which enhances skin blood flow (beta blockers doing the exact opposite, e.g. Propranolol), neuroendocrine control, reflex (e.g. baroreflex) and the nitric oxide system (Hodges, 2009; Johnson, 2010). The entire mechanism by which each of these systems works is highly orchestrated and may be generalized or site specific (e.g. facial flushing). Abnormal vasculature at birth can result in a variety of birthmarks including port wine stains (capillary malformations) which have little dynamic alterations but can be associated with venous anomalies (Enjolras, 2004) and relate to vessel size or a nevus anemicus which relates to increased local vascular tone and can be overcome with adrenergic blockade (Akhami, 1999).

1.2.6 Pigment cells (melanocytes)

Melanocytes are the dendritic pigment cells of the skin that lie in the basal layer of the epidermis after they migrate into the skin from the neural crest (Ernfors, 2010). Migration anomalies are common in individuals of color (e.g. Mongolian spots) and usually resolve once dermal melanocytes make their way into the skin. Once migration is complete in utero and in the first years of life (Silverberg, 2010) pigment production can take place in melanosomes, intracellular organelles. Transfer of melanosomes to the surrounding keratinocytes results in pigmentation of the keratinocytes and the hairs (Schiaffino, 2010). Melanin production occurs with a series of enzymatic alterations starting with L-tyrosine and enzymatic conversion by tyrosinase, a copper-dependent enzyme (Kosmadaki, 2010; Olivares, 2009). Melanin is packaged in melanosomes, specialized organelles of pigmentation that are race and skin tone specific in terms of shape and distribution or density of pigment. Melanin confers the emotionally important pigmentation to the skin and is a UV absorber, providing absorption of mutagenic rays. Melanoma (cancerous, genetically altered melanocyte cells) is a highly lethal skin cancer, therefore pigment cells represent one of the most important areas of skin research in dermatology.

1.2.7 Nervous innervations of the skin

Nervous innervations of the skin is part of the somatosensory system (sensations of the skin, muscle and joints), composed of sympathetic, parasympathetic and sensory innervations. Nervous structure of the skin consists of different sized fibers and sensory endings that allow for greater complexity of sensation with the digits and broader sensation over the proximal skin surfaces. Sensory innervation provides response to touch, pressure, heat, cold and pain (Brodal, 2004). Nerves also control vascular activity in the skin. Sensations such as itching stinging and burning may result from multiple nerve fibers being innervated simultaneously, hence the individual differences in perception and response.

Derived from neural crest, nerves are developed in utero; however, sensory nerve endings may progressively become more active with pediatric development and early brain development. The neuronal cells of the skin, like the brain, require fats for development and can be impaired in the setting of certain genetic diseases e.g. Riley Day syndrome (familial dysautonomia) (Rubin, 2008). Nerves are long cells with dendritic ends that transmit or respond to release of neuroendocrine substances including neuropeptides (e.g. the calcitonin gene-related peptide, pituitary adenylate cyclase-activating peptide, substance P, vasoactive intestinal peptide, and norepinephrine), neurotrophins, corticotropin-releasing hormone and α-melanocyte-stimulating hormone (Seiffert, 2006; Harvima, 2010), thereby communicating a variety of different sensations. Sensation can be influenced by inflammatory cells (e.g. mast cells) (Seiffert, 2006) and can influence antigen presentation (e.g. Langerhans' cell activity) (Harvima, 2010). Sensation in the skin is very complex and this represents a broad overview of the structure.

Nerve fibers

The nervous system of the skin is a portion of the peripheral nervous system that is a three dimensional scaffold of nerves networking throughout all layers of the skin including the epidermis, which derives from dorsal root ganglia (Vega, 2009). The nerves develop into somatic and autonomic nerves. Each set of nerves are fibrous endings of spinal nerve roots and are organized by dermatomes, a genetically distinct unit of nerves that derive from the same dorsal root ganglia and which are labeled or named based on the nerve root from which they derive.

On the face, these nerves derive from trigeminal roots, V1-3, while on the body from nerves originating in the spinal cord, e.g. L1 for lumbar root one. Some of the autonomic fibers are integrally associated with the vascular channels of the skin or attach to the arrector pili muscles of the forearms, while somatic nerve fibers weave through an independent path.

The role of nervous innervations of the skin has not been fully defined as of yet. Nerves are a two way street in the skin: they can sense a tactile, thermal and painful sensations and can promote changes in blood flow, fluid movement and localized inflammation, partially through a close physical association with mast cells. Mast cells (and other inflammatory cells) can also stimulate pain and inflammation through the release of histamine, tryptase and nerve growth factor which can stimulate C-fibers, the pain fibers of the skin (Seiffert, 2006).

Nerve fibers in the skin are categorized roughly by their thickness and the role they play in skin innervation. Some nerves end in free endings which extend into the upper dermis and lower epidermis and others are integrated with sensory receptors of the dermis. Free nerve endings mediate nociception. i.e. pain reception. Nociception can be high-threshold from intense physical injury e.g. pinching or can be polymodal and triggered by skin damage such as intense thermal stimuli. Finally there are silent nociceptors which respond to inflammatory substance release from chronic inflammation. Inflammatory mediators such as histamine, substance P and bradykinin can stimulate pain perception by nociceptors. Excessive sensitivity to pain exists and can be termed hyperalgesia. (Brodal, 2004)

C fibers are unmyelinated traditional pain fibers, but can also aid in sensation of extreme cold or heat. Myelinated Aδ fibers sense moderate cold temperatures, while Aα and Aβ fibers attach to encapsulated sensory nerve endings (Brodal, 2004).

Receptors of sensation

There are three types of encapsulated skin receptors: mechanoreceptors, thermoreceptors and chemorecepotors. Sensory nerve endings are especially abundant on the fingertips and palmar hands, accounting for superior sensation in these regions.

Meissner and Pacinian corpuscles are special sensory organs composed of a central end-tip neuronal cell surrounded by peri-axonic cells contiguous with the nerve cell including Schwann cells and endoneurial-perineurial related cells. Pacinian corpuscles are encapsulated mechanoreceptors and Meissner corpuscles are non-encapsulated and associate with Aα and Aβ nerve fibers on intermediate sized sensory nerves (Legat, 2009). Afferent polyinnervation may be noted in these receptors, e.g. C fibers around the tips of Meissner's corpuscles which may mediate nociception (27). However, Meissner's corpuscles are traditionally rapidly adapting receptors of touch in superficial dermis, while Ruffini's corpuscles are deeper receptors mediating sense of stretch and deep movement of the dermal tissues. Other slow adapting receptors include Merkel's receptors. Pacinian corpuscles are low threshold mechanoreceptors that mediate specific localized pressure/touch perception (Brodal, 2004).

1.2.8 Immunologic cells of the skin

The skin is the largest interface of the body with the world and therefore requires immune cells to maintain homeostasis. The cutaneous immune system consists of circulating immune cells that come to the skin when specific factors cause margination in the blood vessels, diapedesis through the vessels and tissue homing/movement, e.g. the movements of neutrophils into the skin in acne. These cells come to serve a function based on release of homing substances and local need. Some monocytic stem cells differentiate into dendritic cells of the skin.

There is a cutaneous arm of the immune system where the cells live in the epidermis, dermis and subcutaneous fat. The former 2 cell types include Langerhans' cells and dermal dendritic cells which act as antigen presenting cells and are the first line of defense against cutaneous insults. These cells play an important function in atopy and allergic contact dermatitis as well through aberrant function.

Langerhans cells

Langerhans cells are dendritic antigen presenting cells of hematopoietic origin present normally in the epidermis and hair follicle (Merad, 2008). Although sparse in number, their dendritic prominences weave through the epidermis in a broad, continuous network (28). When allergens or pathogens appear at the epidermis, the Langerhans cells recognize these haptens and travel to the lymph nodes for antigen presentation (Tuchinda, 2010). Langerhans cells have specific cellular markers including CD 45, MHC Class II molecules, CD1a, S100 and langerin, which help to identify this subset of cells. On electron microscopy, the identification of Birbeck granules is distinctive to Langerhans cells, and is felt to be the organelle where antigens through which antigens are routed. Langerin may play a role in this routing process. Langerin has been found *in vitro* to have an important ability to prevent HIV1 uptake into T cells, but may not induce CD8+ T cells against certain cytolytic viruses, e.g. HSV1 and Vaccinia (Merad, 2008). The human keratinocytes may cause alteration in Langerhans cell function through expression of certain cellular markers e.g. RANKL during inflammation (Tuchinda, 2010).

Dermal dendrocytes

Dermal dendrocytes are a broad group of antigen presenting cells located in the dermis and serving specific function in exposures to pathogens and in mounting response to vaccination. A large proportion of these are langerin positive cells (Merad, 2008).

Skin associated lymphoid tissue

Skin associated lymphoid tissue is a combination of lymphocytes that come to the dermis for pathogen removal and lymphatic drainage of the skin. This system is of crucial importance in the metastatic spread of melanoma.

1.3 Function of the skin as an organ

1.3.1 Flora of the skin

The skin of a newborn is sterile, but over the first years of life, normal bacterial and fungal constituents of the skin and the hair follicle colonize the skin, forming the microflora, a barrier against pathogens. Particularly important in this process are the bacteria and yeast of the superficial hair follicle, namely the gram positive *Staphylococcus epidermidis* (Cogen, 2008) and *Malassezia* species. *Staphylococcus epidermidis* is a gram positive coccus found in cluster, representing 90% of the skin flora. In immunocompromised individuals this pathogen can be introduced systemically and can be virulent due to biofilm formation. In infancy colonization with the latter may initiate inflammation and cause seborrheic dermatitis or infantile acne.

The *Propionibacterium acnes* an aero-tolerant gram positive bacillus colonizes the mid-portion of the hair follicle and the sebaceous glands, therefore the organism does not begin to play a role in microflora until puberty . In puberty overgrowth of this organism is part of the root cause of acne vulgaris.

Other organisms live in the hair follicle and sebaceous gland including *Demodex folliculorum* and *Demodex brevis*, that can overgrow and cause a rosacea-like dermatitis in immunocompromised individuals (Zhao, 2010).

1.3.2 Innate immune protection and regulation

Innate immune protection includes other specific defenses, including the acidic pH of the skin, i.e. the acid mantle that protects against pathogenic colonization, the production of antimicrobial peptides, lysozymes, proteases, cytokines and chemokines that act as substances of innate immunity reducing risk of infectious generalization (e.g. eczema herpeticum) and attracting immune response when required (Cogen, 2008). Particularly, the acid mantle has a protective role against biofilm formation by bacteria. In the skin, *Staphylococcus epidermidis* is known to produce certain antimicrobial substances including bacteriocins (Cogen, 2008).

1.4 Conclusions

The skin has a broad complexity that ranges from innate host factors to interactions with commensal organisms. The number of elements that are required for healthy skin and normal function are so great that imbalance is a commonplace event, resulting in skin disease states.

References

Ahkami, R.N. and Schwartz, R.A., 1999. Nevus anemicus. Dermatology 198, 327-329.

Brodal, P., 2004. The Central Nervous Systen Structure and Function, Third Ed. Oxford University Press, New York, NY, USA, 138-142 pp.

Cogen, A.L., Nizet, V. and Gallo, R.L., 2008. Skin microbiota: a source of disease or defence? British Journal of Dermatology 158, 442-455.

Enjolras, O., Chapot, R. and Merland, J.J., 2004. Vascular anomalies and the growth of limbs: a review. Journal of Pediatric Orthopaedics. Part B 13, 349-357.

Ernfors, P., 2010. Cellular origin and developmental mechanisms during the formation of skin melanocytes. Experimental Cell Research 316, 1397-407.

Gesase, A.P. and Satoh, Y., 2003. Apocrine secretory mechanism: recent findings and unresolved problems. Histology and Histopathology 18, 597-608.

Hamdy, O., Porramatikul, S. and Al-Ozairi, E., 2006. Metabolic obesity: the paradox between visceral and subcutaneous fat. Current Diabetes Review 2, 367-373.

Harvima, I.T., Nilsson, G. and Naukkarinen, A., 2010. Role of mast cells and sensory nerves in skin inflammation. Giornale Italiano di Dermatologia e Venereologia 145, 195-204.

Hodges, G.J. and Johnson, J.M., 2009. Adrenergic control of the human cutaneous circulation. Applied Physiology, Nutrition, and Metabolism 34, 829-839.

Jaks, V., Kasper, M. and Toftgård, R., 2010. The hair follicle-a stem cell zoo. Experimental Cell Research 316, 1422-1428.

Johnson, J.M. and Kellogg Jr., D.L., 2010. Thermoregulatory and thermal control in the human cutaneous circulation. Frontiers in Bioscience (Scholars Edition) 2, 825-853.

Kosmadaki, M.G., Naif, A. and Hee-Young, P., 2010. Recent progresses in understanding pigmentation. Giornale Italiano di Dermatologia e Venereologia 145, 47-55.

Legat, F.J. and Wolf, P., 2009. Cutaneous sensory nerves: mediators of phototherapeutic effects? Frontiers in Bioscience 14, 4921-4931.

Merad, M., Ginhoux, F. and Collin, M., 2008. Origin, homeostasis and function of Langerhans cells and other langerin-expressing dendritic cells. Nature Reviews Immunology 8, 935-947.

Moll, R., Divo, M. and Langbein, L., 2008. The human keratins: biology and pathology. Histochemistry and Cell Biology 129, 705-733.

Olivares, C. and Solano, F., 2009. New insights into the active site structure and catalytic mechanism of tyrosinase and its related proteins. Pigment Cell & Melanoma Research 22, 750-760.

Omary, M.B., Ku N.O., Strnad P. and Hanada, S., 2009. Toward unraveling the complexity of simple epithelial keratins in human disease. Journal of Clinical Investigation 119, 1794-1805.

Pincelli, C. and Marconi, A., 2010. Keratinocyte stem cells: friends and foes. Journal of Cell Physiology 225, 310-315.

Polcari, I.C., and Stein, S.L. 2010. Panniculitis in childhood. Dermatologic Therapy 23, 356-367.

Protonotarios, N. and Tsatsopoulou, A., 2006. Naxos disease: cardiocutaneous syndrome due to cell adhesion defect. Orphanet Journal of Rare Diseases 13, 4.

Ramot, Y., Paus, R., Tiede, S. and Zlotogorski, A., 2009. Endocrine controls of keratin expression. Bioessays 31, 389-399.

Rubin, B.Y. and Anderson, S.L., 2008. The molecular basis of familial dysautonomia: overview, new discoveries and implications for directed therapies. NeuroMolecular Medicine 10, 148-156.

Sandjeu, Y. and Haftek, M., 2009. Desmosealin and other components of the epidermal extracellular matrix. Journal of Physiology and Pharmacology 60S4, 23-30.

Schiaffino, M.V., 2010. Signaling pathways in melanosome biogenesis and pathology. The International Journal of Biochemistry and Cell Biology 42, 1094-1104.

Schneider, M.R. and Paus, R., 2010. Sebocytes, multifaceted epithelial cells: lipid production and holocrine secretion. The International Journal of Biochemistry and Cell Biology 42, 181-185.

Seiffert, K. and Granstein, R.D., 2006. Neuroendocrine regulation of skin dendritic cells. Annals of the New York Academy of Sciences 1088, 195-206.

Shibasaki, M. and Crandall, C.G., 2010. Mechanisms and controllers of eccrine sweating in humans. Frontiers in Bioscience (Scholars Edition) 2, 685-696.

Silverberg, N.B., 2009. Pediatric Dermatology in Children of Color. Available at http://www.accessdermatology.com/pediatric-dermatology-in-children-of-color.aspx. Accessed 31 December 2010.

Tuchinda, P. and Gaspari, A.A., 2010. Langerhans cells in allergic contact dermatitis. Giornale Italiano di Dermatologia e Venereologia 145, 747-762.

Vega, J.A., García-Suárez, O., Montaño, J.A., Pardo, B. and Cobo, J.M., 2009. The Meissner and Pacinian sensory corpuscles revisited new data from the last decade. Microscopy Research and Technique 72, 299-309.

Zhao, Y.E., Wu, L.P., Peng, Y. and Cheng, H., 2010. Retrospective analysis of the association between Demodex infestation and rosacea. Archives of Dermatology 146, 896-902.

General aspects of skin, nutrition and diet

Key facts

- The term 'gnotobiotic' is derived from Greek *gnostos*, which means 'known', and *bios*, which means 'life'. Gnotobiotic animals are therefore those in which only known strains of microorganisms are present.
- Gnotobiotic animals can be obtained from germ-free animals by the intentional contamination with specific bacteria.
- Germ-free animals are delivered by Caesarean section and reared in sterilized conditions, so that they are completely free of microorganisms.
- The techniques of gnotobiotic experiments were developed nearly 50 years ago. In the gut bacterial study, 'conventional gut mice', which were obtained on oral administration of faeces from the identical strain of normal mouse, are compared with germ-free littermates.
- Using these techniques, numerous studies have suggested that commensal bacteria are key factors for maintaining host homeostasis. For example, a germ-free mouse exhibits immune dysfunction towards microorganisms, although it is known that germ-free mice live 1.5-fold longer than mice with commensal bacteria when reared in the identical environment.

Summary points

- Phenols, including phenol and *p*-cresol, are protein metabolites produced by gut bacteria.
- Phenols are regarded as bioactive toxins and biomarkers of the gut environment.
- Phenols disturb differentiation of monolayer-cultured keratinocytes.
- Phenols produced by gut bacteria enter the circulation and accumulate in skin, where they disrupt keratinocyte differentiation and cause skin dullness in hairless mice.
- Keratinocyte differentiation and skin dryness fluctuated following changes in serum *p*-cresol levels in a human prebiotic beverage administration trial.
- From the results of *in vitro* and *in vivo* experiments, and human trial, it is reasonable to conclude that phenols produced by gut bacteria cause skin problems (dullness and dryness) through disruption of normal keratinocyte differentiation.
- Synbiotics are a promising treatment approach for maintaining healthy skin through decreasing *p*-cresol levels in serum.

2. Gut bacteria and skin health

R. Iizuka
Yakult Central Institute for Microbiological Research, 1796 Yaho, Kunitachi-shi, Tokyo 186-8650, Japan; ryoko-iiduka@yakult.co.jp

Abstract

Recently, gut commensal bacteria and their metabolites have been identified as important factors impacting host homeostasis. This chapter focuses on the relationship between phenolic metabolites produced by gut bacteria and the differentiation of host skin keratinocytes. In *in vitro* experiments, we have shown that phenols (phenol and *p*-cresol) disturb the differentiation of monolayer-cultured keratinocytes (Section 2.1). In hairless mice, a tyrosine-enriched diet caused increased levels of phenols in caecal contents, serum and flank skin, and imparted a yellowish dullness to skin, as determined by increased color meter b* values. Additional *in vivo* experiments were performed by challenging gnotobiotic mice with phenol-producing *Morganella morganii* and phenol-non-producing *Escherichia coli*. The skin of *M. morganii*-colonized gnotobiotic mice exhibited higher b* values and disturbed keratinocyte differentiation compared to *E. coli*-colonized gnotobiotic mice (Section 2.2). To examine the effects of gut bacteria on skin condition in humans, we conducted a prebiotic-beverage administration trial involving 19 healthy female volunteers who refrained from consuming pro- or prebiotics during a 3-week restriction period, and were then administered one daily prebiotic beverage for 3 weeks. Although serum *p*-cresol levels significantly increased during the restriction period, the levels markedly decreased on prebiotic administration. In addition, keratinocyte differentiation and skin conductance, an indicator of skin moisture level, significantly declined during the restriction period, but recovered following prebiotic administration (Section 2.3). Our results suggest that phenols produced by gut bacteria are absorbed, distributed by the circulatory system, and accumulate in skin, where they subsequently cause skin dullness and dryness through disruption of normal keratinocyte differentiation. Finally, we discuss approaches for decreasing phenolic metabolites through diet modification, and describe a pilot study involving the intake of a synbiotic beverage, which appears to be a promising treatment for maintaining healthy skin via decreasing *p*-cresol level (Section 2.4).

Keywords: gut bacterial metabolites, phenol, *p*-cresol, keratinocyte differentiation

Abbreviations

CKD	Chronic kidney disease
D5a	Phenol-non-producing bacterium strain (*Escherichia coli*)
HPLC	High-performance liquid chromatography
TD4	Phenol-producing bacterium strain (*Morganella morganii*)

2.1 Gut bacteria and their metabolites

2.1.1 Gut commensal bacteria

Recently, a number of exciting articles have suggested that a wide variety of bacterial species inhabiting the human gut affect host homeostasis. A few noteworthy studies warrant mention: (1) Sudo *et al.* (2004) provided convincing evidence that gut bacteria influence host stress responses; (2) Ley *et al.* (2006) reported the startling finding that gut bacteria are associated with obesity; and (3) Ivanov *et al.* (2009) revealed that colonization of the mouse small intestine with a single segmented filamentous bacterial species is sufficient to induce Th17 cells, which play key roles in protecting against infections and are also intimately related to autoimmune diseases.

The regulation of host systems by gut bacteria is perhaps not surprising, considering the fact that over 100 trillion bacteria inhabit the adult gut, whereas the entire adult human body itself is only composed of 60 trillion cells. These gut bacteria are predominantly located in the colon, and are estimated to comprise 500 different species of bacteria. The majority of bacterial species, which include bacteroides, bifidobacteria, eubacteria, fusobacteria, clostridia and lactobacilli, are anaerobes, while the remaining <1%, including *Enterobacteriaceae*, enterococci, staphylococci, streptococci and fungal species, are facultative anaerobes.

Bacterial colonization of the gut begins at birth and drastically slows by the end of the weaning period. Although it is generally considered that marked individual variation exists in gut bacterial profiles, the bacterial composition of adult guts is considered to be fairly stable. However, diet can influence the composition of gut microbiota, as demonstrated by the findings of the following interesting study. Filippo *et al.* (2010) compared the faecal bacterial composition of 15 European (Florence, Italy) children, who consumed a typical western diet, and that of 14 rural African (Mossi ethnic village of Burkina Faso) children, whose diet was much lower in fat and animal protein, and richer in fibre and plant polysaccharides than the western diet. Notably, although the gut bacterial compositions of the healthy Burkina Faso and Italian children differed significantly, the diversity was mainly observed in elder children fed a solid diet, with the five youngest children (two Italian and three Burkina Faso children) who were predominantly breast-fed having similar sub-clusters of colonizing bacteria. In addition, the guts of the youngest Burkina Faso and Italian children contained abundant *Bifidobacterium* bacteria, which are well-known to be associated with breast-feeding in developed countries. Taken together, the results from this study clearly suggest that diet has a more dominant influence on gut bacterial composition than ethnicity or climatic environment.

2.1.2 Gut bacterial metabolites

The effect of gut bacteria on host homeostasis in large part depends on the production of bacterial metabolites, which in turn, is influenced by host food intake (Macfarlane *et al.*, 2007). Carbohydrates that are not digested or absorbed in the upper gut are used for growth by carbohydrate-assimilating bacteria, which metabolize the carbohydrates into several end-products, including hydrogen, methane and short-chain fatty acids, such as butyrate, propionate and acetate. The generated metabolites, particularly short-chain fatty acids, are considered to be beneficial to the host. For example, butyrate is an energy source for colonic epithelium, while acidification of the gut environment by acetate prevents infection by pathogens. These representative examples highlight the mutualistic relationship between gut bacteria and hosts.

In contrast to carbohydrate metabolites, gut bacterial protein metabolites tend to have undesirable effects on hosts (Figure 2.1). Proteins and peptides that enter the gut either through food intake or endogenous sources, such as host tissues and secretions, are depolymerised into component amino acids by host pancreatic and bacterial proteases. The available amino acids in the colon

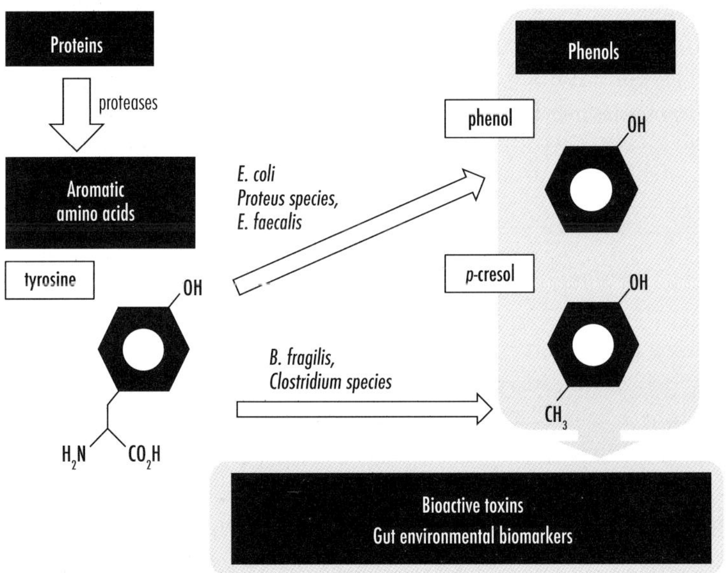

Figure 2.1. Phenols produced by gut bacteria. The aromatic amino acid tyrosine is metabolized to phenol by facultative anaerobic bacteria, such as *Escherichia coli, Proteus* species, and *Enterococcus faecalis*, but is also metabolized to p-cresol by strictly anaerobic bacteria, such as *Bacteroides fragilis* and *Clostridium* species. The generated phenols are regarded as bioactive toxins and biomarkers of the gut environment.

support the growth of protein-assimilating bacteria and are converted into various metabolites, including ammonia, amines and thiols, during bacterial metabolic processes. Moreover, aromatic amino acids in the colon are metabolized into toxic compounds, particularly indoles and phenols, which include phenol and *p*-cresol, by gut bacterial enzymes. An historic and proactive study demonstrated that tyrosine tends to be metabolized to phenol by facultative anaerobic bacteria, such as *Escherichia coli, Proteus* species and *Enterococcus faecalis*, whereas strictly anaerobic bacteria, such as *Bacteroides fragilis* and *Clostridium* species, predominantly metabolize tyrosine to *p*-cresol (Bone *et al.*, 1976). The other main aromatic amino acid, tryptophan, is mainly metabolized to indoles by *E. coli* (Smith *et al.*, 1996).

Although a certain quantity of metabolites is likely excreted in faeces, several recent studies have suggested that a substantial amount of phenols are rapidly absorbed by the colon. Absorbed phenols and indoles are thought to be conjugated to sulphate in the liver and colonic mucosa (De Loor *et al.*, 2005), from where they are subsequently distributed by the circulatory system, and are ultimately excreted in the urine.

Strengthening the view that phenols and indole metabolites are directly produced by the activity of gut bacteria, a recent study found that although these metabolites are present in the blood of conventional mice, phenols and indoles were undetectable in the blood of germ-free mice. Furthermore, 1.4-2.4-fold higher levels of aromatic amino acids, including tyrosine and tryptophan, are found in the blood of germ-free mouse compared to those of conventional mice (Wikoff *et al.*, 2009).

2.1.3 Phenols as bioactive toxins

Owing to their high bioactivity, phenols are characterized as uremic toxins in CKD patients (Evenepoel *et al.*, 2009). In particular, the toxicity of *p*-cresylsulphate, a sulphate-conjugated form of *p*-cresol, has been well studied in CKD. Prior to 2005, numerous *in vitro* studies had suggested that *p*-cresol affects host homeostasis, such as immune dysfunction (Vanholder *et al.*, 1995) and endothelial barrier function (Cerini *et al.*, 2004), and it was speculated that the toxicity of this metabolite might account for the high mortality of CKD patients. In 2005, however, several groups revealed that *p*-cresol predominantly exists in the form of *p*-cresylsulphate in the human body. Although one report has indicated that differences exist between the *in vitro* bioactivity of *p*-cresol and *p*-cresylsulphate (Schepers, 2007), it has since been confirmed that *p*-cresylsulphate is also associated with cardiovascular diseases and mortality in CKD patients (Liabeuf *et al.*, 2009).

2.1.4 Phenols as biomarkers of the gut environment

Phenols can serve as reliable biomarkers of the gut environment, because levels of phenolic metabolites are influenced by the frequency and condition of bowel movements. For example, Nakabayashi *et al.*, (2010) reported a significant inverse rank correlation between serum *p*-cresol (which may have been de-conjugated from *p*-cresylsulphate during sample preparation) levels and scores related to stool form (1: very hard stool - 6: watery stool) and ease of defecation (1: difficult, 2: easy, and 3: very easy).

Among Japanese women, there is popular opinion that persistent constipation correlates with skin problems. Indeed, results obtained from a questionnaire survey suggest that women who suffer from abnormal bowel movements (low frequency, increased strain, foul-smelling stools, etc.) also tend to suffer from subjective skin problems, such as dryness and dullness (unpublished questionnaire survey data for 600 Japanese women conducted in 2007).

From these background studies, we proposed that phenols produced by gut bacteria accumulate during constipation and are possible causes of skin dryness and dullness.

2.1.5 *In vitro* effects of phenols

To investigate the potential mechanisms by which phenolic metabolites affect skin, we performed an *in vitro* experiment with monolayer-cultured keratinocytes (Iizuka *et al.*, 2009b). The differentiation of keratinocytes can be evaluated by monitoring the expression levels of keratin 10 protein, an isoform of the keratins, which are major structural proteins synthesized in keratinocytes during proliferation and differentiation. Keratin 10 is expressed at an early-differentiation stage and plays a role in strengthening the barrier function of skin (Ekanayake-Mudiyanselarge *et al.*, 1998).

In our *in vitro* experiment, monolayers of normal human keratinocytes that were cultured for 3 days in medium supplemented with 0.15 mM Ca^{2+} were shown by western blot analysis to express high levels of keratin 10 protein. To examine the effect of phenols on keratin 10 expression, we supplemented culture media with either 20 nmol/ml phenol or *p*-cresol, a concentration that was determined through preliminary toxicity assessments using monolayer-cultured keratinocytes and is speculated to be the upper limit in healthy humans serum. After 3 days of exposure, WST-8 assays, which detect the dehydrogenase (a key enzyme for cell survival) activity of target cells, revealed that neither phenol nor *p*-cresol affected cell viability (Figure 2.2A). Further, western blot analyses demonstrated that exposure to either phenol or *p*-cresol significantly reduced Keratin 10 expression compared to saline-treated control cells (Figure 2.2B). These results suggest that phenols disturb the early stage of keratinocyte differentiation at concentrations that are not toxic to cells. Notably, we also determined that the exposure of keratinocytes to a physiological dose of indole (up to 1 μmol/ml) had no effect on Keratin 10 expression (unpublished data).

From these observations, we speculated that if phenols produced by gut bacteria under unhealthy gut conditions accumulated in the skin, undesirable effects on the skin would result from disrupted keratinocyte differentiation. To confirm this hypothesis, we performed several *in vivo* studies using mice, in addition to human trials.

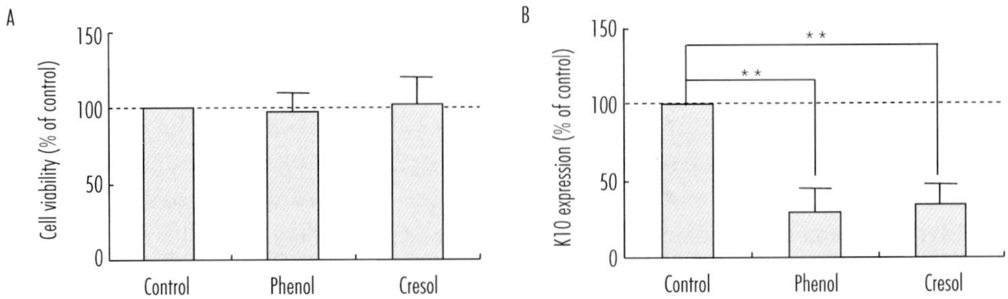

Figure 2.2. *In vitro* effects of phenols on human keratinocyte differentiation. Monolayer keratinocytes supplemented with 20 nmol/ml phenol (Phenol), *p*-cresol (Cresol) or equal volumes of sterilized water (Control) were cultured for 3 days, and then WST-8 assays (A) and western bolt analysis (B) were performed as described in Iizuka *et al.* (2009b). Values are presented as the mean (indicated by columns) ± standard deviation (indicated by error bars). Statistically significant differences were analyzed by the Tukey test, and are indicated as **P<0.01(n=3). (Iizuka *et al.*, 2009b).

2.2 Mice studies

As mentioned above, although it is speculated that constipation leads to adverse skin conditions, little empirical evidence linking the gut environment to skin conditions has been reported. We could find only a single report suggesting that improvement of the gut environment by administration of *Bacillus coagulans* led to a decrease in the number of skin comedones (Ara *et al.*, 2002). We speculate that one reason for the sparsity of evidence may be the lack of appropriate animal models. As a principal cause of constipation is due to diminished gut peristalsis, constipation animal models have been generated using anti-peristalsis drugs, such as loperamide, clonidine and atropine (Kojima *et al.*, 2009). However, such acute constipation models may not be suitable for evaluating the chronic effects of undesirable gut environment on skin characteristics, since skin turn-over requires approximately 1 month. Therefore, we attempted to construct a mouse model with a chronically unhealthy gut environment, such as a high-phenol mouse model, through diet modulation. To facilitate examination of skin conditions, we used Hos:hr-1 hairless mice (Nippon SLC, Shizuoka, Japan), which only have mutations in genes involved in hair growth. Although the external appearance of these mice appears similar to Nude mice, all physiological aspects of Hos:hr-1 hairless mice, including immune system function, are completely normal.

2.2.1 Protein loading experiment

Using Hos:hr-1 hairless mice, we first attempted to construct an experimental mouse model with high levels of gut protein loading, since a protein-rich diet is one of the major causes of an unhealthy human gut environment (Evenepoel *et al.*, 1999). When 7-week-old female hairless mice were fed a protein-enriched diet including 40% casein for 4 weeks, they did not exhibit an increase in levels of faecal or serum phenols and/or indoles compared to mice fed a basal diet including 20% casein.

Notably, increasing the dose of casein in the diet to greater than 40% significantly decreased the food intake and weight of mice. In addition, we noted that 40% casein fed-mice appeared to have significantly fewer wrinkles and more fine-grained skin than basal diet fed-mice, a result which is in line with the knowledge that appropriate protein intake promotes skin health. Thus, we failed to construct an unhealthy gut environment mice model by protein loading.

2.2.2 Tyrosine loading experiment

Although protein loading did not lead to an unhealthy gut environment, we succeeded in inducing a chronically unhealthy gut environment in hairless mice by tyrosine loading according to the protocol described in Iizuka *et al.* (2009a). Briefly, twelve 8-week-old female hairless mice were randomly assigned to two groups of six mice, which were then freely fed either basal (modified AIN-93G) or tyrosine-enriched diets containing 5% tyrosine. During the 21-day experiment, neither body weight nor food intake levels showed significant differences between the two groups. On day 21, after skin characteristics of the flanks were measured, caecal contents, blood, flank skin and livers were collected and homogenized (skin and livers), and levels of phenols were then determined by fluorescence HPLC using a modified method of Niwa *et al.* (1993).

Figure 2.3A. shows the levels of phenols in each of the examined samples as determined by HPLC. During sample preparation, it is likely that sulphate-conjugates were de-conjugated by the acidic and high-temperature conditions; thus, the measured phenol and *p*-cresol levels include their ex-conjugated forms. After 21 days of a tyrosine-rich diet, nearly all phenol levels of the caecal contents, serum, skin and livers were significantly higher in the tyrosine-enriched diet group than those in the basal diet group. Interestingly, it was observed that the phenol levels in skin were much higher than those in livers (phenol: 10.92 vs. 2.36 nmol/g, *p*-cresol: 21.97 vs. 4.51 nmol/g respectively). As skin has a relatively high lipid content, lipid-soluble phenols may preferentially accumulate in this organ.

The fact that substantial levels of phenols accumulate in skin places this organ at higher risk to the adverse effects of phenols. Indeed, we found that on examination of flank skin using a color meter, the b* values of mice fed a tyrosine-enriched diet were significantly higher than those of the basal diet mice group ($P<0.01$) (Figure 2.4A). As b* values are a useful parameter to determine skin dullness, these results suggest the possibility that phenols produced in the gut cause skin dullness.

2.2.3 Gnotobiotic mouse experiment

To more directly examine the effects of phenols produced by gut bacteria on skin, we next performed an experiment with gnotobiotic mice according to the protocol described in Iizuka *et al.* (2009a). As preliminary experiments involving the intraperitoneal administration of phenols showed that phenol, more dramatically than *p*-cresol, caused severe effects on the skin of mice, we attempted to isolate phenol-producing gut bacteria from mouse faeces. Briefly, the faeces of mice fed basal and tyrosine-enriched diets were homogenized and seeded on DHL agar medium, and several of the resulting colonies were subcultured twice to ensure purity and viability. The isolated

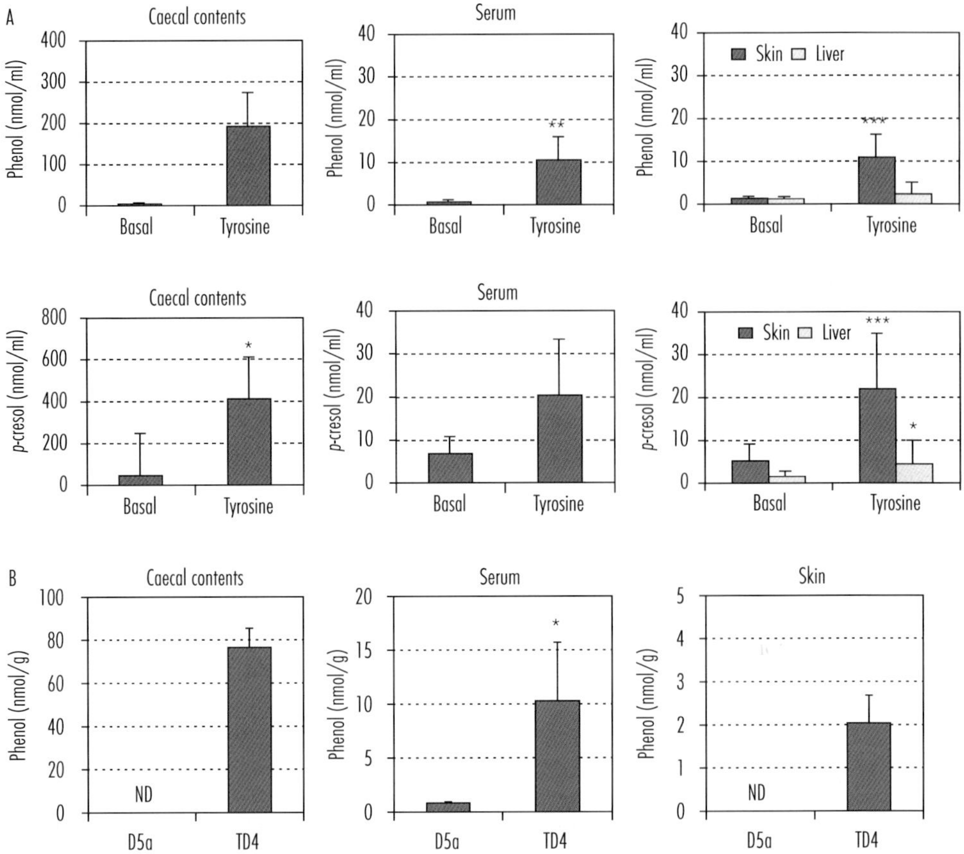

Figure 2.3. Levels of phenols in mouse measured by fluorescence high-performance liquid chromatography, as described in Iizuka *et al.* (2009a). (A) Levels of phenols of mice fed either a basal (Basal) or tyrosine-enriched diet (Tyrosine) are plotted. (B) Phenol levels of D5a strain (D5a)- and TD4 strain (TD4)-gnotobiotic mice are plotted. ND: not detected. Statistically significant differences were analysed by Student's *t* test, and are indicated as $^{*}P<0.05$, $^{**}P<0.01$, $^{***}P<0.001$ (n=5). (Data from Iizuka *et al.*, 2009a)

bacterial strains were then seeded on tyrosine medium, which contained 0.01% tyrosine, but was deficient in other nitrogen and carbon sources (1.0% casein hydrolysate and 1.0% yeast extract), to further selectively isolate bacteria with tyrosine assimilation ability. From this screening procedure, phenol-producing strain TD4 and phenol non-producing strain D5a were isolated, which were identified as *Morganella morganii* (GenBank accession no. AY464464) and *E. coli* (GenBank accession no. U00096), respectively, using standard sequencing methods of 16S rDNA. *M. morganii* is a closely-related species to *Proteus mirabilis*, which is one major phenol-producing residents of the animal gut. *M. morganii* is also often detected in the human gut, and is a cause of urinary-tract infections (Delomenie *et al.*, 2001). Although several *E. coli* strains are known to produce phenol in the animal gut, strain D5a lacks phenol-producing ability. We also determined

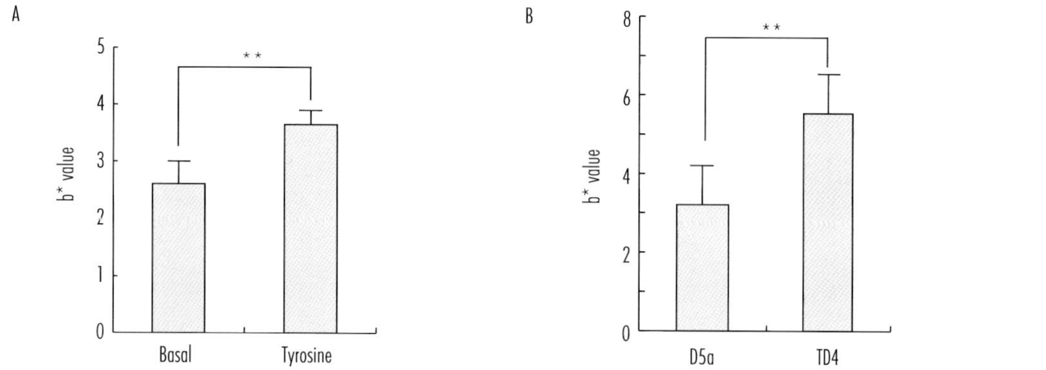

Figure 2.4. Determination of mouse flank skin dullness. The b* values of flank skin in (A) basal-diet (Basal) and tyrosine-enriched-diet (Tyrosine) group mice in the tyrosine-loading experiment, and (B) D5a strain- and TD4 strain-gnotobiotic mice fed a tyrosine-rich diet, were measured using a CM-s600d spectrophotometer (Minolta, Osaka, Japan), as described in Iizuka *et al.* (2009a). Statistically significant differences were analysed by the Student's t test, and are indicated as ***P*<0.01 (n=5). (Iizuka *et al.*, 2009a)

that neither TD4 nor D5a produced *p*-cresol, while both strains produced ammonia, and only D5a produced indole.

A bacterial challenge experiment was performed with 16-week-old female germ-free hairless mice (GF: Hr-1; bred at the Yakult Institute of Laboratory Animals) which were randomly assigned to two groups of five mice. Each group of mice was then selectively administered an oral dose of 10^{10} colony forming units/0.2 ml/mouse of either TD4 or D5a. After 3 weeks of consuming a tyrosine-enriched diet, skin characteristics, and phenol levels of caecal contents, serum and skin were measured as in the tyrosine-loading experiment. Consequently, markedly higher phenol levels were detected in the caecal contents, serum and skin of TD4 mice than in D5a mice (Figure 2.3B). In addition, examination of flank skin revealed that the b* values of TD4 mice skin were significantly higher than those of D5a mice (*P*<0.01) (Figure 2.4B).

Since phenols were shown to disturb keratinocyte differentiation in our *in vitro* experiments with monolayer-cultured keratinocytes (Figure 2.2B), we speculated that one of the causes of skin dullness is abnormal keratinocyte differentiation. To evaluate the differentiation of keratinocytes in the skin of TD4- and D5a-gnotobiotic mice, we examined the size of corneocytes in the *stratum corneum* derived by the tape-stripping method. It is well-known that normal differentiation produces hexagonal-shaped and cobblestone-like-arranged corneocytes, while abnormal differentiation produces deformed and irregularly-arranged corneocytes (Figure 2.5A). Thus, corneocyte morphology is a good marker of the status of keratinocyte differentiation, with decreased corneocyte size in particular serving as a useful marker of abnormal differentiation (Lee *et al.*, 1983). In our gnotobiotic mice experiment, imaging analyses revealed that the average size of TD4 mice corneocytes was significantly smaller than that of control D5a mice cells (Figure 2.5B).

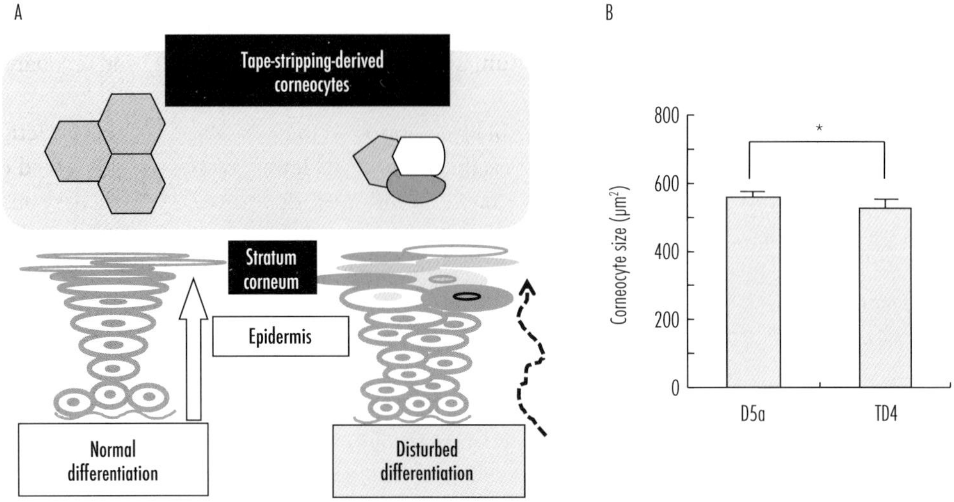

Figure 2.5. Evaluation of keratinocyte differentiation in gnotobiotic mouse. (A) Corneocytes can be non-invasively obtained by a tape-stripping method and used for the diagnosis of skin disorders. (B) Tape-stripping-derived corneocytes from the secondary *stratum corneum* layer of D5a- and TD4-gnotobiotic mice were imaged, and the size of corneocytes was then determined using an image analyser, as described in Iizuka *et al.* (2009a). Significant differences were analysed by the Student's *t* test, and is indicated as $*P<0.05$ (n=5). (Figure 2.5B, Iizuka *et al.*, 2009a)

Taken together, the results of our mice studies reasonably suggest that phenols produced by gut bacteria disturb normal keratinocyte differentiation and cause skin dullness.

2.3 Human trial

We were next interested in determining whether the identical phenomenon observed in mice is also applicable to humans. To investigate this, we performed a prebiotic administration trial to determine if skin conditions are directly affected by serum levels of phenols. For this human trial, we selected a prebiotic beverage containing 3.0 g galacto-oligosaccharides, which had been previously shown to depress amino acid fermentation, including *p*-cresol production, in the human gut (Ito *et al.*, 1993).

Nineteen healthy female volunteers with ages ranging from 21-39 years (mean age 31.9±5.4 years) participated in the trial. All participants had not received any type of medical treatment for 1 month prior to or during the trial, and had a defecation frequency exceeding four times per week. Questionnaire results indicated that all participants had frequently taken probiotics and prebiotics. However, during an initial 3-week restriction period, all participants refrained from consuming any pro- or prebiotics. During the following 3-week administration period, all participants consumed one daily prebiotic beverage (HiLine®; Yakult Honsha Co., Ltd, Tokyo,

Japan), and continued to refrain from any other pro- or prebiotic treatments. On days 0, 21 and 43, inner forearm skin characteristics and serum levels of phenols were measured and compared.

Consequently, we detected that serum *p*-cresol levels appeared to be regulated by the restriction and administration of pro- and prebiotics. Specifically, *p*-cresol levels significantly increased on restriction of pro- and prebiotic administration, but significantly decreased on administration of the prebiotic beverage, indicating improvement of the gut environment (Figure 2.6).

In the analyses of skin characteristics, two parameters indicated that the differentiation of keratinocytes had fluctuated in response to pro- and prebiotic administration. First, corneocyte size, which is a good marker of keratinocyte differentiation (as described above), significantly decreased during the restriction period, indicating that keratinocyte differentiation was adversely affected, but markedly increased on prebiotic administration, indicating that normal differentiation was restored (Figure 2.7A). Second, the cathepsin L-like activity in the *stratum corneum* also significantly decreased in the restriction period, but was restored on prebiotic administration (Figure 2.7B). Cheng *et al.* (2006) reported that cathepsin L is an activator of transglutaminase 3, which is an important factor for constructing normal *stratum corneum*. We detected cathepsin L-like activity in the *stratum corneum* using the non-invasive and simple tape-stripping method to isolate corneocytes. The reduction of cathepsin L-like activity in *stratum corneum* during the restriction period also suggests that keratinocyte differentiation was disturbed, but could be restored by the administration of prebiotics.

We also performed several physiological measurements of participants' skin using a color meter, but failed to detect changes in skin dullness based on the measured b* values, as was observed in the mice studies. We speculate that this discrepancy resulted from larger variations

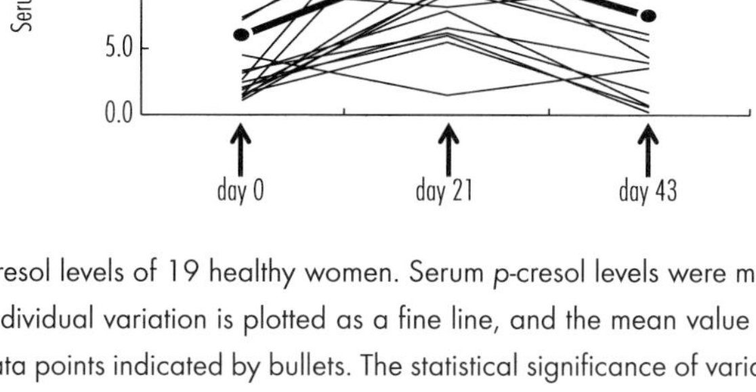

Figure 2.6. Serum *p*-cresol levels of 19 healthy women. Serum *p*-cresol levels were measured as described in Iizuka *et al.* (2009b). Individual variation is plotted as a fine line, and the mean value is plotted as a thick line with each of the three data points indicated by bullets. The statistical significance of variation was determined by Wilcoxon's signed rank sum test, and is indicated as *** $P<0.001$. (Based on Iizuka *et al.*, 2009b).

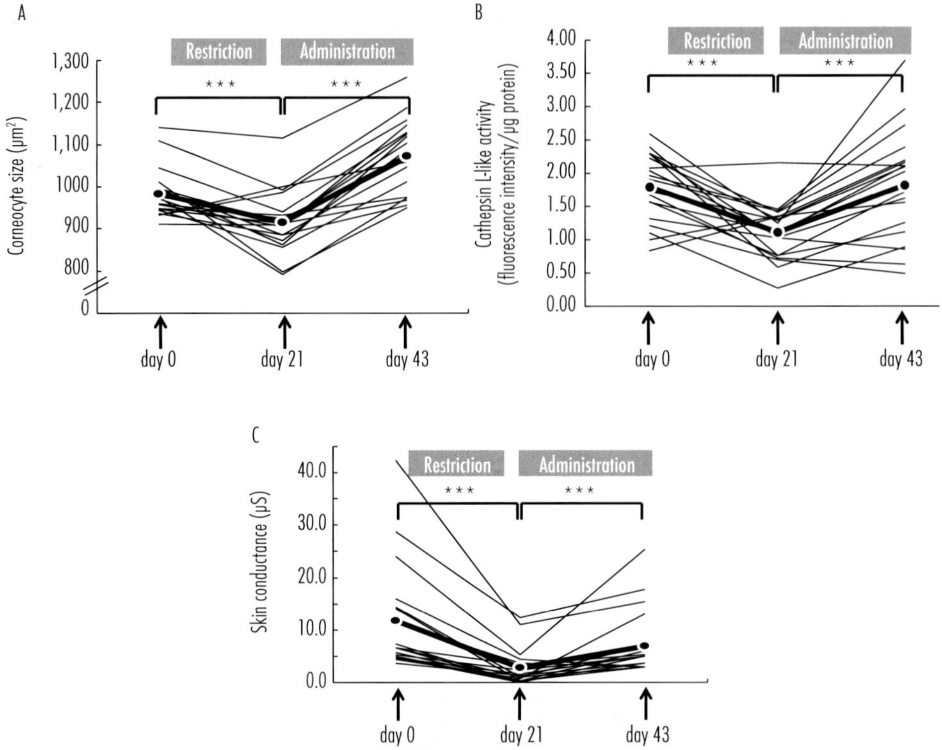

Figure 2.7. Inner forearm skin conditions of 19 healthy women. The size of second-layer corneocytes (A) or the cathepsin L-like protease activity in third-layer corneocytes (B) were measured as described in Iizuka *et al.* (2009b). (C) Skin conductance of the inner forearm was measured using a skin conductance meter (SKICON-200; I.B.S. Co., Ltd., Shizuoka, Japan). Individual variation is plotted as a fine line, and the mean value is plotted as a thick line for each of the three data points indicated by bullets. The statistical significance of variation was determined by Wilcoxon's signed rank sum test, and is indicated as ****P*<0.001. (Figure 2.5A and 2.5B: based on Iizuka *et al.*, 2009b).

in human skin color than those in skin of inbred mouse strains. However, we succeeded in detecting marked fluctuations in skin conductance. The normal differentiation of keratinocytes is important for maintaining the skin barrier function, which protects against trans-epidermal water loss. Therefore, abnormal keratinocyte differentiation results in increased trans-epidermal water loss, leading to skin dryness (Ekanayake-Mudiyanselarge, 1998). Skin dryness can be evaluated by measuring the relative levels of skin conductance using a skin conductance meter, with moisturized and dry skin exhibiting high and low conductance, respectively. In this trial, skin conductance significantly decreased during the restriction period, indicating that the skin suffered dryness, but markedly increased in the administration period, indicating that the dryness was improved on consumption of the prebiotic beverage (Figure 2.7C).

The results of the human trial are summarized as follows: when serum *p*-cresol levels increased, likely due to the restriction of pro- and prebiotics, keratinocyte differentiation was disturbed, resulting in skin dryness. Conversely, the reduction of serum *p*-cresol levels through the administration of prebiotics restored normal keratinocyte differentiation and skin moisture levels.

On consideration of the results from the *in vitro* and *in vivo* mice experiments, and human trial, we conclude that phenols produced by gut bacteria are absorbed, distributed by the circulation system, and accumulate in skin, where they subsequently cause skin problems (dullness and dryness) through the disruption of normal keratinocyte differentiation (Figure 2.8). Despite this apparent association between gut bacteria and skin condition, many unresolved questions remain; particularly, the mechanism by which phenols disturb keratinocyte differentiation is unknown. Before this question can be answered, it is necessary to determine whether phenols in the skin exist in sulphate-conjugated or de-conjugated forms, since differences in bioactivity are expected between these forms, as described above. We speculate that phenols are de-conjugated in the skin, as abundant enzymes capable of de-conjugating phenols, known as sulphatases, are likely present, which are derived from host skin itself and/or from 'skin commensal bacteria'. Further

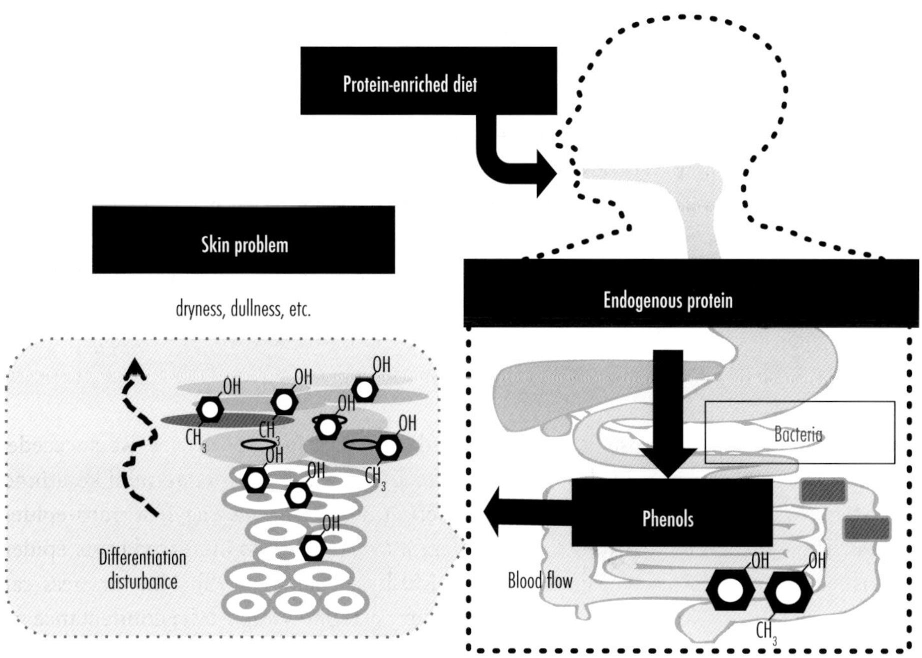

Figure 2.8. Proposed relationships between gut bacteria and skin health. Proteins which enter the gut from food intake and endogenous sources are subsequently fermented to phenols by gut bacteria. The phenols are absorbed, distributed by the circulation system, and accumulate in the skin, where they lead to skin dullness and dryness through the disruption of normal keratinocyte differentiation.

investigation will reveal not only the correlation between distant gut bacteria and host skin, but also determine the involvement of neighbouring skin commensal bacteria on skin health.

2.4 Decreasing levels of phenols via nutrition to maintain skin health

As described in Section 2.1, the gut environment, including levels of phenols, is dynamically influenced by diet; therefore, we can contrive a variety of diet-modification approaches to decrease phenols production and maintain skin health. Limiting the consumption of excess proteins might be a suitable method, although it is important to remember that protein is also needed for maintaining skin health. Through several human trials involving diet-specific questionnaires, we have found that participants who frequently consume tyrosine-enriched foods, such as bamboo shoots, cod roe, cheese and dried whitebait (young sardines) tend to have high serum phenol levels. Although larger population surveys are required to confirm this finding, we speculate that the gut of these individuals is more suitable for colonization by phenol-producing bacteria. Together with the knowledge obtained from tyrosine-loading mouse experiments, limiting the intake of tyrosine-enriched foods might be another approach to promote skin health, although again, an appropriate balance is necessary, as these foods also contain beneficial nutrients for health. Other effective approaches may involve increasing the intake of uncooked vegetables (Ling *et al.*, 1992) and resistant starch (Birkett *et al.*, 1996), which have been shown to lower the faecal concentrations of phenols in humans.

In addition to diet-modification, the results of our human trial suggest that the consumption of prebiotics decreases *p*-cresol levels and helps maintain skin health. At least three possible mechanisms may explain this finding: (1) *p*-cresol-producing activities of gut bacteria are reduced by prebiotics; (2) the population of *p*-cresol-producing bacteria is reduced by a corresponding increase in non-*p*-cresol-producing bacteria that can assimilate prebiotics; and (3) the frequency of bowel movements is increased by prebiotics and/or dietary fibre, which results in reduced *p*-cresol concentrations in the gut. Regardless of the exact mechanism, prebiotics appear to be highly effective for improving the gut environment and decreasing levels of serum phenols.

Several additional studies highlight the potential of probiotics for modifying levels of phenols in the gut environment. For example, De Preter *et al.* (2007) showed the marvellous efficacy of *Bifidobacterium breve* for decreasing *p*-cresol levels in healthy humans. Based on this finding, we conducted a double-blind, placebo-controlled trial with a synbiotic beverage consisting of *B. breve*-fermented milk containing galacto-oligosaccharides. In the trial, which involved 40 healthy women between the ages of 23 and 75 years, the serum *p*-cresol levels of the group administered the synbiotic beverage were significantly reduced compared to the placebo control group. In addition, cathepsin L-like activity, a cornification marker, and moisture content in the *stratum corneum* of the test group had a tendency to be higher than those of the placebo group (Kano *et al.*, 2010). These promising results support the need for further investigation to more clearly demonstrate that synbiotics lead to improved skin health through decreasing levels of phenols.

References

Ara, K., Meguro, S., Hase, T., Tokimitsu, I., Otsuji, K., Kawai, S., Ito, S. and Iino, H., 2002. Effect of spore-bearing lactic acid-forming bacteria (*Bacillus coagulans* SANK 70258) administration on the intestinal environment, defection frequency, fecal characteristics and dermal characteristics in humans and rats. Microbial Ecology in Health and Disease 14, 4-13.

Birkett, A., Muir, J., Phillips, J., Jones, G. and O'Dea, K., 1996. Resistant starch lowers fecal concentrations of ammonia and phenols in humans. American Journal of Clinical Nutrition 63, 766-772.

Bone, E., Tamm, A. and Hill, M., 1976. The production of urinary phenols by gut bacteria and their possible role in the causation of large bowel cancer. The American Journal of Clinical Nutrition 29, 1448-1454.

Cerini, C., Dou, L., Anfosso, F., Sabatier, F., Moal, V., Glorieux, G., De Smet, R., Vanholder, R., Dignat-George, F., Sampol, J., Berland, Y. and Brunet, P., 2004. *P*-cresol, a uremic retention solute, alters the endothelial barrier function *in vitro*. Journal of Thrombosis and Haemostasis 92, 140-150.

Cheng, T., Hitomi, K., Van Vlijmen-Willems, I.M.J.J., De Jongh, G.J., Yamamoto, K., Nishi, K., Watts, C., Reinheckel, T., Schalkwijk, J. and Zeeuwen, P.L.J.M., 2006. Cystatin M/E is a high affinity inhibitor of cathepsin V and cathepsin L by a reactive site that is distinct from the legumain-binding site: a novel clue for the role of cystatin M/E in epidermal cornification. Journal of Biological Chemistry 281, 15893-15899.

Delomenie, C., Fouix, S., Longuemaux, S., Brahimi, N., Bizet, C., Picard, B., Denamur, E. and Dupret, J.M., 2001. Identification and functional characterization of arylamine *N*-actetyltransferases in eubacteria: evidence for highly selective acetylation of 5-aminosalicylic acid. Journal of Bacteriology 183, 3417-3427.

De Loor, H., Bammens, B., Evenepoel, P., Preter, V.D. and Verbeke, K., 2005. Gas chromatographic-mass spectorometric analysis for measurement of *p*-cresol and its conjugated metabolites in uremic and normal serum. Clinical Chemistry 51, 1535-1538.

De Preter, V., Vanhoutte, T., Huys, G., Swings, J., De Vuyst, L., Rutgeerts, P. and Verbeke, K., 2007. Effects of *Lactobacillus casei* Shirota, *Bifidobacterium breve*, and oligofructose-enriched inulin on colonic nitrogen-protein metabolism in healthy humans. American Journal of Physiology-Gastrointestinal and Liver Physiology 292, G358-G368.

Ekanayake-Mudiyanselarge, S., Aschauer, H., Schmook, F.P., Jensen, J.M., Meingassner, J.G. and Proksch, E., 1998. Expression of epidermal keratins and the cornified envelope protein involucrin is influenced by permeability barrier disruption. Journal of Investigative Dermatology 111, 517-523.

Evenepoel, P., Claus, D., Geypens, B., Hiele, M., Geboes, K., Rutgeers, P. and Ghoos, Y., 1999. Amount and fate of egg protein escaping assimilation in the small intestine of humans. American Journal of Physiology 277, G935-G943.

Evenepoel, P., Meijers, B.K.I., Bammens, B.R.M. and Verbeke, K., 2009. Uremic toxins originating from colonic microbial metabolism. Kidney International 76, s12-s19.

Filippo, C.D., Cavalieri, D., Paola, M.D., Ramazzotti, M., Poullet, J.B., Massart, S., Collini, S., Pieraccini, G. and Lionetti, P., 2010. Impact of diet in shaping gut microbiota revealed by a comparative study in children from Europe and rural Africa. In: Proceedings of the National Academy of Sciences of the United States of America 107, pp. 14691-14696.

Iizuka, R., Kawakami, K., Izawa, N. and Chiba, K., 2009a. Phenols produced by gut bacteria affect the skin in hairless mice. Microbial Ecology in Health and Disease 21, 50-56.

Iizuka, R., Kawakami, K. and Chiba, K., 2009b. Gut bacteria producing phenols disturb keratinocyte differentiation in human skin. Microbial Ecology in Health and Disease 21, 221-227.

Ito, M., Kimura, M., Deguchi, Y., Miyamori-Watabe, A., Yajima, T. and Kan, T., 1993. Effects of transgalactosylated disaccharides on the human intestinal microflora and their metabolism. Journal of Nutritional Science and Vitaminology 39, 279-288.

Ivanov, I.I., Atarashi, K., Manel, N., Brodie, E.L., Shima, T., Karaoz, U., Wei, D., Goldfarb, K.C., Santee, C.A., Lynch, S.V., Tanoue, T., Imaoka, A., Itoh, K., Takeda, K., Umesaki, Y., Honda, K. and Littman, D.R., 2009. Induction of intestinal Th17 cells by segmented filamentous bacteria. Cell 139, 1-14.

Kano, M., Masuoka, N., Kaga, C., Iizuka, R., Manabe, K., Sone, T., Ooeda, K., Nonaka, C., Miyazaki, K. and Ishikawa, F., 2010. A double-blind, placebo-controlled, randomized trial of fermented milk containing bifidobacteria to verify the effect of gut bacteria producing phenols on skin conditions in healthy Japanese women. The Japan Society for Lactic Acid Bacteria 2010 annual meeting, July 26-27, Sendai Japan, p. 43.

Kojima R., Doihara, H., Nozawa, K., Kawabata-Shoda, E., Yokoyama, T. and Ito, H., 2009. Characterization of two models of drug-induced constipation in mice and evaluation of mustard oil in these models. Pharmacology 84, 227-233.

Lee, S., Park, Y.K., Kim, Y.K. and Kang, J.S., 1983. An experimental study on corneocytes of acutely and chronically irritated skin. Achievement of Dermatological Research 275, 49-52.

Ley, R.E., Turnbaugh, P.J., Klein, S. and Gordon, J.I., 2006. Microbial ecology: human gut microbes associated with obesity. Nature 444, 1022-1023.

Liabeuf, S., Barreto, D.V., Barreto, F.C., Meert, N., Glorieux, G., Schepers, E., Temmar, M., Choukroun, G., Vanholder, R., Massy, Z.A. and on behalf of the European Ureaemic Toxin Work Group (EUTox), 2009. Free *p*-cresylsulphate is a predictor of mortality in patients at different stages of chronic kidney disease. Clinical Journal of American Society of Nephrology 4, 1932-1938.

Ling, W.H. and Hänninen, O., 1992. Shifting from a conventional diet to an uncooked vegan diet reversibly alters fecal hydrolytic activities in humans. Journal of Nutrition 122, 924-930.

Macfarlane, G.T. and Macfarlane, S., 2007. Models for intestinal fermentation: association between food components, delivery systems, bioavailability and functional interactions in the gut. Current Opinion in Biotechnology 18, 156-162.

Nakabayashi, I., Nakamura, M., Kawakami, K., Ohta, T., Kato, I., Uchida, K. and Yoshida, M., 2010. Effects of synbiotic treatment on serum level of *p*-cresol in haemodialysis patients: a preliminary study. Nephrology Dialysis Transplantation 16, 1094-1098.

Niwa, T., 1993. Phenol and *p*-cresol accumulated in uremic serum measured by HPLC with fluorescence detection. Clinical Chemistry 39, 108-111.

Schepers, E., Meert, N., Glorieux, G., Goeman, J., Van der Eycken, J. and Vanholder R., 2007. *P*-cresylsulphate, the main *in vivo* metabolite of *p*-cresol activates leucocyte free radical production. Nephrology Dialysis Transplantation 22, 592-596.

Smith, E.A. and Macfarlane, G.T., 1996. Enumeration of human colonic bacteria producing phenolic and indolic compounds: effects of pH, carbohydrate availability and retention time on dissimilatory aromatic amino acid metabolism. Journal of Applied Bacteriology 81, 288-302.

Sudo, N., Chida, Y., Aiba, Y., Sonoda, J., Oyama, N., Yu, X.N., Kubo, C. and Koga, Y., 2004. Postnatal microbial colonization programs the hypothalamic-pituitary-adrenal system for stress response in mice. The Journal of Physiology 558, 263-275.

Vanholder, R., De Smet, R., Waterloos, M.A., Van Landschoot, N., Vogeleere, P., Hoste, E. and Ringoir, S., 1995. Mechanisms of uremic inhibition of phagocyte reactive species production: Characterization of the role of *p*-cresol. Kidney International 47, 510-517.

Wikoff, W.R., Anfora, A.T., Liu, J., Schultz, P.G., Lesley, S.A., Peters, E.C. and Siuzdak, G., 2009. Metabolomics analysis reveals large effects of gut microflora on mammalian blood metabolites. In: Proceedings of the National Academy of Sciences of the United States of America 106, pp. 3698-3703.

Key facts

- All mammals have a water permeability barrier in the epidermis of the skin that prevents water loss from the body.
- This barrier consists of dead skin cells surrounded by layers of lipids.
- The barrier lipids consist of a mixture of cholesterol, free fatty acids and special ceramides, among which is a unique ceramide containing linoleic acid.
- Linoleic acid can not be formed by mammals but has to be ingested with the food.
- Linoleic acid is a (n-6)-fatty acid, also called ω6-fatty acid.
- Linoleic acid is a polyunsaturated fatty acid, which in the body can be converted to a number of other polyunsaturated fatty acid, all belonging to the linoleic acid-family or also called (n-6) fatty acids.
- Polyunsaturated fatty acids are fatty acids with two or more chemical double bonds, and they are typically found in many vegetable oils.
- Linoleic acid is an essential fatty acid that has to be consumed in the diet.
- Prolonged deficiency of linoleic acid results in a defective water permeability barrier of the skin, resulting in increased water loss over the skin.
- Humans who ingest less than 2 g/day of linoleic acid for many months may risk development of a defective water permeability barrier.
- Humans in the western societies consume between 8-25 g/day of linoleic acid, which is found in most vegetable oils and other food items.
- Thus, deficiency of linoleic acid in humans is practically non-existing.
- Linoleic acid and its endogenous derivative arachidonic acid are most probably also involved in many other both beneficial and harmful functions within the skin, among which are skin cell differentiation, inflammation, immune reactions and cancer.
- However, our scientific knowledge is still too scattered to know whether dietary or topical supplementation with specific fatty acid of the linoleic acid-family can have any beneficial effect on various skin functions.

Summary points

- Linoleic acid serves via its incorporation into special ceramides an essential function in maintaining the water permeability barrier of the skin.
- The dietary requirement of linoleic acid for this essential function is below the 2% of dietary energy intake, and lack of linoleic acid in the diet of humans in the western societies is practically non-existing.
- Derivatives of linoleic acid and arachidonic acid most probably serve several important functions in maintaining a healthy skin.
- Our knowledge is still too scattered to suggest that dietary supplementation or topical application of specific (n-6)-fatty acids will have any beneficial effect.

3. The fatty acids and the skin: a focus on the n-6 family of unsaturated fatty acids

H.S. Hansen
Department of Pharmacology & Pharmacotherapy, Faculty of Pharmaceutical Sciences, University of Copenhagen, Universitetsparken 2, 2100 Copenhagen, Denmark; hsh@farma.ku.dk

Abstract

Linoleic acid (18:2(n-6)) is an essential fatty acid, and it is the parent fatty acid for all other fatty acids belonging to the n-6 fatty acid family, e.g. γ-linolenic acid (18:3(n-6), dihomo-γ-linolenic acid (20:3(n-6), arachidonic acid (20:4(n-6)), and adrenic acid (22:4(n-6)). The traditional symptoms seen after prolonged dietary deficiency of (n-6)-fatty acids, e.g. increased trans-epidermal water loss, scaly skin, hair loss and poor growth, can all be ascribed to a lack of linoleic acid in special O-acylated ceramides, which constitute part of the lipid structure of the lamellar membranes between the cornocytes in *stratum corneum* of the epidermis. These extracellular lamellar membranes form together with the cornocytes a significant part of the water permeability barrier of the skin. Other deficiency symptoms, e.g. kidney failure, may possibly be explained by specific functions of arachidonic acid and/or its derivatives. For young growing animals, including humans, a dietary intake of 1 energy% of linoleic acid should be enough to prevent the appearance of these deficiency symptoms. Linoleic acid is found in high amounts in most vegetable oils like safflower oil, sunflower oil, and corn oil while olive oil and rapeseed oil contain lesser amounts. The estimated dietary requirement for linoleic acid is less than 2 energy%, but the dietary intake of linoleic acid in Western societies is generally much higher (4-12 energy%). Thus, deficiency of linoleic acid is practically non-existing in the populations of the well-nourished Western societies. Beside these essential functions, metabolites of linoleic acid and arachidonic acid seems also be involved in regulating different biochemical processes important for skin function, e.g. skin inflammation, sebum formation and cell proliferation and differentiation. However, our scientific knowledge concerning these areas is at present too scattered to suggest specific use of any (n-6)-fatty acid for treatment of various skin diseases.

Keywords: linoleic acid, arachidonic acid, trans-epidermal water loss, essential fatty acids, endocannabinoids, prostaglandins, leukotrienes

Abbreviations

COX	Cyclooxygenase
EFA	Essential fatty acid
LOX	Lipoxygenase
PUFA	Polyunsaturated fatty acid
XX	Enzymes in endocannabinoid and leukotriene formation

3.1 Introduction

The (n-6) fatty acids are essential dietary components for all mammals and birds, whereas their nutritive importance for fish and insects is less clear. Burr and Burr discovered in 1929-1930 that young growing rats would develop scaly skin, hair loss and poor growth on a fat-free diet, and these deficiency symptoms (Table 3.1) could be cured by adding linoleic acid or other (n-6) fatty acids to the diet (Hansen, 1986). Burr and Burr coined the name EFAs for the group of (n-6) fatty acids, which can be formed in the mammalian body from dietary linoleic acid. In chemical nomenclature, fatty acids are numbered from the carboxylic group and the two double bonds of linoleic acid (Figure 3.1) are thus positioned at carbon atom nr 9 and carbon atom nr 12. The double bond at carbon nr 12 cannot be formed by any mammals or birds, and it is important for the essentiality of linoleic acid. Plants have enzymes that can catalyse formation of fatty acids with this double bond. Linoleic acid can be elongated by enzymes in mammals and birds to C20 fatty acids by addition of two carbon atoms to the carboxylic end of the molecule, and the structural position of this important double bond will thus be found at carbon nr 14. However, if one counts from the methyl-end of the molecule (atom number n (or ω)), then the important double bond will still have the position n-6 in both a C18 fatty acid and a C20 fatty acid. (n-6) Fatty acids are also called ω6 fatty acids.

The symptoms described in Table 3.1 will eventually appear when young rats are raised on a fat-free diet. The earliest symptoms (increased transepidermal water loss and increased water consumption) are seen after 4-6 weeks. Caudal necrosis, kidney degeneration and impaired

Table 3.1. Classical symptoms of essential fatty acid-deficiency in rats.

Scaly skin	Impairment of growth and weight gain
Caudal necrosis	Increased trans-epidermal water loss
Inflamed hind feet	Increased water consumption
Dandruff	Kidney degeneration
Loss of hair	Impaired reproductive function and sterility (male and female)
Haemorrhagic skin scores	Increased susceptibility to bacteria

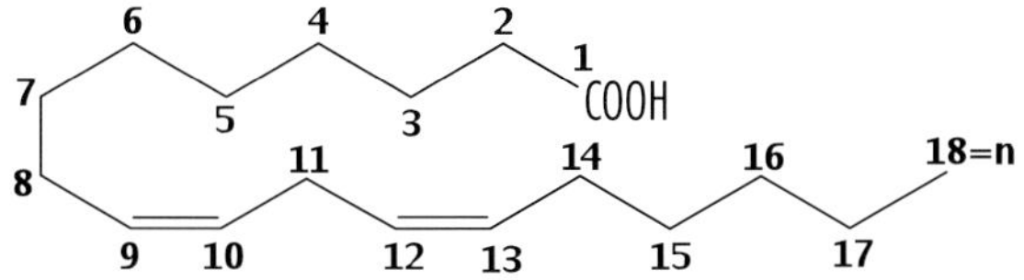

Figure 3.1. Chemical structure of linoleic acid (18:2 (n-6)).

fertility appear much later. The symptoms can be prevented by giving linoleic acid (less than 2 energy%) in the diet.

The other group of essential fatty acids is the (n-3) fatty acids (also called ω3 fatty acids) that have an additional double bond at position n-3. The (n-3) fatty acids and (n-6) fatty acids can not be inter-converted in mammals or birds, and the (n-3) fatty acids also serve quite different function in the body. α-Linolenic acid (18:3(n-3)) is the parent fatty acid of the (n-3)-fatty acid family. The essentiality of the (n-3) fatty acids can most likely be ascribed to docosahexaenoic acid (22:6(n-3), which seems to be of importance for prober brain function (Lauritzen *et al.*, 2001). Docosahexaenoic acid also seems to have essential functions for the fertility of male rats (Roqueta-Rivera *et al.*, 2010).

3.2 (n-6) fatty acid turnover and nutrition

Adult humans as well as adult experimental animals, which consume a varied diet, usually have so large stores of linoleic acid in the adipose tissue that it is nearly impossible to induce EFA-deficiency-symptoms when consuming a diet devoid of linoleic acid. Most of the linoleic acid consumed with the diet is degraded and used for energy production, just as any other dietary fatty acid. Linoleic acid may also serve some structural function in phospholipids in the cell membranes. A small part of the linoleic acid consumed can be elongated (i.e. adding more carbon atoms) and further desaturated (i.e. introduction of new double bonds) by enzymes found in the human body (Figure 3.2). The intermediate fatty acids not shown on the figure, i.e. γ-linolenic acid (18:3(n-6)) and dihomo-γ-linolenic acid (20:3(n-6)), are found only in minor amounts in human tissue, and they are generally thought of as just being steps in the formation of arachidonic acid (20:4(n-6)). Adrenic acid (22:4(n-6)) is considered a sort of storage form of arachidnoic acid because it can be retro-converted into arachidonic acid by enzymes. Arachidonic acid is relatively abundant especially in the phospholipids that are found in cellular membranes, and it is precursor for many hormone-like signaling molecules, e.g. prostaglandins and leukotrienes. It shall be liberated from the phospholipids by phospholipase enzymes before it can be converted into these signaling molecules. Omnivorous humans also consume small amounts of arachidonic acid with the diet, since arachidonic acid is found in small amounts in all animal products. Arachidonic acid is not found in higher plants. Arachidonic acid serves many important functions in the body, e.g. as a structural element in membrane phospholipids, and as precursor for signalling

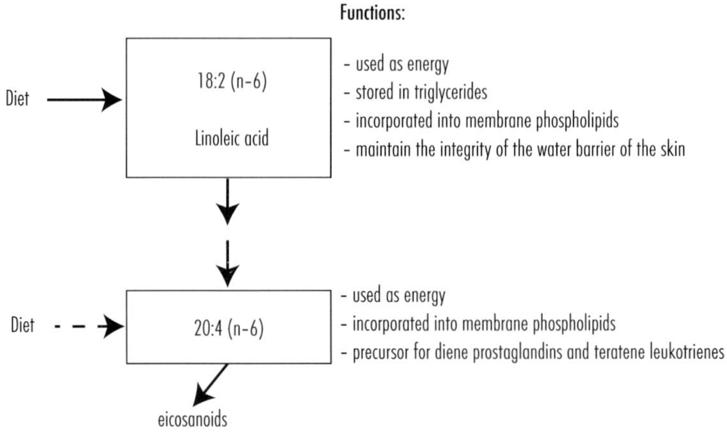

Figure 3.2. Turnover and some biological functions of the two major (n-6)-fatty acids, linoleic acid and arachidonic acid, in humans. A high dietary intake of linoleic acid (18:2(n-6)) and a corresponding large pool of linoleic acid esterified in body lipids are typical for humans in most Western societies. A very minor amount of linoleic acid can be elongated and desaturated by human enzymes to form arachidonic acid (20:4(n-6), which is found in a much smaller pool in the body, primarily in phospholipids in cell membranes. Linoleic acid is also found in some special ceramides of the skin (see Figure 3.4), where it serves an essential function in maintaining the water permeability barrier of the skin. Omnivorous humans have a small dietary intake of arachidonic acid, derived from animal products. The very small daily endogenous production of signalling eicosanoids (i.e. prostaglandins, leukotrienes, and other oxidized arachidnoic acid derivatives) is far smaller than the turnover of arachidnoic acid, either formed endogenously from linoleic acid or available in the diet. Besides the eicosanoids, arachidonic acid is also a precursor for signaling endocannabinoids, which probably also are formed in minute amounts. (After Hansen and Artmann, 2008)

molecules like prostaglandins, leukotrienes, and various epoxy- and hydroxyl-derivatives as well as endocannabinoids (Figure 3.3) (Hansen and Artmann, 2008; Khanapure *et al.*, 2007; Murphy and Gijon, 2007; Serhan *et al.*, 2008). Usually, they are as a unified group called eicosanoids. This group of signalling molecules represents a huge number of different oxygenated derivatives, formed from arachidonic acid by several different human enzymes, and these oxygenated derivatives all seems to have some biological activity (i.e. possible signalling function) that can be considered beneficial or harmful. Linoleic acid can also be converted to a number of oxygenated derivatives, e.g. 9-hydroxy-octadecanoic acid, by some of the same enzymes. The enzymes involved are COX, LOX, and cyt P450 enzymes, as well as enzymes involved in endocannabinoid formation. The signalling molecules functions as mediators in various physiological processes (e.g. parturition and kidney function) and patophysiological processes (e.g. inflammation, immune responses, and cancer). Kidney failure as well as male infertility, described as two of the late appearing symptoms of deficiency of linoleic acid (Table 3.1), may be caused by lack of prostaglandins derived from arachidonic acid. Leukotrienes are implicated in the pathogenesis of many inflammation-driven diseases, including skin diseases. Conversion of arachidonic acid to leukotrienes can promote inflammation, but hydroxylation of leukotrienes by cyt P450 enzymes,

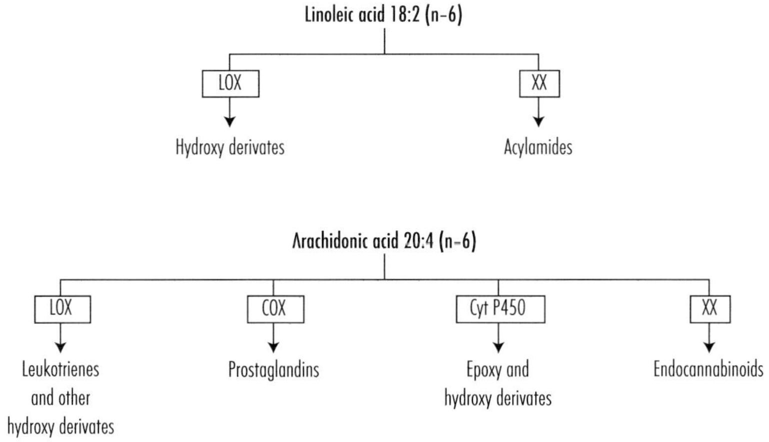

Figure 3.3. Bioactive derivatives of linoleic acid and arachidonic acid.

e.g. CYP4F3A, can inactivate pro-inflammatory leukotrienes in the skin (Kalsotra *et al.*, 2008). Furthermore, some arachidonic acid derivatives are anti-inflammatory (e.g. epoxyeicosatrienoic acid and lipoxin A) while others are pro-inflammatory (Kris-Etherton *et al.*, 2010). This illustrates the complexity in the understanding of the biological functions of these biotransformations of arachidonic acid and linoleic acid. Endocannabinoids derived from arachidonic acid also fulfill many diverse biological functions in the human body (Maccarrone *et al.*, 2010), and recent research has found that endocannabinoids can reduce contact dermatitis in animal models (Karsak *et al.*, 2007).

When essential fatty acids are deficient in the diet (i.e. both (n-6)- and (n-3)-fatty acids) for a prolonged time, an unusual fatty acid called Mead acid, 20:3(n-9), will be formed endogenously, and an increased ratio of the content of this fatty acid to the content of arachidonic acid in the tissues is a recognized biochemical marker for EFA-deficiency (Holman, 1971; Jeppesen *et al.*, 1998). It is not clearly known how this ratio is related to all the different symptoms of EFA-deficiency described in Table 3.1. This ratio is normally below 0.4 in liver and heart tissues and can in rats rise to above 5.0 after several month on an EFA-deficient diet (Holman, 1971).

Linoleic acid is found in high amounts in most vegetable oils like safflower oil, sunflower oil, and corn oil, while lesser amounts are found in olive oil and rapeseed oil. Generally the dietary intake of linoleic acid by humans in the Western societies is between 8-25 g/day, and the intake of arachidonic acid is between 100-300 mg/day for those who eat animal products like meat and fish (Hansen and Artmann, 2008). Vegetarians generally have very high intake of linoleic acid and a very low intake of arachidnoic acid. The daily production of the arachidonic acid-derived signalling molecules within the body is probably less that 10 mg/day. The dietary requirement of linoleic acid for avoiding EFA-deficiency symptoms is below 2 energy% (= ca 4 g/day), and a few of these deficiency symptoms (Table 3.1) have only been seen in very few special cases in humans, especially in some patients receiving parental nutrition (Jeppesen *et al.*, 1998).

Authorities recommend that dietary linoleic acid intake should be 5-10 energy% from the point of view that displacing dietary saturated fat with dietary polyunsaturated fat (i.e. high in linoleic acid) will decrease coronary heart disease events (Kris-Etherton *et al.*, 2010). Some people have argued that the a higher linoleic acid intake (i.e. above the required 1-2 energy%) will result in higher tissue levels of arachidonic acid and thus in a pro-inflammatory state, since arachidnoic acid is precursor for many pro-inflammatory signalling molecules. However, there is no clear evidence that a higher dietary intake of linoleic acid will increase tissue levels of arachidonic acid (Hansen and Artmann, 2008; Kris-Etherton *et al.*, 2010), and as described above oxidative derivatives of arachidonic acid can have both pro-inflammatory and anti-inflammatory properties. Thus, there is no clear indication that a high linoleic acid intake will promote pro-inflammatory conditions.

Some skin diseases, e.g. atopic dermatitis, has been thought to benefit from a dietary supplement of (n-6)-fatty acids like γ-linolenic acid (18:3(n-6)). This fatty acid is usually not present in human food items, but it is found in higher amounts in evening primrose oil and especially in borage oil. Some people have argued that an intake of γ-linolenic acid could result in increased endogenous formation of anti-inflammatory eicosanoids. However, results from studies on dietary supplementation of humans with γ-linolenic acid have been conflicting and confusing. Two recent reviews conclude that there is no or only marginal beneficial clinical effect of supplementing γ-linolenic acid to patients with atopic dermatitis (Bayles and Usatine, 2009; Foster *et al.*, 2010). However, atopic dermatitis seems in many ways to be connected to a defective epidermal barrier function that again may be caused by multiple different biochemical mechanisms (Elias, 2010).

3.3 Linoleic acid and epidermal water permeability barrier

Young rats maintained from weaning on a diet deficient in linoleic acid will within 4-6 weeks develop disturbed water balance due to an increased trans-epidermal water loss (Hansen, 1986). Secondary to the energy loss caused by the increased water evaporation from the skin, the rats also have a poor growth and weight gain. Figure 3.4 illustrates this difference in growth of a control rat and of an EFA-deficient rat of equal age (nr 10 and nr 58), both from the strain of hairless rats. From weaning the rats have been provided with a diet devoid of essential fatty acids (nr 58) or a control diet (nr 10) for 8 weeks. This increase in trans-epidermal water loss can be alleviated by supplementing the rats with small amounts of linoleic acid or of arachidonic acid (Hansen, 1986). However, following these supplementations it is only linoleic acid that is incorporated into the special *O*-acylated ceramides of the skin (Figure 3.5), which are known to serve an important function in maintaining the integrity of the water permeability barrier of the skin (Uchida and Holleran, 2008). In the tissues of EFA-deficient rats, dietary arachidonic acid can be retro-converted by β-oxidation and elongation to linoleic acid, which then is incorporated into the special *O*-acylated ceramides of the skin (Hansen *et al.*, 1986). The pathway of formation of linoleic acid from arachidonic acid seems to be: 20:4(n-6) → 18:4(n-6) → 16:3(n-6) → 14:2(n-6) → 16:2(n-6) → 18:2(n-6). Thus, it is only linoleic acid that by its incorporation fulfills an essential function in maintaining the integrity of the water permeability of the skin, and dietary arachidonic

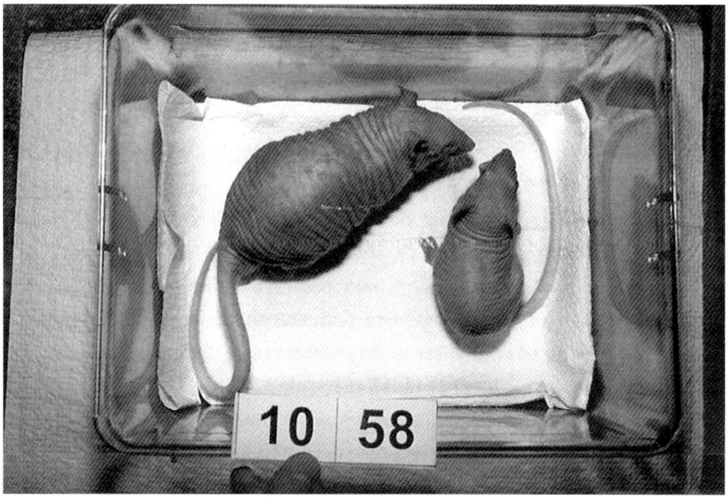

Figure 3.4. An example of a linoleic acid-deficient rat and a control rat, both of the strain of hairless rats. The rat to the left (nr 10) is the control rat, and it has a normal body weight and weak yellow pigmentation on the skin. The rat to the right (nr 58) has received the essential fatty acid-deficient diet, and it shows poor growth (low body weight), increased trans-epidermal water loss, and no pigmentation of the skin. (Jensen *et al.*, 2002)

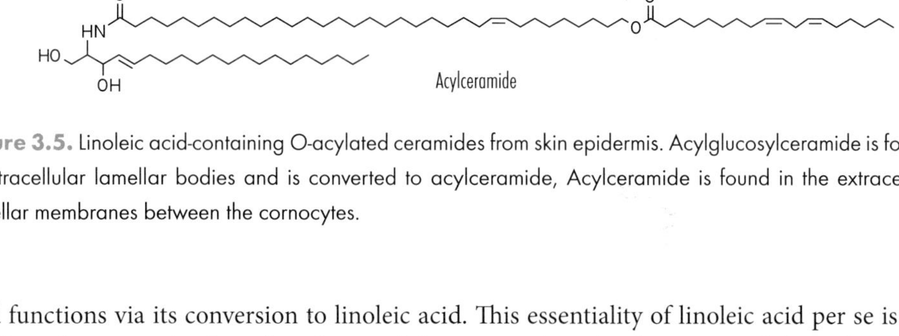

Figure 3.5. Linoleic acid-containing O-acylated ceramides from skin epidermis. Acylglucosylceramide is formed in intracellular lamellar bodies and is converted to acylceramide, Acylceramide is found in the extracellular lamellar membranes between the cornocytes.

acid functions via its conversion to linoleic acid. This essentiality of linoleic acid per se is also confirmed in mice that have knock out of the gene coding for the enzyme delta-6-desaturase (also called FADS2). This enzyme is responsible for the first desaturation step in the enzymatic conversion of linoleic acid to arachidonic acid. These delta-6-desaturase-null mice are not able of generating arachidonic acid from dietary linoleic acid and they do not show the typical EFA-deficiency symptoms of scaly skin, poor growth and increased trans-epidermal water loss on a normal rat chow that is extremely low in arachidonic acid (Stoffel *et al.*, 2008; Stroud *et al.*, 2009), indicating that it is linoleic acid that is important for the epidermal water permeability barrier.

The epidermal water permeability barrier is localized to the outermost layers of epidermis consisting of dead cornocytes embedded in lipids that forms lamellar membranes between the cornocytes. The lipid lamellar membranes consists of a mixture of free fatty acid, cholesterol and several different species of ceramides, including some very unusual *O*-acylated long-chain ceramides having mainly linoleic acid as the *O*-acylated fatty acid (Figure 3.5) (Feingold, 2007). The formation of these epidermal *O*-acylated ceramides requires several metabolic steps, including synthesis of very long-chain fatty acids, omega-hydroxylation of the fatty acids, and esterification of the omega-hydroxy group with primarily linoleic acid. It appears that during differentiation and apoptosis of keratinocytes in the upper *stratum granulosum*, the keratinocytes release the content of their lamellar bodies to the extracellular environment, and this content then eventually form the extracellular lamellar membranes around the dead cornocytes (Figure 3.6). Lipoxygenase metabolites of linoleic acid, e.g. 13-hydoxy(S)-octadecadienoic acid, may stimulate keratinocyte differentiation via the NF-κB signaling pathway (Ogawa *et al.*, 2010). Acylglucosylceramide, which was discovered in human skin in 1978 (Gray *et al.*, 1978), together with other glucosylceramides seems to be formed within the lamellar bodies of the keratinocyte from where the lamellar bodies

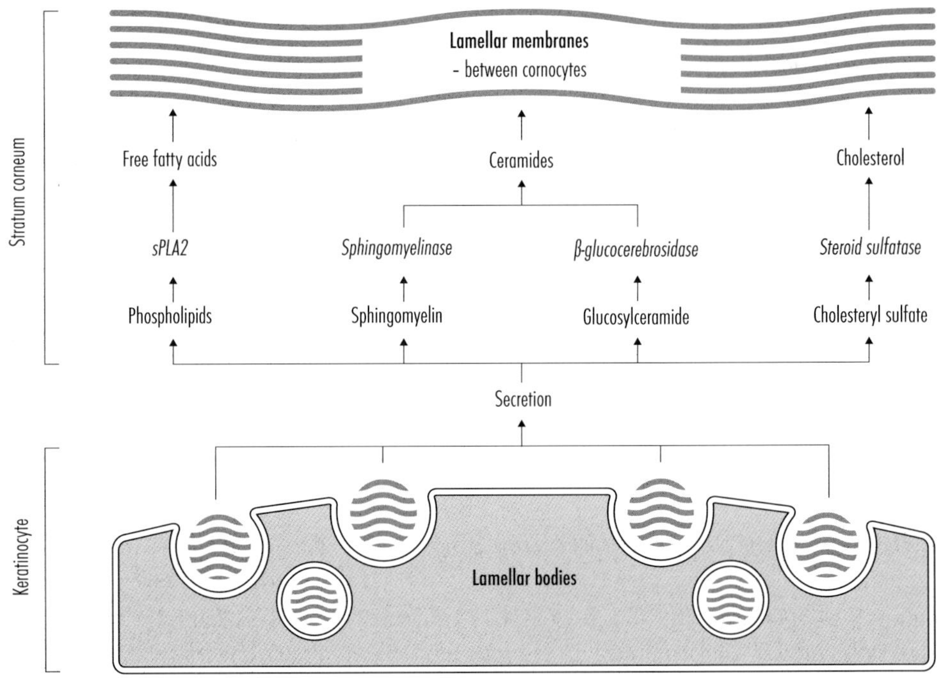

Figure 3.6. The content of lamellar bodies in dying keratinocytes is secreted and processed to become extracellular lamellar membranes, which mainly consists of ceramides, free fatty acids cholestrol. Several enzymes are involved in this process, e.g. secretory phospholipase A_2 ($sPLA_2$) that hydrolyses phospholipids, β-glucosidase that hydrolyses glucosylceramide, sphingomyelinase that hydrolyses sphingomyelin and steroid sulfatase that hydrolyses cholesteryl sulfate. The extracellular lamellar membranes that are positioned between the dead keratinocytes/cornocytes represent a major part of the water permeability barrier of the skin.

eventually are secreted. Following secretion, all glucosylceramides are cleaved by the enzyme β-glucosidase resulting in formation of long-chain ceramides including acylceramide, which has mainly linoleic acid as the *O*-acylated acyl group. Furthermore, phospholipids are degraded to free fatty acids, sphingomyelin to ceramide, and cholesteryl sulphate to cholesterol, and this group of lipids then form the extracellular membranes (Groen *et al.*, 2010).

Linoleic acid used for formation of acylglucosylceramide has been thought to come from hydrolysis of epidermal phospholipids but new research suggest that epidermal triacylglycerol is an important source since a defect in the enzyme CGI-58 results in defective barrier function probably due to lack of linoleic acid (Uchida *et al.*, 2010). Inactivating mutations in the genes for the enzymes 12R-lipoxygenase and epidermal-type Lipoxygenase-3 are causally linked to the development of autosomal recessive congenital ichtyosis, an inherited skin disease characteriszed by dry, thickened and scaly skin as a result of impairment of barrier formation (Juanes *et al.*, 2009), These lipoxygenase enzymes may form epoxy-alcohol dervatives of linoleic acid that in some unknown way may be important for the formation of the water permeability barrier of the skin. As keratinocytes terminally differentiate into dead cornocytes, their plasma membranes are replaced by cornified proteins whereto long-chain omega-hydroxylated fatty acids are esterified. These omega-hydroxylated fatty acids that seems to originate from degraded acylglucosylceramide form a covalently-attached cornocyte lipid envelope to which the extracellular lamellar membranes can adhere to (Uchida and Holleran, 2008). In EFA-deficient rats it is oleic acid instead of linoleic acid that is found in acylglucosylceramide and acylceramide and in some unknown way this results in a defective water permeability barrier (Hansen, 1986). In studies with EFA-deficient rats, dietary α-linolenic acid has not been able to correct the epidermal water permeability barrier (Hansen and Jensen, 1987).

Thus, it is clear that linoleic acid fulfill an important function for maintaining the integrity of the water permeability barrier of the skin, and this function is a major explanation for the essentiality of (n-6)-fatty acids. However, several other biological functions can be ascribed to oxidized derivatives of linoleic acid and arachidonic acid, and some examples are shortly mentioned in the next section.

3.4 Other skin functions of (n-6) fatty acids

Sebum is a complex mixture of lipids that are synthesized in the sebaceous gland and excreted on the skin surface. The sebum mixture consists of triacylglycerol, diacylglycerol, wax esters, squalene and its oxygenated derivatives, cholesteryl esters, and free fatty acids, of which ca. 3.5% is linoleic acid (Camera *et al.*, 2010). It is unclear whether this small amount of linoleic acid has any importance for the function of the skin. However, β-oxidation of dietary linoleic acid in the sebaceous gland may serve as a source of carbon atoms for formation of sebum lipids, e.g. wax esters and squalene (Ottaviani *et al.*, 2010). This is in parallel to observations in the brain where dietary linoleic acid also may serve as a source of carbon atoms in the synthesis of cholesterol and other lipids (Menard *et al.*, 1998).

Loss of hair is one of the symptoms of EFA-deficiency (Table 3.1). One signaling lipid involved in controlling hair growth seems to be lysophosphatidic acid. Lysophosphatidic acid containing polyunsaturated fatty acids may via a G-protein-coupled receptor (called p2y5) control hair growth as it is known that human hair growth deficiency is linked to a genetic effect in the enzyme lipase H, also called phosphatidic acid-specifik phospholipase A2 which generates such bioactive lysophosphatidic acids (Shinkuma *et al.*, 2010). It is at present unclear whether it is important that lysophosphatidic acid must contain an acylated polyunsaturated fatty acid (e.g. linoleic acid or arachidonic acid), or whether an acylated monounsaturated fatty acid is sufficient for the lysophosphatidic acid to have its receptor-stimulating activity. Thus, it is at present also unclear whether the hair loss of EFA-deficient rats can be associated with decreased signaling through the p2y5 receptor.

As discussed above, delta-6-desaturase null-mice that are unable to form arachidonic acid and eicosanoids do not show the usual symptoms of EFA-deficiency (Stoffel *et al.*, 2008; Stroud *et al.*, 2009). However, one of the two research groups working with delta-6-desaturase null-mice reported that the mice had a slight skin irritation that eventually resulted in an obsessive scratching behavior (Stroud *et al.*, 2009). Whether this skin irritation is caused by lack of formation of prostaglandins (Stroud *et al.*, 2009) or endocannabinoids (Karsak *et al.*, 2007) is not clear. Obviously, arachidonic acid metabolites have immune- or allergy-related functions in skin.

In recent years, skin whitening products have become the best selling skin care products in Asia. Different ingredients in the skin care products have been claimed to have a skin whitening effect. One of these ingredients/compounds is linoleic acid. Topically applied linoleic acid has been shown to have a whitening effect on UVB-induced hyper-pigmentated guinea pig skin and possibly also on UVB-treated human skin. The effect is suggested to be caused by inhibition of a tyrosinase enzyme, which is involved in formation of melanin that is the pigment of the skin (Shigeta *et al.*, 2004). However, prober double-blind placebo-controlled human studies are needed for verifying these initial results.

References

Bayles, B. and Usatine, R., 2009. Evening primrose oil. American Family Physician 80, 1405-1408.

Camera, E., Ludovici, M., Galante, M., Sinagra, J.L. and Picardo, M., 2010. Comprehensive analysis of the major lipid classes in sebum by rapid resolution high-performance liquid chromatography and electrospray mass spectrometry. Journal of Lipid Research 51, 3377-3388.

Elias, P.M., 2010. Therapeutic Implications of a Barrier-based Pathogenesis of Atopic Dermatitis. Annals of Dermatology 22, 245-254.

Feingold, KR., 2007. Thematic review series: skin lipids. The role of epidermal lipids in cutaneous permeability barrier homeostasis. Journal of Lipid Research 48, 2531-2546.

Foster, R.H., Hardy, G. and Alany, R.G., 2010. Borage oil in the treatment of atopic dermatitis. Nutrition 26, 708-718.

Gray, G.M., White, R.J. and Majer, J.R., 1978. 1-(3′-O-acyl)-b-glucosyl-N-dihydroxypentatriacontadienoylsphingosine, a major component of the glucosylceramides of pig and human epidermis. Biochimica et Biophysca Acta 528, 127-137.

Groen, D., Gooris, G.S. and Bouwstra, J.A., 2010. Model membranes prepared with ceramide EOS, cholesterol and free fatty acids form a unique lamellar phase. Langmuir 26, 4168-4175.

Hansen, H.S., 1986. The essential nature of linoleic acid in mammals. Trends in Biocheical Sciences 11, 263-265.

Hansen, H.S. and Artmann, A., 2008. Endocannabinoids and nutrition. Journal of Neuroendocrinology 20 Suppl 1, 94-99.

Hansen, H.S. and Jensen, B., 1987. Why is linoleic acid essential? Further studies of fatty acid specificity in correction of transepidermal water loss. In: Lands W.E.M. (ed.). Proceedings of the AOCS Short Course on Polyunsaturated Fatty Acids and Eicosanoids. American Oil Chemists Society, Champaign, IL, USA, pp. 549-554.

Hansen, H.S., Jensen, B. and Von Wettstein-Knowles, P., 1986. Apparent *in vivo* retroconversion of dietary arachidonic to linoleic acid in essential fatty acid-deficient rats. Biochimica et Biophysica Acta 878, 284-287.

Holman, R.T. , 1971. Essential fatty acid deficiency. Progess in the Chemistry of Fats and Lipids 9, 279-348.

Jensen, M., Groth, L., Holmer, G., Hansen, H.S. and Fullerton, A., 2002. The potential of the essential fatty acid-deficient hairless rat as a psoriasis screening model for topical anti-proliferative drugs. Skin Pharmacology and Applied Skin Physiology 15, 401-413.

Jeppesen, P.B., Hoy, C.E. and Mortensen, P.B., 1998. Essential fatty acid deficiency in patients receiving home parenteral nutrition. American Journal of Clinical Nutrition 68, 126-133.

Juanes, S.D., Epp, N., Latzko, S., Neumann, M., Furstenberger, G., Hausser, I., Stark, H.J. and Krieg, P., 2009. Development of an Ichthyosiform Phenotype in Alox12b-Deficient Mouse Skin Transplants. Journal of Investigational Dermatology 129, 1429-1436.

Kalsotra, A., Du, L., Wang, Y., Ladd, P.A., Kikuta, Y., Duvic, M., Boyd, A.S., Keeney, D.S. and Strobel, H.W., 2008. Inflammation resolved by retinoid X receptor-mediated inactivation of leukotriene signaling pathways. FASEB Journal 22, 538-547.

Karsak, M Gaffal, E., Date, R., Wang-Eckhardt, L., Rehnelt, J., Petrosino, S., Starowicz, K., Steuder, R., Schlicker, E., Cravatt, B., Mechoulam, R., Buettner, R., Werner, S., Di Marzo, V., Tuting, T. and Zimmer, A., 2007. Attenuation of allergic contact dermatitis through the endocannabinoid system. Science 316, 1494-1497.

Khanapure, S.P., Garvey, D.S., Janero, D.R. and Letts, L.G., 2007. Eicosanoids in inflammation: biosynthesis, pharmacology, and therapeutic frontiers. Current Topics in Medicinal Chemistry 7, 311-340.

Kris-Etherton, P., Fleming, J. and Harris, W.S., 2010. The debate about n-6 polyunsaturated fatty acid recommendations for cardiovascular health. Journal of American Dieticians Association 110, 201-204.

Lauritzen, L., Hansen, H.S., Jorgensen, M.H. and Michaelsen, K.F., 2001. The essentiality of long chain n-3 fatty acids in relation to development and function of the brain and retina. Progess in Lipid Research 40, 1-94.

Maccarrone, M., Gasperi, V., Catani, M.V., Diep, T.A., Dainese, E., Hansen, H.S. and Avigliano, L., 2010. The endocannabinoid system and its relevance for nutrition. Annual Review of Nutrition 30, 423-440.

Menard, C.R., Goodman, K.J., Corso, T.N., Brenna, J.T. and Cunnane, S.C., 1998. Recycling of carbon into lipids synthesized de novo is a quantitatively important pathway of a-[U-^{13}C]linolenate utilization in the developing rat brain. Journal of Neurochemistry 71, 2151-2158.

Murphy, R.C. and Gijon, M.A., 2007. Biosynthesis and metabolism of leukotrienes. Biochemical Journal 405, 379-395.

Ogawa, E., Owada, Y., Ikawa, S., Adachi, Y., Egawa, T., Nemoto, K., Suzuki, K., Hishinuma, T., Kawashima, H., Kondo, H., Muto, M., Aiba, S. and Okuyama, R., 2010. Epidermal FABP (FABP5) Regulates Keratinocyte Differentiation by 13(S)-HODE-Mediated Activation of the NF-kappaB Signaling Pathway. Journal of Investigational Dermatology 131, 604-612.

Ottaviani, M., Camera, E. and Picardo, M., 2010. Lipid mediators in acne. Mediators of Inflammation Volume 2010, 6 p. Available at: http://www.hindawi.com/journals/mi/2010/858176/.

Roqueta-Rivera, M., Stroud, C.K., Haschek, W.M., Akare, S.J., Segre, M., Brush, R.S., Agbaga, M.P., Anderson, R.E., Hess, R.A. and Nakamura, M.T., 2010. Docosahexaenoic acid supplementation fully restores fertility and spermatogenesis in male delta-6 desaturase-null mice. Journal of Lipid Research 51, 360-367.

Serhan, C.N., Chiang, N. and Van Dyke, T.E., 2008. Resolving inflammation: dual anti-inflammatory and pro-resolution lipid mediators. Nature Review of Immunology 8, 349-361.

Shigeta, Y., Imanaka, H., Ando, H., Ryu, A., Oku, N., Baba, N., Makino, T., 2004. Skin whitening effect of linoleic acid is enhanced by liposomal formulations. Biological and Pharmaceutical Bulletin 27, 591-594.

Shinkuma, S., Akiyama, M., Inoue, A., Aoki, J., Natsuga, K., Nomura, T., Arita, K., Abe, R., Ito, K., Nakamura, H., Ujiie, H., Shibaki, A., Suga, H., Tsunemi, Y., Nishie, W. and Shimizu, H., 2010. Prevalent LIPH founder mutations lead to loss of P2Y5 activation ability of PA-PLA1alpha in autosomal recessive hypotrichosis. Human Mutations 31, 602-610.

Stoffel, W., Holz, B., Jenke, B., Binczek, E., Gunter, R.H., Kiss, C., Karakesisoglou, I., Thevis, M., Weber, A.A., Arnhold, S. and Addicks, K., 2008. Delta6-Desaturase (FADS2) deficiency unveils the role of omega3- and omega6-polyunsaturated fatty acids. EMBO Journal 27, 2281-2292.

Stroud, C.K., Nara, T.Y., Roqueta-Rivera, M., Radlowski, E.C., Lawrence, P., Zhang, Y., Cho, B.H., Segre, M., Hess, R.A., Brenna, J.T., Haschek, W.M. and Nakamura, M.T., 2009. Disruption of FADS2 gene in mice impairs male reproduction and causes dermal and intestinal ulceration. Journal of Lipid Research 50, 1870-1880.

Uchida, Y., Cho, Y., Moradian, S., Kim, J., Nakajima, K., Crumrine, D., Park, K., Ujihara, M., Akiyama, M., Shimizu, H., Holleran, W.M., Sano, S. and Elias, P.M., 2010. Neutral lipid storage leads to acylceramide deficiency, likely contributing to the pathogenesis of Dorfman-Chanarin syndrome. Journal of Investigational Dermatology 130, 2497-2499.

Uchida, Y. and Holleran, W.M., 2008. Omega-O-acylceramide, a lipid essential for mammalian survival. Journal of Dermatological Sciences 51, 77-87.

Key facts

- A substantial proportion of the population of USA, European countries and Japan appears not to meet currently recommended daily intake of vitamin B6.
- Vitamin B6 has an anti-dermatitis factor, and contact allergy to vitamin B6 is generally rare.
- Vitamin B6 has the anti-oxidative, anti-glycation, anti-inflammation and anti-angiogenesis activities.
- Vitamin B6 has protective roles against several diseases including cardiovascular diseases, diabetes, brain diseases and colon cancer.
- For skin cancer, higher consumption of vitamin B6 enhances UV-induced skin tumorigenesis, but not in carcinogen-induced skin tumorigenesis.
- Vitamin B6 has a phototoxic activity under UV exposure.

Summary points

- Vitamin B6 was discovered as an anti-dermatitis factor, and has been believed to be essential for skin development and maintenance.
- Vitamin B6 inadequacy appears to be a global problem, and recently much attention has been paid to the role of vitamin B6 in the development of certain diseases, including cardiovascular disease, diabetes and colon cancer.
- Recent studies have underlined that higher intake of vitamin B6 enhances UV-irradiated skin tumorigenesis, but not carcinogen induced skin tumorigenesis.
- There is evidence that pharmacological dose of vitamin B6 causes phototoxicity under UV-irradiation.
- Thus, it is our recommendation that under strong sunlight, higher dose or abuse of vitamin B6 for skin maintenance should be avoided despite its essentiality for skin health.

4. Role of vitamin B6 in skin health and diseases

N. Kato
Laboratory of Molecular Nutrition, Graduate School of Biosphere Science, Hiroshima University, Higashi-Hiroshima 739-8528, Japan; nkato@hiroshima-u.ac.jp

Abstract

Vitamin B6 has been believed to be essential for skin development and maintenance. Additionally, vitamin B6 has long been considered to have an important role in amino acids metabolism as a coenzyme. Vitamin B6 deficiency has been known to be associated with dermatitis, while contact allergy to vitamin B6 is generally rare. There is accumulating evidence that dietary vitamin B6 has protective roles in cardiovascular disease and diabetes. Recent studies have further highlighted the important protective role of vitamin B6 in carcinogenesis, especially colon carcinogenesis. For skin cancer, a recent study has underlined that dietary supplemental vitamin B6 to a low vitamin B6 diet enhanced UV-irradiated skin tumorigenesis in hairless mice. However, dietary vitamin B6 caused no influence on carcinogen-induced skin tumorigenesis in mice. Topical application of vitamin B6 has been reported to exaggerate UV-irradiated skin phototoxicity. The toxic properties of irradiated vitamin B6 compounds have been also demonstrated for human fibroblasts. Thus, there is concern that excessive dose or abuse of vitamin B6 might cause adverse effect on skin health under certain conditions such as strong sunlight despite its essential roles for skin maintenance.

Keywords: vitamin B6, dermatitis, cancer, skin, phototoxicity

Abbreviations

AGEs	Advanced glycation endproducts
AOM	Azoxymethane
DMBA	Dimethylbezanthracene
PL	Pyridoxal
PLP	Pyridoxal 5'-phosphate
PM	Pyridoxamine
PN	Pyridoxine
ROS	Reactive oxygen species
TBARS	Thiobarbituric acid-reactive substances
TPA	12-*O*-tetradecanoylphorbol-13-acetate
UV	Ultraviolet

4.1 Introduction

In 1932, Paul György discovered vitamin B6 as an anti-dermatitis factor (György and Eckardt, 1932). In the ensuing decades, vitamin B6 in the form of pyridoxal 5'-phosphate had been discovered to serve as a coenzyme in many reactions of human intermediary metabolisms. Tremendous advances have been made in the field of vitamin B6 enzymology during more than half century (Merrill and Henderson, 1987). Beyond the role of vitamin B6 as a coenzyme, recent discovery of potential functions of vitamin B6 such as antioxidant (as radical scavenger), anti-glycation and anti-inflammation activities re-awakened interest in the protective roles of vitamin B6 in certain diseases, including vascular disease, diabetes and cancers (Clayton, 2006). On the other hand, recent growing evidence has suggested that excessive dose of vitamin B6 can cause adverse effect on skin health despite its anti-dermatitis role. This review surveys the utilization of B6 in the treatment of skin health and diseases.

4.2 B6 vitamers and their nutrition

All three forms of vitamin B6, PN, PL and PM, are precursors of an activated compound known as PLP, which plays a vital role as the co-factor of a large number of essential enzymes in the human body (Figure 4.1). Enzymes dependent on PLP focus a wide variety of chemical reactions mainly involving amino acids. Vitamin B6 is obtained from various dietary sources (cereals, meat, fish, poultry, soy and some fruits such as bananas and avocado, etc.), although bioavailability can vary considerably. Foods contain three types of B6 vitamers, PN, PM, and PL. PN is plentiful in plants and readily converted to other forms of vitamin B6 compounds after absorption from the human intestine. PLP accounts for at least 60% of the circulating vitamin B6 in humans. Pharmacologic formulations of vitamin B6 include PN and PN hydrochloride in various dosages.

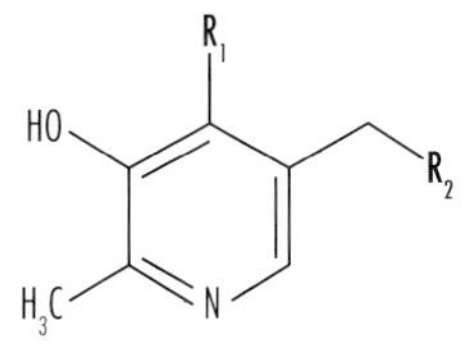

Vitamin B6 compound	R_1	R_2
PL	$-CHO$	$-OH$
PN	$-CH_2OH$	$-OH$
PM	$-CH_2NH_2$	$-OH$
PLP	$-CHO$	$-O-P(=O)(OH)-OH$

Figure 4.1. Chemical structures of vitamin B6 compounds.
PL: pyridoxal, PN: pyridoxine, PM: pyridoxamine and PLP: pyridoxal 5′-phosphate.

Higher intakes of vitamin B6 are required in pregnancy and lactation and possibly also in the elderly (Merrill and Henderson, 1987). Alcohol consumption and high protein intake elevate the requirement of vitamin B6. A substantial proportion of the population of USA, European countries and Japan appears not to meet currently recommended daily intake. Despite the close link of vitamin B6 with human health, vitamin B6 inadequacy is not widely recognized as a problem. There is little evidence that pharmacological doses of vitamin B6 have any beneficial effect. Extremely high intakes (in excess of 500 mg/day) have been reported to cause neurological damage (Clayton, 2006).

Vitamin B6 deficiency is associated with elevation in blood homocysteine, a risk factor for cardiovascular disease, by impairing homocysteine metabolism. Low vitamin B6 status is linked to inflammation, a risk factor of cardiovascular disease (Friso *et al.*, 2001). PN has been reported to have a potent radical scavenger activity (Bilski *et al.*, 2000). PM has an inhibitory activity for glycation (Voziyan and Hudson, 2005). AGEs are marker and predictors of progression of diabetic complications. Large doses of PM appear to be beneficial for the treatment of diabetic nephropathy.

4.3 Vitamin B6 and dermatitis

Skin lesions were the first symptoms identified with vitamin B6 deficiency in rats (György and Eckardt, 1932). Seborrheic dermatitis was subsequently observed in vitamin B6 deficiency in humans (Prasad *et al.*, 1983). The dermatitis caused by this vitamin deficiency has been suggested to be due to impaired collagen biosynthesis (Prasad *et al.*, 1983; Merrill and Henderson, 1987). A role of vitamin B6 in the anti-inflammatory response might also partially relate to the dermatitis

(Friso *et al.*, 2001). PLP is important as a cofactor in the decarboxylation reactions by which skin produces biogenic amines such as catecholamines, histamine and serotonin (Slominski and Wortsman, 2000). Contact allergy to vitamin B6 is generally rare, and there have been a few reports for hypersensitivity to vitamin B6 (Yoshikawa *et al.*, 1985; Friedman *et al.*, 1986; Camarasa *et al.*, 1990).

4.4 Vitamin B6 and cancers

4.4.1 Skin cancer

Vitamin B6 is important for skin development and maintenance. Vitamin B6 deficiency is associated with skin dermatitis. Additionally, vitamin B6 has been reported to have a strong antioxidant effect (Bilski *et al.*, 2000). Therefore, adequate intake of vitamin B6 has been believed to be essential for skin health and maintenance.

UV radiation, particularly UVB wavelengths (280-320 nm), results in an increased generation of ROS that overwhelms the antioxidant defense mechanisms of the skin. ROS are associated with the initiation, promotion and progression of skin cancer. Vitamin B6 has a strong antioxidant effect like a singlet oxygen quencher (Bilski *et al.*, 2000) and prevents lipid peroxidation (Kannan and Jain, 2004). From these facts, vitamin B6 might be beneficial for the skin exposed to UV. Thus, Lu *et al.* (2008) have examined the effect of dietary vitamin B6 on UVB-induced skin tumorigenesis. Surprisingly, they have found the exacerbation of UV-induced skin tumorigenesis by dietary vitamin B6. The data of tumor incidence and tumor multiplicity indicated that dietary supplementation of vitamin B6 to a low vitamin B6 diet enhances UV-induced skin tumorigenesis in hairless mice treated with DMBA (Figure 4.2). In their experiment, PN hydrochloride (HCl) was supplemented to the basal diet at the dose of 1 mg/kg (the minimum level required for preventing the growth depression by this vitamin deficiency), 7 mg/kg (the recommended level) or 35 mg/kg. The skin levels of oxidative stress markers such as lipid peroxides (TBARS) and protein carbonyls (a marker of protein oxidation) were unaffected by dietary level of vitamin B6 (Table 4.1). Vitamin B6 has gained widespread acceptance as being nontoxic, with large doses being given to treat various disease states. Their study indicates that compared with the low vitamin B6 diet, UV-induced skin tumorigenesis is enhanced by the 35 mg PN HCl/kg diet, which is far lower than any acute toxic level (>500 mg PN HCl/kg diet). This adverse effect of dietary vitamin B6 on UV-induced tumorigenesis appears to be similar to the exacerbation of UV-induced carcinogenesis by treatment of β-carotene (a singlet oxygen quencher). Very recently, Kato *et al.* have studied the influence of dietary level of vitamin B6 on DMBA-treated, TPA-induced skin tumorigenesis in mice (unpublished data, 2010). There was no significant difference in the tumor incidence and number among the groups fed 1, 7 and 35 mg PN HCl/kg diet in TPA-induced skin tumorigenesis (Table 4.2). Thus, a higher incidence of skin tumor with a higher vitamin B6 intake appears to be observed in the UV-induced, but not in the TPA-induced skin tumorigenesis. Taken together, these results suggest that an interaction between vitamin B6 and UV-radiation might be responsible for higher carcinogenic process in skin.

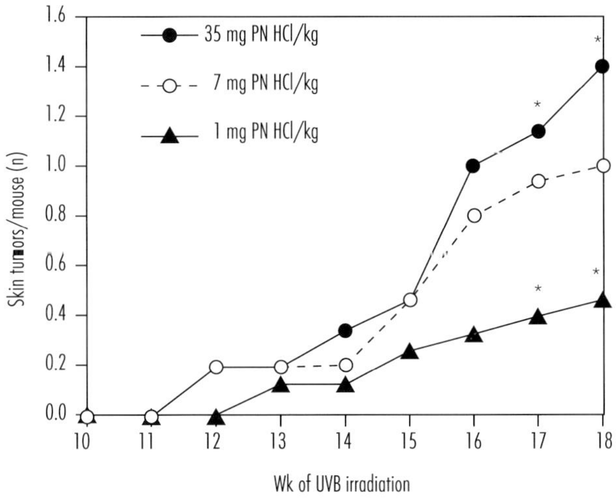

Figure 4.2. Effect of dietary level of vitamin B6 on the cumulative skin tumor multiplicity (skin tumors/mouse) in DMBA-UV treated mice. Values from 15 mice per group are shown. *Significantly different (Pooled SEM = 0.170 and 0.134 at 17 weeks and 18 weeks, respectively, Scheffe's multiple-range test, $P<0.05$). (From Lu *et al.*, 2008: with permission from Center for Academic Publications Japan)

Table 4.1. Effect of dietary vitamin B6 on skin oxidative stress markers, thiobarbituric acid-reactive substances (TBARS) and protein carbonyls and PLP status in DMBA-UV treated hairless mice (From Lu *et al.* 2008: with permission from Center for Academic Publications Japan).

	Dietary pyridoxine HCl		
	1 mg/kg	**7 mg/kg**	**35 mg/kg**
Skin TBARS (nmol/mg protein)	2.35±0.50	1.95±0.32	1.16±0.25
Skin carbonyls (nmol/mg protein)	4.04±0.61	3.92±0.71	4.74±0.79
Skin PLP (nmol/g tissue)	2.61±0.10	2.46±0.18	2.66±0.06
Serum PLP (nmol/l)	$124±8^{b}$	$157±10^{a}$	$151±7^{ab}$

The data indicate means ± SE (n=5). Means in a row not sharing a superscript letter are significantly different (Scheffe's multiple-rang test, $P<0.05$).

Vitamin B6 had been discovered as an anti-dermatitis factor, and has long been believed to be essential for skin development and maintenance. On the basis of such information, vitamin B6 has been often incorporated into various supplements and skin creams for skin care and maintenance. However, it is possible that dietary supplemental vitamin B6 potentiates UV-induced skin tumorigenesis at the dose being far lower than acute toxic level. This raises concern that higher dose of vitamin B6 might be harmful for the skin under strong UV exposure.

Table 4.2. Effect of dietary vitamin B6 on the incidence (%) and number (n) of skin tumors in DMBA-TPA treated mice.

Dietary pyridoxine HCl	7 weeks		9 weeks		11 weeks	
	%	N	%	N	%	N
1 mg/kg	15[1]	0.75±0.10*	45	4.0±1.2	85	11.1±2.2
7 mg/kg	15	0.75±0.10	50	2.3±1.0	85	10.1±2.1
14 mg/kg	5	0.25±0.10	60	2.5±0.5	90	10.0±1.5
35 mg/kg	19	0.95±0.05	75	4.5±2.1	85	15.3±2.6

[1]The data indicate the percentage (%) of mice with tumors from 20 mice per group, and means ± SE (n=20) of the number (n) of tumors per mice.

4.4.2 Colon and other cancers

There have been many epidemiological studies indicating an inverse association between vitamin B6 status and the risk of colon cancer. Larsson *et al.* (2010) have recently reviewed and conducted meta-analysis of several previous studies. Their results showed that the blood PLP levels were inversely associated with the risk of colorectal cancer in the meta-analysis. Consistent with these studies, Komatsu *et al.* (2001) strikingly demonstrated that compared with a low vitamin B6 diet, a moderate dose of dietary vitamin B6 caused a marked reduction in colon tumorigenesis in mice received azoxymethane. It has been suggested that the anti-colon tumor effect is mediated by suppression in colon epithelium cell proliferation, oxidative stress, inflammation and angiogenesis (Komatsu *et al.*, 2003).

There have been limited studies of epidemiological studies on the relation between B6 status and other cancers. For lung cancer, case control study and cohort study in European countries have both indicated a significant association between serum B6 status and incidence of the cancer (Hartman *et al.*, 2001; Johansson *et al.*, 2010). There has been controversy over the role of vitamin B6 in breast cancer (Zhang *et al.*, 2008, 2003) and prostate cancer (Weinstein *et al.*, 2003; Weinstein *et al.*, 2006). Studies on the relation of vitamin B6 status and other cancers, including bladder, pancreatic, ovarian, gastric, bladder, endometrical, oral cancers, etc., are sparse and currently under intense investigation.

4.5 Vitamin B6 and phototoxicity

There have been a few reports of photosensitivity due to PN HCl in pharmacological doses. In 1986, Friedman and coworkers (1986) reported a phototoxic reaction with pyridoxine abuse. Murata *et al.* (1998) have described photosensitive dermatitis caused by PN HCl. It has been reported that vitamin B6 has a strong cytotoxic effect after UV-irradiation in cell culture system of human skin fibroblasts (Wondrak *et al.*, 2004). These studies suggest that UV-induced vitamin B6 cytotoxicity is caused by toxic photoproducts resulting from irradiated vitamin B6. Wondrak *et al.* (2004) have reported that the cytotoxic effects of UV-irradiated B6 vitamers include inhibition of cell proliferation, elevation of protein photocross-linking, peptide photooxidation and intracellular peroxide formation. The addition of several types of antioxidants do not revert UV-irradiated PN effects, implying that ROS formation appears not to be essential for skin phototoxicity of B6 vitamers (Wondrak *et al.*, 2004). Although the mechanism of phototoxicity of vitamin B6 is still unclear, the higher development of skin tumors with higher intake of vitamin B6 mentioned above might be ascribed to the phototoxicity of vitamin B6.

References

Bajaj, A.K., Rastogi, S., Misra, A., Misra, K. and Bajaj, S., 2001. Occupational and systemic contact dermatitis with photosensitivity due to vitamin B6. Contact Dermatitis 44, 184.

Bilski, P., Li M.Y., Ehrenshaft, M., Daub, M.E. and Chignell C.F., 2000. Vitamin B6 (pyridoxine) and its derivatives are efficient singlet oxygen quenchers and potential fungal antioxidants. Photochemistry and Photobiology 71, 129-134.

Black, H.S., 2004. Pro-carcinogenic activity of beta-carotene, a putative systemic photoprotectant. Photochemical and Photobiological Sciences 3, 753-758.

Camarasa, J.G., Serra-Baldrich, E. and Lluch, M., 1990. Contact allergy to vitamin B6. Contact Dermatitis 23, 115.

Clayton, P.T., 2006. B6-responsive disorders: A model of vitamin dependency. Journal of Inherited Metabolic Disease 29, 317-326.

Cohen, P.A. Schneidman, K., Ginsberg-Fellner, F., Sturman, J.A., Knittle, J. and Gaull G.F., 1973. High pyridoxine diet in the rat: possible implications for megavitamin therapy. The Journal of Nutrition 103, 143-151.

DiSorbo, D.M. and Nathanson, L., 1983. High-dose pyridoxal supplemented culture medium inhibits the growth of a human malignant melanoma cell line. Nutrition and Cancer 5, 10-15.

Friedman, M.A. Resnick, J.S. and Baer, R.L., 1986. Subepidermal vesicular dermatosis and sensory peripheral neuropathy caused by pyridoxine abuse. Journal of the American Academy of Dermatology 14, 915-917.

Friso, S., Jacques, P.F., Wilson, P.W.F., Rosenberg, I.H. and Selhub, J., 2001. Low circulating vitamin B6 is associated with elevation of the inflammation marker C-reactive protein independently of plasma homocysteine levels. Circulation 103, 2788-2791.

György P. and Eckardt R.E., 1932. Vitamin B6 and skin lesions. Nature 144, 512.

Hartman, T.J., Woodson, K., Stolzenberg-Solomon, R., Virtamo, J., Selhub, J., Barrett, J.M. and Albanes, D., 2001. Association of the B-vitamins pyridoxal 5'-phosphate (B(6)), B(12), and folate with lung cancer risk in older men. American Journal of Epidemiology 153, 688-694.

Johansson, M., Relton, C., Ueland, P.M., Vollset, S.E., Midttun, Ø., Nygård, O., Slimani, N., Boffetta, P., Jenab, M., Clavel-Chapelon, F., Boutron-Ruault, M.C., Fagherazzi, G., Kaaks, R., Rohrmann, S., Boeing, H., Weikert, C., Bueno-de-Mesquita, H.B., Ros, M.M., Van Gils, C.H., Peeters, P.H., Agudo, A., Barricarte, A., Navarro, C., Rodríguez, L., Sánchez, M.J., Larrañaga, N., Khaw, K.T., Wareham, N., Allen, N.E., Crowe, F., Gallo, V., Norat,

T., Krogh, V., Masala, G., Panico, S., Sacerdote, C., Tumino, R., Trichopoulou, A., Lagiou, P., Trichopoulos, D., Rasmuson, T., Hallmans, G., Riboli, E., Vineis, P. and Brennan, P., 2010. Serum B vitamin levels and risk of lung cancer. Journal of American Medical Association 303, 2377-2385.

Kannan, K. and Jain, S.K., 2004. Effect of vitamin B6 on oxygen radicals, mitochondrial membrane potential, and lipid peroxidation in H_2O_2-treated U937 monocytes. Free Radical Biology and Medicine 36, 423-428.

Komatsu, S. Watanabe, H., Oka, K., Tsuge, H., Nii, H. and Kato, N., 2001. Vitamin B-6-supplemented diets compared with a low vitamin B-6 diet suppress azoxymethane-induced colon tumorigenesis in mice by reducing cell proliferation. The Journal of Nutrition 131, 2204-2207.

Komatsu, S. Yanaka, N., Matsubara, K. and Kato, N., 2003. Antitumor effect of vitamin B6 and its mechanisms. Biochimica et Biophysica Acta 1647, 127-130.

Larsson, S.C., Orsini, N. and Wolk, A., 2010. Vitamin B6 and risk of colorectal cancer, A meta-analysis of prospective studies. Journal of American Medical Association 303, 1077-1083.

Lu, T., Xu, Y., Monttinen, E.S. and Kato, N., 2008. Supplemental vitamin B6 to a low vitamin B6 diet exaggerates UVB-induced skin tumorigenesis in DMBA-treated hairless mice. Journal Nutritional Science and Vitaminology 54, 262-265.

Merrill Jr., A.H. and Henderson, J.M., 1987. Diseases associated with defects in vitamin B6 metabolism and nutrition. Annual Reviews of Nutrition 7, 137-156.

Murata, Y. Kumano, K., Ueda, T., Araki, N., Nakamura, T. and Tani, M., 1998. Photosensitive dermatitis caused by pyridoxine hydrochloride. Journal of the American Academy of Dermatology 39, 314-317.

Prasad, R., Lakshmi, A.V. and Bamji, M.S., 1983. Impaired collagen maturity in vitamins B-2 and B-6 deficiency - probable molecular basis of skin lesions. Biochemical Medicine 30, 333-341.

Slominski, A. and Wortsman, J., 2000. Neuroendocrinology of the skin. Endocrine Reviews 21, 457-487.

Tworoger, S.S. Hecht, J.L., Giovannucci E. and Hankinson, S.E., 2006. Intake of folate and related nutrients in relation to risk of epithelial ovarian cancer. American Journal of Epidemiology 163, 1101-1111.

Voziyan, P.A. and Hudson, B.G., 2005. Pyridoxamine as a multifunctional pharmaceutical: targeting pathogenic glycation and oxidative damage. Cellular and Molecular Life Sciences 62, 1671-1681.

Weinstein, S.J., Hartman, T.J., Stolzenberg-Solomon, R., Pietinen, P., Barrett, M.J., Taylor, P.R., Virtamo, J. and Albanes, D., 2003. Null association between prostate cancer and serum folate, vitamin B(6), vitamin B(12), and homocysteine. Cancer Epidemiology Biomarkers and Prevention 12, 1271-1272.

Weinstein, S.J., Stolzenberg-Solomon, R., Pietinen, P., Taylor, P.R., Virtamo, J. and Albanes D., 2006. Dietary factors of one-carbon metabolism and prostate cancer risk. American Journal of Clinical Nutrition 84, 929-935.

Wondrak, G.T., Roberts, M.J., Jacobson, M.K. and Jacobson, E.L., 2004. 3-Hydroxypyridine chromophores are endogenous sensitizers of photooxidative stress in human skin cells. The Journal of Biological Chemistry 279, 30009-30020.

Yoshikawa, K., Watanabe, K. and Mizuno, N., 1985. Contact allergy to hydrocortisone 17-butyrate and pyridoxine hydrochloride. Contact Dermatitis 12, 55-56.

Zhang, S.M., Cook, N.R., Albert, C.M., Gaziano, J.M., Buring, J.E. and Manson, J.E., 2008. Effect of combined folic acid, vitamin B6, and vitamin B12 on cancer risk in women. A randomized trial. Journal of American Medical Association 300, 2012-2021.

Zhang, S.M. Willett, W.C., Selhub, J. Hunter, D.J., Giovannucci, E.L., Holmes, M.D., Colditz, G.A. and Hankinson, S.E., 2003. Plasma folate, vitamin B6, vitamin B12, homocysteine, and risk of breast cancer. Journal of the National Cancer Institute 95, 373-380.

Key facts

- Antioxidants are substances that combine with or neutralize reactive oxygen species (ROS) to prevent oxidative damage to cells and tissues.
- Human skin has endogenous enzymatic antioxidants and exogenous antioxidants which can be applied topically or systemically.
- Endogenous enzymatic antioxidants include superoxide dismutase, catalase, glutathione peroxidase.
- Endogenous nonenzymatic antioxidants including vitamin E, vitamin C, coenzyme Q10, glutathione.
- Endogenous antioxidative systems attenuate the production of ROS or their ensuing damage.
- Exogenous antioxidants have been introduced to mitigate the damage caused by ROS and have been largely used in products to offer complementary protection against skin damage.

Summary points

- Human skin is a continual target for chemical toxicity, due to its constant exposure to xenobiotics.
- Antioxidants are substances that combine with or neutralize ROS to prevent or treat oxidative damage to cells and tissues.
- There are many kinds of antioxidants.
- Well characterized anti-oxidants that may be useful in the treatment of skin conditions, either cosmetically or therapeutically, include vitamin E, ferulic acid, coenzyme Q10 (ubiquinone), lycopene, curcumin, vitamin C (ascorbic acid), glutathione, green tea, silymarin, resveratrol, grape seeds extract, alpha lipoic acid, genistein and melatonin.
- Other antioxidants include extracts or pure compounds of coffee, polypodium leucotomos, pomegranate, pycnogenol, dehydroepiandrosterone, selenium, quercetin and rosemarinic acid.
- Evidence suggests that antioxidants in combination with other agents, i.e. cocktails, may offer superior benefits compared to one antioxidant alone.
- Many studies are at their preclinical or experimental stage and their application to main stream dermatology awaits further confirmatory or clinical studies.

5. Antioxidants and skin: an overview

Y. Wu[1], H.-D. Chen[1], Y.-H. Li[1], X.-H. Gao[1] and V.R. Preedy[2]
[1]Department of Dermatology, No.1 Hospital of China Medical University, 155 North Nanjing Street, Shenyang 110001, China; [2]Nutritional Sciences Division, School of Biomedical & Health Sciences, Kings College London, Franklin-Wilkins Building, 150 Stamford Street, London SE1 9NH, United Kingdom; gaobarry@hotmail.com

Abstract

As the largest and outermost organ of the body, human skin is a continual target for chemical toxicity which has the potential to cause tissue damage. In combination with certain chemicals, the presence of additional or existing oxidative stress may exacerbate the degree of damage, due to photoactivation of the chemical. In addition, oxidative-stress per se is a known trigger for cellular pathology. Antioxidants are substances which prevent oxidative damage to cells and tissues. Endogenous enzymatic antioxidants include superoxide dismutase, catalase, glutathione peroxidase; and nonenzymatic antioxidants include vitamins E and vitamin C, coenzyme Q10, glutathione, etc. However, oxidative imbalance will arise when such cellular defences are not able to counteract the rate at which free radicals are generated. An excess of free radicals will cause structural or functional damage to cellular macromolecules leading to perturbations in metabolism, functional impairment, necrosis or apoptosis. In this chapter, we review studies on the use of antioxidants on the skin.

Keywords: antioxidants, skin, reactive oxygen species

Abbreviations

8-OHdG	8-hydroxy-2'-deoxyguanosine
ALA	α-lipoic acid
ATP	Adenosine triphosphate
BP	Benz(a)pyrene
CAT	Catalase
CEES	2-chloroethyl ethyl sulfide
CL	Chemiluminescense
COX-2	Cyclooxygenase-2
DHLA	Dihydrolipoic acid
DMBA	Dimethylbenz[a]anthracene
DNCB	2,4-dinitrochlorobenzene
DOPA	Dihydroxyphenylalanine
DPPH	1,1-diphenyl-2-picrylhydrazyl
ERK	Extracellular signal-regulated kinase
FAK	Focal adhesion kinase
Grb2	Growth-factor receptor-bound protein 2
GSE	Grape seed extract
GSH-Px	Glutathione peroxidase
GSPs	Grape seed proanthocyanidins
GPx	Glutathione peroxidase
HD	Sulfur mustard
HO-1	Heme oxygenase-1
ICAM-1	Intercellular adhesion molecule-1
ICD	Irritant contact dermatitis
IFN	Interferon
IGF-1	Insulin-like growth factor-1
IKKα	IkappaB kinase alpha
IL	Interleukin
IL-12-KO	IL-12p40 knockout
iNOS	Inducible NO synthase
JNK	Jun N-terminal kinase
LDH	Lactate dehydrogenase
LPS	Lipopolysaccharide
LPO	Lipid peroxides
MAPK	Mitogen-activated protein kinase
MCP	Monocyte chemoattractant protein
MDA	Malondialdehyde
MG	Mono/diglycerides of capric and caprylic acids
MMP	Matrix metalloproteinase
MPO	Myeloperoxidase
NF-κB	Nuclear factor-kappaB

PAF	Platelet-activating factor
PAF-R	PAF-receptor
PARP	Poly ADP ribose polymerase
PCNA	Proliferating cell nuclear antigen
PD	Pyrimidine dimmer
PDGF-BB	Platelet-derived growth factor-BB
PGE_2	Prostaglandin E_2
PKC	Protein kinase C
PMNL	Polymorphonuclear leukocytes
ROS	Reactive oxygen species
SOD	Superoxide dismutase
STAT1	Signal transducer activator of transcription
SVCT-1	Sodium-dependent vitamin C transporter-1
TG	Triglycerides of capric and caprylic acids
TGF-β	Transforming growth factor-beta
TCHQ	Tetrachlorohydroquinone
t-BHP	t-butylhydroperoxide
TPA	12-*O*-tetradecanoylphorbol-13-acetate
TNF-α	Tumor necrosis factor-alpha
UVA	Ultraviolet A
UVB	Ultraviolet B
VEGF	Vascular endothelial growth factor

5.1 Introduction

When O_2 combines with other molecules, an odd number of electrons i.e. ROS may arise. Such ROS include superoxide and hydroxyl radicals and non-radical species such as hydrogen peroxide and singlet oxygen. Even specific chemicals will generate their own reactive species such as the hydroxyethyl radical from ethanol metabolism. In mammalian tissue, leakage of electrons from the respiratory chain of the mitochondria is the main source of ROS during oxidative metabolism. However, constituent ROS generation is often exacerbated by a wide variety of environmental insults including metabolic processing or exposure to environmental toxins, infection or ingested food products. When there is an overproduction of free radicals and cells are not able to neutralize them by their own antioxidant mechanisms, then oxidative stress occurs. ROS are involved in many pathological conditions including photoaging, carcinogenesis and inflammation.

Skin is the largest and outermost organ of the body i.e. approx 1.8 to 2 square meters. Human skin is a continual target for chemical toxicity, due to its constant exposure to xenobiotics. In some circumstances, such as in the presence of certain chemicals, the concomitant presence of oxidative stress will exacerbate chemical toxicity, due to either photoactivation of the chemical or an oxidatively-stressed state in the skin.

Antioxidants are substances that combine with or otherwise neutralize ROS to reduce tissue damage to cells and tissues. Naturally, as with other tissues, human skin has endogenous enzymatic antioxidants including superoxide dismutase, catalase, glutathione peroxidase; and non-enzymatic antioxidants include vitamins E and C, coenzyme Q10, glutathione. However, with ongoing aging and environmental influences, the body's endogenous antioxidative system attenuates and the production of ROS increases. An excess of free radicals will perturb cellular metabolism, affect macromolecules directly and lead to cell damage, functional impairment, necrosis or apoptosis.

Exogenous antioxidants have been investigated to mitigate the damage caused by ROS and have been largely used in commercial products to offer complementary protection to skin. Although some of these antioxidants are not included in the classical definition of nutrients, they have a pivotal role in maintaining the cellular milieu of skin. Here, we review the studies on the use of antioxidants on the skin. It is apparent that many of these studies use cocktails of antioxidants i.e. protective super-imposition.

5.2 Vitamin E

Vitamin E is the generic term for four tocopherols (taken from the Greek words tokos, meaning offspring and phero, meaning to bear) and four tocotrienols. The family of four tocopherols are alpha, beta, gamma, and delta (respectively: α-, β-, γ-, and δ-) tocopherol and a similar notation is used for the four corresponding tocotrienols. Normal human epidermis contains 87% α-tocopherol, 9% γ-tocopherol, 3% γ-tocotrienol and 1% α-tocotrienol (Fuchs *et al.*, 2003). Natural sources of vitamin E include vegetables, oils, seeds, nuts, corn, soy, whole wheat flour, margarine, meats and dairy products. Vitamin E has been used in the healing of wounds of skin, photoprotection such as sunburn, photocarcinogenesis, photoimmunoinhibition, and changes in the dermal matrix i.e. wrinkles. It has also been used in cardiovascular disease, inflammation and immunostimulation (Baumann, 2009).

α-tocopherol is quantitatively the most important vitamin E isoform and comprises the bulk of first line free radical defence in the lipid compartment. It removes free radical intermediates and prevents oxidation reactions, by reacting with lipid radicals produced in the lipid peroxidation chain reaction. Using DPPH, a good correlation has been found between epidermal α-tocopherol, γ-tocopherol and the total vitamin E content with the free radical scavenging activity of the epidermis (Fuchs *et al.*, 2003).

Césarini *et al.* (2003) investigated the capacity of an antioxidant complex containing α-tocopherol together with β-carotene, lycopene, selenium in reducing UV-induced skin damage. After orally administering the cocktail daily for 7 weeks to 25 healthy individuals, UV-induced erythema, p53 expression, sunburnt cells, and lipoperoxide levels were all significantly reduced. This was accompanied by an elevation in the actinic erythema threshold and pigmentation (Césarini *et al.*, 2003). Pre-incubation of HaCaT keratinocytes with α-tocopherol submicron emulsion can reduce UVB-induced damage. After irradiation with 30 mJ/cm^2 and 90 mJ/cm^2 UVB, the cell proliferation rate in the untreated groups decreased by 18% and 40%, respectively, while the cellular viability of the tocopherol-treated group increased by 44% at 24 h (Luo *et al.*, 2007).

5.3 Ferulic acid

Ferulic acid belongs to the family of ***hydroxycinnamic acids***. It is an abundant phenolic phytochemical found in plant cell wall components. Natural sources of ferulic acid are leaves and seeds of many plants, such as cereals, coffee, apples, artichokes, peanuts, oranges, pineapples and wine.

Orally administered ferulic acid completely prevents the formation of skin tumors, reverts the status of phase I and phase II detoxication agents, lipid peroxidaton byproducts and antioxidants to near-normal ranges in 7,12-DMBA-treated mice (Alias *et al.*, 2009). The observation demonstrate that orally administered ferulic acid has potent suppressive effects on cell proliferation during DMBA-induced skin carcinogenesis.

Ferulic acid also has the capacity to prevent UV-induced damage to cells. Ferulic acid is often added as an ingredient to anti-aging supplements. When ferulic acid was incorporated into a formulation of α-tocopherol and/or ascorbic acid, the topical delivery of the vitamins was improved. There was enhanced chemical stability and the photoprotection to solar-simulated irradiation doubled (Lin *et al.*, 2005; Cassano *et al.*, 2009). For example, Murray *et al.* (2008) applied a stable topical formulation (containing 1% α-tocopherol, 15% L-ascorbic acid, and 0.5% ferulic acid) to normal-appearing human skin and a pig skin model. These were then irradiated with solar-simulated UV. The results showed the complex of antioxidants provided substantial UV photoprotection against erythema, sunburnt cells, thymine dimmers, p53 as well as UV-induced cytokine formation including IL-1α, IL-6, IL-8, and IL-10, and TNF-α (Murray *et al.*, 2008).

5.4 Coenzyme Q10 (ubiquinone)

Coenzyme Q10 (ubiquinone, ubidecarenone, coenzyme Q) is a 1,4-benzoquinone, where Q refers to the quinone chemical group, and 10 refers to the number of isoprenyl chemical subunits in its tail. This oil-soluble, vitamin-like substance is found in all cells as part of the electron transfer chain responsible for energy production and provides 95% of the body's energy in the form of ATP. There are three redox states of coenzyme Q10: fully oxidized (ubiquinone), semiquinone (ubisemiquinone), and fully reduced (ubiquinol). A natural resource of coenzyme Q10 is fish and shellfish although animal's foods such heart will have very high concentrations reflecting mitochondrial levels in such tissue (Crane, 2001). It has been used to reduce wrinkles (changes in the dermal matrix), reduce cell death and DNA oxidative stress, increase ATP systhesis, and suppress collagenase following UVA irradiation (Muta-Takada *et al.*, 2009). For example, coenzyme Q10 accelerates production of basement membrane components, i.e. laminin 332 and type IV and VII collagens, in keratinocytes and fibroblasts, respectively. However, it has no effect on type I collagen production in fibroblasts. Coenzyme Q10 also shows protective effects against cell death induced by ROS in keratinocytes (Muta-Takada *et al.*, 2009).

Primary coenzyme Q10 deficiencies cause variable defects of ATP synthesis and oxidative stress in cultured skin fibroblasts. In genes responsible for coenzyme Q10 biosynthesis, there are two mutations: coenzyme Q2 and PDSS2. PDSS2 mutant fibroblasts have only 12% of the normal amount of coenzyme Q10 and markedly reduced ATP synthesis. Coenzyme Q2 mutant fibroblasts, on the other hand, have 30% coenzyme Q10 with partial defects in ATP synthesis, as well as significantly increased ROS production and oxidation of lipids and proteins (Quinzii *et al.*, 2008). Residual coenzyme Q10 in concentrations of 10-15% and >60% are not associated with significant ROS production, whereas 30-50% residual coenzyme Q10 is accompanied by increased ROS production and cell death (Quinzii *et al.*, 2010).

A functional loss of mitochondria represents an inherent part of the cutaneous aging process. Significant age-dependent differences were seen in mitochondrial function of keratinocytes isolated from skin biopsies of young and old donors (Prahl *et al.*, 2008). Coenzyme Q10 positively influences age-affected cellular metabolism and counteracts signs of aging starting at the cellular level (Prahl *et al.*, 2008). Topical application of coenzyme Q10 penetrates into the viable layers of the epidermis, and reduces the level of oxidation and wrinkle depth. Coenzyme Q10 was determined to be effective in ameliorating UVA mediated oxidative stress in human keratinocytes, activation of specific phosphotyrosine kinases, prevention of oxidative DNA damage and suppression of the expression of collagenase in human dermal fibroblasts following UVA irradiation (Hoppe *et al.*, 1999). In the presence of coenzyme Q10, UVB-induced IL-6 production of normal human keratinocyte and MMP-1 production of fibroblasts significantly decreased. A 1% coenzyme Q10 cream for five months reduced wrinkle score grade observed by a dermatologist. Coenzyme Q10 may inhibit the production of IL-6 as well as stimulating fibroblasts in dermis in a paracrine manner, leading to rejuvenation of wrinkled skin (Inui *et al.*, 2008).

5.5 Lycopene

Lycopene (from the Greek word lykopersikon, meaning tomato) is a bright red carotene and carotenoid pigment. The natural resources are red fruits and vegetables, such as tomatoes, pink grapefruit, watermelon, and apricots. After absorbing from the stomach, lycopene is transported in the blood and accumulates in the liver, adrenal glands, and testes. Lycopene has been used to prevent carcinogenesis, cardiovascular diseases and aging.

Lycopene may act as an inhibitor of tumor cells. In one study, lycopene was shown to inhibit PDGF-BB-induced signalling and cell migration in human cultured skin fibroblasts (Wu *et al.*, 2007). Trapping of PDGF by lycopene compromised melanoma-induced fibroblast migration and attenuated signalling transduction in fibroblasts (Wu *et al.*, 2007). In functional studies, lycopene inhibited melanoma-induced fibroblast migration in a noncontact coculture system and attenuated signalling in fibroblasts simulated by melanoma-derived conditioned medium (Chiang *et al.*, 2007).

Microemulsion formulations are efficient and safe systems to increase lycopene delivery to the skin and the antioxidant activity in tissue. In one study, lycopene was incorporated into two microemulsions containing propylene glycol but different oil phases: mono/diglycerides of capric and caprylic acids or triglycerides of the same fatty acids. Both MG- or TG-containing microemulsions markedly increased lycopene penetration in the *stratum corneum* (6- and 3.6-fold, respectively) and in viable layers of porcine ear skin. The antioxidant activity of skin treated with MG-containing microemulsion was found to be 10-fold higher than untreated skin. The cytotoxicity of MG-containing microemulsion in cultured fibroblasts was similar to propylene glycol (considered safe) and significantly less than of sodium lauryl sulfate (a moderate-to-severe irritant) (Lopes *et al.*, 2010).

5.6 Curcumin

Curcumin is a dietary pigment from the plant *Curcuma longa*. It exists in two tautomeric forms, keto and enol. The enol form is more energetically stable in the solid phase and in liquid phase. Curcumin reacts with boric acid to form a red colored compound. It is a phytochemical with diverse antioxidant, anti-inflammatory and anti-proliferative properties.

Curcumin has the photodynamic ability to induce apoptosis. It modulates transmembrane signal transduction via PKC to affect TPA-induced biochemical and molecular alterations in mouse skin. A single topical application of TPA (5 nmol) is able to increase the translocation of PKC from the cytosolic to the particulate fraction of cells. Pre-treatment with curcumin (10 μmol) decreases the TPA-induced translocation of PKC isozymes (α, β, γ, ε, eta). It also blunts the TPA-induced levels of MAPK and transcription factors (c-jun, c-fos and NF-κB), and downstream target proteins associated with cell proliferation (cyclin D1 and ornithine decarboxylase), cell death (Bax and Bcl2), inflammation (cyclooxygenase 2 (COX 2) and prostaglandin E_2 (PGE_2)) and oxidative stress (8-hydroxy-2'-deoxyguanosine (8-OHdG)) (Garg *et al.*, 2008).

Curcumin, even at low concentrations (0.2-1 μg/ml), has an antiproliferative effect when applied in combination with UVA or visible light (Dujic *et al.*, 2007). Curcumin can induce apoptosis in human skin keratinocytes as represented by an increase in fragmented cell nuclei, release of cytochrome c from mitochondria, activation of caspases-9 and -8, and inhibition of NF-κB activity, extracellular regulated kinases 1/2 and protein kinase B (Dujic *et al.*, 2007).

Curcumin also accelerates wound healing in rats. Curcumin treated skin wounds were found to heal much faster as indicated by improved rates of epitheliazation, wound contraction and increased tensile strength (Panchatcharam *et al.*, 2006). Curcumin increases cellular proliferation and collagen synthesis at the wound site, as evidenced by increase in DNA, total protein and type

III collagen content of wounded skin tissues. Concomitantly there are decreases in the levels of LPO, increases in the levels of SOD, CAT and GSH-Px (Panchatcharam *et al.*, 2006).

5.7 Vitamin C (ascorbic acid)

Vitamin C is a water-soluble antioxidant and well known for its role in preventing scurvy. The natural dietary sources are citrus fruit, black currants, red peppers, and leafy green vegetables. It has been used as an antioxidant. It is also required for collagen synthesis and has anti-inflammatory activities. The cosmetic uses of vitamin C include melasma, postlaser erythema and stretch marks (Figure 5.1 and 5.2).

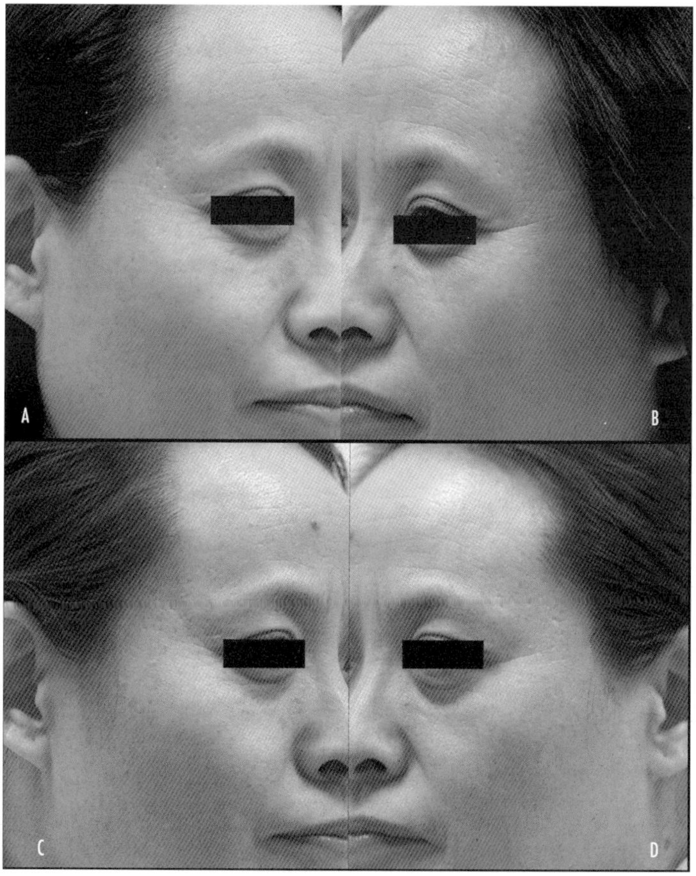

Figure 5.1. Effect of topical 23.8% L-ascorbic acid serum with iontophoresis on canthal wrinkles and pigmented spots. A 50-year-old woman was treated with topical L-ascorbic acid serum with iontophoresis on one side of the face once a day for 2 weeks and the other side was spared of treatment as self control. At baseline, untreated side (A) and treated side (B) presented similar photoaging signs. At the time follow-up (after treatment), canthal wrinkles and pigmented spots significantly reduced on the treated side (C) and remained unchanged on untreated side (D).

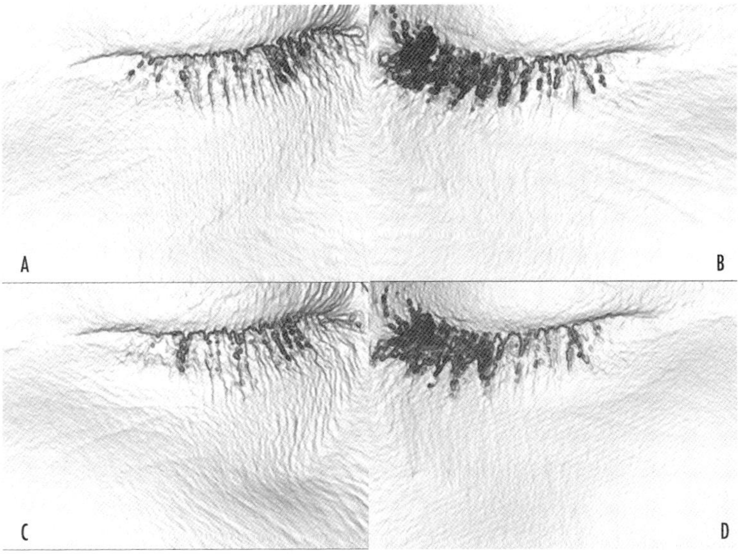

Figure 5.2. Effects of topical 23.8% L-ascorbic acid with iontophoresis on skin roughness and wrinkle measured by (PRIMOS) 3D. A 50-year-old woman was treated with topical L-ascorbic acid serum with iontophoresis on one side of the face once a day for 2 weeks and the other side was spared of treatment as self control. At baseline, untreated side (A) and treated side (B) presented similar roughness and wrinkle severity. At the time follow-up (after treatment), roighness and wrinkle measured by (PRIMOS) 3D remained unchanged on the untreated side (C) and significantly reduced on the treated side (D).

The lipid mediator PAF in UVB-mediated systemic immunosuppression is known to be a major cause of skin cancers. UVB irradiation of human epidermal KB cells results in increased levels of ROS and PAF-R agonistic activity (Yao *et al.*, 2009). Pretreatment of KB cells with vitamin C was shown to inhibit these effects. Vitamin C also abolishes UVB-mediated immunosuppression and blocks production of PAF-R agonists from UVB-irradiated mouse skin (Yao *et al.*, 2009). Another study indicated that vitamin C regulates UVB-induced skin inflammation via the modulation of chemokine production. Vitamin C uptake increases in UVB-irradiated keratinocytes through the translocation of SVCT-1 from the cytosol to the membrane. Vitamin C treatment suppresses chemokine mRNA overexpression, such as IL-8 and MCP-1, indicating vitamin C regulates the inflammatory responses in the skin via the down regulation of IL-8 and MCP-1 production (Kang *et al.*, 2007).

When vitamin C and vitamin E coexist, they demonstrate synergistic effects (Figure 5.3 and 5.4). The proposed mechanism of action is that when vitamin E intercepts a radical, thus forming a complex α-tocopheroxyl-radical, it can be reduced back to α-tocopherol by vitamin C or other agents, thus attenuating the propagation of free radical reactions. Vitamin C prevents the pro-oxidant activity of vitamin E by decreasing the activity of the tocopheroxyl radical to α-tocopherol, thereby contributing to increased total antioxidant status and reducing oxidative stress (Chen *et al.*, 2001).

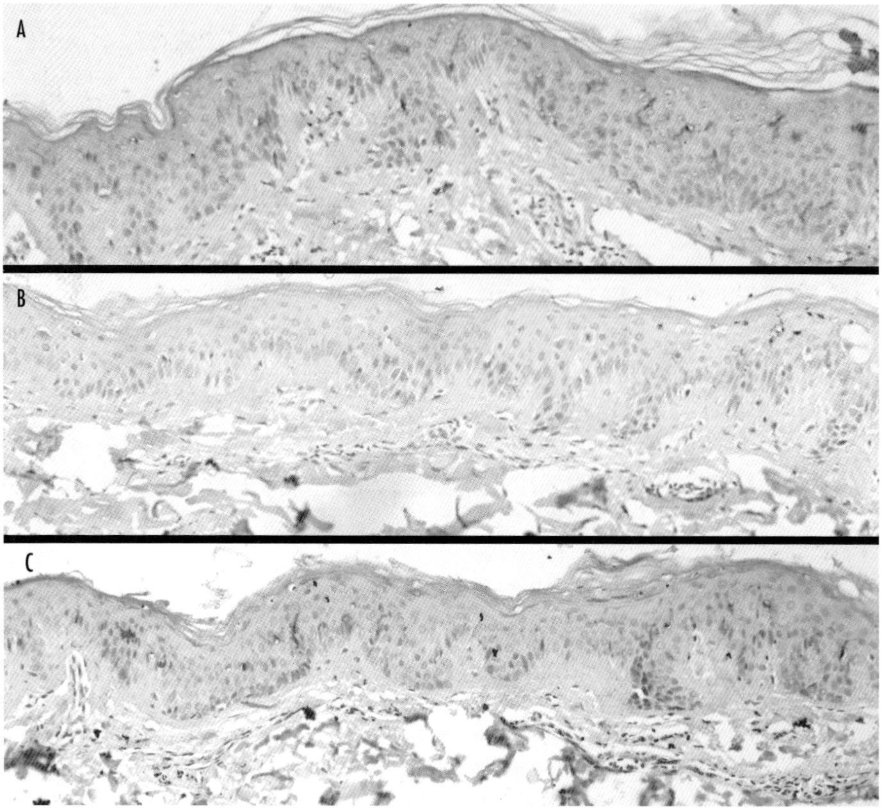

Figure 5.3. Effects of topical antioxidant complex on CD1a positive langerhans cells induced by solar simulator ultraviolet irradiation (Immunohistochemisty staining ×200). The antioxidant complex contained ascorbic acid, tocopherol, ferulic acid, sodium hyaluronate, sodium hydroxide, dipropylene glycol, glycerin laureth-23, phenoxyethanol. (A) UV+antioxidant complex; (B) positive control (UV only); (C) negative control (no UV, no products). After UV irradiation, CD1a positive langerhans cells decreased significantly compared to negative control. Antioxidant complex showed protective effects from the depletion of CD1a positive langerhans cells.

5.8 Glutathione

GSH is the most abundant antioxidant in the network of antioxidants, produced from the amino acids glutamic acid, cysteine, and glycine. It prevents damage to important cellular components caused by reactive species such as free radicals and peroxides.

Exposure to the chemical warfare agent HD is linked to oxidative stress and skin injury. GSH can ameliorate skin injury due to exposure to the HD analog, CEES (Tewari-Singh *et al.*, 2010). In mouse JB6 and human HaCaT epidermal keratinocytes, exogenous GSH (1 or 10 mM) exhibits both protective and therapeutic effects on CEES-induced decreases in cell viability and DNA synthesis, and S and G2-M phase arrest in cell cycle progression. Pre-treatment of mice with 300 mg/kg GSH via oral gavage 1 h prior to topical application of CEES results in significant

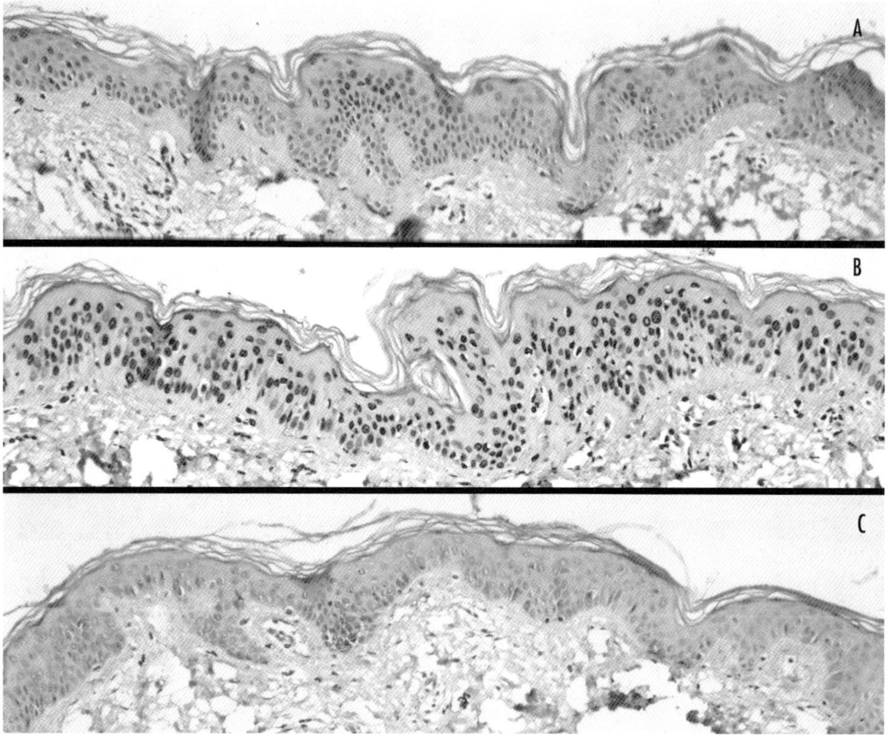

Figure 5.4. Effects of topical antioxidant complex on thymine dimmers induced by solar simulated ultraviolet irradiation (Immunohistochemisty staining ×200). The antioxidant complex contained ascorbic acid, tocopherol, ferulic acid, sodium hyaluronate, sodium hydroxide, dipropylene glycol, glycerin laureth-23, phenoxyethanol. (A) UV+antioxidant; (B) positive control (UV only); (C) negative control (no UV, no products). After UV irradiation, thymine dimmers increased significantly compared to negative control. Antioxidant complex showed protective effects on the reduced Thymine dimmers.

protection against CEES-induced increases in skin bi-fold and epidermal thickness, apoptotic cell death and MPO activity (Tewari-Singh *et al.*, 2010).

Aside from its many ascribed biological functions, GSH has also been implicated in skin lightening. Proposed mechanisms of action include: (1) direct inactivation of the enzyme tyrosinase (the rate-limiting melanogenic enzyme) by binding with the copper-containing active site of the enzyme; (2) mediating the switch mechanism from eumelanin to phaeomelanin production; (3) quenching of free radicals and peroxides that contribute to tyrosinase activation and melanin formation; and (4) modulation of depigmenting abilities of melanocytotoxic agents (Villarama *et al.*, 2005).

GSH may promote the survival rate of random skin flaps and the possible mechanism may be due to the fact that GSH can reduce the level of oxygen free radicals and lipid peroxidation, as well as lessen neutrophil infiltration (Jing *et al.*, 2007). In one animal experiment on twenty rats, skin flaps were made on the back and GSH (250 mg/kg) was injected into the abdominal cavity. The mean survival rate of the skin flap and SOD activity were significantly higher, while MDA level was obviously lower, than the controls. Histological observations also showed reduced neutrophil infiltration, and superior angiogenesis, fibroblasts, fair cells and well presented cutaneous glands with the GSH treatment (Jing *et al.*, 2007).

5.9 Green tea

Green tea is made from unfermented leaves of the plant *Camellia sinensis* and is rich in polyphenols called catechins, which are potent antioxidant flavonoids. Epigallocatechin gallate is particularly abundant in green tea and biologically active. Green tea has been used in the inhibition of tyrosinase activity, inhibition of melanin transfer, anti-inflammation and anti-carcinogenesis.

Green tea polyphenols are able to alleviate the UVB-induced destructive morphological changes in HaCat cells, including shedding of cell membrane microvilli, degeneration of nucleus and nucleols and changes in mitochondrial size and internal cristae. It is considered that green tea polyphenols offers protection against UVB-induced stress via both interacting with UVB-induced reactive oxygen species and attenuating mitochondrion-mediated apoptosis (Wu *et al.*, 2009).

Consumption of green tea polyphenols in drinking water prevents photocarcinogenesis in mice. Administration of green tea polyphenols (0.2%, w/v) in drinking water in one study resulted in a reduction in the levels of markers of inflammation (COX-2, PGE_2, PCNA, and cyclin D1) and proinflammatory cytokines (TNF-α, IL-6, and IL-1β) in chronically UVB-exposed skin and skin tumors of wild-type mice. These effects were less effective in IL-12p40 knockout (IL-12-KO) mice (Meeran *et al.*, 2009). UVB-induced DNA damage was shown to be resolved rapidly in wild-type mice treated with green tea polyphenols and was less pronounced in IL-12-KO mice (Meeran *et al.*, 2009). Oral administration of green tea polyphenols reduces UVB-induced tumor incidence (35%), tumor multiplicity (63%), and tumor growth (55%), expression of MMP-2 and MMP-9, and enhanced expression of tissue inhibitor of MMPs in tumors. Green tea polyphenols also reduces expression of CD31, VEGF and PCNA (Meeran *et al.*, 2006).

5.10 Silymarin

Silymarin is a naturally-occurring polyphenolic flavonoid compound. Silymarin is derived from the seeds of the milk thistle plant. It has inhibitory effects on melanogenesis in a spontaneously immortalized mouse melanocyte cell line, Mel-Ab. In one study (Choo *et al.*, 2009), silymarin was shown to significantly prevent melanin production in a dose-dependent manner with an IC50 value of 28.2 μg/ml, without effects on cell viability (Choo *et al.*, 2009). Also, silymarin inhibits the L-DOPA oxidation activity of the rate-limiting melanogenic enzyme tyrosinase in cell based-systems but it does not directly affect cell-free tyrosinase activity (Choo *et al.*, 2009). Furthermore, silymarin decreases the expression of tyrosinase protein (Choo *et al.*, 2009).

Silymarin is also effective against burn-induced oxidative damage and morphological alterations in rat skin. In one study, Wistar albino rats were exposed to a 90 °C bath for 10 seconds to induce a burn injury. Both local and systemic silymarin treatments significantly reversed high level of TNF-α and LDH levels in blood caused by the burn, and reduced MPO activity and luminol-lucigenin CL in the skin at 48 h after the burn (Toklu *et al.*, 2007).

Topical application of silymarin reduces chemical-induced ICD. Silymarin blocks neutrophil accumulation into the ear induced by irritants. Silymarin also inhibits DNCB-induced expression of TNF-α, keratinocyte, ICAM-1 and E-selectin in mouse ear. Moreover, TNF-α, IL-1β, IL-8, TNF-α- and DNCB-induced NF-κB activation are suppressed by silymarin treatment in HaCaT (Han *et al.*, 2007).

5.11 Resveratrol

Resveratrol is a polyphenolic phytoalexin compound produced naturally by several plants. Resveratrol is found in the skin of red grapes and is a constituent of red wine. Resveratrol has also been produced by chemical synthesis and by biotechnological synthesis. Resveratrol exerts antioxidant, metabolism-regulating and pro-apoptotic/anti-cancer effects and has been suggested to protect skin from ultraviolet (UV)-induced photodamage and photoaging (Baumann, 2009) (Figure 5.5 and 5.6).

p66Shc is a member of the Src homologue and collagen homologue family with redox activity. It has striking interactions with resveratrol. In HaCaT cells, resveratrol induces dose- and time-dependent growth arrest, p66Shc-Ser36 phosphorylation, ERK1/2 phosphorylation and AKT dephosphorylation. Resveratrol-induced p66Shc-Ser36 phosphorylation is dependent on ERK1/2 activation. These resveratrol-induced molecular effects are associated with reduced adhesion and reversible growth arrest rather than cell death pathways (Fabbrocini *et al.*, 2010).

Exposure of A431 carcinoma cells to UVB radiation or resveratrol can inhibit cell proliferation and induce apoptosis. The combination of resveratrol and UVB on A431 cells disrupts the NF-κB pathway by blocking phosphorylation of serine 536 and inactivating NF-κB. Resveratrol and UVB treatment also decreases the phosphorylation of tyrosine 701 of the important transcription factor STAT1, and also inhibits the metastatic protein LIMK1, which reduces the motility of A431 cells (Roy *et al.*, 2009). SIRT1 is a member of a highly conserved gene family called sirtuins which encode nicotinamide adenine dinucleotide NAD^+-dependent deacetylases for numerous protein targets involved in various cellular pathways, including stress responses, apoptosis and axonal degeneration. The SIRT1 activator, resveratrol, protects against UV- and H_2O_2-induced cell death by inhibiting UV- and H_2O_2-induced p53 acetylation (Cao *et al.*, 2009).

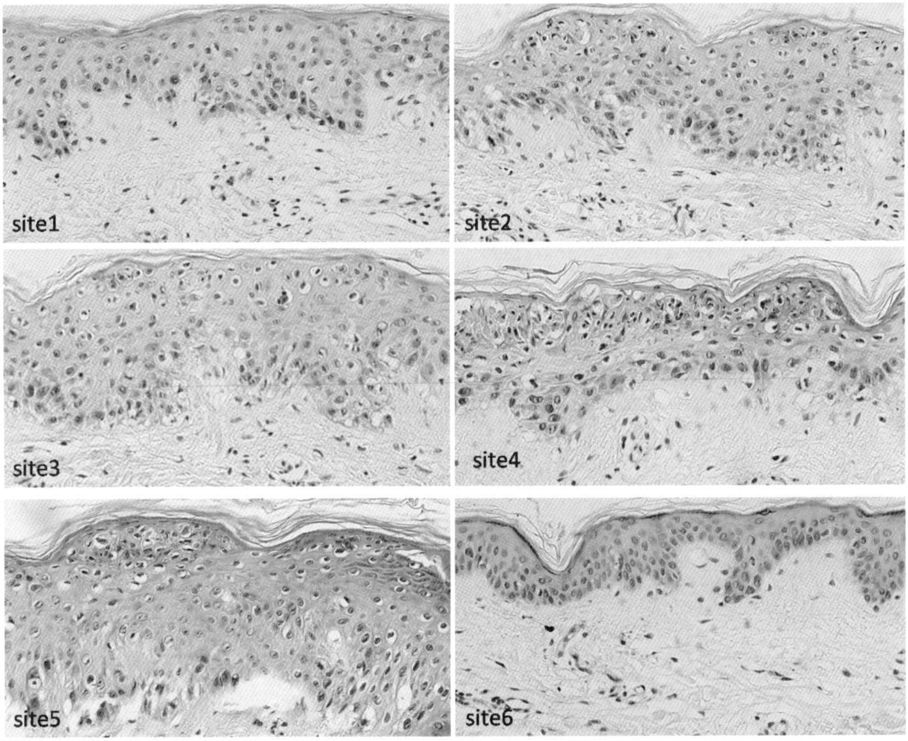

Figure 5.5. Effect of resveratrol on the number of sunburnt cells incuced by solar simulated ultraviolet irradiation (HE staining, ×400). site1: resveratrol+antioxidant+UV; site2: antioxidant+UV; site3: resveratrol+UV; site4: base+UV; site5: UV only; site6: negative control. After solar simulated UV irradiation, a large number of sunburnt cells can be observed in the epidermis of sites 4 and 5. Resveratrol+antioxidant and resveratrol alone effectively inhibit the sunburnt cell formation induced by solar simulator UV irradiation. The antioxidant contains contained ascorbyl phosphate (0.1%), tocopherol acetate (0.5%), echinacea pallida extract (0.01%), chamomile extract (0.12%) and caffeine (0.18%).

5.12 Grape seeds extract

GSEs are industrial derivatives from whole grape seeds. GSEs contain polyphenolic proanthocyanidins and procyanidins, which have strong antioxidant effects. It can facilitate skin wound healing and protect collagen and elastin from degradation. GSEs show tyrosinase-inhibiting activity, and have been used in antiaging and skin-lightening cosmetics.

Oral administration of proanthocyanidin-rich GSE is effective in lightening the UV-induced pigmentation of guinea pig skin (Yamakoshi *et al.*, 2003). GSE decreases the number of 3,4-DOPA-positive melanocytes as well as 8-OHdG-positive, Ki-67-positive, PCNA-positive melanin-containing cells in the basal epidermal layer, and inhibits the activity of tyrosinase and melanogenesis. Thus, the skin lightening effect of GSE may be related to the inhibition of

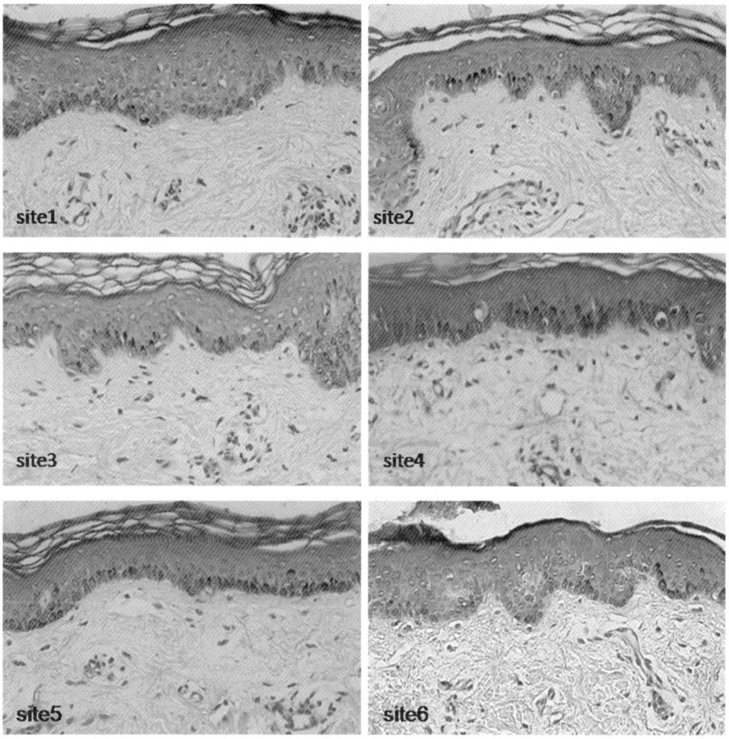

Figure 5.6. Effect of resveratrol on the number of melanin induced by solar simulated ultraviolet irradiation (Fontana Manna staining ×400). site1: resveratrol+antioxidant+UV; site2: antioxidant+UV; site3: resveratrol+UV; site4: base+UV; site5: UV only; site6: negative control. After solar simulated UV irradiation, the measured percentage and average optical density of melanin increased significantly on sites 4 and 5, much higher than those on site1, 2 and 3. The antioxidant contains contained ascorbyl phosphate (0.1%), tocopherol acetate (0.5%), echinacea pallida extract (0.01%), chamomile extract (0.12%) and caffeine (0.18%).

melanin synthesis by tyrosinase in melanocytes and the ROS related proliferation of melanocytes (Yamakoshi *et al*,. 2003).

GSPs have photoprotective effects on UVB-induced oxidative stress and activation of MAPK and NF-κB signaling pathways both in mice skin and in normal human epidermal keratinocytes (Sharma *et al.*, 2007). Treatment with GSPs inhibits UVB-induced depletion of antioxidant defence components, such as glutathione peroxidase, catalase, superoxide dismutase and glutathione, and UVB-induced H_2O_2, lipid peroxidation, protein oxidation, nitric oxide, and DNA damage. GSPs also inhibits phosphorylation of ERK1/2, c-Jun-NH(2)-kinase, p38 proteins of the MAPK family and the expression of PCNA, cyclin D1, inducible nitric oxide synthase, and COX-2. Moreover, GSPs inhibits UVB-induced activation of NF-κB/p65 through inhibition of degradation of IkappaBα and activation of IKKα (Sharma *et al.*, 2007).

Clinically, GSE is effective in reducing the hyperpigmentation of women with chloasma. In one study, the first 6 months of GSE intake was shown to improve or slightly improve chloasma in 10 of the 12 women and following 5 months of intake improved or slightly improved chloasma in 6 of the 11 candidates (Yamakoshi *et al.*, 2004). L* values (lightening) increased and the melanin-index significantly decreased after GSE intake (Yamakoshi *et al.*, 2004).

Topical application of GSE represents a feasible and productive approach to support dermal wound healing. GSE treatment is associated with a more well-defined hyperproliferative epithelial region, higher cell density, enhanced deposition of connective tissue, and improved histological architecture. GSE treatment also accelerates skin wound contraction and closure, increases VEGF and tenascin expression in wound edge tissue, and enhances the oxidizing environment at skin wound sites (Khanna *et al.*, 2002).

5.13 Alpha lipoic acid

ALA is an organosulfur compound derived from octanoic acid. It is a universal lipid- and water-soluble antioxidant and is synthesized in bacteria, yeast, plants, and mammals. It can be used in superficial chemical peeling and acts as an anti-inflammatory agent to reduce post-laser erythema, anti-photoaging and anti-photocarcinogenesis (Baumann, 2009).

ALA enhances the biosynthesis of new collagen in normal human dermal fibroblasts. ALA effectively increases the expression and subsequently the deposition of type I collagen and facilitates the expression of a collagen-processing enzyme, prolyl-4-hydroxylase (Tsuji-Naito *et al.*, 2010). Pretreatment of SB431542, a specific TGF-β receptor type I kinase inhibitor, blocks ALA-mediated Smad 2/3 phosphorylations and both type I collagen and prolyl-4-hydroxylase expression. Levels of TGF-β secretion after 4 hr of treatment with ALA are not remarkably elevated (Tsuji-Naito *et al.*, 2010).

ALA creams in 5% concentrations have been used for treating facial photoaging. Assessments such as self-evaluation by the test subjects, clinical evaluation, photographic evaluation and laser profilometry after using ALA cream show a statistically significant improvement of photoaging. Laser profilometry, an objective method, shows an significant decrease in skin roughness of 51% after ALA cream treatment (Beitner, 2003).

DHLA is a reduced form of ALA. DHLA can be a possible chemopreventive agent in tumorigenesis. DHLA/ALA significantly inhibits LPS-induced NO and PGE(2) formation in RAW 264.7 cells. Meanwhile, treatment with DHLA/LA suppresses the expression of iNOS protein, significantly inhibits the priming and activation stages of skin inflammation induced by TPA application. Furthermore, DHLA inhibits DMBA (0.3 micromol)/TPA (2.0 nmol)-induced skin tumor formation by reducing the tumor incidence and tumor multiplicity (Ho *et al.*, 2007).

5.14 Genistein

Isoflavones, such as genistein and daidzein, are one main group of phytoestrogens. The natural sources are a number of plants including lupin, fava beans, soybeans, kudzu, and psoralea. Isoflavones have antioxidative and photoprotective effects.

Genistein is the most abundant isoflavone of the soy derived phytoestrogen compounds. It can be used to treat or prevent inflammation, photoaging, and photocarcinogenesis. Genistein also prevents certain cardiovascular conditions and osteoporosis. It provides protective effects on collagen biosynthesis in t-BHP-treated fibroblasts via prevention of disturbances in the IGF-I receptor-mediated, ERK1/ERK2-associated signaling pathway evoked by the oxidant (Sienkiewicz *et al.*, 2008). Genistein, at 1 μM, counteracts the inhibition of collagen biosynthesis evoked by t-BHP in fibroblasts. At higher concentrations of 10 μM it exerts significantly diminished protective effects on collagen biosynthesis in fibroblasts, while at 100 μM it inhibits this process. The protective effects of genistein are not related to the modulation of prolidase activity or the expression of the β1-integrin receptor, FAK pp125, Src or Grb2 (Sienkiewicz *et al.*, 2008).

Genistein has photoprotective efficacy on reconstituted skin induced by acute UVB irradiation. It preserves cutaneous proliferation and repair mechanisms, as evidenced by the preservation of proliferating cell populations with increasing genistein concentrations and noticeable paucity in PCNA immunoreactivity in the absence of genistein. It also inhibits UV-induced DNA damage, evaluated with PD immunohistochemical expression profiles (Moore *et al.*, 2006). Moreover, genistein possesses tyrosine kinase inhibitory and/or antioxidant activities for photoaging prevention. Topical genistein inhibits UV induction of extracellular signal-regulated kinase, cJun N-terminal protein kinase, cJun-driven enzyme collagenase and epidermal growth factor receptor tyrosine kinase activities. However, genistein has no effect on UV-induced erythema (Kang *et al.*, 2003).

5.15 Melatonin

Melatonin is a naturally compound found in animals, plants, and microbes. It is a hormone secreted by the pineal gland. Melatonin can directly neutralize a number of ROS, stimulate several antioxidant enzymes, reduce UV-induced erythema, modulate the expression of apoptosis and alleviate sleep disturbances. The mechanism of biological effects is via activation of melatonin receptors or protection of nuclear and mitochondrial DNA.

In the skin, melatonin scavenges and inactivates free radicals arising from UV irradiation. Melatonin represents a substance which protects cells from UVA and UVB action *in vitro* and *in vivo* experiments. Preincubation with melatonin can lead to the normalization of the decreased UV-induced mitochondrial membrane potential. Such effects are followed by suppression of the activation of mitochondrial pathway-related initiator caspase 9 (casp-9), but not of death receptor-dependent casp-8. Melatonin also down-regulates casp-3/casp-7 and reduces PARP activation (Fischer *et al*,. 2008).

One study added melatonin to a culture medium 30 minutes before exposure of keratinocytes and fibroblasts to UVA (15 J/cm^2) and UVB (30 mJ/cm^2, 60 mJ/cm^2) (Izykowska *et al*,. 2009). Melatonin at 10^{-3} M increased the number of surviving keratinocytes and at 10^{-6} M increased the number of surviving fibroblasts exposed to UVB (30 mJ/cm^2, 60 mJ/cm^2). In addition, at 10^{-6} M, melatonin protected keratinocytes exposed to the UVA dose of 30 mJ/cm^2. In these studies melatonin at 10^{-3} M exerted a protective effect on both types of cells irradiated with UVA (15 J/cm^2) (Izykowska *et al.*, 2009).

Melatonin can also inhibit skin carcinogenesis induced by benz(a)pyrene in mice (Deriabina *et al.*, 2010). In one study, the clean-shaven backs of SHR male mice were painted with 0.2 ml of 0.05% acetone solution of BP twice a week. At the same time, mice received melatonin 2 mg/l in their drinking water at nighttime. Six months later, skin tumor frequency decreased by 67% and squamous cell carcinoma decreased by 37%. Treatment with melatonin was followed by significantly lower tumor multiplicity and smaller size, longer latency period and survival of tumor-bearers. Melatonin treatment was also shown to significantly lower MDA content and catalase of blood serum (Deriabina *et al.*, 2010).

Melatonin exerts positive effects on skin wound healing, whether it is administered topically or systemically. When melatonin was deprived through pinealectomy, wound healing was prolonged. Fifteen days after pinealectomy, a bipediculed flap was formed on the back of the rats and then excisional skin wounds were produced. Following treatment with melatonin for 7 days, the wound surface areas were measured and wound tissues were removed. In experimental animals deprived of melatonin through pinealectomy, hydroxyproline, SOD and GSH-Px levels decreased, whereas wound surface areas MDA levels increased. Melatonin administered systemically or topically decreases MDA level and increases SOD and GSH-Px enzyme levels (Ozler *et al.*, 2010).

Melatonin also suppresses DNFB-induced atopic dermatitis-like skin lesions in NC/Nga mice (Kim *et al.*, 2009). In one study, the topical administration of melatonin to DNFB-treated NC/Nga mice was found to inhibit ear thickness increases and the skin lesions induced by DNFB. Furthermore, IL-4 and IFN-γ secretion by activated CD4(+) T cells from the draining lymph nodes of DNFB-treated NC/Nga mice were also significantly inhibited by melatonin, and total IgE levels in serum were reduced (Kim *et al.*, 2009).

5.16 Other antioxidants

Other antioxidants include extracts or pure compounds of coffee, polypodium leucotomos, pomegranate, pycnogenol, dehydroepiandrosterone, selenium, quercetin, rosemarinic acid.

Acknowledgements

This paper was partly supported by Programme for Changjiang Scholars and Innovative Teams in Universities, Ministry of Education, China (IRT0760).

References

Alias, L.M., Manoharan, S., Vellaichamy, L., Balakrishnan, S. Ramachandran, C.R., 2009. Protective effect of ferulic acid on 7,12-dimethylbenz[a]anthracene-induced skin carcinogenesis in Swiss albino mice. Experimental Toxicologic Pathology 61, 205-214.

Beitner H., 2003. Randomized, placebo-controlled, double blind study on the clinical efficacy of a cream containing 5% alpha-lipoic acid related to photoageing of facial skin. British Journal of Dermatology 149, 841-849.

Baumann, L., 2009. Cosmetic dermatology: principles and practice, second edition. McGraw-Hill Medical Publishers, New York, NY, USA.

Cao, C., Lu, S., Kivlin, R., Wallin, B., Card, E., Bagdasarian, A., Tamakloe, T., Wang, W.J., Song, X., Chu, W.M., Kouttab, N., Xu, A. and Wan, Y., 2009. SIRT1 confers protection against UVB- and H_2O_2-induced cell death via modulation of p53 and JNK in cultured skin keratinocytes. Journal of Cellular and Molecular Medicine 13, 3632-3643.

Cassano, R., Trombino, S., Muzzalupo, R., Tavano, L., and Picci, N., 2009. A novel dextran hydrogel linking trans-ferulic acid for the stabilization and transdermal delivery of vitamin E. European Journal of Pharmuetics and Biopharmeutics 72, 232-238.

Césarini, J.P., Michel, L., Maurette, J.M., Adhoute, H. and Béjot, M., 2003. Immediate effects of UV radiation on the skin: modification by an antioxidant complex containing carotenoids. Photodermatology, Photoimmunology and Photomedicine 19, 182-189.

Chen, X., Touyz, R.M., Park, J.B. and Schiffrin, E.L., 2001. Antioxidant effects of vitamins C and E are associated with altered activation of vascular NADPH oxidase and superoxide dismutase in stroke-prone SHR. Hypertension. 38, 606-611.

Chiang, H.S., Wu, W.B., Fang, J.Y., Chen, D.F., Chen, B.H., Huang, C.C., Chen, Y.T. and Hung, C.F., 2007. Lycopene inhibits PDGF-BB-induced signaling and migration in human dermal fibroblasts through interaction with PDGF-BB. Life Sciences 81, 1509-1517.

Choo, S.J., Ryoo, I.J., Kim, Y.H., Xu, G.H., Kim, W.G., Kim, K.H., Moon, S.J., Son, E.D., Bae, K. and Yoo, I.D., 2009. Silymarin inhibits melanin synthesis in melanocyte cells. Journal of Pharmacy and Pharmacology 61, 663-667.

Crane, F.L., 2001. Biochemical functions of coenzyme Q10. Journal of American College of Nutrition 20, 591-598.

Deriabina, O.N., Plotnikova, N.A. and Anisimov, V.N., 2010. Melatonin and metformin inhibit skin carcinogenesis induced by benz(a)pyrene in mice. Vopr Onkol. 56, 583-587.

Dujic, J., Kippenberger, S., Hoffmann, S., Ramirez-Bosca, A., Miquel, J., Diaz-Alperi, J., Bereiter-Hahn, J., Kaufmann, R. and Bernd, A., 2007. Low concentrations of curcumin induce growth arrest and apoptosis in skin keratinocytes only in combination with UVA or visible light. Journal of Investigative Dermatology 127, 1992-2000.

Fabbrocini, G., Kisslinger, A., Iannelli, P., Vitale, N., Procaccini, C., Sparanceo, G., Chieffi, P., Ayala, F., Mencini, F.P. and Tramontano, D., 2010. Resveratrol regulates p66Shc activation in HaCaT cells. Experiamental Dermatology 19, 895-903.

Fischer, T.W., Zmijewski, M.A., Wortsman, J. and Slominski, A., 2008. Melatonin maintains mitochondrial membrane potential and attenuates activation of initiator (casp-9) and effector caspases (casp-3/casp-7) and PARP in UVR-exposed HaCaT keratinocytes. Journal of Pineal Research 44, 397-407.

Fuchs, J., Weber, S., Podda, M., Groth, N., Herrling, T., Packer, L., Kaufmann, R., 2003. HPLC analysis of vitamin E isoforms in human epidermis: correlation with minimal erythema dose and free radical scavenging activity. Free Radic Biol Med.34, 330-336.

Garg, R., Ramchandani, A.G. and Maru, G.B., 2008. Curcumin decreases 12-O-tetradecanoylphorbol-13-acetate-induced protein kinase C translocation to modulate downstream targets in mouse skin. Carcinogenesis. 29, 1249-1257.

Han, M.H., Yoon, W.K., Lee, H., Han, S.B., Lee, K., Park, S.K., Yang, K.H., Kim, H.M. and Kang, J.S., 2007. Topical application of silymarin reduces chemical-induced irritant contact dermatitis in BALB/c mice. International Immunopharmacology 7, 1651-1658.

Ho, Y.S., Lai, C.S., Liu, H.I., Ho, S.Y., Tai, C., Pan, M.H. and Wang, Y.J., 2007. Dihydrolipoic acid inhibits skin tumor promotion through anti-inflammation and anti-oxidation. Biochemical Pharmacology 73, 1786-1795.

Hoppe, U., Bergemann, J., Diembeck, W., Ennen, J., Gohla, S., Harris, I., Jacob, J., Kielholz, J., Mei, W., Pollet, D., Schachtschabel, D., Sauermann, G., Schreiner, V., Stäb, F. and Steckel, F., 1999. Coenzyme Q10, a cutaneous antioxidant and energizer. Biofactors. 9, 371-378.

Inui, M., Ooe, M., Fujii, K., Matsunaka, H., Yoshida, M. and Ichihashi, M., 2008. Mechanisms of inhibitory effects of CoQ10 on UVB-induced wrinkle formation *in vitro* and *in vivo*. Biofactors. 32, 237-243.

Izykowska, I., Cegielski, M., Gebarowska, E., Podhorska-Okolow, M., Piotrowska, A., Zabel, M. and Dziegiel, P., 2009. Effect of melatonin on human keratinocytes and fibroblasts subjected to UVA and UVB radiation *In vitro*. In Vivo. 23, 739-745.

Jing, W., Cen, Y. and Zhang, Y., 2007. Experimental study on effect of reduced glutathione on random flap survival in rats. Zhongguo Xiu Fu Chong Jian Wai Ke Za Zhi. (Chinese journal of reparative and reconstructive surgery) 21, 909-912.

Kang, J.S., Kim, H.N., Jung da, J., Kim, J.E., Mun, G.H., Kim, Y.S., Cho, D., Shin, D.H., Hwang, Y.I. and Lee, W.J., 2007. Regulation of UVB-induced IL-8 and MCP-1 production in skin keratinocytes by increasing vitamin C uptake via the redistribution of SVCT-1 from the cytosol to the membrane. Journal of Investigative Dermatology 127, 698-706.

Kang, S., Chung, J.H., Lee, J.H., Fisher, G.J, Wan, Y.S., Duell, E.A. and Voorhees, J.J., 2003. Topical N-acetyl cysteine and genistein prevent ultraviolet-light-induced signaling that leads to photoaging in human skin *in vivo*. Journal of Investigative Dermatology 120, 835-841.

Khanna, S., Venojarvi, M., Roy, S., Sharma, N., Trikha, P., Bagchi, D., Bagchi, M., Sen and C.K., 2002. Dermal wound healing properties of redox-active grape seed proanthocyanidins. Free Radical Biology and Medicine 33, 1089-1096.

Kim, T.H., Jung, J.A., Kim, G.D., Jang, A.H., Ahn, H.J., Park, Y.S. and Park, C.S., 2009. Melatonin inhibits the development of 2,4-dinitrofluorobenzene-induced atopic dermatitis-like skin lesions in NC/Nga mice. Journal of Pineal Research 47, 324-329.

Lin, F.H., Lin, J.Y., Gupta, R.D., Tournas, J.A., Burch, J.A., Selim, M.A., Monteiro-Riviere, N.A., Grichnik, J.M., Zielinski, J. and Pinnell, S.R., 2005. Ferulic acid stabilizes a solution of vitamins C and E and doubles its photoprotection of skin. Journal of Investigative Dermatology 125, 826-832.

Lopes, L.B., VanDeWall, H., Li, H.T., Venugopal, V., Li, H.K., Naydin, S., Hosmer, J., Levendusky, M., Zheng, H., Bentley, M.V., Levin, R. and Hass, M.A., 2010. Topical delivery of lycopene using microemulsions: enhanced skin penetration and tissue antioxidant activity. Journal of Pharmaceutical Sciences 99, 1346-1357.

Luo, D., Lin, X.F., Min, W., Ma, Q.H., Gu, N., Jin, S.L. and Wang, D.G., 2007. Photoprotection by tocopherol submicron emulsion against UV-mediated damage in HaCaT cells. Methods and Findings in Experimental and Clinical Pharmacology 29, 185-189.

Meeran, S.M., Akhtar, S. and Katiyar, S.K., 2009. Inhibition of UVB-induced skin tumor development by drinking green tea polyphenols is mediated through DNA repair and subsequent inhibition of inflammation. Journal of Investigative Dermatology 129, 1258-1270.

Meeran, S.M., Mantena, S.K., Elmets, C.A. and Katiyar, S.K., 2006. (-)-Epigallocatechin-3-gallate prevents photocarcinogenesis in mice through interleukin-12-dependent DNA repair. Cancer Research 66, 5512-5520.

Moore, J.O., Wang, Y., Stebbins, W.G., Gao, D., Zhou, X., Phelps, R., Lebwohl, M. and Wei, H., 2006. Photoprotective effect of isoflavone genistein on ultraviolet B-induced pyrimidine dimer formation and PCNA expression in human reconstituted skin and its implications in dermatology and prevention of cutaneous carcinogenesis. Carcinogenesis. 27, 1627-1635.

Murray, J.C., Burch, J.A., Streilein, R.D., Iannacchione, M.A., Hall, R.P. and Pinnell, S.R., 2008. A topical antioxidant solution containing vitamins C and E stabilized by ferulic acid provides protection for human skin against damage caused by ultraviolet irradiation. Journal of the American Academy of Dermatology 59, 418-425.

Muta-Takada, K., Terada, T., Yamanishi, H., Ashida, Y., Inomata, S., Nishiyama, T. and Amano, S., 2009. Coenzyme Q10 protects against oxidative stress-induced cell death and enhances the synthesis of basement membrane components in dermal and epidermal cells. Biofactors. 35, 435-441.

Ozler, M., Simsek, K., Ozkan, C., Akgul, E.O., Topal, T., Oter, S. and Korkmaz, A., 2010. Comparison of the effect of topical and systemic melatonin administration on delayed wound healing in rats that underwent pinealectomy. Scandinavian Journal of Clinical and Laboratory Investigation 70, 447-452.

Panchatcharam, M., Miriyala, S., Gayathri, V.S. and Suguna, L., 2006. Curcumin improves wound healing by modulating collagen and decreasing reactive oxygen species. Molecular and Cellular Biochemistry 290, 87-96.

Prahl, S., Kueper, T., Biernoth, T., Wöhrmann, Y., Münster, A., Fürstenau, M., Schmidt, M., Schulze, C., Wittern, K.P., Wenck, H., Muhr, G.M. and Blatt, T., Aging skin is functionally anaerobic: importance of coenzyme Q10 for anti aging skin care. Biofactors 2008. 32, 245-255.

Quinzii, C.M., López, L.C., Von-Moltke, J., Naini, A., Krishna, S., Schuelke, M., Salviati, L., Navas, P., DiMauro, S. and Hirano, M., 2008. Respiratory chain dysfunction and oxidative stress correlate with severity of primary CoQ10 deficiency. FASEB Journal. 22, 1874-1885.

Quinzii, C.M., López, L.C., Gilkerson, R.W., Dorado, B., Coku, J., Naini, A.B., Lagier-Tourenne, C., Schuelke, M., Salviati, L., Carrozzo, R., Santorelli, F., Rahman, S., Tazir, M., Koenig, M., DiMauro, S. and Hirano, M., 2010. Reactive oxygen species, oxidative stress, and cell death correlate with level of CoQ10 deficiency. FASEB Journal 24, 3733-3743.

Roy, P., Madan, E., Kalra, N., Nigam, N., George, J., Ray, R.S., Hans, R.K., Prasad, S. and Shukla, Y., 2009. Resveratrol enhances ultraviolet B-induced cell death through nuclear factor-kappaB pathway in human epidermoid carcinoma A431 cells. Biochememical and Biophysical Research Communications 384, 215-220.

Sharma, S.D., Meeran, S.M. and Katiyar S.K., 2007. Dietary grape seed proanthocyanidins inhibit UVB-induced oxidative stress and activation of mitogen-activated protein kinases and nuclear factor-kappaB signaling in in vivo SKH-1 hairless mice. Molecular Cancer Therapeutics 6, 995-1005.

Sienkiewicz, P., Surazyński, A., Pałka, J. and Miltyk, W., 2008. Nutritional concentration of genistein protects human dermal fibroblasts from oxidative stress-induced collagen biosynthesis inhibition through IGF-I receptor-mediated signaling. Acta Poloniae Pharmaceutica 65, 203-211.

Tewari-Singh, N., Agarwal, C., Huang, J., Day, B.J., White, C.W. and Agarwal, R., 2011. Efficacy of glutathione in ameliorating sulfur mustard analog-induced toxicity in cultured skin epidermal cells and in SKH-1 mouse skin *in vivo*. Journal of Pharmacology and Experimental Therapeutics 336, 450-459.

Toklu, H.Z., Tunali-Akbay, T., Erkanli, G., Yüksel, M., Ercan, F. and Sener, G., 2007. Silymarin, the antioxidant component of Silybum marianum, protects against burn-induced oxidative skin injury. Burns. 33, 908-916.

Tsuji-Naito, K., Ishikura, S., Akagawa, M. and Saeki, H., 2010. α-Lipoic acid induces collagen biosynthesis involving prolyl hydroxylase expression via activation of TGF-β-Smad signaling in human dermal fibroblasts. Connective Tissue Research 51, 378-387.

Villarama, C.D. and Maibach, H.I., 2005. Glutathione as a depigmenting agent: an overview. International Journal of Cosmetic Science 27, 147-153.

Wu, L.Y., Zheng, X.Q., Lu, J.L. and Liang, Y.R., 2009. Protective effect of green tea polyphenols against ultraviolet B-induced damage to HaCaT cells. Human Cell. 22, 18-24.

Wu, W.B., Chiang, H.S., Fang, J.Y. and Hung, C.F., 2007. Inhibitory effect of lycopene on PDGF-BB-induced signalling and migration in human dermal fibroblasts: a possible target for cancer. Biochemical Society Transactions 35, 1377-1378.

Yamakoshi, J., Otsuka, F., Sano, A., Tokutake, S., Saito, M., Kikuchi, M. and Kubota, Y., 2003. Lightening effect on ultraviolet-induced pigmentation of guinea pig skin by oral administration of a proanthocyanidin-rich extract from grape seeds. Pigment Cell Research 16, 629-638.

Yamakoshi, J., Sano, A., Tokutake, S., Saito, M., Kikuchi, M., Kubota, Y., Kawachi, Y. and Otsuka, F., 2004. Oral intake of proanthocyanidin-rich extract from grape seeds improves chloasma. Phytotherapy Research 18, 895-899.

Yao, Y., Wolverton, J.E., Zhang, Q., Marathe, G.K., Al-Hassani, M., Konger, R.L. and Travers, J.B., 2009. Ultraviolet B radiation generated platelet-activating factor receptor agonist formation involves EGF-R-mediated reactive oxygen species. Journal of Immunology 182, 2842-2848.

Micronutrients

Key facts

- The discovery of vitamin D production by the skin and by plants following UVB treatment played a key role in preventing/treating the epidemic of rickets that developed during the industrial revolution (Hess *et al.*, 1921).
- The vitamin D receptor (VDR) is found in most tissues, not just those in bone, intestine, and kidney indicating a role for vitamin D in the function of most cells (Bouillon *et al.*, 2008).
- The kidney is the major source of circulating $1,25(OH)_2D$, but a number of tissues including the epidermis have this capability; the regulation of the extrarenal CYP27B1 differs from that of CYP27B1 in the kidney (Bikle, 2009).

Summary points

- The skin produces vitamin D and converts it to its active form $1,25(OH)_2D$.
- $1,25(OH)_2D$ is the principal ligand for the VDR, which also found in the skin.
- The VDR is a transcription factor regulating the expression of a number of genes including the induction of genes important for the differentiation process and inhibition of genes involved with proliferation.
- These actions of $1,25(OH)_2D$/VDR require coactivators, which govern different functions of the VDR providing some explanation for the sequential action of $1,25(OH)_2D$/VDR in regulating the differentiation process.
- The regulation of hair follicle cycling by VDR does not require $1,25(OH)_2D$.
- Hairless (Hr) is an essential coregulator of hair follicle cycling with VDR.
- VDR protects the skin from tumor development, in part due to its suppression of both the hedgehog and β-catenin pathways.

6. The skin and vitamin D

D.D. Bikle
Veterans Affairs Medical Center and University of California San Francisco, 4150 Clement St (111N), San Francisco, CA 94121; daniel.bikle@ucsf.edu

Abstract

The skin is the major source of vitamin D for the body. It contains ample stores of its precursor 7-dehydrocholesterol that under the influence of ultraviolet radiation (spectrum 280-320) opens the B ring leading to pre vitamin D3 that isomerizes to vitamin D3. Vitamin D2 is formed in similar fashion from the plant sterol ergosterol. Vitamin D when converted to its active metabolites is an important regulator not only of bone mineral homeostasis, but of a wide range of non classical actions including that of differentiation and proliferation of numerous cell types, hormone secretion, and immune function. Moreover, the keratinocytes of the skin are unique in being not only the primary source of vitamin D for the body, but in possessing the enzymatic machinery to metabolize vitamin D to active metabolites (in particular $1,25(OH)_2D$) and the vitamin D receptor (VDR) that enables the keratinocytes to respond to the $1,25(OH)_2D$ they produce. Numerous functions of the skin are regulated by vitamin D and/or its receptor. These include inhibition of proliferation, stimulation of differentiation including formation of the permeability barrier, promotion of innate immunity, regulation of the hair follicle cycle, and suppression of tumor formation. Regulation of these actions is exerted by a number of different coregulators including the coactivators DRIP and SRC, a less well known inhibitor, hairless (Hr), and β-catenin. Different coregulators appear to be involved in different VDR regulated functions. This review will examine the various functions of vitamin D and its receptor in the skin, and to the extent known explore the mechanisms by which these functions are regulated.

Keywords: CYP27B1, differentiation, skin cancer, innate immunity

Abbreviations

$1,25(OH)_2D$	1,25 dihydroxyvitamin D
25OHD	25 hydroxyvitamin D
BCC	Basal cell carcinoma
BCNS	Basal cell nevus syndrome
CBP	CREB binding protein
CTS	Connective tissue sheath
DMBA	7,12-dimethyl benzanthracene
DRIP	Vitamin D receptor interacting protein
Dvl	Disheveled
GSK	3β-glycogen synthase kinase-3β
HAT	Histone acetyl transferase
Hh	Hedgehog
Hr	Hairless
HVDRR	hereditary vitamin D resistant rickets
IRS	Inner root sheath
LEF	Lymphoid enhancer factor
MeT	Methyl transferase
ORS	Outer root sheath
Ptch	Patched
SB	*Stratum basale*
SC	*Stratum corneum*
SCC	Squamous cell carcinoma
SG	*Stratum granulosum*
Shh	Sonic hedgehog
Smoh	Smoothened
SRC	Steroid receptor coactivator
SS	*Stratum spinosum*
Sufu	Suppressor of fused
TCF	T cell factor
THR	Thyroid hormone receptor
TLR	Toll like receptor
UV	Ultraviolet
VDR	Vitamin D receptor
VDRE	Vitamin D response element

6.1 Introduction

The epidermis is the major source of vitamin D for the body. Under the influence of sunlight (ultraviolet radiation, action spectrum 280-320nM or UVB) 7-dehydrocholesterol in the epidermis is converted to vitamin D3. Vitamin D2 is formed in similar fashion from the plant sterol ergosterol. Vitamin D when converted to its active metabolites, in particular 1,25 dihydroxyvitamin D (1,25(OH)$_2$D), is an important regulator not only of bone mineral homeostasis, but of a wide range of non classical actions including that of differentiation and proliferation of numerous cell types, hormone secretion, and immune function. These nutritional aspects of vitamin D will not be covered in detail in this review. Rather the focus will be on the role of vitamin D signaling in the skin. Interest in this subject emanates from the finding that the major cell of the epidermis, the keratinocyte, not only produces vitamin D but possesses the enzymes (CYP27A and CYP27B1) to further metabolize vitamin D to its active form 1,25(OH)$_2$D) (Bikle *et al.*, 1986) and the receptor (VDR) capable of responding to the 1,25(OH)$_2$D so produced (Bikle *et al.*, 1993). 1,25(OH)$_2$D, acting through the VDR, regulates epidermal proliferation in the basal layer (SB) and promotes the sequential differentiation of keratinocytes as they form the upper layers of the epidermis. Loss of VDR or loss of the capacity to produce 1,25(OH)$_2$D (CYP27B1 mutations/deletion) disrupts differentiation of the epidermis and results in hyperproliferation of the basal layers. The keratinocytes lining the outer layer of the hair follicle (ORS) also possess VDR (Bikle *et al.*, 2006). Loss of VDR function either by inactivating mutations or bioengineered deletions leads to loss of hair follicle cycling and alopecia (Sakai and Demay, 2000). In this case, it is less obvious that the VDR requires 1,25(OH)$_2$D for its activity in that deletion of CYP27B1 does not produce alopecia (Bikle *et al.*, 2004). Whether regulation of hair follicle cycling is truly a ligand independent action of VDR or whether an as yet to be identified other ligand is required is unclear. VDR also functions as a tumor suppressor, a function seen in other epithelial tissues such as the colon, breast and prostate. Mice lacking the VDR are susceptible to tumor formation in the skin induced either by chemical means or by UVR (Ellison *et al.*, 2008; Zinser *et al.*, 2002). As for hair follicle cycling, the role of 1,25(OH)$_2$D in this tumor suppressor function is not clear, although some studies support a protective role. The specificity of VDR action within the skin for the different functions it regulates is attributed at least in part to the different coregulators that modulate its genomic actions (Oda *et al.*, 2007). In the proliferating keratinocytes of the epidermis and hair follicle, the DRIP complex (VDR interacting protein complex) also known as Mediator is the dominant coregulator. In the more differentiated keratinocytes of the epidermis, the SRC complexes (SRC 2 and 3) dominate VDR function. In the hair follicle, the coregulator Hr plays an important role. For 1,25(OH)$_2$D regulated VDR actions, Hr acts as a cosuppressor (Xie *et al.*, 2006). But its interaction with VDR in regulating hair follicle cycling, a 1,25(OH)$_2$D independent action of VDR, is less clear. In this review we will examine these different actions of vitamin D and its receptor, emphasizing the many roles vitamin D plays in regulating epidermal proliferation and differentiation, hair follicle cycling, and tumorigenesis.

6.2 Vitamin D regulation of epidermal proliferation and differentiation

The epidermis is composed of four layers of keratinocytes at different stages of differentiation (Figure 6.1) (review in Bikle and Pillai, 1993). The SB rests on the basal lamina separating the dermis and epidermis. Within this layer are the stem cells. These cells proliferate, providing the cells for the upper differentiating layers. They contain an extensive keratin network comprised of keratins K5 and K14. By a process that we are only beginning to understand, cells migrate upward

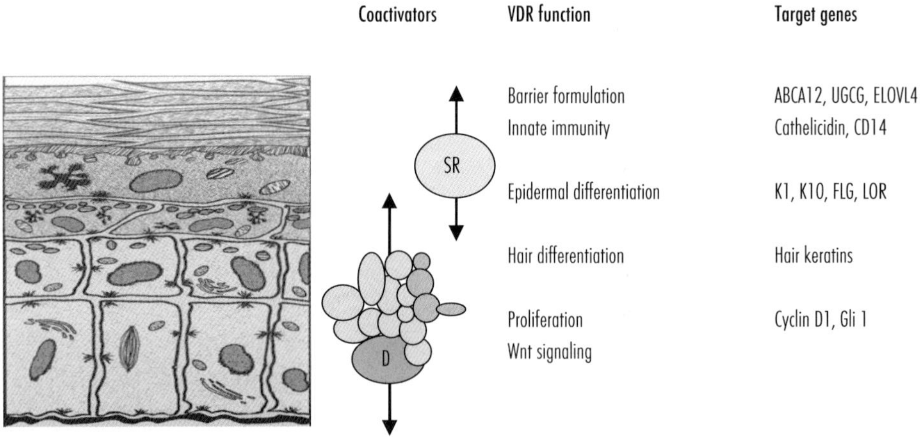

Figure 6.1. The different layers of the epidermis, and the functions within those layers regulated by VDR and its coactivators. The basal layer of the epidermis (stratum basale) contains the stem cells that through proliferation provide the cells for the upper layers. As the keratinocytes leave the basal layer differentiation takes place with K1, K10, involucrin, and transglutaminase being expressed in the stratum spinosum, filaggrin and loricrin being expressed in the stratum granulosum. Lamellar bodies forming in the stratum granulosum inject their lipid content into the intercellular spaces between the stratum granulosum and *stratum corneum* to provide the water proofing for the permeability barrier. DRIP205 is most highly expressed in the stratum basale and spinosum where it participates with VDR in regulating proliferation. SRC3 on the other hand is found in highest concentration in the stratum granulosum where it participates with VDR in the regulation of terminal differentiation.

from this basal layer, acquiring the characteristics of a fully differentiated corneocyte, which is eventually sloughed off. The layer above the basal cells is the SS. These cells initiate the production of the keratins K1 and K10, which are the keratins characteristic of the more differentiated layers of the epidermis. Cornified envelope precursors such as involucrin also appear in the spinous layer as does the enzyme transglutaminase K, responsible for the ε-(γ-glutamyl)lysine cross-linking of these substrates into the insoluble cornified envelope. The SG, lying above the spinous layer, is characterized by electron-dense keratohyalin granules. These are of two types. The larger of the two granules contains profilaggrin, the precursor of filaggrin, a protein thought to facilitate the aggregation of keratin filaments; the smaller granule contains loricrin, a major component of the cornified envelope. The granular layer also contains lamellar bodies, lipid-filled structures that fuse with the plasma membrane, divesting their contents into the extracellular space between the SG and SC where the lipid contributes to the permeability barrier of skin. As the cells pass from the granular layer to the SC, they undergo destruction of their organelles with further maturation of the cornified envelope into an insoluble, highly resistant structure surrounding the keratin-filaggrin complex and linked to the extracellular lipid milieu. The outer layer of the epidermis provides not only a barrier to water loss (permeability barrier) but a barrier to invasion by infectious organisms via its expression of the innate immune system. In particular disruption of the barrier triggers the induction of defensins such as cathelicidin that provide the initial defense in killing such organisms.

$1,25(OH)_2D$ increases the expression of involucrin, transglutaminase, loricrin, and filaggrin and increased cornified envelope formation (Bikle *et al.*, 1991; Hawker *et al.*, 2007; Hosomi *et al.*, 1983; McLane *et al.*, 1990; Pillai and Bikle, 1991; Smith *et al.*, 1986) while inhibiting proliferation. The antiproliferative effects are accompanied by a reduction in the mRNA levels for c-myc (Matsumoto *et al.*, 1990) and cyclin D1 and an increase in the cell cycle inhibitors $p21^{cip}$ and $p27^{kip}$. In addition, $1,25(OH)_2D$ and its receptor regulate the processing of the long-chain glycosylceramides that are critical for permeability barrier formation (Oda *et al.*, 2009) and induce the receptors, TLR2 and its coreceptor CD14, that initiate the innate immune response in skin (Schauber *et al.*, 2007). Activation of these receptors leads to the induction of CYP27B1, which in turn induces cathelicidin resulting in the killing of invasive organisms (Schauber *et al.*, 2006, 2007). Mice lacking the VDR or the enzyme (CYP27B1) producing its ligand $1,25(OH)_2D$ show defective epidermal differentiation manifesting as reduced levels of involucrin and loricrin and loss of keratohyalin granules (Bikle *et al.*, 2006; Xie *et al.*, 2002), decreased lipid content of the lamellar bodies leading to a defective permeability barrier (Oda *et al.*, 2009), and defective response in the innate immune system to wounding (Schauber *et al.*, 2006).

6.3 Role of VDR coactivators in epidermal proliferation and differentiation

The process of epidermal differentiation is sequential. $1,25(OH)_2D$ and VDR regulate all steps from the control of proliferation in the SB, to the regulation of K1, K10, involucrin, and transglutaminase production in the SS, to the regulation of loricrin and filaggrin production in the SG, to the synthesis of lipids required for the permeability barrier in the SC, and to the development of the innate immune system (Bikle *et al.*, 1993; Hawker *et al.*, 2007; Schauber *et al.*, 2007). How does this occur? Although CYP27B1 and VDR are found in highest concentration in the SB, they are both distributed throughout the epidermis (Milde *et al.*, 1991; Stumpf *et al.*, 1984; Zehnder *et al.*, 2001), so this does not provide an obvious explanation for the sequential induction of genes involved in the differentiation process. However, VDR requires the binding of coactivators to stimulate transcription. The two major coactivator complexes in the epidermis are DRIP (Mediator) and SRC (McKenna *et al.*, 1999; Oda *et al.*, 2003). We (Oda *et al.*, 2003) observed that in proliferating keratinocytes DRIP was the major coactivator complex binding to VDR, whereas the SRC complex dominated VDR binding in differentiated keratinocytes (Oda *et al.*, 2003, 2007). These results are consistent with our finding that in the epidermis DRIP205 (Med1) is expressed in highest concentration in the SB and SS, whereas SRC3 is expressed in highest concentration in the SG (Schauber *et al.*, 2008). The DRIP complex is anchored to VDR via DRIP205 (Rachez *et al.*, 1999, 2000a). The SRC complex is anchored to VDR with one of three homologous proteins, SRC1, 2, and 3 (Leo and Chen, 2000), but only SRC2 and 3 are found in keratinocytes (Oda *et al.*, 2003). These coactivator complexes interact with the C terminal (AF2) domain of VDR following ligand binding via LxxLL motifs (NR boxes). They do not bind to VDR at the same time, competing as they do for the same region of the VDR (Rachez *et al.*, 2000a). The SRC coactivators have 3 NR boxes, whereas DRIP205 has 2. Our recent studies (Teichert *et al.*, 2009) indicate that VDR binds most strongly to the 2nd and 3rd NR box of SRC1 and 2, the 3rd NR box of SRC3, and the 2nd NR box of DRIP205. Different nuclear hormone receptors differ in the affinity for the different NR boxes of the different coactivators suggesting some degree of specificity (Acevedo *et al.*, 2004). SRC recruits CBP or P300 and other HATs and MeTs to the VDR resulting in a multisubunit complex (Christakos *et al.*, 2003). The HAT and MeT activity of the SRC complex is thought to destabilize the interaction between DNA and the histone core, enabling transcription to occur. The DRIP complex does not have HAT or MeT activity but functions, at least in part, through recruitment of RNA polymerase II to the transcription start site (Rachez *et al.*, 2000a,b). Some studies suggest that the order of coactivator binding to its nuclear hormone receptor is sequential with different kinetics, generally with SRC binding preceding and being required for DRIP binding (Acevedo *et al.*, 2004). Other studies indicate that the specificity of coactivator binding to VDR depends on the gene being regulated (Carvallo *et al.*, 2008; Issa *et al.*, 2002), the ligand being evaluated (Bouillon *et al.*, 2005), and the cellular context (Maeda *et al.*, 2002; Peleg *et al.*, 2003). Our data, summarized below, indicate that the sequential action of $1,25(OH)_2D$ and its receptor on keratinocyte differentiation is due to the differential expression and distribution of these coactivators according to the differentiation status of the cell coupled with the selectivity of genes regulated by VDR for one or the other of the coactivator complexes.

DRIP205 is expressed in proliferating keratinocytes, and its expression decreases with differentiation, as the expression of SRC3 is increased (Oda *et al.*, 2003). Knockdown of DRIP205 using siRNA results in increased keratinocyte proliferation, similar to that seen by knocking down VDR itself, but knockdown of SRC3 does not show such an effect (Oda *et al.*, 2007). Inhibition of DRIP205 binding to VDR in proliferating keratinocytes blocks VDR transcriptional activity with a VDRE reporter construct, but such inhibition is not seen in differentiated keratinocytes (Oda *et al.*, 2003). However, this inhibition of transcriptional activity turns out to be gene specific. For example, knockdown of DRIP205 using siRNA methodology has a greater impact on keratins 1, 10, and involucrin expression than does knockdown of SRC3, although depletion of both coactivators profoundly reduces loricrin and filaggrin expression (Hawker *et al.*, 2007). On the other hand SRC3 knockdown, not DRIP205 knockdown, reduces glucosylceramide production and lamellar body formation similar to that of VDR knockdown (Oda *et al.*, 2009), and prevents $1,25(OH)_2D$ induced cathelicidin expression (Schauber *et al.*, 2008). Thus, our hypothesis is that SRC facilitates the ability of VDR to regulate the more differentiated functions of the keratinocyte, whereas DRIP facilitates the ability of VDR to regulate proliferation and early keratinocyte differentiation, although some overlap in coactivator function is observed.

6.4 Role of VDR in hair follicle cycling

The hair follicle cycle is divided into three main stages: anagen, catagen, and telogen (Figure 6.2). Anagen is the stage of hair follicle growth; catagen is the stage of regression; telogen is the resting stage. Only the proximal or dermal portion of the hair follicle cycles; the distal or epidermal portion does not. The duration of these stages in a given species varies from location to location on the body and between genders. Furthermore, there are two types of cycles: developmental and postnatal. The developmental cycle is initiated during embryogenesis. The follicle develops from specific regions of the epidermis called placodes. The follicle is induced to grow by its interaction

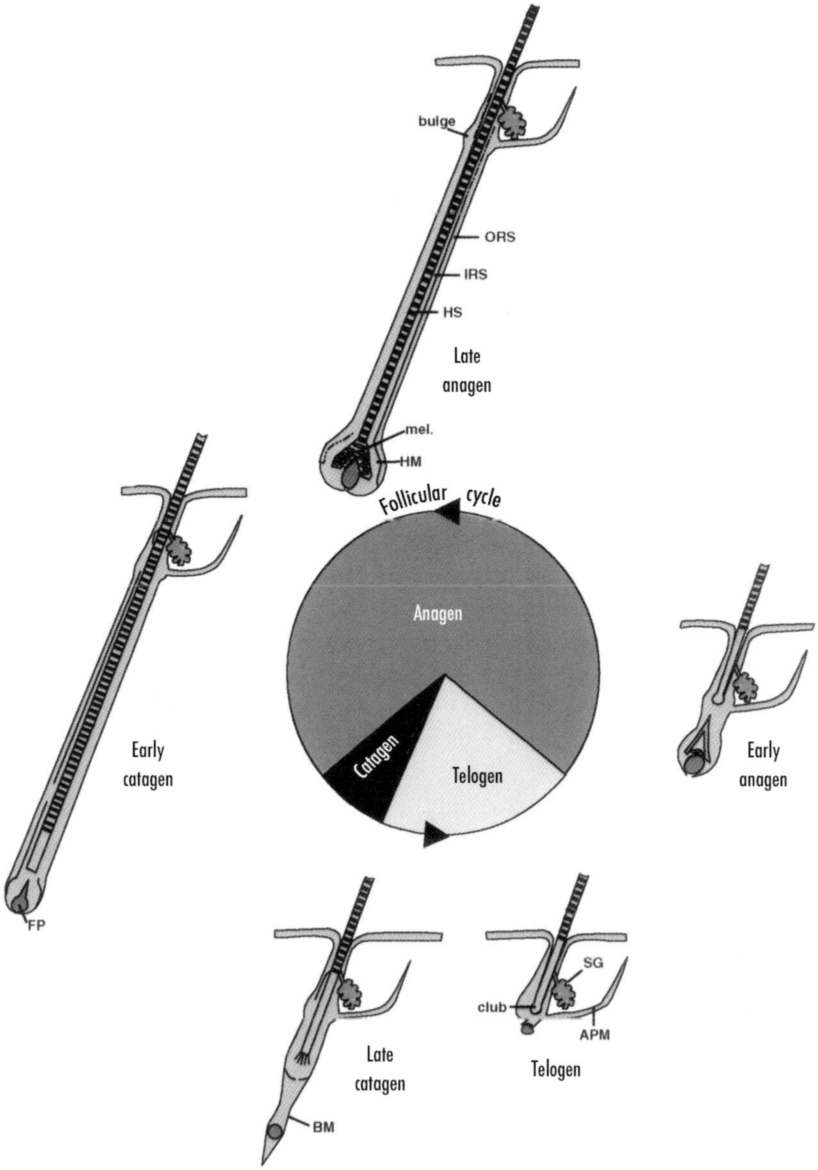

Figure 6.2. Hair follicle cycling. The initial developmental phase of hair follicle development terminates with catagen and the first telogen, after which repetitive cycles of anagen (the growth phase), catagen (the regression phase), and telogen (the resting phase) occur throughout the life span of the animal. In general the hair follicle spends most of its time in anagen. But cycle duration varies according to location, gender, age, species. The bulge is the source of stem cells for the regenerating hair follicle, responding to signals from the dermal or follicular papilla (FP). ORS: outer root sheath, IRS: inner root sheath, HS: hair shaft, mel: melanin for the hair shaft, HM: hair matrix, BM: basement membrane, SG: sebaceous gland, APM: arrector pili muscle. (After Stenn and Paus, 2001)

with a collection of specialized mesenchymal cells in the dermis called the dermal papilla. Wnt signaling (β-catenin) appears to be necessary to maintain the ability of the dermal papilla to stimulate hair follicle growth (Kishimoto *et al.*, 2000; Shimizu and Morgan, 2004). Following the developmental cycle, which leads to the initial coat of hair, the follicle undergoes repetitive cycling until senescence. The length of the hair is dependent on the duration of anagen. During this stage the follicle grows through the dermis into the subcutaneous tissue. As the follicle develops different cell layers appear. The ORS is a direct extension of the *stratum basale*, and separates the hair follicle from the surrounding CTS. From outside in are found the companion layer, the three layers of the IRS – Henle's layer, Huxley's layer, cuticle of the IRS – and the hair shaft itself including the cuticle of the shaft, shaft cortex, and shaft medulla. Stem cells in the bulge are capable of generating all cells in the hair follicle and epidermis (Morris *et al.*, 2004). The keratins produced by the cells of the IRS and hair shaft differ from those expressed by epidermal keratinocytes (Langbein *et al.*, 2003). Of particular interest is these hair keratins have β-catenin/lef1 binding sites in their promoters that regulate their expression (Zhou *et al.*, 1995). Following anagen, the follicle enters catagen during which massive apoptosis occurs primarily in the cells of the proximal follicle (the dermal portion), and the hair shaft produced during anagen is generally shed. At the end of catagen the follicle enters telogen, the resting phase. Duration of telogen is highly variable. A new cycle then begins with anagen. The juxtaposition of the dermal papilla to the bulge is critical for this process to begin, and it is associated with increased proliferation of stem cells in the bulge with migration of cells from the bulge into the hair bulb to restart the growth of the hair follicle. The regulatory elements that control the transition from one stage to the next are not well understood.

6.4.1 VDR is required for hair follicle cycling

Alopecia is a well-known part of the phenotype of many patients with mutations in their VDR (Hochberg *et al.*, 1985; Marx *et al.*, 1986), a syndrome currently known as HVDRR. Vitamin D deficiency per se or CYP27B1 mutations are not associated with alopecia, indicating that the regulation of hair follicle cycling requires VDR but not its principal ligand 1,25$(OH)_2$D. VDR null mice develop their first coat of hair normally, but re-initiation of anagen following the first cycle or after depilation is impaired (Sakai *et al.*, 2000). Reconstitution of the VDR to the VDR null mouse skin using a keratinocyte specific promoter reverses the defect in hair growth without reversing the metabolic defects of skeletal growth retardation, hypocalcemia, and rickets otherwise associated with the VDR null condition (Chen *et al.*, 2001; Kong *et al.*, 2002). On the other hand, correction of the metabolic abnormalities with a high calcium diet prevents the rickets and hyperparathyroidism but does not prevent the alopecia (Li *et al.*, 1998). Furthermore, it is the lack of VDR in the keratinocyte as opposed to the dermal papilla that is critical. Dermal papilla cells obtained from either VDR null or wildtype mice can initially induce hair growth in a hair reconstitution assay when mixed with epidermal keratinocytes obtained from wildtype or VDR null mice, but if the hair grown with keratinocytes from VDR null mice is then depilated, anagen will not be reinitiated regardless of the source of dermal papilla cells (Sakai *et al.*, 2000).

6.4.2 Role of Hr in the regulation of hair follicle cycling and its interactions with VDR

Hr mutations in both mice (Panteleyev *et al.*, 1999) and humans (Ahmad *et al.*, 1998; Miller *et al.*, 2001) result in phenocopies of the VDR null mouse and some human VDR mutations, respectively, with regard to the morphologic changes observed in hair follicle cycling. The dissociation of the dermal papilla from the hair bulb by the end of catagen is thought to account for the failure to initiate the subsequent anagen in both Hr mutant and VDR null mice (Bikle *et al.*, 2006; Huelsken *et al.*, 2001; Panteleyev *et al.*, 1999). The distal (epidermal) portion of the hair follicle including the sebaceous gland as well as the interfollicular epidermis is less impacted (DasGupta *et al.*, 2002; Huelsken *et al.*, 2001; Panteleyev *et al.*, 1999; Xie *et al.*, 2002). We have found VDR and Hr in the nuclei of keratinocytes in the *stratum basale* and ORS (Bikle *et al.*, 2006). However, we found little or no VDR in the IRS and hair bulb or cells of the dermal papilla and CTS, whereas we did find Hr in those locations (Bikle *et al.*, 2006). As will be discussed subsequently, deletion of the transcriptional domain of β-catenin results in a similar phenotype (DasGupta *et al.*, 2002; Huelsken *et al.*, 2001).

Hr has characteristics of a coregulator in that it resides in the nucleus; its structure contains a nuclear localization signal, a putative zinc finger, and three LXXLL motifs (Djabali *et al.*, 2001) like that found in coactivators that interact with nuclear hormone receptors such as VDR as well as ΦXXΦΦ motifs (Φ=hydrophobic amino acid) similar to regions in corepressors like SMRT and NCoR responsible for the binding of these corepressors to nuclear hormone receptors. In the brain, Hr has been suggested as a corepressor of the THRb in that Hr can bind to THRb and inhibit its transcriptional activity (Thompson *et al.*, 1997). However, Hr does not appear to regulate thyroid hormone action in the keratinocyte (Engelhard *et al.*, 2004). Rather, VDR appears to be the target (Xie *et al.*, 2006). Hsieh *et al.* (Hsieh *et al.*, 2003) demonstrated that Hr could bind to VDR in COS cells. They noted that Hr bound to VDR in the same region predicted for corepressor binding, and different from the C-terminal region to which coactivators bind. The region of Hr responsible for VDR binding contains one LXXLL motif, but also a ΦXXΦΦ motif. However, only mutations in the ΦXXΦΦ motif altered binding to VDR (Hsieh *et al.*, 2003). Furthermore, when we tested both motifs separately for their binding to VDR, the ΦXXΦΦ motif had the higher affinity regardless of the presence or absence of 1,25(OH)$_2$D (Teichert *et al.*, 2009). We have shown that the endogenous VDR binds to endogenous Hr in keratinocytes (Xie *et al.*, 2006). Binding of Hr to VDR inhibited 1,25(OH)$_2$D stimulation of a CYP24A1 (24-hydroxylase) promoter construct containing the VDRE of this vitamin D target gene. Overexpression of Hr blocks the ability of 1,25(OH)$_2$D to induce differentiation markers in keratinocytes, whereas inhibition of Hr expression enhances the stimulation by 1,25(OH)$_2$D of these markers (Xie *et al.*, 2006). The Hr null animal demonstrates upregulation of differentiation markers in the epidermis (Zarach *et al.*, 2004) consistent with a corepressor role for Hr in vitamin D regulated epidermal differentiation. Antibodies to Hr enhance the binding of VDR to VDREs in vitamin D target genes in gel retardation assays (Xie *et al.*, 2006) suggesting that Hr binding to VDR blocks its binding to VDREs. 1,25(OH)$_2$D displaces Hr from VDREs as it recruits the coactivators DRIP205 and SRC3 to these same VDREs (Xie *et al.*, 2006). Thus at least for 1,25(OH)$_2$D stimulated actions of VDR, Hr is a cosuppressor.

6.5 Role of β-catenin as VDR coregulator

Wnt signaling via activation of β-catenin has a complex role in VDR (Figure 6.3). Wnt ligands bind to their seven-transmembrane Frizzled receptors and an LRP5 or LRP6 co-receptor leading to phosphorylation of Dvl resulting in disruption of the axin/APC complex and inhibition of the kinase activity of glycogen synthase kinase-3β (GSK-3β), which otherwise phosphorylates the serine(s) within exon 3 of β-catenin facilitating its degradation by the E3 ubiquitin ligase. Thus wnt signaling increases the availability of β-catenin in the cytoplasm, which can then bind to transcription factors of the TCF and LEF families to promote expression of genes such as cyclin D1 and c-myc (He *et al.*, 1998) important for proliferation. β-catenin also forms part of the adherens junction complex with E-cadherin where it may play an important role in keratinocyte differentiation (Xie and Bikle, 2007). Tyrosine phosphorylation of E-cadherin, as occurs after calcium administration to keratinocytes, promotes the binding of β-catenin and other catenins to the adherens junction complex (Bienz, 2005; Xie and Bikle, 2007) making it less available for transcriptional activity. Cells differ markedly in the components of the β-catenin signaling pathway utilized. This is well illustrated in the keratinocytes of the hair follicle and interfollicular epidermis (DasGupta *et al.*, 2002; Huelsken *et al.*, 2001) LEF1 is the

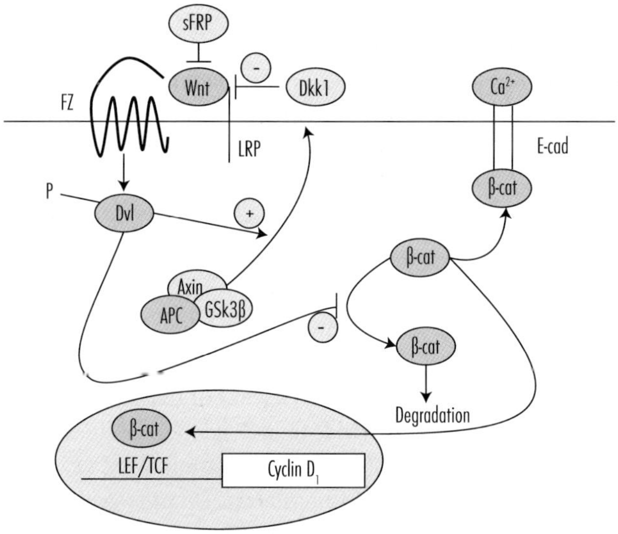

Figure 6.3. The canonical wnt signaling pathway. Wnts bind to their frizzled receptors (FZ) and coreceptors LRP in the membrane. This binding can be blocked by dickkopf (Dkk) or soluble frizzled related proteins (sFRP). Activation of FZ by wnt results in phosphorylation of Dvl, which induces the disruption of the axin/APC/GSK-3βcomplex and recruitment of axin to the membrane. This complex when active phosphorylates β-catenin, leading to its proteosomal degradation. However, following wnt stimulation β-catenin is no longer degraded and can enter the nucleus where in combination with members of the LEF/TCF family can induce expression of its target genes such as cyclin D1. β-catenin also binds to the E-cadherin complex in the plasma membrane, a complex stabilized by calcium and induced by 1,25$(OH)_2$D.

dominant transcription partner for β-catenin in the dermal portion of the hair follicle, which has little E-cadherin. The epidermal keratinocyte, on the other hand, has little LEF1 but a lot of E-cadherin especially in the differentiated layers (Stenn *et al.*, 2001). Over expression and/or activating mutations in the β-catenin pathway lead to skin tumors, in this case pilomatricomas or trichofolliculomas (hair follicle tumors) (Chan *et al.*, 1999; Gat *et al.*, 1998; Xia *et al.*, 2006) indicative of the hyperproliferative response to β-catenin in these cells. Activating mutations of specific serines within exon 3, or deletion of exon 3, block phosphorylation of β-catenin by GSK-3β, phosphorylation which otherwise leads to its proteosomal degradation. As a result β-catenin levels increase in the nucleus where its transcriptional activity is exerted in association with members of the LEF/TCF family of transcription factors.

In colon cancer cells VDR has been shown to bind to β-catenin, and reduce its transcriptional activity in a ligand dependent fashion (Palmer *et al.*, 2001). Furthermore, in these cells $1,25(OH)_2D$ has been shown to increase E-cadherin expression, such that β-catenin is redistributed from the nucleus to the plasma membrane where it forms a complex with E-cadherin and other catenins at adherens junctions (Shah *et al.*, 2003). However, the suppression of β-catenin signaling by $1,25(OH)_2D$ does not necessarily require E-cadherin (Shah *et al.*, 2006). Rather β-catenin binds to VDR in its AF-2 domain, binding that enhances the ability of $1,25(OH)_2D$ to activate the transcriptional activity of the VDR (Shah *et al.*, 2006) but blocks the transcriptional activity of β-catenin. Mutations in the AF-2 domain of VDR that block coactivator binding do not necessarily block β-catenin binding (Shah *et al.*, 2006). Whether β-catenin binding alters DRIP205 or SRC3 binding to this same region has not been determined. Palmer *et al.* (Palmer *et al.*, 2008) evaluated the interaction between VDR and β-catenin in transcriptional regulation in keratinocytes, and identified putative response elements for VDR and β-catenin/LEF in a number of genes. These interactions were either positive or negative, depending on the gene being evaluated. The hypothesis put forward is that genes in which the interaction is positive (i.e. stimulated transcription) benefits from β-catenin acting as a coactivator for VDR on VDREs, whereas in situations where the interaction is negative (ie. suppression of transcription) VDR prevents β-catenin from binding to TCF/LEF required for transcription of those genes. We (Oda and Bikle, personal communication) have found in keratinocytes that knockdown of VDR reduces E-cadherin expression and formation of the β-catenin/E-cadherin membrane complex resulting in increased β-catenin transcriptional activity, whereas $1,25(OH)_2D$ administration has the opposite effect. This was associated with increased (with VDR and DRIP205 knockdown) or decreased (with $1,25(OH)_2D$ administration) keratinocyte proliferation and cyclin D1 expression, respectively. Other studies have demonstrated that VDR potentiates, not inhibits, β-catenin transcriptional activity. Cianferotti *et al.* (2007) found a reduction in proliferation of keratinocytes in the dermal portion of the hair follicle (below the bulge) in VDR null mice, and no stimulation of proliferation when β-catenin was overexpressed in these cells in contrast to the stimulation of proliferation in control animals. Thus VDR/β-catenin interactions can be positive or negative, and this depends on the gene/cell/function being evaluated.

6.6 VDR as a tumor suppressor in the skin

Over 1 million skin cancers occur annually in the United States, 80% of which are BCC (16% SCC, 4% melanomas), making it by far the most common cancer (Greenlee *et al.*, 2001). UVR is the major etiologic agent. UV wavelengths shorter than 280nm (UVC) are absorbed by the ozone layer and do not reach the earth. UV wavelengths longer than 320nm (UVA) have limited ability to induce the characteristic mutations in DNA seen in epidermal cancers. Thus UVB with a spectrum between 280-320nm is the major cause of these cancers (Freeman *et al.*, 1989). The principle genotoxic lesions induced by UVR are CPDs and pyrimidine(6-4)pyrimidone photoproducts, which if not repaired result in C to T or CC to TT mutations, the UVB 'signature' lesion (Hussein, 2005). Such mutations in p53 are common (50-90%) in both BCC and SCC (Brash *et al.*, 1991; Daya-Grosjean and Sarasin, 2005; Ziegler *et al.*, 1993, 1994) as well as in actinic keratoses, the precursor lesions to SCC (Bito *et al.*, 1995). Precursor lesions for BCC have not been identified, but BCC are thought to arise from interfollicular basal cells, hair follicles, and sebaceous glands. Mutations in ras are much more common in SCC than BCC (Reifenberger *et al.*, 2005), whereas mutations in the Hh signaling pathway, in particular in Ptch1, characterize BCC (Aszterbaum *et al.*, 1998, 1999; Hahn *et al.*, 1996; Johnson *et al.*, 1996), but can also be found in SCC in patients also susceptible to BCC (Ping *et al.*, 2001).

Recent studies indicate that vitamin D plays a protective role in the skin with respect to carcinogenesis. Zinser *et al.* (2002) treated VDR null mice bearing medroxyprogesterone pellets with two oral administrations of DMBA at 5.5 and 7 wks, a protocol designed to induce breast cancers. No breast tumors were observed at least initially. Instead they found that 85% of the VDR null mice developed skin tumors within two months. No tumors (breast or skin) were found in the wildtype controls. The tumors were mostly sebaceous, squamous, and follicular papillomas, but several BCC were observed. No SCC were reported. These results have been confirmed using topical administration of DMBA/TPA, although only papillomas were seen in the VDR null mice unlike RXRα null mice which developed both BCC and SCC (Indra *et al.*, 2007). The more recent study by Ellison *et al.* (2008) likewise confirmed the results by Zinser *et al.*, demonstrated that mice lacking CYP27B1 (and so lacking 1,25$(OH)_2$D) were not prone to DMBA induced tumors, and further demonstrated that the VDR null mice were also more susceptible to tumor formation induced by UVR. We have confirmed these results, and like DMBA induced tumor formation we showed that CYP27B1 null mice are not more susceptible to UVR induced tumor formation than wildtype controls (Teichert and Bikle, personal communication). UVR induced tumors include SCC and BCC. The appearance of BCC in these studies is surprising since the typical malignancy induced in mouse skin by UVR, ionizing radiation, or chemical carcinogens is SCC not BCC (Daya-Grosjean *et al.*, 2005) providing a clue that tumors in these mice involve the Hh pathway.

6.6.1 The hedgehog pathway in epidermal tumor formation

The Hh pathway is depicted in Figure 6.4. The appearance of BCC is characteristic of tumors formed when Hh signaling is disrupted (Regl *et al.*, 2004a). We (Teichert and Bikle, personal communication) have found that the epidermis and epidermal portion (utricles) of the hair

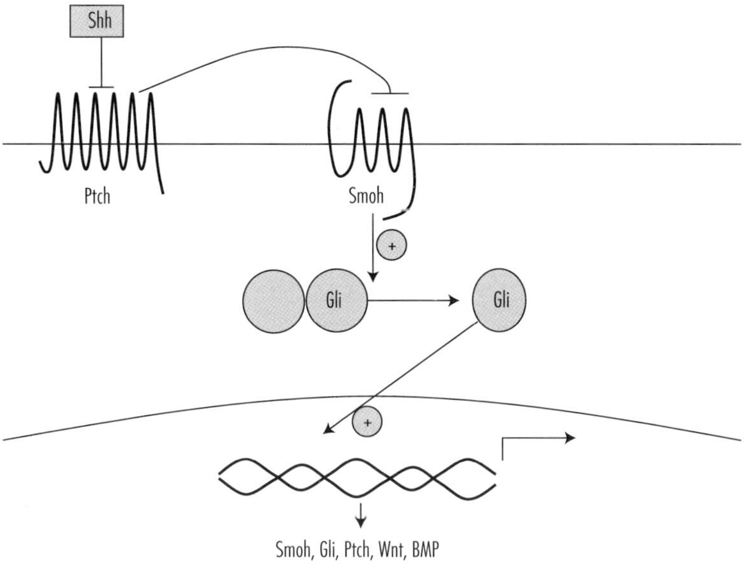

Figure 6.4. The Sonic hedgehog (Shh) signaling pathway. In the absence of Shh, Ptch1 suppresses signaling by smoothened (Smoh). Binding of Shh to Ptch1 relieves this inhibition. Activation of Smoh leads to the activation and translocation of transcription factors of the Gli family into the nucleus, with subsequent changes in gene expression.

follicles of adult VDR null animals overexpress elements of the Hh signaling pathway unlike the dermal portion of the hair follicle, suggesting that one of the causes of the increased susceptibility of the epidermis to malignant transformation is due to a loss of $1,25(OH)_2D$ and/or VDR regulation of Hh signaling. Of interest is that vitamin D may regulate this pathway not only via the genomic actions of $1,25(OH)_2D$ acting through its receptor, VDR, but also by direct inhibition by vitamin D independent of its receptor. The latter possibility stems from recent observations that vitamin D itself as well as its precursor 7-dehydrocholesterol can bind to and inhibit the actions of Smoh, a critical step in Hh signaling (Bijlsma *et al.*, 2006).

Appreciation of the pivotal role of the Hh signaling pathway in BCC development began with the identification of the Ptch1 gene as the site of mutations underlying the rare autosomal dominant heritable BCNS (Gorlin Syndrome), one of whose cardinal features is a high susceptibility to the development of BCCs (Aszterbaum *et al.*, 1998; Hahn *et al.*, 1996). The BCCs in these subjects frequently lose function of the inherited wildtype Ptch1 allele, leaving the tumor cells functionally null of Ptch1. Subsequently it has become clear that essentially all BCCs, whether arising in patients with BCNS or sporadically, have mutations in Ptch1 or other alterations in Hh signaling (Aszterbaum *et al.*, 1999). This appreciation has resulted in the development of the Ptch1+/− (Gorlin) mouse as the first practical model of murine BCCs (Regl *et al.*, 2004a). Treatment of these mice (unlike treatment of Ptch1 wildtype mice) with UVR or ionizing radiation produces BCC as well as SCC (Regl *et al.*, 2004a).

Ptch1 is the membrane receptor for Shh (Figure 6.4). In the absence of Shh, Ptch1 inhibits the function of another membrane protein Smoh. Shh reverses this inhibition freeing Smoh to enable the activation of a family of transcription factors Gli1, Gli2, and Gli3. Sufu may maintain these transcription factors in the cytoplasm and/or limit their activity in the nucleus such that the loss of Sufu leads to increased Hh signaling (Barnfield *et al.*, 2005; Svard *et al.*, 2006). Gli1 and 2 overexpression in keratinocytes can increase the expression of each other as well as Ptch1, the anti apoptotic factor bcl2, cyclins D1 and D2, E2F1, cdc45 (all of which promote proliferation) while suppressing genes associated with keratinocyte differentiation such as K1, K10, involucrin, loricrin and the VDR (Grachtchouk *et al.*, 2000; Nilsson *et al.*, 2000; Regl *et al.*, 2002, 2004a,b). $1,25(OH)_2D$ and VDR, on the other hand, have the opposite action on these genes. Mice overexpressing Gli1, Gli2, or Shh in their basal keratinocytes (Grachtchouk *et al.*, 2000; Nilsson *et al.*, 2000; Oro *et al.*, 1997) or grafted with human keratinocytes overexpressing Shh (Fan *et al.*, 1997) develop BCC like lesions. Furthermore, BCC show overexpression of Ptch1, Smoh, Gli1 and Gli2 (Bonifas *et al.*, 2001; Tojo *et al.*, 1999). Gli2 deletion partially rescues Ptch1 null animals, and Gli2 is required for Shh signaling in hair follicle development (Eichberger *et al.*, 2004). Shh, Ptch1, Ptch2, Gli1 and Gli2 have consensus sequences for VDRE in their promoters (Palmer *et al.*, 2008; Wang *et al.*, 2005) suggesting that $1,25(OH)_2D$/VDR may regulate their expression directly. In fact we (Teichert and Bikle, personal communication) have found such regulation.

6.6.2 β-catenin signaling in epidermal tumor formation

The appearance of hair follicle elements in many of the tumors forming in the VDR null mouse skin also suggests that the β-catenin pathway is activated. Like the Hh pathway, over expression and/or activating mutations in the β-catenin pathway lead to skin tumors, in this case pilomatricomas or trichofolliculomas (hair follicle tumors) as previously mentioned (Chan *et al.*, 1999; Gat *et al.*, 1998; Xia *et al.*, 2006). Palmer *et al.* (2008) evaluated the interaction between VDR and β-catenin in transcriptional regulation, and identified putative response elements for VDR and β-catenin/LEF in a number of genes including Shh, Ptch1 and 2, Gli1 and 2. Furthermore, they found that the ability of β-catenin overexpression to induce trichofolliculomas was blocked by an analog of $1,25(OH)_2D$, and in the absence of VDR, BCC were induced rather than trichofolliculomas.

6.7 Applications to other areas of skin health, care, and treatments

The ability of plants and animals to produce vitamin D goes way back in evolutionary times (Bikle, 2010). The role of vitamin D in primitive organisms such as plankton may have involved protecting the DNA from damage by solar (UVB) radiation. As animals (and plants) developed in complexity, the vitamin D produced in the skin and leaves took on the function of regulating calcium fluxes. This function ensures sufficient calcium for the skeleton by increasing intestinal calcium absorption but it also provides a mechanism for the vast array of physiologic processes utilizing calcium signaling regulated by vitamin D. The skin is an excellent example for vitamin D regulation of a number of different processes including the control of epidermal differentiation and proliferation, hair follicle cycling, and protection against the DNA damage induced by solar exposure. Clinically this has led to the use of vitamin D analogs in a number of skin conditions such as psoriasis and potentially could lead to prevention of photoaging and cancer development.

6.8 Conclusions

Vitamin D and its receptor have many roles in the skin. Some of these roles – induction of genes required for differentiation, suppression of genes involved with proliferation – appear to require both $1,25(OH)_2D$ and VDR. Other roles such as regulation of hair follicle cycling and tumor suppression do not appear to be ligand dependent, or at least $1,25(OH)_2D$ dependent roles of VDR. Different coactivator complexes including DRIP and SRC modulate the actions of VDR, and the choice of coactivator complex in many cases is gene specific. Regulation of proliferation is dependent on DRIP, whereas more differentiated functions including innate immunity and permeability barrier formation are SRC dependent. Hr is a coregulator with profound actions in hair follicle cycling. Although Hr blocks $1,25(OH)_2D$ regulated VDR functions, its role in VDR regulated hair follicle cycling is less clear. β-catenin interactions with VDR function can enhance VDR induction of some genes, whereas VDR can suppress β-catenin induction of other genes. Many genes in the skin contain both putative VDREs and LEF/TCF sites. Like β-catenin the Hh pathway is critical for hair follicle cycling, and again like β-catenin, the Hh pathway is regulated differently in the interfollicular epidermis than it is in the proximal hair follicle. Both β-catenin and Hh pathways are likely mediators of the predisposition of VDR null animals to epidermal tumor formation. At this point we are a long ways away from understanding how vitamin D via its active metabolite(s) and receptor regulates all the various functions in the skin that it appears to regulate. But what we have learned indicates that the skin is a fertile area for understanding the mechanisms by which vitamin D regulates so many different physiologic processes.

Acknowledgements

The author acknowledges the administrative support of Ms Teresa Tong and the financial support from the Veterans Affairs Medical Center and NIH grant AR050023.

References

Acevedo, M.L., Lee, K.C., Stender, J.D., Katzenellenbogen, B.S. and Kraus, W.L., 2004. Selective recognition of distinct classes of coactivators by a ligand-inducible activation domain. Molecular Cell 13, 725-738.

Ahmad, W., Faiyaz ul Haque, M., Brancolini, V., Tsou, H.C., ul Haque, S., Lam, H., Aita, V.M., Owen, J., deBlaquiere, M., Frank, J., Cserhalmi-Friedman, P.B., Leask, A., McGrath, J.A., Peacocke, M., Ahmad, M., Ott, J. and Christiano, A.M., 1998. Alopecia universalis associated with a mutation in the human hairless gene. Science 279, 720-724.

Aszterbaum, M., Rothman, A., Johnson, R.L., Fisher, M., Xie, J., Bonifas, J.M., Zhang, X., Scott, M.P. and Epstein Jr., E.H., 1998. Identification of mutations in the human PATCHED gene in sporadic basal cell carcinomas and in patients with the basal cell nevus syndrome. Journal of Investigative Dermatology 110, 885-888.

Aszterbaum, M., Epstein, J., Oro, A., Douglas, V., LeBoit, P.E., Scott, M.P. and Epstein Jr., E.H., 1999. Ultraviolet and ionizing radiation enhance the growth of BCCs and trichoblastomas in patched heterozygous knockout mice. Nature Medicine 5, 1285-1291.

Barnfield, P.C., Zhang, X., Thanabalasingham, V., Yoshida, M. and Hui, C.C., 2005. Negative regulation of Gli1 and Gli2 activator function by Suppressor of fused through multiple mechanisms. Differentiation 73. 397-405.

Bienz, M., 2005. beta-Catenin: a pivot between cell adhesion and Wnt signaling. Current Biology 15, R64-R67.

Bijlsma, M.F., Spek, C.A., Zivkovic, D., Van de Water, S., Rezaee, F. and Peppelenbosch, M.P., 2006. Repression of smoothened by patched-dependent (pro-)vitamin D3 secretion. PLoS Biology 4, e232.

Bikle, D.D., Nemanic, M.K., Whitney, J.O. and Elias, P.W., 1986. Neonatal human foreskin keratinocytes produce 1,25-dihydroxyvitamin D3. Biochemistry 25, 1545-1548.

Bikle, D.D., Pillai, S. and Gee, E., 1991. Squamous carcinoma cell lines produce 1,25 dihydroxyvitamin D, but fail to respond to its prodifferentiating effect. Journal of Investigative Dermatology 97, 435-441.

Bikle, D.D. and Pillai, S., 1993. Vitamin D, calcium, and epidermal differentiation. Endocrine Review 14, 3-19.

Bikle, D.D., Elalieh, H., Chang, S., Xie, Z. and Sundberg, J.P., 2006. Development and progression of alopecia in the vitamin D receptor null mouse. Journal of Cellular Physiology 207, 340-353.

Bikle, D.D., 2009. Extra renal synthesis of 1,25 dihydroxyvitamin D and its Health Implications. Clinical Review in Bone and Mineral Metabolism 7, 114-125.

Bikle, D.D., 2010. Vitamin D: an ancient hormone. Experimantal Dermatology 20, 7-13.

Bito, T., Ueda, M., Ahmed, N.U., Nagano, T. and Ichihashi, M., 1995. Cyclin D and retinoblastoma gene product expression in actinic keratosis and cutaneous squamous cell carcinoma in relation to p53 expression. Journal of Cutaneous Pathology 22, 427-434.

Bonifas, J.M., Pennypacker, S., Chuang, P.T., McMahon, A.P., Williams, M., Rosenthal, A., De Sauvage, F.J. and Epstein Jr., E.H., 2001. Activation of expression of hedgehog target genes in basal cell carcinomas. Journal of Investigative Dermatology 116, 739-742.

Bouillon, R., Verlinden, L., Eelen, G., De Clercq, P.V., M., Mathieu, C. and Verstuyf, A., 2005. Mechanisms for the selective action of Vitamin D analogs. Journal of Steroid Biochemistry and Molecular Biology 97, 21-30.

Bouillon, R., Carmeliet, G., Verlinden, L., Van Etten, E., Verstuyf, A., Luderer, H.F., Lieben, L., Mathieu, C. and Demay, M., 2008. Vitamin D and human health: lessons from vitamin D receptor null mice. Endocrine Reviews 29, 726-776.

Brash, D.E., Rudolph, J.A., Simon, J.A., Lin, A., McKenna, G.J., Baden, H.P., Halperin, A.J. and Ponten, J., 1991. A role for sunlight in skin cancer: UV-induced p53 mutations in squamous cell carcinoma. Proceedings of the National Academy of Sciences of the United States of America 88, 10124-10128.

Carvallo, L., Henriquez, B., Paredes, R., Olate, J., Onate, S., Van Wijnen, A.J., Lian, J.B., Stein, G., Stein, J.L. and Montecino, M., 2008. 1,25-dihydroxy vitamin D3-enhanced expression of the osteocalcin gene involves increased promoter occupancy of basal transcription regulators and gradual recruitment of the 1,25-dihydroxy vitamin D3 receptor-SRC-1 coactivator complex. Journal of Cellular Physiology 214, 740-749.

Chan, E.F., Gat, U., McNiff, J.M. and Fuchs, E., 1999. A common human skin tumour is caused by activating mutations in beta-catenin. Nature Genetics 21, 410-413.

Chen, C.H., Sakai, Y. and Demay, M.B., 2001. Targeting expression of the human vitamin D receptor to the keratinocytes of vitamin D receptor null mice prevents alopecia. Endocrinology 142, 5386-5389.

Christakos, S., Dhawan, P., Liu, Y., Peng, X. and Porta, A., 2003. New insights into the mechanisms of vitamin D action. Journal of Cell Biochemistry 88, 695-705.

Cianferotti, L., Cox, M., Skorija, K. and Demay, M.B., 2007. Vitamin D receptor is essential for normal keratinocyte stem cell function. Proceedings of the National Academy of Sciences of the United States of America 104, 9428-9433.

DasGupta, R., Rhee, H. and Fuchs, E., 2002. A developmental conundrum: a stabilized form of beta-catenin lacking the transcriptional activation domain triggers features of hair cell fate in epidermal cells and epidermal cell fate in hair follicle cells. Journal of Cell Biology 158, 331-344.

Daya-Grosjean, L. and Sarasin, A., 2005. The role of UV induced lesions in skin carcinogenesis: an overview of oncogene and tumor suppressor gene modifications in xeroderma pigmentosum skin tumors. Mutatation Research 571, 43-56.

Djabali, K., Aita, V.M. and Christiano, A.M., 2001. Hairless is translocated to the nucleus via a novel bipartite nuclear localization signal and is associated with the nuclear matrix. Journal of Cell Science 114, 367-376.

Eichberger, T., Regl, G., Ikram, M.S., Neill, G.W., Philpott, M.P., Aberger, F. and Frischauf, A.M., 2004. FOXE1, a new transcriptional target of GLI2 is expressed in human epidermis and basal cell carcinoma. Journal of Investigative Dermatology 122, 1180-1187.

Ellison, T.I., Smith, M.K., Gilliam, A.C. and Macdonald, P.N., 2008. Inactivation of the Vitamin D Receptor Enhances Susceptibility of Murine Skin to UV-Induced Tumorigenesis. Journal of Investigative Dermatology 128, 2508-2517.

Engelhard, A. and Christiano, A.M., 2004. The hairless promoter is differentially regulated by thyroid hormone in keratinocytes and neuroblastoma cells. Experimental Dermatology 13, 257-264.

Fan, H., Oro, A.E., Scott, M.P. and Khavari, P.A., 1997. Induction of basal cell carcinoma features in transgenic human skin expressing Sonic Hedgehog. Natural Medicine 3, 788-792.

Freeman, S.E., Hacham, H., Gange, R.W., Maytum, D.J., Sutherland, J.C. and Sutherland, B.M., 1989. Wavelength dependence of pyrimidine dimer formation in DNA of human skin irradiated *in situ* with ultraviolet light. Proceedings of the National Academy of Sciences of the United States of America 86, 5605-5609.

Gat, U., DasGupta, R., Degenstein, L. and Fuchs, E., 1998. De Novo hair follicle morphogenesis and hair tumors in mice expressing a truncated beta-catenin in skin. Cell 95, 605-614.

Grachtchouk, M., Mo, R., Yu, S., Zhang, X., Sasaki, H., Hui, C.C. and Dlugosz, A.A., 2000. Basal cell carcinomas in mice overexpressing Gli2 in skin. Nature Genetics 24, 216-617.

Greenlee, R.T., Hill-Harmon, M.B., Murray, T. and Thun, M., 2001. Cancer statistics, 2001. CA Cancer Journal for Clinicians 51, 15-36.

Hahn, H., Wicking, C., Zaphiropoulous, P.G., Gailani, M.R., Shanley, S., Chidambaram, A., Vorechovsky, I., Holmberg, E., Unden, A.B., Gillies, S., Negus, K., Smyth, I., Pressman, C., Leffell, D.J., Gerrard, B., Goldstein, A.M., Dean, M., Toftgard, R., Chenevix-Trench, G., Wainwright, B. and Bale, A.E., 1996. Mutations of the human homolog of Drosophila patched in the nevoid basal cell carcinoma syndrome. Cell 85, 841-851.

Hawker, N.P., Pennypacker, S.D., Chang, S.M. and Bikle, D.D., 2007. Regulation of Human Epidermal Keratinocyte Differentiation by the Vitamin D Receptor and its Coactivators DRIP205, SRC2, and SRC3. Journal of Investigative Dermatology 127, 874.

He, T.C., Sparks, A.B., Rago, C., Hermeking, H., Zawel, L., da Costa, L.T., Morin, P.J., Vogelstein, B. and Kinzler, K.W., 1998. Identification of c-MYC as a target of the APC pathway. Science 281, 1509-1512.

Hess, A.F. and Unger, L.F., 1921. Cure of infantile rickets by sunlight. Journal of the American Medical Association 77, 39.

Hochberg, Z., Gilhar, A., Haim, S., Friedman-Birnbaum, R., Levy, J. and Benderly, A., 1985. Calcitriol-resistant rickets with alopecia. Archives of Dermatology 121, 646-647.

Hosomi, J., Hosoi, J., Abe, E., Suda, T. and Kuroki, T., 1983. Regulation of terminal differentiation of cultured mouse epidermal cells by 1 alpha,25-dihydroxyvitamin D3. Endocrinology 113, 1950-1957.

Hsieh, J.C., Sisk, J.M., Jurutka, P.W., Haussler, C.A., Slater, S.A., Haussler, M.R. and Thompson, C.C., 2003. Physical and functional interaction between the vitamin D receptor and hairless corepressor, two proteins required for hair cycling. Journal of Biological Chemistry 278, 38665-386674.

Huelsken, J., Vogel, R., Erdmann, B., Cotsarelis, G. and Birchmeier, W., 2001. beta-Catenin controls hair follicle morphogenesis and stem cell differentiation in the skin. Cell 105, 533-545.

Hussein, M.R., 2005. Ultraviolet radiation and skin cancer: molecular mechanisms. Journal of Cutaneous Pathology 32, 191-205.

Indra, A.K., Castaneda, E., Antal, M.C., Jiang, M., Messaddeq, N., Meng, X., Loehr, C.V., Gariglio, P., Kato, S., Wahli, W., Desvergne, B., Metzger, D. and Chambon, P., 2007. Malignant Transformation of DMBA/TPA-Induced Papillomas and Nevi in the Skin of Mice Selectively Lacking Retinoid-X-Receptor alpha in Epidermal Keratinocytes. Journal of Investigative Dermatology 127, 1250-1260.

Issa, L.L., Leong, G.M., Sutherland, R.L. and Eisman, J.A., 2002. Vitamin D Analogue-Specific Recruitment of Vitamin D Receptor Coactivators. Journal of Bone and Mineral Research 17, 879-890.

Johnson, R.L., Rothman, A.L., Xie, J., Goodrich, L.V., Bare, J.W., Bonifas, J.M., Quinn, A.G., Myers, R.M., Cox, D.R., Epstein Jr., E.H. and Scott, M.P., 1996. Human homolog of patched, a candidate gene for the basal cell nevus syndrome. Science 272, 1668-1671.

Kishimoto, J., Burgeson, R.E. and Morgan, B.A., 2000. Wnt signaling maintains the hair-inducing activity of the dermal papilla. Genes and Development 14, 1181-1185.

Kong, J., Li, X.J., Gavin, D., Jiang, Y. and Li, Y.C., 2002. Targeted expression of human vitamin d receptor in the skin promotes the initiation of the postnatal hair follicle cycle and rescues the alopecia in vitamin D receptor null mice. Journal of Investigative Dermatology 118, 631-638.

Langbein, L., Rogers, M.A., Praetzel, S., Winter, H. and Schweizer, J., 2003. K6irs1, K6irs2, K6irs3, and K6irs4 represent the inner-root-sheath-specific type II epithelial keratins of the human hair follicle. Journal of Investigative Dermatology 120, 512-522.

Leo, C. and Chen, J D., 2000. The SRC family of nuclear receptor coactivators. Genes 245, 1-11.

Li, Y.C., Amling, M., Pirro, A.E., Priemel, M., Meuse, J., Baron, R., Delling, G. and Demay, M.B., 1998. Normalization of mineral ion homeostasis by dietary means prevents hyperparathyroidism, rickets, and osteomalacia, but not alopecia in vitamin D receptor-ablated mice. Endocrinology 139, 4391-4396.

Maeda, Y., Rachez, C., Hawel, I., L., Byus, C.V., Freedman, L.P. and Sladek, F.M., 2002. Polyamines Modulate the Interaction between Nuclear Receptors and Vitamin D Receptor-Interacting Protein 205. Molecular Endocrinology 16, 1502-1510.

Marx, S.J., Bliziotes, M.M. and Nanes, M., 1986. Analysis of the relation between alopecia and resistance to 1,25-dihydroxyvitamin D. Clinical Endocrinology (Oxf) 25, 373-381.

Matsumoto, K., Hashimoto, K., Nishida, Y., Hashiro, M. and Yoshikawa, K., 1990. Growth-inhibitory effects of 1,25-dihydroxyvitamin D3 on normal human keratinocytes cultured in serum-free medium. Biochemical and Biophysical Research Communications 166, 916-923.

McKenna, N.J., Lanz, R.B. and O'Malley, B.W., 1999. Nuclear receptor coregulators: cellular and molecular biology. Endocrine Reviews 20, 321-344.

McLane, J.A., Katz, M. and Abdelkader, N., 1990. Effect of 1,25-dihydroxyvitamin D3 on human keratinocytes grown under different culture conditions. In Vitro Cellular and Developmental Biology 26, 379-387.

Milde, P., Hauser, U., Simon, T., Mall, G., Ernst, V., Haussler, M.R., Frosch, P. and Rauterberg, E.W., 1991. Expression of 1,25-dihydroxyvitamin D3 receptors in normal and psoriatic skin. Journal of Investigative Dermatology 97, 230-239.

Miller, J., Djabali, K., Chen, T., Liu, Y., Ioffreda, M., Lyle, S., Christiano, A.M., Holick, M. and Cotsarelis, G., 2001. Atrichia caused by mutations in the vitamin D receptor gene is a phenocopy of generalized atrichia caused by mutations in the hairless gene. Journal of Investigative Dermatology 117, 612-617.

Morris, R.J., Liu, Y., Marles, L., Yang, Z., Trempus, C., Li, S., Lin, J.S., Sawicki, J.A. and Cotsarelis, G., 2004. Capturing and profiling adult hair follicle stem cells. Nature Biotechnology 22, 411-417.

Nilsson, M., Unden, A.B., Krause, D., Malmqwist, U., Raza, K., Zaphiropoulos, P.G. and Toftgard, R., 2000. Induction of basal cell carcinomas and trichoepitheliomas in mice overexpressing GLI-1. Proceedings of the National Academy of Sciences of the United States of America 97, 3438-3443.

Oda, Y., Sihlbom, C., Chalkley, R.J., Huang, L., Rachez, C., Chang, C.P., Burlingame, A.L., Freedman, L.P. and Bikle, D.D., 2003. Two distinct coactivators, DRIP/mediator and SRC/p160, are differentially involved in vitamin D receptor transactivation during keratinocyte differentiation. Molecular Endocrinology 17, 2329-2339.

Oda, Y., Ishikawa, M.H., Hawker, N.P., Yun, Q.C. and Bikle, D.D., 2007. Differential role of two VDR coactivators, DRIP205 and SRC-3, in keratinocyte proliferation and differentiation. Journal of Steroid Biochemistry and Molecular Biology 103, 776-780.

Oda, Y., Uchida, Y., Moradian, S., Crumrine, D., Elias, P. and Bikle, D., 2009. Vitamin D receptor and coactivators SRC 2 and 3 regulate epidermis-specific sphingolipid production and permeability barrier formation. Journal of Investigative Dermatology 129, 1367-1378.

Oro, A.E., Higgins, K.M., Hu, Z., Bonifas, J.M., Epstein Jr., E.H. and Scott, M.P., 1997. Basal cell carcinomas in mice overexpressing sonic hedgehog. Science 276, 817-821.

Palmer, H.G., Gonzalez-Sancho, J.M., Espada, J., Berciano, M.T., Puig, I., Baulida, J., Quintanilla, M., Cano, A., De Herreros, A.G., Lafarga, M. and Munoz, A., 2001. Vitamin D(3) promotes the differentiation of colon carcinoma cells by the induction of E-cadherin and the inhibition of beta-catenin signaling. Journal of Cell Biology 154, 369-387.

Palmer, H.G., Anjos-Afonso, F., Carmeliet, G., Takeda, H. and Watt, F.M., 2008. The Vitamin D Receptor Is a Wnt Effector that Controls Hair Follicle Differentiation and Specifies Tumor Type in Adult Epidermis. PLoS ONE 3, e1483.

Panteleyev, A.A., Botchkareva, N.V., Sundberg, J.P., Christiano, A.M. and Paus, R., 1999. The role of the hairless (hr) gene in the regulation of hair follicle catagen transformation. American Journal of Pathology 155, 159-171.

Peleg, S., Ismail, A., Uskokovic, M. and Avnur, Z., 2003. Evidence for tissue- and cell-type selective activation of the vitamin D receptor by Ro-26-9228, a noncalcemic analog of vitamin D3. Journal of Cellular Biochemistry 88, 267-273.

Pillai, S. and Bikle, D.D., 1991. Role of intracellular-free calcium in the cornified envelope formation of keratinocytes: differences in the mode of action of extracellular calcium and 1,25 dihydroxyvitamin D3. Journal of Cellular Physiology 146, 94-100.

Ping, X.L., Ratner, D., Zhang, H., Wu, X.L., Zhang, M.J., Chen, F.F., Silvers, D.N., Peacocke, M. and Tsou, H.C., 2001. PTCH mutations in squamous cell carcinoma of the skin. Journal of Investigative Dermatology 116, 614-616.

Rachez, C., Lemon, B.D., Suldan, Z., Bromleigh, V., Gamble, M., Naar, A.M., Erdjument-Bromage, H., Tempst, P. and Freedman, L.P., 1999. Ligand-dependent transcription activation by nuclear receptors requires the DRIP complex. Nature 398, 824-828.

Rachez, C., Gamble, M., Chang, C.P., Atkins, G.B., Lazar, M.A. and Freedman, L.P., 2000a. The DRIP complex and SRC-1/p160 coactivators share similar nuclear receptor binding determinants but constitute functionally distinct complexes. Molecular and Cellular Biology 20, 2718-2726.

Rachez, C. and Freedman, L.P., 2000b. Mechanisms of gene regulation by vitamin D(3) receptor: a network of coactivator interactions. Gene 246, 9-21.

Regl, G., Neill, G.W., Eichberger, T., Kasper, M., Ikram, M.S., Koller, J., Hintner, H., Quinn, A.G., Frischauf, A.M. and Aberger, F., 2002. Human GLI2 and GLI1 are part of a positive feedback mechanism in Basal Cell Carcinoma. Oncogene 21, 5529-5539.

Regl, G., Kasper, M., Schnidar, H., Eichberger, T., Neill, G.W., Ikram, M.S., Quinn, A.G., Philpott, M.P., Frischauf, A.M. and Aberger, F., 2004a. The zinc-finger transcription factor GLI2 antagonizes contact inhibition and differentiation of human epidermal cells. Oncogene 23, 1263-1274.

Regl, G., Kasper, M., Schnidar, H., Eichberger, T., Neill, G.W., Philpott, M.P., Esterbauer, H., Hauser-Kronberger, C., Frischauf, A.M. and Aberger, F., 2004b. Activation of the BCL2 promoter in response to Hedgehog/GLI signal transduction is predominantly mediated by GLI2. Cancer Research 64, 7724-7731.

Reifenberger, J., Wolter, M., Knobbe, C.B., Kohler, B., Schonicke, A., Scharwachter, C., Kumar, K., Blaschke, B., Ruzicka, T. and Reifenberger, G., 2005. Somatic mutations in the PTCH, SMOH, SUFUH and TP53 genes in sporadic basal cell carcinomas. British Journal of Dermatology 152, 43-51.

Sakai, Y. and Demay, M.B., 2000. Evaluation of keratinocyte proliferation and differentiation in vitamin D receptor knockout mice. Endocrinology 141, 2043-2049.

Schauber, J., Dorschner, R.A., Yamasaki, K., Brouha, B. and Gallo, R.L., 2006. Control of the innate epithelial antimicrobial response is cell-type specific and dependent on relevant microenvironmental stimuli. Immunology 118, 509-519.

Schauber, J., Dorschner, R.A., Coda, A.B., Buchau, A.S., Liu, P.T., Kiken, D., Helfrich, Y.R., Kang, S., Elalieh, H.Z., Steinmeyer, A., Zugel, U., Bikle, D.D., Modlin, R.L. and Gallo, R.L., 2007. Injury enhances TLR2 function and antimicrobial peptide expression through a vitamin D-dependent mechanism. Journal of Clinical Investigation 117, 803-811.

Schauber, J., Oda, Y., Buchau, A.S., Steinmeyer, A., Zugel, U., Bikle, D.D. and Gallo, R.L., 2008. Histone acetylation in keratinocytes enables control of the expression of cathelicidin and CD14 by 1,25-dihydroxyvitamin D3. Journal of Investigative Dermatology 128, 816-824.

Shah, S., Hecht, A., Pestell, R. and Byers, S.W., 2003. Trans-repression of beta-catenin activity by nuclear receptors. Journal of Biological Chemistry 278, 48137-48145.

Shah, S., Islam, M.N., Dakshanamurthy, S., Rizvi, I., Rao, M., Herrell, R., Zinser, G., Valrance, M., Aranda, A., Moras, D., Norman, A., Welsh, J. and Byers, S.W., 2006. The molecular basis of vitamin D receptor and beta-catenin crossregulation. Molecular Cell 21, 799-809.

Shimizu, H. and Morgan, B.A., 2004. Wnt signaling through the beta-catenin pathway is sufficient to maintain, but not restore, anagen-phase characteristics of dermal papilla cells. Journal of Investigative Dermatology 122, 239-245.

Smith, E.L., Walworth, N.C. and Holick, M.F., 1986. Effect of 1 alpha,25-dihydroxyvitamin D3 on the morphologic and biochemical differentiation of cultured human epidermal keratinocytes grown in serum-free conditions. Journal of Investigative Dermatology 86, 709-714.

Stenn, K.S. and Paus, R., 2001. Controls of hair follicle cycling. Physiological Reviews 81, 449-494.

Stumpf, W.E., Clark, S.A., Sar, M. and DeLuca, H.F., 1984. Topographical and developmental studies on target sites of 1,25 (OH)2 vitamin D3 in skin. Cell and Tissue Research 238, 489-496.

Svard, J., Heby-Henricson, K., Persson-Lek, M., Rozell, B., Lauth, M., Bergstrom, A., Ericson, J., Toftgard, R. and Teglund, S., 2006. Genetic elimination of Suppressor of fused reveals an essential repressor function in the mammalian Hedgehog signaling pathway. Developmental Cell 10, 187-197.

Teichert, A., Arnold, L.A., Otieno, S., Oda, Y., Augustinaite, I., Geistlinger, T.R., Kriwacki, R.W., Guy, R.K. and Bikle, D.D., 2009. Quantification of the vitamin D receptor-coregulator interaction. Biochemistry 48, 1454-1461.

Thompson, C.C. and Bottcher, M.C., 1997. The product of a thyroid hormone-responsive gene interacts with thyroid hormone receptors. Proceedings of the National Academy of Sciences of the United States of America 94, 8527-8532.

Tojo, M., Mori, T., Kiyosawa, H., Honma, Y., Tanno, Y., Kanazawa, K.Y., Yokoya, S., Kaneko, F. and Wanaka, A., 1999. Expression of sonic hedgehog signal transducers, patched and smoothened, in human basal cell carcinoma. Pathology International 49, 687-694.

Wang, T.T., Tavera-Mendoza, L.E., Laperriere, D., Libby, E., MacLeod, N.B., Nagai, Y., Bourdeau, V., Konstorum, A., Lallemant, B., Zhang, R., Mader, S. and White, J.H., 2005. Large-scale in silico and microarray-based identification of direct 1,25-dihydroxyvitamin D3 target genes. Molecular Endocrinology 19, 2685-2695.

Xia, J., Urabe, K., Moroi, Y., Koga, T., Duan, H., Li, Y. and Furue, M., 2006. beta-Catenin mutation and its nuclear localization are confirmed to be frequent causes of Wnt signaling pathway activation in pilomatricomas. Journal of Dermatology Science 41, 67-75.

Xie, Z., Komuves, L., Yu, Q.C., Elalieh, H., Ng, D.C., Leary, C., Chang, S., Crumrine, D., Yoshizawa, T., Kato, S. and Bikle, D.D., 2002. Lack of the vitamin D receptor is associated with reduced epidermal differentiation and hair follicle growth. Journal of Investigative Dermatology 118, 11-16.

Xie, Z., Chang, S., Oda, Y. and Bikle, D.D., 2006. Hairless suppresses vitamin D receptor transactivation in human keratinocytes. Endocrinology 147, 314-323.

Xie, Z. and Bikle, D.D., 2007. The recruitment of phosphatidylinositol 3-kinase to the E-cadherin-catenin complex at the plasma membrane is required for calcium-induced phospholipase C-gamma1 activation and human keratinocyte differentiation. Journal of Biological Chemistry 282, 8695-8703.

Zarach, J.M., Beaudoin, G.M., 3rd, Coulombe, P.A. and Thompson, C.C., 2004. The co-repressor hairless has a role in epithelial cell differentiation in the skin. Development 131, 4189-41200

Zehnder, D., Bland, R., Williams, M.C., McNinch, R.W., Howie, A.J., Stewart, P.M. and Hewison, M., 2001. Extrarenal expression of 25-hydroxyvitamin d(3)-1 alpha-hydroxylase. Journal of Clinical Endocrinology and Metabolism 86, 888-894.

Zhou, P., Byrne, C., Jacobs, J. and Fuchs, E., 1995. Lymphoid enhancer factor 1 directs hair follicle patterning and epithelial cell fate. Genes and Development 9, 700-713.

Ziegler, A., Jonason, A.S., Leffell, D.J., Simon, J.A., Sharma, H.W., Kimmelman, J., Remington, L., Jacks, T. and Brash, D.E., 1994. Sunburn and p53 in the onset of skin cancer. Nature 372, 773-776.

Ziegler, A., Leffell, D.J., Kunala, S., Sharma, H.W., Gailani, M., Simon, J.A., Halperin, A.J., Baden, H.P., Shapiro, P.E., Bale, A.E. and Brash, D.E., 1993. Mutation hotspots due to sunlight in the p53 gene of nonmelanoma skin cancers. Proceedings of the National Academy of Sciences of the United States of America 90, 4216-4220.

Zinser, G.M., Sundberg, J.P. and Welsh, J., 2002. Vitamin D(3) receptor ablation sensitizes skin to chemically induced tumorigenesis. Carcinogenesis 23, 2103-2109.

Key facts on oxidative stress

- Reactive oxygen species (ROS) are continuously produced in the mitochondria of living cells and can be detrimental by damaging lipids, proteins and the DNA.
- An imbalance in the cellular redox state, where the production of ROS overwhelms the cellular antioxidant capacity, results in the condition termed oxidative stress.
- Oxidative stress is thought to be involved in the aetiology of a wide variety of diseases, including atherosclerosis, diabetes, neurodegenerative diseases, chronic inflammatory diseases, cancer and in aging.

Key facts on the human requirement for vitamin C

- Higher vertebrates are able to synthesize vitamin C from D-glucose in the liver or kidney, but humans, other primates, guinea pigs and fruit bats lack the last enzyme involved in the synthesis of ascorbic acid (gulonolactone oxidase), which means that the vitamin is derived exclusively from the diet.
- Vitamin C is a cofactor for several enzymes participating in the post-translational hydroxylation of collagen, in the biosynthesis of carnitine, in the conversion of the neurotransmitter dopamine to norepinephrine, in peptide amidation and in tyrosine metabolism.
- The prolonged deprivation of vitamin C generates defects in the post-translational modification of collagen that cause scurvy.
- Descriptions of scurvy were common during the late 15th century, when the disease affected sailors deprived from fruits and vegetables for extended periods of time, causing large personnel losses.

Summary points

- Vitamin C contributes to the formation of skin barrier function by enhancing epidermal differentiation.
- Vitamin C promotes higher rates of collagen synthesis and cell proliferation in skin fibroblasts, and is required for efficient wound healing.
- Vitamin C is capable of reducing DNA damage and erythema formation, which is attributed to the scavenging of reactive species generated as a direct or indirect result of exposure to ultraviolet radiation.
- Clinical studies showed that vitamin C increased the synthesis of collagen, reduced facial wrinkles, increased the number of dermal papillae and improved the overall aspect of the skin.
- Vitamin C displays a skin lightening effect via the inhibition of tyrosinase activity in melanocytes.
- The effectiveness of topical application of vitamin C is limited due to its reduced stability in aqueous solution and poor penetration of the skin.
- In recent years, different stable derivatives of ascorbic acid have been synthesized and some have already been employed in a variety of cosmetic and pharmaceutical formulations.
- To improve stability and increase skin permeation, vitamin C has been successfully formulated in multiple emulsions and microemulsions and encapsulated in chitosan microspheres and nanoparticles.
- The combined topical application of vitamins C and E increases the photoprotective effect synergistically.

7. Vitamin C, gene expression and skin health

T.L. Duarte[1] *and I.F. Almeida*[2]
[1]*Iron and Innate Immunity Group, Instituto de Biologia Molecular e Celular, Rua do Campo Alegre, 823, 4150-180 Porto, Portugal;* [2]*Departamento de Ciências do Medicamento, Laboratório de Tecnologia Farmacêutica, Faculdade de Farmácia da Universidade do Porto, Rua Aníbal Cunha 164, 4099-030 Porto, Portugal; tduarte@ibmc.up.pt*

Abstract

Vitamin C (ascorbic acid) is an essential micronutrient, required for multiple biological functions. In this chapter, we review *in vitro* and *in vivo* evidence that vitamin C plays an important role in skin biology, contributing to the formation of efficient skin barrier and to fibroblast proliferation/ migration during wound healing. Vitamin C is also widely used in cosmetic and dermatological products since it is reported to have a number of beneficial effects on the skin. As an antioxidant it can scavenge deleterious free radicals and other oxidizing species that contribute to the process of skin aging. By stimulating collagen synthesis it improves the elasticity of the skin, thereby reducing fine lines and wrinkles. Furthermore, its ability to suppress skin pigmentation makes it a useful whitening agent. Ascorbic acid is, however, an unstable compound in aqueous solution, and several water- and lipid-soluble derivatives were synthesized to improve stability and increase skin penetration. The effectiveness of different ascorbic acid derivatives that are currently employed in dermatological preparations and cosmetics, in particular magnesium L-ascorbyl-2-phosphate and ascorbyl palmitate, is discussed. Finally, we compare the efficacy of different routes of administration of vitamin C, alone or in combination with other antioxidants, with a focus on newly developed topical delivery systems.

Keywords: ascorbic acid, antioxidant, photoprotection, wound healing, topical delivery

Abbreviations

DDR2	Discoidin domain receptor 2
HIF-1	Hypoxia-inducible factor 1
LUV	Large unilamellar vesicle
MLV	Multilamellar vesicle
NLC	Nanostructured lipid carrier
OLV	Oligolamellar vesicle
ROS	Reactive oxygen species
SLN	Solid lipid nanoparticle
SUV	Small unilamellar vesicle
UV	Ultraviolet
UVA	Ultraviolet A
UVB	Ultraviolet B

7.1 Introduction

The skin serves as a protective barrier against external insults. In practice, this means that the skin itself is continuously exposed to external injury. One of the most important stress factors is exposure to sunlight, especially solar UV radiation can damage the DNA directly as a consequence of the absorption of UVB by pyrimidine bases, leading to the formation of mutagenic pyrimidine dimer lesions, and indirectly through the formation of ROS. Moreover, UV damage induces an inflammatory response, which recruits lymphocytes and neutrophils that produce ROS. Photosensitization reactions, often involving the formation of singlet oxygen, are also known to occur in the skin *in vivo*. Exposure to sunlight is thus both a risk factor for skin cancers and a major contributor to skin aging, which is often accompanied by the formation of wrinkles, loss of elasticity, skin fragility and slower wound healing. DNA damage is a major contributor to skin aging and cancer, as supported by the fact that patients (and laboratory animals) defective for DNA repair are often more susceptible to skin cancer.

Physiological antioxidants such as α-tocopherol (vitamin E) and ascorbic acid (vitamin C) protect the skin from the damaging effects of sunlight. Whilst vitamin E protects lipid structures including membranes, vitamin C protects the aqueous environment. One reasonable strategy to prevent ROS-mediated damage to the skin would thus be to support its antioxidant defence systems with exogenous antioxidants. To meet the demands of consumers, who are increasingly aware of the need to protect against excessive exposure to sunlight and/or to prevent skin aging, product manufacturers often add antioxidant supplements (tocopherols, vitamin C and flavonoids) as protective agents to skin formulations. Some skin care companies offer stabilized and/or slow-releasing derivatives of the antioxidant molecules, which oxidize less rapidly and thus afford a sustained protective action.

Vitamin C is found in the composition of many skin care products due to its antioxidant properties. It is a potent reducing agent and a scavenger of free radicals. The redox metabolism of vitamin C has been described in many review articles (e.g. Duarte and Lunec, 2005). Briefly, the mono-anion form (ascorbate) is the predominant chemical species at physiological pH. Ascorbate readily undergoes two consecutive, yet reversible, one-electron oxidations to generate dehydroascorbate and an intermediate, the ascorbate free radical (Figure 7.1). The latter is, however, a relatively unreactive free radical, with a reduction potential considerably low compared to the α-tocopherol radical, the glutathione radical and virtually all reactive oxygen and nitrogen species that are thought to be involved in human disease. These properties make ascorbate an efficient electron donor in many biological redox reactions, capable of replacing potentially highly damaging radicals by the poorly reactive ascorbate radical, and of recycling other important antioxidant molecules such as α-tocopherol or glutathione from their respective radical species.

7.2 Vitamin C and skin biology

Human skin consists of two tissue layers, namely a keratinized stratified epidermis and an underlying thick layer of collagen-rich dermal connective tissue. Vitamin C is reported to contribute to the formation of skin barrier function by enhancing epidermal differentiation. Studies performed on postconfluent cultured human keratinocytes demonstrated that vitamin C stimulates ceramide synthase, the final step in ceramide synthesis, and increases the production of selected ceramide and glucosylceramide species, resulting in a spectrum of sphingolipids as observed *in vivo*. In addition, vitamin C increases several structural markers of permeability barrier formation (e.g. covalently bound lipids, mature lamellar bodies, secretion, and postsecretory membrane structural reorganization) (Uchida *et al.*, 2001). Likewise, incubation of cultured skin substitutes in medium containing vitamin C stimulates keratinocyte proliferation, results in more complete basement membrane development and promotes formation of epidermal barrier (Boyce *et al.*, 2002).

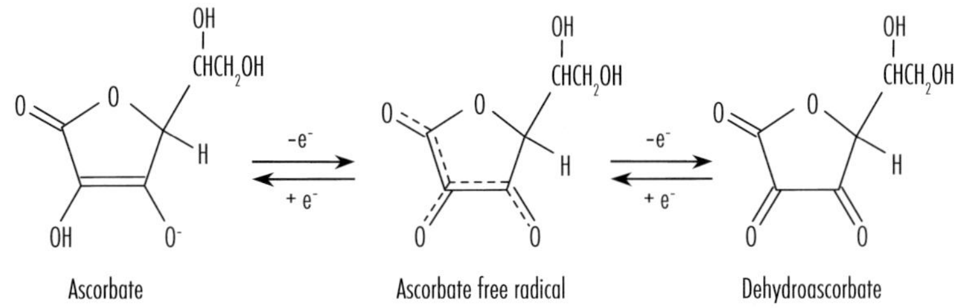

Figure 7.1. Redox forms of ascorbic acid. At physiological pH, most ascorbic acid is found as the monovalent anion ascorbate. The one electron oxidation of ascorbate yields the ascorbate free radical, with one electron delocalized between three oxygen atoms. Further oxidation generates dehydroascorbate, an unstable molecule that decomposes if not reduced back to ascorbate.

The most widely recognized role of vitamin C in the skin is, however, its participation in collagen maturation. Vitamin C is a cofactor required for the enzymatic activity of prolyl hydroxylase, which hydroxylates prolyl residues in procollagen and elastin, thus playing an important role in the structure of the dermis. A number of studies have also associated vitamin C with a higher proliferation rate in skin fibroblasts (reviewed by Duarte and Lunec, 2005). Dermal fibroblasts are the major source of extracellular connective tissue matrix and they play an important role in wound healing. They are recruited to the wound area by inflammatory cells, invade lesions, and play an important role in promoting reepithelialization and reestablishing tissue integrity. The genome-wide effects of vitamin C on gene expression in primary dermal fibroblasts were recently studied (Duarte *et al.*, 2009). Fibroblast cultures were grown to postconfluence and incubated with a stable vitamin C derivative (magnesium L-ascorbyl-2-phosphate) for 5 days, with daily repletion. In agreement with earlier reports that vitamin C stimulates post-confluent proliferation of skin fibroblasts, the most striking effect observed in the gene expression profiling study was the induction of a large set of genes that are involved in DNA replication or in the progression through the G_2/M phases of the mitotic cell cycle. Moreover, vitamin C increased the percentage of quiescent fibroblasts that were stimulated to synthesize DNA and proliferate after serum stimulation. *In vivo*, dermal fibroblasts are exposed to serum during the recolonization of a wound. Growth factors released at a wound site can act as mitogens or as chemotactic factors for dermal fibroblasts. In response, fibroblasts begin to proliferate and eventually migrate into the wound clot. In this respect, the work of Duarte *et al.* (2009) suggests that vitamin C may favour wound healing by stimulating reentry of quiescent fibroblasts into the cell cycle and also by promoting cell migration. Indeed, fibroblasts incubated in the presence of vitamin C expressed higher levels of genes involved in cell motility and matrix remodelling during wound healing (e.g. urokinase-type plasminogen activator, hyaluronan-mediated motility receptor, and interleukin-6). Moreover, vitamin C-treated fibroblasts showed markedly faster wound invasion *in vitro* (Figure 7.2). This supports previous observations of impaired wound healing in vitamin C-deficient humans and guinea pigs (as discussed by Duarte *et al.*, 2009). A study performed in mice, which unlike humans or guinea pigs are able to synthesize vitamin C, has also shown that vitamin C enhances wound healing. In the study, vitamin C increased the degree of wound contraction and reduced the mean wound healing time (Jagetia *et al.*, 2007). Faster wound closure was also reported in a study where cultured skin substitutes incubated with vitamin C were grafted to athymic mice (Boyce *et al.*, 2002).

When vitamin C is present in the medium, fibroblasts are able to repair 8-oxoguanine lesions with a faster kinetics (Duarte *et al.*, 2009). Importantly, 8-oxoguanine is not only one of the most abundant DNA lesions formed during oxidative stress, it is also mutagenic causing GC to TA transversions and is implicated in carcinogenesis. Since the Base Excision Repair pathway (i.e. the cellular pathway that performs the excision of damaged DNA bases) utilizes the DNA replication machinery, it is likely that vitamin C, by stimulating cell cycle progression, may simultaneously increase the cellular DNA repair capacity. In the context of wound healing, this might provide fibroblasts in proximity to the wound greater protection from ROS-induced DNA damage arising from the activity of inflammatory cells. This hypothesis warrants further validation.

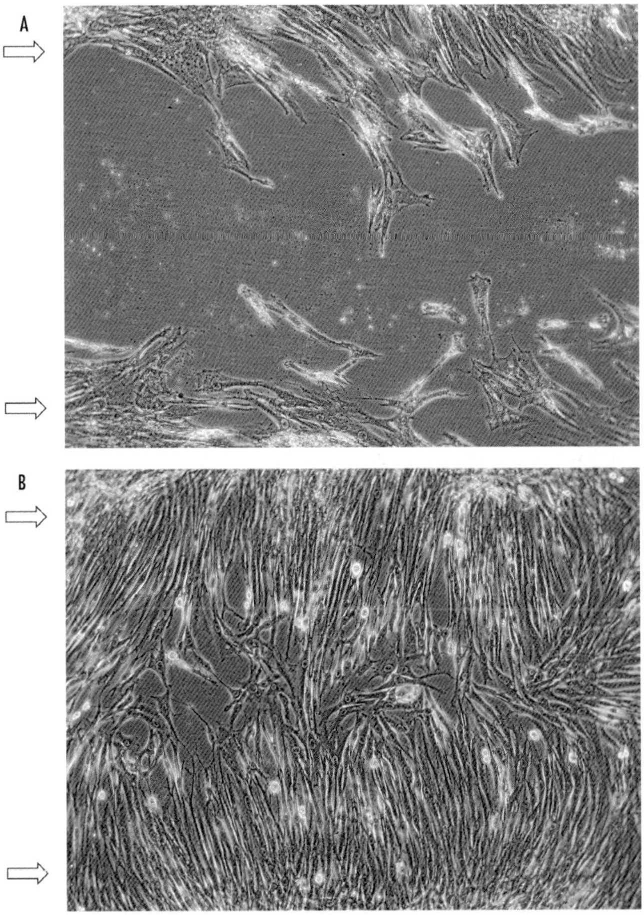

Figure 7.2. Vitamin C promotes the migration of skin fibroblasts toward the wound area in the *in vitro* wound healing assay. Confluent monolayers of fibroblasts were incubated with growth medium alone (A) or in the presence of 100 μM magnesium L-ascorbyl-2-phosphate (B) for 3 days. The postconfluent cell layer was wounded by gently scratching with a micropipette tip and incubated in fresh growth medium (A) or in medium containing L-ascorbyl-2-phosphate (B) for 24 h. The arrows indicate the wound edge. (Adapted from Duarte *et al.*, 2009: with permission from Elsevier)

What has not been established yet is exactly how vitamin C regulates fibroblast proliferation and motility, i.e. what signalling pathways are activated. Dermal fibroblasts express two transmembrane receptors, α1β1 integrin receptor and DDR2, which can transduce signals in response to collagen, resulting in the downstream activation of intracellular signalling pathways that modulate fibroblast proliferation and migration. Therefore, it is possible that vitamin C may activate gene expression indirectly through its ability to stimulate collagen synthesis (discussed in Duarte *et al.*, 2009). To the best of our knowledge, this possibility has not been addressed to date. Several studies have highlighted, however, the ability of vitamin C to inhibit the response to hypoxic stress by regulating the transcription factor HIF-1 signalling. Vitamin C is required for the optimal activity of Fe^{2+}- and 2-oxoglutarate-dependent hydroxylases that in normoxia promote the degradation HIF-1 in the proteasome (reviewed by Mandl *et al.*, 2009). Ascorbate stimulates enzyme activity presumably by keeping the iron center in the reduced state. Human skin fibroblasts cultured in vitamin C deficient media thus have a basal level of HIF-1 protein that disappears when the medium is supplemented with physiological concentrations of the vitamin (Vissers *et al.*, 2007). Consequently, vitamin C-treated fibroblasts express lower levels of genes that are targeted by HIF-1 and that are involved in processes such as angiogenesis, cell proliferation, glucose uptake, glycolysis, erythropoiesis and iron homeostasis (Duarte *et al.*, 2009). The induction of HIF-1 in response to hypoxia or $CoCl_2$ is also down-modulated by intracellular ascorbate (Vissers *et al.*, 2007).

7.3 Photoprotection

Exposure to simulated solar light or UV depletes ascorbate in the skin (Shindo *et al.*, 1993; Burke, 2004), which suggests that the topical administration of vitamin C may have a beneficial impact on skin health. Accordingly, many reports have associated the topical application of vitamin C prior to UV irradiation with the prevention of different UV-induced deleterious effects. Whilst these effects remain to be thoroughly assessed in humans, studies with animal models demonstrated a reduction in DNA damage and erythema formation (Darr *et al.*, 1992; Lin *et al.*, 2003). In pig skin, vitamin C is equally effective in protecting against both UVB (290-320 nm) and UVA (320-400 nm) (Darr *et al.*, 1992). As vitamin C does not act as sunscreen, the protective effects that have been reported to date are attributed to its antioxidant activity. It is apparent, however, that photoprotection is only achieved with rather high concentrations of the vitamin (10-15%) (Darr *et al.*, 1992; Lin *et al.*, 2003).

Whilst the photoprotective effect of topical antioxidants applied before UV exposure has been well recognized, the beneficial effect of these compounds when administered after irradiation is less obvious. A study reported that vitamin C administration to hairless mice immediately after exposure significantly delayed skin tumor formation and hyperplasia induced by chronic exposure to UVB (Kobayashi *et al.*, 1998). On the other hand, no photoprotective effect was observed in a human trial (Dreher *et al.*, 1999). Noteworthy, the ascorbic acid concentration used in this study (5%) was lower than the concentrations reported to be effective in other trials. Another possible explanation for the low efficacy of vitamin C post-irradiation could be the fact that the UV-induced ROS formation and subsequent ROS-mediated attack towards skin biomolecules leading to acute skin damage is a very rapid process. Since antioxidants applied after irradiation would likely not be present at the site of action (e.g. superficial skin layers) in relevant amounts when oxidative stress is generated, they would not decrease UV-induced erythema formation.

7.4 Skin whitening and anti-aging effects

Skin pigmentation is mainly determined by the amount of melanin present in the surface of the skin. Vitamin C is reported to display a skin lightening effect via the inhibition of tyrosinase activity in melanocytes, which reduces the synthesis of melanin. For that reason, vitamin C is a commonly employed whitening agent in cosmetics. Its beneficial role in hyperpigmentation disorders has been reported in human trials (e.g. Espinal-Perez *et al.*, 2004).

Premature senescence of the skin is a well-documented consequence of frequent exposures to intense or repeated UV irradiation. Solar lentigo and the loss of collagen are two common features in photoaged skin, which has prompted the use of vitamin C in many formulations as an anti-aging ingredient. The potential anti-aging effect of vitamin C has also been investigated in several human studies. A wide range of doses of the vitamin was reported (3-15%), and the number of volunteers enrolled was generally small (n=10-25). However, the studies were carefully designed and included a placebo, often a hydrating cream. Studies were usually split-face and double blind, which reassures the relevance of the reported effects. The duration of the studies ranged between 1 and 6 months. Clinical grading, bioengineering evaluation, patient self-assessment and histological examination were used to document the anti-aging effects. A number of beneficial effects were associated with vitamin C treatment, including the reduction of facial wrinkles and increase in the number of dermal papillae (Raschke *et al.*, 2004), improvement in the overall aspect of the skin as determined by dermatologists (Fitzpatrick and Rostan, 2002; Humbert *et al.*, 2003), increased Grenz zone collagen, and increased staining for type I collagen mRNA (Fitzpatrick and Rostan, 2002). Examples of the ability of topical administration of vitamin C preparations to reverse skin photodamage are presented in Figures 7.3 and 7.4. Overall, these studies support the original hypothesis that vitamin C displays anti-aging properties in the skin.

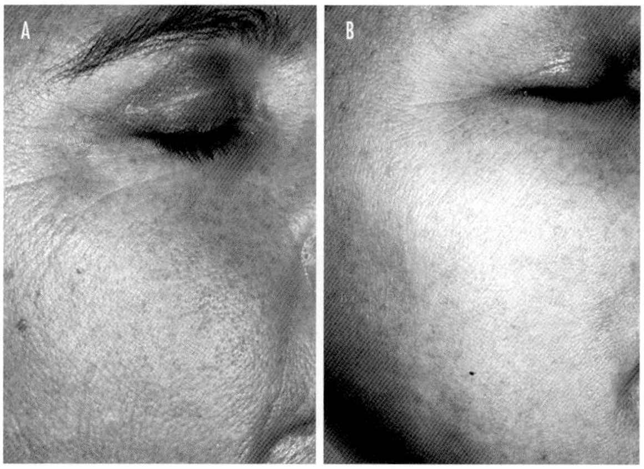

Figure 7.3. Reversal of facial photodamage with topical vitamin C administration. (A) Significant facial photodamage with textural irregularity and deep lines prior to therapy. (B) Visible improvement in texture and lines after 90 days of topical application of a vitamin C preparation containing 10% ascorbic acid and 7% tetrahexyldecyl ascorbate, a lipid soluble analogue with increased ability to penetrate to the dermis. (Adapted from Fitzpatrick and Rosta, 2002: with permission from John Wiley & Sons, Inc.)

A

B

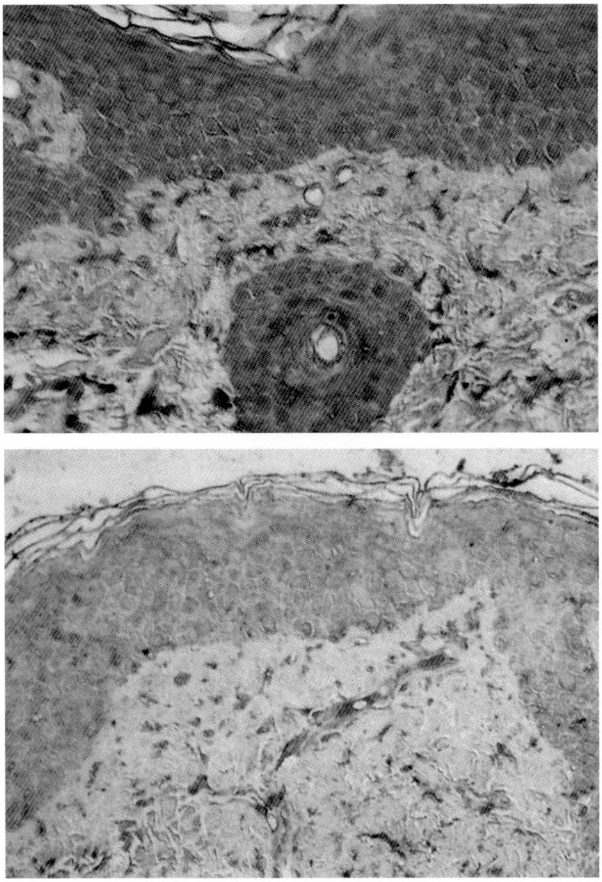

Figure 7.4. Increased collagen synthesis in vitamin C-treated skin. Vitamin C was applied topically for 90 days to one side of the face of a patient with facial photodamage, whereas the other side of the face received a placebo. Immunohistologic staining for mRNA for collagen type I was more evident in a biopsy of the topical vitamin C-treated side (A) than in a biopsy of the placebo side (B). (Adapted from Fitzpatrick and Rostan, 2002: with permission from John Wiley & Sons, Inc.)

7.5 Constraints to vitamin C effectiveness

7.5.1 Stable vitamin C derivatives

Vitamin C is widely used topically and orally for treating skin disorders. Whilst it is found in many skincare products, the level of effectiveness can vary. Effectiveness of topical formulations incorporating vitamin C is challenged by its intrinsic lack of stability and its poor penetration of the skin (e.g. Kameyama *et al.*, 1996). L-ascorbic acid is readily oxidized to dehydroascorbic acid and decomposes in aqueous solution. Oxidation not only leads to the loss of the active substance but may also cause fast product yellowing, which affects consumer acceptability. A number of different strategies can be followed to improve stability, such as exclusion of oxygen during production, oxygen impermeable packaging, encapsulation, low pH, minimization of water content, and inclusion of other antioxidants (Bisset, 2006). A promising approach to avoid the intrinsic lack of stability of L-ascorbic acid is the synthesis of different stable derivatives (Figure 7.5). Two water-soluble molecules (magnesium and sodium L-ascorbyl-2-phosphates, Figure 7.5A and 7.5B) and a lipid-soluble molecule (ascorbyl palmitate, Figure 7.5C) have been employed in a variety of cosmetic and pharmaceutical formulations, primarily as antioxidants.

Magnesium L-ascorbyl-2-phosphate has been used in many cell culture studies as a water-soluble, long-acting and possibly non-toxic vitamin C analogue. It protects human fibroblasts from *in vitro* cellular aging and is superior to L-ascorbic acid in enhancing proliferation, collagen accumulation and extracellular matrix formation (reviewed by Duarte and Lunec, 2005). In humans, magnesium L-ascorbyl-2-phosphate is absorbed percutaneously and retained in the skin, reducing skin hyperpigmentation in some patients (Kameyama *et al.*, 1996). It delays the onset of skin tumors and hyperplasia induced by chronic exposure to UV radiation in hairless mice (Kobayashi *et al.*, 1998).

Ascorbyl palmitate is an ester formed from ascorbic acid and palmitic acid resulting in a lipid-soluble vitamin C derivative. Its effectiveness in the reduction of UV-induced free radical formation was confirmed *in vivo* (Jurkovic *et al.*, 2004). A new lipophilic vitamin C derivative, tetra-isopalmitoyl ascorbic acid (Figure 7.5D), was synthesized by esterification of all four intramolecular hydroxyl groups of an ascorbic acid molecule with branched chain fatty groups that are more fluid than the corresponding straight fatty chain, resulting in a liquiform compound.

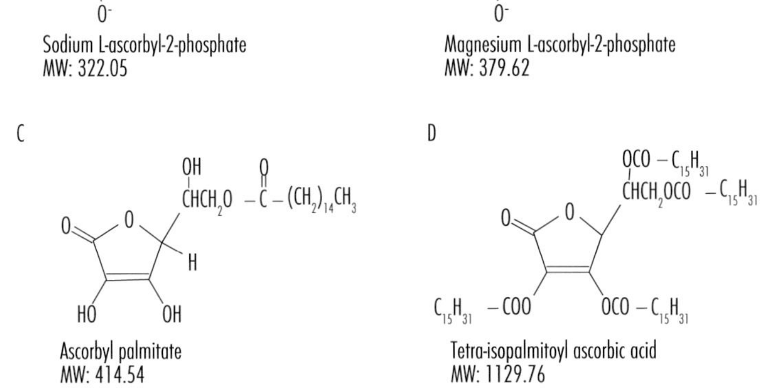

Figure 7.5. Chemical structure of common vitamin C derivatives.

Tetra-isopalmitoyl ascorbic acid protects cultured keratinocytes from UVA-induced oxidative stress and enhances type-IV collagen production in dermal fibroblasts more efficiently than ascorbic acid (Xiao *et al.*, 2009). It was also demonstrated to be an efficient precursor of vitamin C in a reconstructed skin model and to suppress UV-induced skin pigmentation when topically applied immediately after irradiation (Ochiai *et al.*, 2006).

7.5.2 Topical delivery systems

Several antioxidant molecules are labile to degradation in the presence of oxygen, water and light, and thus the use of a delivery system will increase their stability and enhance their performance (reviewed by Kaur *et al.*, 2007). Among the colloidal carriers, liposomes, niosomes and tranfersomes are particularly interesting regarding their application on the skin. In the nanometer range, SLNs and NLCs are appealing choices for cosmetic formulators, especially for skincare products. Special emulsions such as microemulsions and multiple emulsions have also been tested as antioxidant skin delivery systems. Liposomes are spherical vesicles with a membrane composed of a phospholipid and cholesterol bilayer, which encloses an aqueous core. Liposomes are classified into three categories on the basis of their size and lamellarity (number of bilayers): SUVs or OLVs, LUVs and MLVs. The similarity of the bilayer structure with biological membranes is considered an advantage regarding their application on the skin. Niosomes are essentially non-ionic surfactant based multilamellar or unilamellar vesicles in which an aqueous solution of solute(s) is entirely enclosed by a membrane composed of surfactant macromolecules organized as bilayers. Transfersomes are special vesicular particles, consisting of at least one inner aqueous compartment surrounded by a lipid bilayer. These vesicles are elastic and thus sufficiently deformable to penetrate pores much smaller than their own size, which distinguish them from liposomes. All of the above mentioned vesicles are appropriate to incorporate both hidrophilic and lipophlic compounds. Lipid nanoparticles with a solid matrix can be composed either of SLNs, or of a blend of a solid lipid with a liquid lipid (oil) (NLCs). These lipid particles form a mono-layered film on the skin that has an occlusive action and contributes to skin hydration. These carriers are better suited for well lipid-miscible lipophilic active compounds.

As discussed in the previous section, the effectiveness of topical formulations incorporating vitamin C is mainly challenged by its intrinsic lack of stability and its poor penetration of the skin. In recent years, this has prompted efforts to utilize different delivery systems. Ascorbic acid has been successfully formulated in multiple emulsions (Farahmand *et al.*, 2006) and in microemulsions (Pakpayat *et al.*, 2009) showing improved stability. Encapsulation in chitosan microspheres (Desai *et al.*, 2006) and nanoparticles (Jang and Hyeon, 2008) was also successfully attempted. Ascorbate derivatives have also been incorporated into colloidal carriers. Ascorbyl palmitate has been incorporated into SLNs and NLCs with increased stability (Üner *et al.*, 2005). The effectiveness of the lipid particles loaded with ascorbyl palmitate and sodium ascorbyl phosphate against free radical formation has been demonstrated in UV-A irradiated pig skin (Kristl *et al.*, 2005) and was shown to depend significantly on the carrier system (i.e. the type of microemulsion and its concentration), rather than on the time of application (Jurkovic *et al.*, 2004).

7.5.3 Oral administration

Vitamin C is water-soluble and is well absorbed from the gastrointestinal tract. Oral administration of vitamin C could thus be an alternative to topical application, benefiting from increased stability and less dependence on the vehicle. However, plasma levels saturate at about 100 μM, due to efficient vitamin C excretion in the urine. Since the mean plasma vitamin C levels for healthy, well-nourished, non-smoking individuals are 50-60 μM, high supplemental doses can only cause modest increases in plasma vitamin C. Not surprisingly, the level of vitamin C attained in the skin with oral vitamin C is 20-40 times lower than that achievable by topical application (Burke, 2004). This could possibly justify the absence of a photoprotective effect in humans after oral administration of vitamin C even at high doses (Mireles-Rocha *et al.*, 2002).

7.5.4 Combination therapies

The effectiveness of topical vitamin C formulations could be improved through inclusion of other antioxidants (Bisset, 2006). In human and pig skin, the combined topical application of vitamins C and E was shown to increase the photoprotective effects of the vitamins when comparing with the monotherapies (i.e. the application of a single antioxidant) (Darr *et al.*, 1996; Lin *et al.*, 2003). The synergistic effect is attributed to the ability of vitamin C to regenerate the tocopheryl radical formed following α-tocopherol oxidation. The combined use of vitamin C and sunscreens was also studied. In an animal study, topical vitamin C provided additional protection against acute UVB damage when combined with a UVB sunscreen and, when formulated with an UVA sunscreen, an apparently greater than additive protection was observed (Darr *et al.*, 1996).

Acknowledgements

TLD is grateful to the 'Ciência Programme', supported by POPH - QREN - Tipologia 4.2 - Promoção do Emprego Científico, by 'Fundo Social Europeu' and by national funds from MCTES, Portugal.

References

Bisset, D.L., 2006. Anti-aging skin care formulations. In: Draelos, Z. and Thaman, L. (eds.) Cosmetic formulation of skin care products. Cosmetic science and technology. Taylor & Francis, New York, London, pp. 167-186.

Boyce, S.T., Supp, A.P., Swope, V.B. and Warden, G.D., 2002. Vitamin C regulates keratinocyte viability, epidermal barrier, and basement membrane *in vitro*, and reduces wound contraction after grafting of cultured skin substitutes. Journal of Investigative Dermatology 118, 565-572.

Burke, K.E., 2004. Photodamage of the skin: protection and reversal with topical antioxidants. Journal of Cosmetic Dermatology 3, 149-155.

Darr, D., Combs, S., Dunston, S., Manning, T. and Pinnell, S., 1992. Topical vitamin C protects porcine skin from ultraviolet radiation-induced damage. British Journal of Dermatology 127, 247-253.

Darr, D., Dunston, S., Faust, H. and Pinnell, S., 1996. Effectiveness of antioxidants (vitamin C and E) with and without sunscreens as topical photoprotectants. Acta Dermato-Venereologica 76, 264-268.

Desai, K.G., Liu, C. and Park, H.J., 2006. Characteristics of vitamin C encapsulated tripolyphosphate-chitosan microspheres as affected by chitosan molecular weight. Journal of Microencapsulation 23, 79-90.

Dreher, F., Denig, N., Gabard, B., Schwindt, D.A. and Maibach, H.I., 1999. Effect of topical antioxidants on UV-induced erythema formation when administered after exposure. Dermatology 198, 52-55.

Duarte, T.L., Cooke, M.S. and Jones G.D., 2009. Gene expression profiling reveals new protective roles for vitamin C in human skin cells. Free Radical Biology and Medicine 46, 78-87.

Duarte, T.L. and Lunec, J., 2005. When is an antioxidant not an antioxidant? A review of novel actions and reactions of vitamin C. Free Radical Research 39, 671-686.

Espinal-Perez, L.E., Moncada, B. and Castanedo-Cazares, J.P., 2004. A double-blind randomized trial of 5% ascorbic acid vs. 4% hydroquinone in melasma. International Journal of Dermatology 43, 604-607.

Farahmand, S., Tajerzadeh, H. and Farboud, E.S., 2006. Formulation and evaluation of a vitamin C multiple emulsion. Pharmaceutical Development and Technology 11, 255-261.

Fitzpatrick, R.E. and Rostan, E.F., 2002. Double-blind, half-face study comparing topical vitamin C and vehicle for rejuvenation of photodamage. Dermatologic Surgery 28, 231-236.

Humbert, P.G., Haftek, M., Creidi, P., Lapiere, C., Nusgens, B., Richard, A., Schmitt, D., Rougier, A. and Zahouani, H., 2003. Topical ascorbic acid on photoaged skin. Clinical, topographical and ultrastructural evaluation: double-blind study vs. placebo. Experimental Dermatology 12, 237-244.

Jagetia, G.C., Rajanikant, G.K. and Rao, K.V., 2007. Ascorbic acid increases healing of excision wounds of mice whole body exposed to different doses of gamma radiation. Burns 33, 484-494.

Jang, K.I. and Hyeon, G.L., 2008. Stability of chitosan nanoparticles for L-ascorbic acid during heat treatment in aqueous solution. Journal of Agricultural and Food Chemistry 56, 1936-1941.

Jurkovic, P., Sentjurc, M., Kristl, J., Pecar, S. and Gasperlin, M., 2004. Comparison of two ascorbic acid derivatives effectiveness for scavenging ultraviolet-induced free radicals in the skin. Journal of Drug Delivery Science and Technology 14, 229-233.

Kameyama, K., Sakai, C., Kondoh, S., Yonemoto, K., Nishiyama, S., Tagawa, M., Murata, T., Ohnuma, T. and Quigley, J., 1996. Inhibitory effect of magnesium L-ascorbyl-2-phosphate (VC-PMG) on melanogenesis *in vitro* and *in vivo*. Journal of the American Academy of Dermatology 34, 29-33.

Kaur, I.P., Kapila, M. and Agrawal, R., 2007. Role of novel delivery systems in developing topical antioxidants as therapeutics to combat photoageing. Ageing Research Reviews 6, 271-288.

Kobayashi, S., Takehana, M., Kanke, M., Itoh, S. and Ogata, E., 1998. Postadministration protective effect of magnesium-L-ascorbyl-phosphate on the development of UVB-induced cutaneous damage in mice. Photochemistry and Photobiology 67, 669-675.

Kristl, J., Gombač, K. and Šentjurc, M., 2005. Effectiveness of lipid nanoparticles with derivatives of ascorbic acid for skin protection from free radicals after UV-A irradiation. European Journal of Pharmaceutical Sciences 25, S140-142.

Lin, J.Y., Selim, M.A., Shea, C.R., Grichnik, J.M., Omar, M.M., Monteiro-Riviere, N.A. and Pinnell, S.R., 2003. UV photoprotection by combination topical antioxidants vitamin C and vitamin E. Journal of the American Academy of Dermatology 48, 866-874.

Mandl, J., Szarka, A. and Bánhegyi, G., 2009. Vitamin C: update on physiology and pharmacology. British Journal of Pharmacology 157, 1097-1110.

Mireles-Rocha, H., Galindo, I., Huerta, M., Trujillo-Hernandez, B., Elizalde, A. and Cortes-Franco, R., 2002. UVB photoprotection with antioxidants: effects of oral therapy with d-alpha-tocopherol and ascorbic acid on the minimal erythema dose. Acta Dermato-Venereologica 82, 21-24.

Ochiai, Y., Kaburagi, S., Obayashi, K., Ujiie, N, Hashimoto, S., Okano, Y., Masaki, H., Ichihashi, M. and Sakurai, H., 2006. A new lipophilic pro-vitamin C, tetra-isopalmitoyl ascorbic acid (VC-IP), prevents UV-induced skin pigmentation through its anti-oxidative properties. Journal of Dermatological Science 44, 37-44.

Pakpayat, N., Nielloud, F., Fortune, R., Tourne-Peteilh, C., Villarreal, A., Grillo, I. and Bataille, B., 2009. Formulation of ascorbic acid microemulsions with alkyl polyglycosides. European Journal of Pharmaceutics and Biopharmasceutics 72, 444-452.

Raschke, T., Koop, U., Dusing, H.J., Filbry, A., Sauermann, K., Jaspers, S., Wenck, H. and Wittern, K.P., 2004. Topical activity of ascorbic acid: from *in vitro* optimization to *in vivo* efficacy. Skin Pharmacology and Physiology 17, 200-206.

Shindo, Y., Witt, E. and Packer, L., 1993. Antioxidant defense mechanisms in murine epidermis and dermis and their responses to ultraviolet light. Journal of Investigative Dermatology 100, 260-265.

Uchida, Y., Behne, M., Quiec, D., Elias, P.M. and Holleran, W.M., 2001. Vitamin C stimulates sphingolipid production and markers of barrier formation in submerged human keratinocyte cultures. Clinical Journal of Investigative Dermatology 117, 1307-1313.

Üner, M., Wissing, S.A., Yener, G. and Müller, R.H., 2005. Solid lipid nanoparticles (SLN) and nanostructured lipid carriers (NLC) for application of ascorbyl palmitate. Pharmazie 60, 577-582.

Vissers, M.C.M., Gunningham, S.P., Morrison, M.J., Dachs, G.U. and Currie, M.J., 2007. Modulation of hypoxia-inducible factor-1 alpha in cultured primary cells by intracellular ascorbate. Free Radical Biology & Medicine 42, 765-772.

Xiao, L., Kaneyasu, K., Saitoh, Y., Terashima, Y., Kowata, Y. and Miwa, N., 2009. Cytoprotective effects of the lipoidic-liquiform pro-vitamin C tetra-isopalmitoyl-ascorbate (VC-IP) against ultraviolet-A ray-induced injuries in human skin cells together with collagen retention, MMP inhibition and p53 gene repression. Journal of Cellular Biochemistry 106, 589-598.

Key facts

- Very few drugs can passively diffuse across the skin, at therapeutically useful rates, because of *stratum corneum* barrier properties.
- To overcome the *stratum corneum* barrier, various mechanisms have been investigated, including the use of chemical or physical enhancers and transdermal delivery systems that have also the potential of overcoming the skin barrier and have been reported to enhance permeability of drug through the *stratum corneum* barrier.
- Transdermal delivery of drugs through the skin to the systemic circulation provides a convenient route of administration for a variety of clinical indications.
- Transdermal delivery has a variety of advantages compared with the oral route. In particular, it is used when there is a significant first-pass effect of the liver that can prematurely metabolize drugs.
- Transdermal delivery systems have the advantage of maintaining constant drug plasma levels and providing it release for long periods of time (up to one week).
- In addition, transdermal delivery systems are non-invasive and can be self-administered. They also improve patient compliance and the systems are generally inexpensive.

Summary points

- Vitamin E, a natural biological antioxidant, has been shown to protect the skin against cell damage following exposure to UV light.
- Vitamin E-based formulations may be employed as protective agents against UV light-induced cell damage. Barrier properties of the skin and the poor stability of the vitamin E, due to direct exposure to UV light, limit its use and deposition on the skin.
- It is possible make use of chemicals and methods, that reduce the barrier capability of the *stratum corneum*, such as chemical enhancers, physical and electrical methods that permit the vitamin E release.
- Pro-drugs and several drug delivery systems (DDS) such as liposomes, polymeric nanoparticles, microemulsions, hydrogels, lipogels, and nanofibers can be useful to favor vitamin E transdermal delivery.

8. Strategies for vitamin E transdermal delivery

S. Trombino
University of Calabria, Department of Pharmaceutical Science,Edificio Polifunzionale, Arcavacata di Rende, Via Pietro Bucci, 87036, Arcavacata di Rende, Cosenza, Italy; sonia.trombino@unical.it

Abstract

Long term UV exposure is known to induce cell damage which may cause skin carcinogenesis. Vitamin E (α-tocopherol), a natural biological antioxidant, has been shown to protect against UV light-induced cell damage, to treat skin diseases and several types of cancer or to decrease oxidative stress. For this reason vitamin E based formulations, may be employed as protective agents. To obtain an effective photoprotection, a significant amount of antioxidant needs to arrive at the site, but skin barrier properties and vitamin E poor stability, due to direct exposure to UV light, limit its use and deposition in the skin. Consequently, the choice of an appropriate carrier is particularly important for enhancing vitamin E penetration through the *stratum corneum* and improving its efficacy. There are numerous strategies to facilitate the transdermal delivery of vitamin E. Chemicals substances and methods that modify the properties of *stratum corneum* can be used. In addition, during the last years, particular increasing attention has been given to drug delivery systems that have proved efficacy in enhancing pharmacological effects, vitamin E photoprotection and favouring its transdermal delivery.

Keywords: α-tocopherol, UV light exposure, transdermal release, drug delivery systems

Abbreviations

α-T	α-tocopherol
α-TAc	α-tocopherol acetate
α-TOS	α-tocopherol succinate
δ-TG	δ-tocopherol glucoside
DDS	Drug delivery systems
DMSO	Dimethyl sulfoxide
IPM	Isopropyl mlyristate
ME	Microemulsion
o/w	Oil in water
Pluronic F-88	Polyoxyethylene-olypropylene block copolymer
SLN	Solid lipid nanoparticle
Span 80	Sorbitan monooleate
Tween 80	Polyoxyethylene (20) sorbitan monooleate
UV	Ultraviolet
UVB	Ultraviolet B
w/o	Water in oil

8.1 Introduction

The skin is exposed to several environmental, chemical, and physical agents, such as ultraviolet light, air pollutants, and chemical oxidants that cause erythema, edema, skin thickening, wrinkling and increase risk of skin cancer or precursor lesions (Fuchs *et al.*, 2001). The oxidants may play a key role in skin cancer, aging and other skin diseases. Most of the facial skin is exposed to a variety of environmental oxidants and consequently requires a higher antioxidant protection.

Vitamin E is the most potent lipid-soluble antioxidant *in vivo* (Ingold *et al.*, 1987) that plays an essential function in skin defence delaying aging progression. In particular this vitamin stabilizes biological membranes, especially those containing large amounts of polyunsaturated fatty acids (Rangarajan and Zatz, 1999). Furthermore, because of its emollient properties and non-irritant nature, this molecule shows high compatibility with the skin. Ekanayake-Mudiyanselage *et al.* (2005) verified that even the use of rinsing products containing α-tocopherol at concentrations less than 0.2%, can lead to significantly increased levels of vitamin E in the *stratum corneum* of human skin and protect against lipid peroxidation *in vivo*. Therefore, topical formulations containing α-tocopherol at concentrations ranging from 0.1% to 1% significantly reduce erythema, edema, and skin hypersensitivity associated with UVB radiation resulting from solar exposure (Trevithick *et al.*, 1993).

On the other hand the vitamin E administration is limited because of its physicochemical properties. In fact, in order to obtain an efficient and accurate transdermal delivery of vitamin E, a few unavoidable problems need to be overcome. In fact the drug is highly viscous oil, practically insoluble in water and readily oxidized by atmospheric oxygen and must be carefully protected from light, heat and prolonged contact with air or UV light (Drott *et al.*, 1991). For this reason the choice of an appropriate carrier to increase vitamin E solubility, protect it from photodegradation and permit its transdermal release, is extremely important.

A good transdermal delivery system must not only offer an adequate drug release from the formulation but, also permit considerable amounts of drugs to overcome the skin barrier, ensure a non-irritancy of the skin, and also guarantee that the drug will not be inactivated on the skin's surface or during the permeation process (Langer, 2000). Furthermore transdermal delivery can count a number of advantages over conventional drug administration methods, including greater convenience and improved patient compliance. By delivering a steady flow of drugs into the blood stream over an extended period of time, transdermal systems can, in addition, elude the first metabolism step through the gastrointestinal tract.

The strategies to facilitate the vitamin E transdermal delivery are numerous. In order to reduce the barrier capability of the *stratum corneum*, chemical enhancers, physical and electrical methods, can be employed. In addition, as reported in literature, pro-drugs and several DDS (Figure 8.1) such as liposomes, polymeric nanoparticles, microemulsions, hydrogels, lipogels, nanofibers can be useful to obtain improved pharmacological effects such as decreased side effects, a controlled drug release and vitamin E photoprotection.

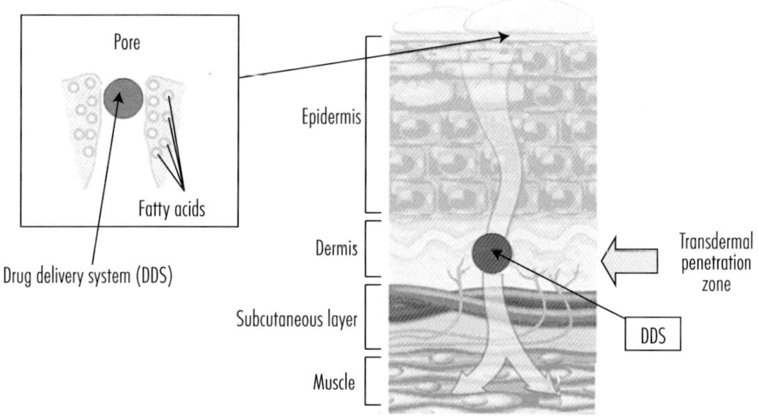

Figure 8.1. Schematic representation of transdermal delivery through DDS. Transdermal delivery systems are able to transfer the drug into the systemic circulation through the skin at a controlled rate. They represent a non-invasive route of drug administration but its application is limited by low skin permeability. Therefore to facilitate drug vehiculation through the transdermal barrier some carriers such as liposomes, microemulsions, solid lipid nanoparticles, etc. are used.

8.2 Chemical enhancers

Chemical enhancers are substances that alter the skin barrier function to permit a faster and better drug permeation through the skin. Many enhancers, such as azone, DMSO, alcohols, fatty acids and terpenes, have been shown to increase permeability by disordering or 'fluidising' the lipid structure of the *stratum corneum* (Benson, 2005). Enhancer molecules can form microcavities within the lipid bilayers hence increasing the diffusion of the drug. In some cases the enhancers penetrate into and mix homogeneously with the lipids.

Trivedi *et al.* (1995) also demonstrated that vitamin E itself can act as a penetration enhancer by intercalating within the lipid bilayer region of the *stratum corneum*, thus altering the characteristics of membrane affecting permeability, presumably by disordering gel phase lipids. Unlike other exposed enhancers, Vitamin E is generally thought to be non-irritating, and additionally, it possesses antioxidant and emollient properties.

8.3 Physical and electrical methods

The electrical methods of penetration enhancement (Jadoul *et al.*, 1999) include iontophoresis (driving charged molecules into the skin by a small direct current: approximately 0.5 mA/cm^2), phonophoresis (cavitation caused by low frequency ultrasound energy increases lipid fluidity), electroporation (application of short micro- to milli-second electrical pulses of approximately 100-1000 V/cm to create transient aqueous pores in lipid bilayers) and photomechanical waves (laser-generated stress waves reported to cause a possible transient permeabilisation of the *stratum corneum*). A number of methods to bypass or remove the *stratum corneum* have also been assessed, such as microneedles (a device containing 400 solid or hollow silicon needles, approximately 150 μm in length, that penetrate through the *stratum corneum* into the upper epidermis), jet-propelled particles (high-velocity jet of compressed gas carrying drug particles) and ablation of the *stratum corneum* (by laser, adhesive tape or chemical peels).

8.4 Pro-drugs

The pro-drug design approach involves the addition of a promoiety to increase the solubility and -tocopherol transport into the *stratum corneum* improving its stability. Upon reaching the viable epidermis, the esterases release the drug by hydrolysis thereby optimising solubility in the aqueous epidermis (Sloan and Wasdo, 2003) (Figure 8.2). Therefore α-tocopherol pro-drug esters, such as α-tocopherol acetate and α-tocopherol succinate etc., are usually used, for cosmetic formulations.

Several studies have examined the photoprotective benefits after the topical application of α-TAc (Figure 8.3). In particular, a recent *in vitro* study on mouse keratinocytes, showed that α-TAc has the potential to protect the skin from UVB-associated epidermal damage (Maalouf *et al.*,

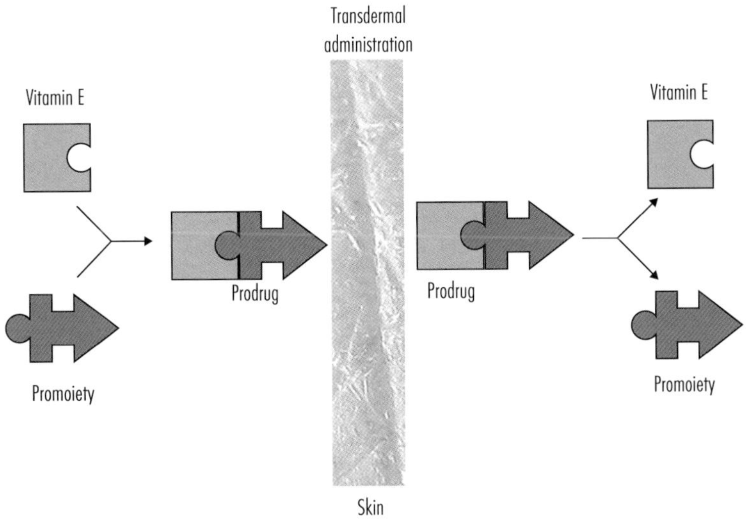

Figure 8.2. Schematic representation of vitamin E release by pro-drug strategy. Prodrug is a substance administered as an inactive form that is then metabolized *in vivo* into the active compound. It could be used to alter the physicochemical properties of drug, in a temporary manner, to increase its usefulness and/or to decrease associated toxicity. Due its instability and physicochemical properties, vitamin E is usually used, as its prodrug ester, for cosmetic formulations and clinical use.

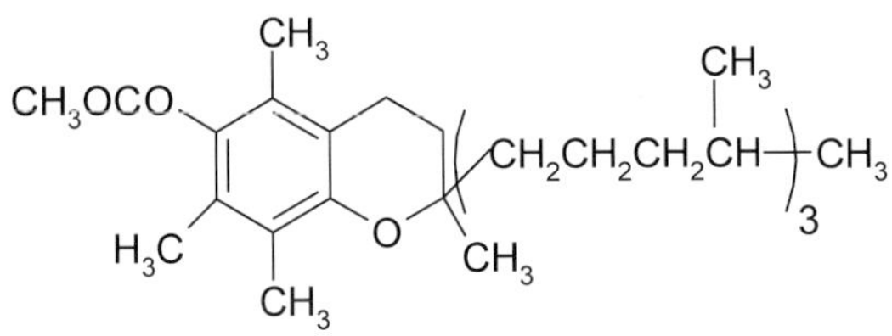

Figure 8.3. α-tocopherol acetate structure. α-tocopherol acetate (α-TAC) is a vitamin E prodrug. The acetate is slowly hydrolyzed, when it is absorbed into the skin, regenerating α-tocopherol and providing protection against UV light.

2002). Rangarajan and Zatz (2001) evaluated the effect of delivery system on the permeation and metabolism of α-TAc in micro-Yucatan pig skin, which closely resembles human skin. In particular various α-TAc formulations, including a simple IPM, o/w emulsion, microemulsions, alcoholic and hydroalcoholic gels were made. They demonstrated the metabolic conversion of α-TAc to α-T in pig skin and showed that permeation of α-TAc and its metabolism depend on the delivery system used. An emulsion system containing IPM emerged as the most desirable formulation in terms of skin delivery of α-TAc.

Wang *et al.* (2006), in a review, proposed a transdermal application to promote the delivery of the anticancer drug α-TOS (Figure 8.4). Intravenous administration of α-TOS in fact was impractical for humans and, an alternative route of administration such as a transdermal one, they thought to be useful and, also, safer.

Mavon and co-workers (2004) presented as prodrug, a gluco-vitamin E conjugate, δ-tocopherol glucoside (δ-TG)δδ (Figure 8.5), synthesized by making use of β-glucocerebrosidase activity for dermatological employ. The glycosidic bond cleavage, catalyzed by β-glucocerebrosidase permits the release of active vitamin E. As same as the acetate form, the gluco-conjugation one improves the stability of vitamin E.

Particularly, the application of δ-TG on the skin, gives a substantial reservoir effect and a slow delivery of active tocopherol, first in the *stratum corneum* and then in other skin layers. More recently, Marra *et al.* (2009) synthesized new pro-vitamins of a-tocopherol that accumulated into the skin in significant amounts and were able to generate vitamin E. The synthesized pro-vitamins resulted more water soluble than the currently used vitamin E acetate, thus allowing the use of more hydrophilic vehicles for skin application.

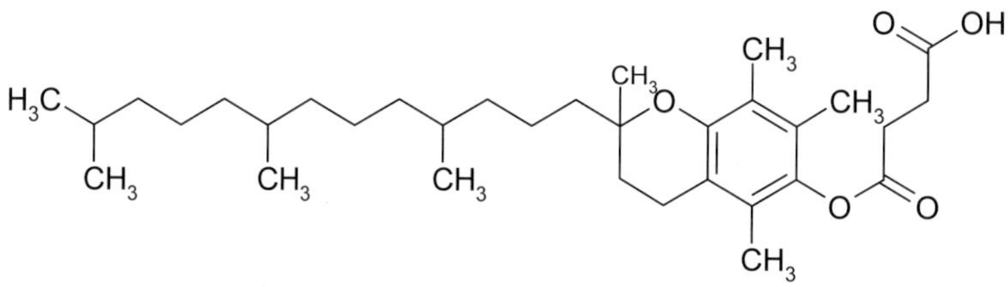

Figure 8.4. a-tocopherol succinate (a-TOS) is vitamin E antineoplastic prodrug that prevents skin cancer and immuno-suppression induced by UV radiation

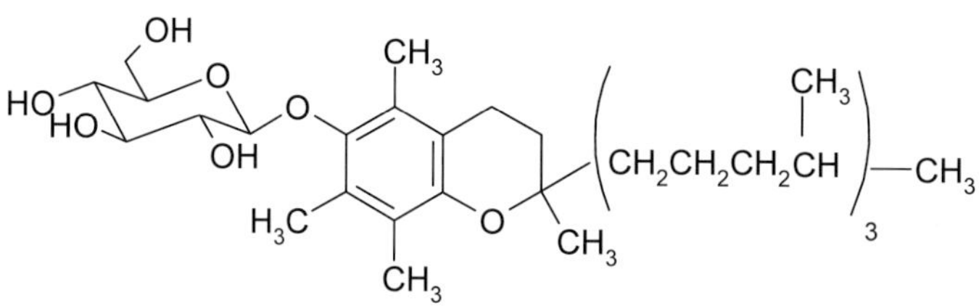

Figure 8.5. δ-tocopherol glucoside (δ-TG) structure. δ-TG is vitamin E prodrug that is metabolized in the skin at a low concentration in a slow and prolonged manner, furnishing a continuous reserve of antioxidant.

8.5 Liposomes

Liposomes have been promoted as another means for enhancing transdermal drug delivery. These are microscopic bilayer vesicles made of phospholipids and cholesterol (Figure 8.6). The liposomes are trapped within the top layer of *stratum corneum* cells and interact with skin lipids to release their drug (Gandhi *et al.*, 2010).

Darr *et al.* (1996) demonstrated that vitamin E liposomes provide time release, target delivery of essential antioxidant power with superior benefits. The frequent use of this liposomes-based formulation, for deep hydration of the skin, could substantially diminish the visible signs of premature aging.

8.5.1 Deformable liposomes

In 2006 Gallarate *et al.* obtained deformable liposomes to increase the vitamin E skin deposition. These particular elastic vesicles containing phosphatidylcholine and usually sodium cholate, Span 80, or Tween 80, can penetrate relatively easily even through pores that are much smaller than their own diameter. For this reason, the active substance included in the vehicle may have high concentration in the deeper layers of the epidermis. In particular, Gallarate and co-workers (2006) obtained the deformable liposomes, using several hydrogenated lecithin-surfactant mixtures that when applied, significantly improved the α-tocopherol *in vitro* skin delivery when compared to non-elastic liposomes and reference solutions.

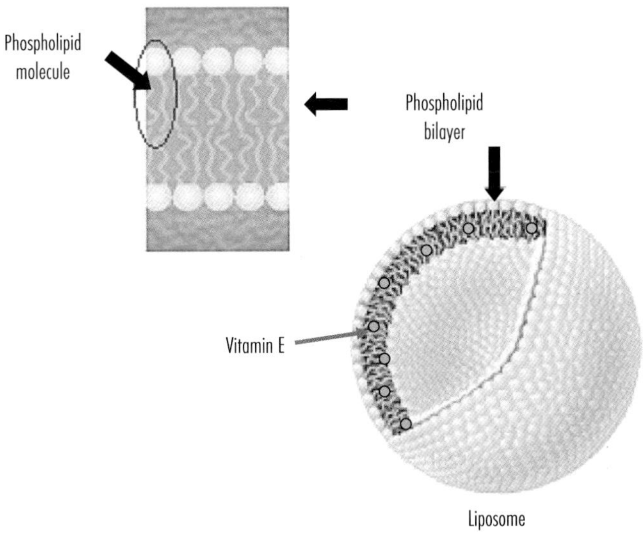

Figure 8.6. Schematic representation of liposome. Liposomes are artificial microscopic vesicles consisting of an aqueous core enclosed in one or more phospholipid layers, used to convey vaccines, drugs, enzymes, or other substances to cell or organs. In the liposomes, vitamin E, due to its lipophilic nature, is located in phospholipid layers.

8.5.2 Ethosomes

Ethosomes are phospholipid soft vesicles composed of safe and natural components approved for pharmaceutical and cosmetic use. This innovative carrier technology is a tailored delivery system designed to meet the necessary criteria for efficient and safe administration of molecules into the various layers of the skin. When ethosomes are applied, owing to their softness, they easily penetrate the *stratum corneum* barrier releasing the active substance into the cells located in the deep skin strata. In 2008 Touitou and Godin obtained an efficient intracellular delivery of vitamin E from ethosomes.

8.6 Polimeric nanoparticles

Polymeric nanoparticles are solid colloidal suspensions with a mean particle size <1 μm (Figure 8.7). These systems can be classified in capsules and spheres as a consequence of their morphology. More specifically, spheres are usually characterized by a porous matrix in which the drug is contained, whereas capsules are formed by a core containing the drug surrounded by a shell.

Several studies on entrapment of α-tocopherol particularly in polymeric nanoparticles are reported in the literature. Delivery of vitamin E entrapped in polymeric nanoparticles shows specific improvements over the delivery of non-entrapped vitamins. The release rate of the vitamin can be controlled and consequently the dose frequency can be reduced. Furthermore, the bioactivity and stability of active substance entrapped in the nanoparticle is protected by encapsulation. Duclairoir *et al.* (2002) entrapped vitamin E in nanoparticles obtained from

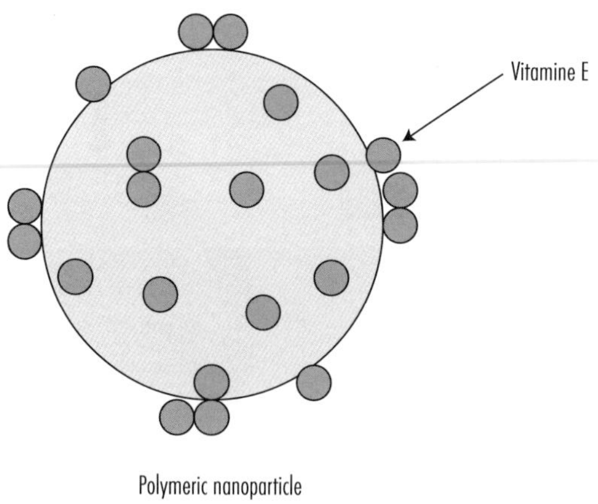

Figure 8.7. Schematic representation of nanoparticle. Polymeric nanoparticles are materials prepared from polymers. The drug is dissolved, entrapped, encapsulated or attached to a nanoparticle and depending upon the method of preparation. Nanoparticle-based drug delivery approaches have the potential for making hydrophobic agents, like vitamin E, dispersible in aqueous media, overcoming thus the problem of poor solubility.

vegetal proteins. In particular gliadins, extracted from wheat gluten, formed nanoparticles, able to interact with epidermal keratin, for therapeutic and cosmetic formulations.

8.7 Microemulsions

Microemulsions are clear, thermodynamically stable dispersions of water and oil, with a mean particle size between 10 to 200 nm, stabilized by an interfacial film of surfactant molecules (Figure 8.8).

Topically applied ME significantly increase skin absorption of drugs and, in some cases, provide its sustained release (Kogan and Garti, 2006). These favourable drug delivery properties of ME are attributed mainly to their excellent solubilizing properties. They can also act as penetration enhancers, depending on the nature of the oil and surfactant constituents.

Rozman and Gasperlin (2009) proposed an interesting strategy for protecting skin from excessive exposure to free radicals through a combined therapy based on at least two antioxidants. In particular o/w, w/o and gel-like ME, all composed of the same ingredients, were selected as carrier systems for dermal delivery of vitamins C and E. Gel-like ME was found to offer the best protection for both vitamins, although other ME also significantly increased their stability compared with the one in solution. In the presence of vitamin C no decrease in vitamin E content occurred. To obtain ME appropriate for dermal use, their viscosity was increased by adding a thickening agent such as colloidal silica. Finally, the stability of both vitamins was examined as

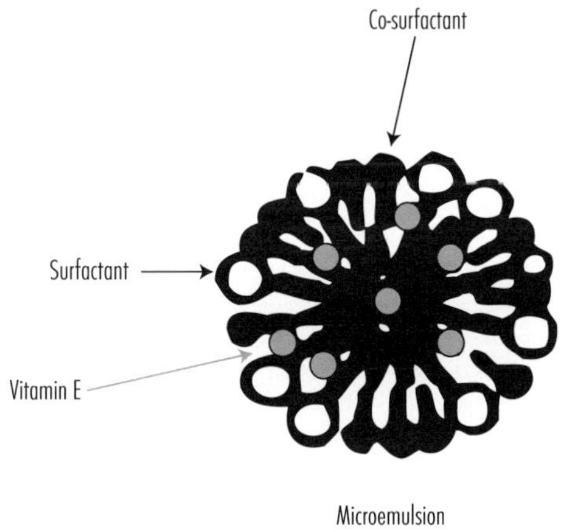

Figure 8.8. Schematic representation of microemulsion. Microemulsions are dispersions of oil and water made with surfactants, frequently in combination with co-surfactants. The use of microemulsions is advantageous not only thanks to the easy and low cost preparation, but also because of the improved bioavailability. The increased absorption of vitamin E in topical applications microemulsion-based is attributed to enhancement of penetration through the skin by this carrier.

a function of thickening agent and of vitamins location in the ME. The addition of thickeners changed the stability of at least one vitamin, but the systems generally still protected vitamins better than solutions.

8.8 Lipid nanoparticles

Solid lipid nanoparticles are nanoparticles with a solid lipid matrix presenting an average diameter in the nanometer range. General ingredients include solid lipids, emulsifiers and water. Lipids comprise triglycerides, partial glycerides, fatty acids, steroids, and waxes (Figure 8.9).

All excipients are generally recognized as safe substances, so a wide variety of compounds can be used for formulation purposes (Trapasso *et al.*, 2009). Due to physicochemical properties of vitamin E, SLNs could be an appropriate system for its stabilization and release. In fact, these carriers could be used to protect the compound from consequential light exposure damage and to improve its pharmacological activity. In addition, the occlusive effect of SLNs leads to an increase of skin hydration and, thus, to wrinkle smoothing, enhancing the penetration of encapsulated compound in specific skin layers.

In 1999 Dingler *et al.* obtained vitamin E loaded SLNs and then evaluated their incorporation into a cream and its effect with regard to occlusion and vitamin E penetration into the *stratum corneum*. Furthermore they studied the physical stability of vitamin E loaded solid lipid nanoparticles

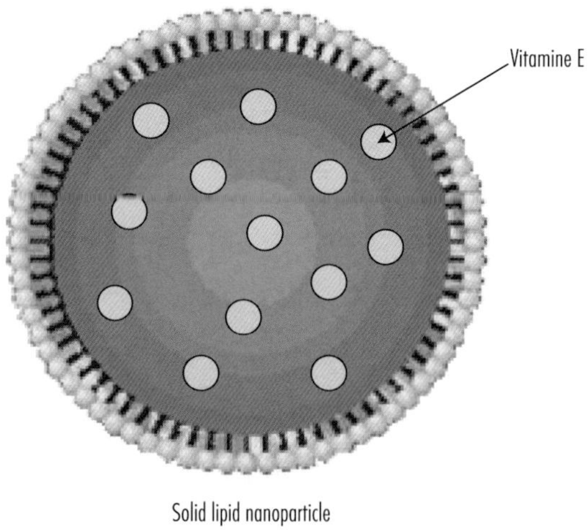

Figure 8.9. Schematic representation of solid lipid nanoparticle. Solid lipid nanoparticles are specific delivery systems made of solid lipids which remain solid at room temperature. Their small particle size ensures close contact to the *stratum corneum* and increase the amount of encapsulated agent penetrating into the viable skin. Furthermore, thanks to their lipidic nature, they represent the ideal carrier for vitamin E transdermal delivery.

in creams. The obtained results showed that these nanoparticles can protect vitamin E against chemical degradation and improve its penetration into the upper layer of the epidermis, mainly into the *stratum corneum*, enhancing its dermatological performance.

To improve the vitamin E stability and skin penetration, Kim and Lee (1999) obtained lecithin lipidic-based nanospheres. In order to increase the stability, the lipid layer surface was coated with several emulsifiers such as Tween 80 and Pluronic F-88. In addition when a non-ionic surfactant was used as a cosurfactant, the lipid nanosperes size was smaller. Then, *in vitro* permeation experiments of vitamin E through rat skin were performed and the obtained results showed that the permeation of the vitamin from lipid nanospheres was greater compared to the free vitamin E ones and varied depending on the composition and size of the nanospheres.

8.9 Lipogels and hydrogels

Another method for vitamin E transdermal delivery is its loading in lipogel and hydrogel formulations.

Lipogels are formulations based on vegetable, mineral or animal oils gellified by gelling agents. Thanks to their lipidic composition they are useful in the cosmetic field because they serve as a barrier between the skin and the outside environment and help to regenerate the *stratum corneum* because of their rich content in fatty acids (i.e. oleic fatty acid) which are usually compatible with the epidermis.

Hydrogels are three-dimensional hydrophilic polymer networks capable of swelling in water or biological fluids and retaining a large amount of fluids in the swollen state (Figure 8.10).

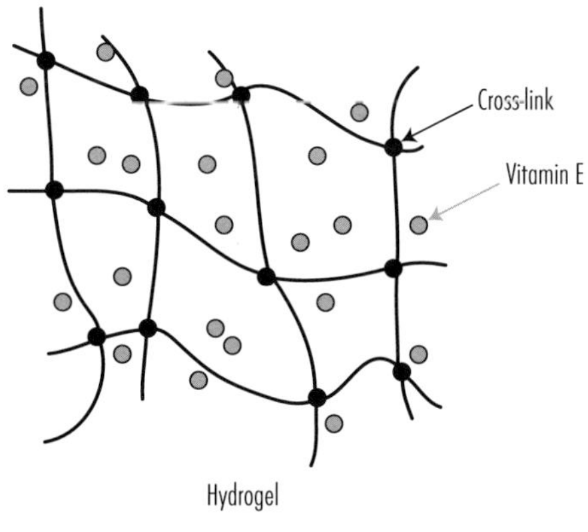

Figure 8.10. Schematic representation of hydrogel. Hydrogels are a class of polymer materials that can absorb large amounts of water without dissolving. This is due to physical or chemical crosslinkage of the hydrophilic polymer chains. A particular antioxidant dextran-based hydrogel showed the ability in preserving vitamin E during its release through the skin.

Their ability to absorb water is due to the presence of hydrophilic groups such as -OH, -CONH-, $-CONH_2$, -COOH, and $-SO_3H$. Due to high water content, hydrogels show physical properties similar to living tissue, soft and rubbery consistency and low interfacial tension with water or biological fluids. The ability of different sized molecules to diffuse into and out of hydrogels permits the use of hydrogels as delivery systems. Gallardo *et al.* (2005) prepared both lipogels and hydrogels, as vitamin E vehicles. In particular, the lipogels, because of their spreadability and persistence on the skin, were an adequate system for cosmetic anti-aging treatment with vitamin E. Furthermore the compatibility of these types of lipophilic vehicles with the skin makes them especially appropriate for vitamin E administration. On the other hand, vitamin E loaded hydrogels were ideal as an efficient protective action from the UV-radiation oxidative damage and they showed good organoleptic properties and good release kinetics.

More recently, Cassano *et al.* (2009) prepared a dextran-based hydrogel, containing ferulic moieties, as a carrier to increase vitamin E deposition, its flux on the skin and to improve further its stability to the light, heat etc. Ferulic acid was chosen, as linking to hydrogel, because it is a potent antioxidant having synergistic effects with other antioxidants and it is able to protect and stabilize them from degradation.

The vitamin E deposition on the skin was evaluated and compared to that of analogous hydrogel not containing ferulic groups. The ferulate hydrogel showed to be a good approach to preserve vitamin E during its release. In particular, the linking of ferulic acid on hydrogel greatly influenced the drug deposition in the ear skin of a rabbit. In fact, an increased drug deposition was observed suggesting that the ferulic groups protect drug during the deposition compared to the analogous hydrogel without ferulic moieties. Therefore, the ferulate hydrogel could be an adequate system for the controlled release of vitamin E in the human skin.

8.10 Nanofibers

Nanofibers are submicron sized fibers with average diameters in sub-micrometer down to nanometer range, obtained through an electrospinning process (Figure 8.11), that exhibit several interesting characteristics such as a high surface area to mass or volume ratio, a small inter-fibrous pore size with high porosity, vast possibilities for surface functionalization, etc. (Deitzel *et al.*, 2001). Thanks to these interesting properties the nanofibers have been developed as carriers for drugs delivery.

In 2007 Taepaiboon and co-workers, produced mats of electrospun cellulose acetate nanofibers as carriers for delivery of vitamin A and vitamin E to the skin and then investigated the *in vitro* release profiles. The obtained results indicated that the vitamin-loaded elecrospun nanofiber mats, could be successfully developed for vitamin E transdermal delivery or for obtaining dermal patches.

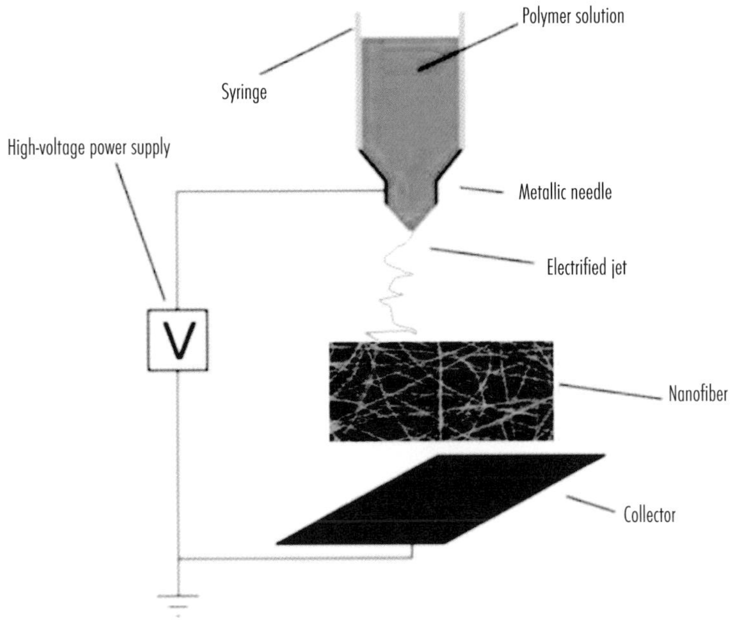

Figure 8.11. Electrospinning process uses an electrical charge for fabricating very fine (typically on the micro or nano scale) fibers from a liquid. Thanks to their dimension and characteristics, the nanofibers represent good candidates as vitamin E transdermal delivery systems.

Acknowledgements

The author thanks Ms Stefania Pasqua, linguistic expert of the Faculty of Pharmacy of the University of Calabria, for her assistance in the drafting of the manuscript.

References

Benson, H.A.E., 2005. Transdermal Drug Delivery: Penetration Enhancement Techniques. Current Drug Delivery 2, 23-33.

Cassano, R., Trombino, S., Muzzalupo, R., Tavano, L. and Picci, N., 2009. A novel dextran hydrogel linking trans-ferulic acid for the stabilization and transdermal delivery of vitamin E. European Journal of Pharmaceutics and Biopharmaceutics 72, 232-238.

Darr, D., Dunston, S., Faust H. and Pinell, S., 1996. Effectiveness of antioxidants (Vitamin C and E) with and without sunscreens as topical photoproctants. Acta Dermato-Venereologica 76, 264-268.

Deitzel, J.M., Kleinmeyer, J., Harris, D., Beck and Tan, N.C., 2001. The effect of processing variables on the morphology of electrospun nanofibers and textiles. Polymer 42, 261-272.

Dingler A, Blum R.P., Niehus H, Muller, R.H. and Gohla, S., 1999. Solid lipid nanoparticles (SLN^{TM}/Lipopearls^{TM}) a pharmaceutical and cosmetic carriers for the application of vitamin E in dermal products. Journal of Microencapsulation 16, 751-767.

Drott, P., Meurling, S. and Meurling, L., 1991. Clinical adsorption and photodegradation of the fat-soluble vitamins A and E. Clinical Nutrition 10, 348-351.

Duclairoir, C., Orecchioni, A.M., Depraetere, P. and Nakache, E., 2002. Alpha-Tocopherol encapsulation and *in vitro* release from wheat gliadin nanoparticles. Journal of Microencapsulation 19, 53-60.

Ekanayake-Mudiyanselage, S., Tavakkol, A., Polefka, T.G., Nabi, Z., Elsner, P. and Thiele, J.J., 2005. Vitamin E delivery to human skin by a rinse-off product: Penetration of alpha-tocopherol versus wash-out effects of skin surface lipids. Skin Pharmacology and Physiology 18, 20-26.

Fuchs, J., Zollner, T.M., Kaufman, R. and Podda, M., 2001. Redox-modulated pathways in inflammatory skin diseases. Free Radical Biology and Medicine 30, 337-353.

Gallardo, V., Muñoz, M. and Ruíz, M.A., 2005. Formulations of hydrogels and lipogels with vitamin E. Journal of Cosmetic Dermatology 4, 187-192.

Gallarate, M., Chirio, D., Trotta, M. and Parlotti, M.E., 2006. Deformable liposomes as topical formulations containing alpha-tocopherol. Journal of Dispersion Science and Technology 27, 703-713.

Gandhi, M., Verma, S. and Shukla, S., 2010. Various approaches to transedermal delivery system. Journal of Natura Conscientia 1, 227-232.

Ingold, K.U., Webb, A.C., Witter, D., Burton, G.W., Metcalfe, T.A. and Muller, D.P., 1987. Vitamin E remains the major lipid-soluble, chain-breaking antioxidant in human plasma even in individuals suffering severe vitamin E deficiency. Archives of Biochemistry and Biophysics 259, 224-225.

Jadoul, A., Bouwstra, J. and Preat, V.V., 1999. Effects of iontophoresis and electroporation on the *stratum corneum* - Review of the biophysical studies. Adanced Drug Delivery Reviews 35, 89-105.

Kim, S.Y. and Lee, Y.M., 1999. Lipid nanospheres containing vitamin A or Vitamin E: evaluation of their stabilites and in vitro skin permeability. Journal of Industry and Engineering Chemistry 5, 306-313.

Kogan, A. and Garti, N., 2006. Microemulsions as transdermal drug delivery vehicles. Advances in Colloid and Interface Science 123, 369-385.

Langer, R., 2000. Transdermal drug delivery: past, progress, current status and future prospects. Advanced Drug Delivery Reviews 56, 557-558.

Maalouf, S., El-Sabban, M., Darwiche, N. and Gali-Muhtasib, H., 2002. Protective effect of vitamin E on ultraviolet B light-induced damage in keratinocytes. Molecular Carcinogenesis 34, 121-130.

Marra, F., Ostacolo, C., Laneri, S., Bernardi, A., Sacchi, A., Padula, C., Nicoli, S. and Santi, P., 2009, Synthesis, Hydrolysis, and Skin Retention of Amino Acid Esters of Alpha-Tocopherol. Journal of Pharmaceutical Science 98, 2364-2377.

Mavon A, Raufast V. and Redouèls D. 2004. Skin absorptionand metabolism of new vitamin E prodrug, δ-tocopherol-glucoside: in vitro evaluation in human skin models. Journal of Controlled Release 100, 221-231.

Rangarajan, M. and Zate, J.L. 1999. Skin delivery of vitamin E. Journal of Cosmetic Science 50, 249-279.

Rangarajan, M. and Zatz, J.L. 2001. Effect of formulation on the delivery and metabolism of a-tocopheryl acetate. Journal of Cosmetic Science 52, 225-236.

Rozman, B. and Gasperlin, M. 2009. Stability of vitamins C and E in topical microemulsions for combined antioxidant therapy. Drug Delivery 14, 235-245.

Sloan, K.B. and Wasdo, S. 2003. Designing for topical delivery: prodrugs can make the difference. Medicinal Research. Reviews 23, 763-793.

Taepaiboon, P., Rungsardthong, U. and Supaphol, P., 2007. Vitamin-loaded electrospun cellulose acetate nanofiber mats as transdermal and dermal therapeutic agents. of vitamin A acid and vitamin E. European Journal of Pharmaceutics and Biopharmaceutics 67, 387-397.

Touitou, E. and Godin, B., 2008. Skin nonpenetrating sunscreens for cosmetic and pharmaceutical formulations. Clinics in Dermatology 26, 375-379.

Trapasso, E. Cosco, D, Celia C., Fresta, M. and Paolino, D., 2009. Retinoids: new use by innovative drug-delivery systems. Expert Opinion On Drug Delivery 6, 465-483.

Trevithick, J.R., Shum, D.T., Redae, S., Mitton, K.P., Norley, C., Karlik, S.J., Croom, A.C. and Schmidt, E.E., 1993. Reduction of sunburn damage to skin by topical application of vitamin-e acetate following exposure to ultraviolet-b radiation-effect of delaying application or of reducing concentration of vitamin-e acetate applied. Scanning Microscopy 7, 1269-1281.

Trivedi, J.S, Krill, S.L and Fort, J.J. 1995. Vitamin-E as a human skin penetration enhancer. European Journal of Pharmaceutical Science 3, 241-243.

Wang, X.F, Dong, L.F., Zhao, Y., Tomasetti, M., Wu, K. and Neuzil, J., 2006. Vitamin E analogues as anticancer agents: Lessons from studies with *α*-tocopheryl succinate. Molecular Nutrition and Food Research 50, 675-685.

Key facts

- Reactive oxygen species include chemically reactive molecules derived from oxygen. Some of those molecules are extremely reactive, such as the hydroxyl radical, while some are less reactive, such as superoxide and hydrogen peroxide. They can readily react with most biomolecules, starting a chain reaction of free radical formation.
- Low-density lipoprotein are sometimes called 'bad' cholesterol. Lipoproteins are made of fat and protein. They carry cholesterol, triglycerides, and other fats, called lipids, in the blood to various parts of the body.
- Protonic nuclear magnetic resonance serves as a great resource in determining the structure of an organic compound by revealing the hydrogen and carbon skeleton.
- Carbon nuclear magnetic resonance plays an important role in determining the structure of unknown organic molecules and the study of organic reactions and processes. The idea and theory behind is the same as with protonic nuclear magnetic resonance just a different nucleus.
- Fourier transform infrared spectroscopy is an analysis technique that provides information about the chemical bonding or molecular structure of materials, whether organic or inorganic.
- Gas Chromatography/Mass Spectrometry is a technique that separates chemical mixtures (the GC component) and identifies the components at a molecular level (the MS component). It is one of the most accurate tools for analyzing environmental samples.
- An international unit is a globally accepted amount of a substance. This measure is used for the fat-soluble vitamins, certain hormones, enzymes, and biologicals such as vaccines.

Summary points

- Vitamin E is an essential nutrient required for human health care.
- Vitamin E is an umbrella term for a group of compounds called tocopherols and tocotrienols.
- α-tocopherol is the name of the most active form of vitamin E in human that is maintained in the blood circulation.
- α-tocopherol both reacts with fatty acid peroxyl radicals, the primary products of lipid peroxidation, intercepting the chain reaction, than shows, *in vitro*, pro-oxidative effects.
- Vitamin E benefits are due to the antioxidant properties: it helps to stabilize cell membranes and protects the biological tissues which are more sensitive to oxidation.
- Vitamin E also protects the skin from deleterious effects due to its exposure to exogeneous toxic agents such as pollulants, chemical and sunrays, preventing the propagation of free-radicals.

9. Vitamin E chemistry, biological activity and benefits on the skin

R. Cassano
Department of Pharmaceutical Sciences, University of Calabria, Edificio Polifunzionale, Via Pietro Bucci, 87036 Arcavacata di Rende (CS) Italy; roberta.cassano@unical.it

Abstract

Vitamin E is the most important fat-soluble antioxidant especially applied in the animal nutrition. Due to its anti-inflammatory effects on the skin it is contained in many cosmetic products. In fact, if topically applied, vitamin E deactives unstable free radicals providing one of its electrons to the electron deficient free radical making it more stable. As a result, it protects the skin from deleterious effects due to its exposure to exogenous toxic agents such as pollulants, chemicals and sun rays, preventing the propagation of free-radicals. Although, mainly acting as an antioxidant, vitamin E can also exert a pro-oxidant activity. Vitamin E also regulates the functions of vitamin A in the body. This is fundamental because vitamin A itself is a dominant vitamin for skin care.

Keywords: tocopherol, tocotrienol, antioxidant, fatty acid, pro-oxidation, health care

Abbreviations

^{1}H-NMR	Protonic nuclear magnetic resonance
^{13}C-NMR	Carbon nuclear magnetic resonance
FT-IR	Fourier transform infrared spectroscopy
GC/MS	Gas chromatography/mass spectrometry
IU	International unit
LDL	Low-density lipoprotein
PKC	Protein kinase C
PUFA	Polyunsaturated fatty acid
RDA	Recommended daily allowance
ROS	Reactive oxygen species
TTP	Tristetraprolin

9.1 Introduction

Vitamins are substances widely distributed in vegetable lipids and in the body fat of animals that play an essential role in metabolic processes. However, animals cannot synthesize them and, for this reason, they must be obtained from plant and animal foods. Vitamin E, a light yellow oil, is found in butter, egg, milk fat, and liver. Good sources of vitamin E are vegetable oils, nut oils, soya beans, avocados and tomatoes. Its most rich resource is the wheat germ oil. The very important vitamin E function is as antioxidant. It protects the cells from attack by ROS reducing the oxidation of lipid membranes and unsaturated fatty acids that will cause the formation of free radicals. These last are harmful molecules that can lead to cellular and tissue damage and then to chronic diseases such as atherosclerosis, arthritis, etc. Then, vitamin E stabilizes cell membranes and protects the tissues more sensitive to oxidation such as the skin. Moreover, it helps the preservation of the vitamin A biological activity, a further very important oil-soluble vitamin, and prevents the oxidation of some hormones, such as those released from the pituitary and adrenal glands. There is the possibility by which vitamin E can exert a pro-oxidant activity due to its ability to reduce transition metal ions to their reduced form that lead to the generation of reactive alkoxyl radicals.

The aim of this chapter is to give a complete overview of the antioxidant and pro-oxidant chemistry of vitamin E. On the other hand will be described the vitamin E involving in physiologic process such as the progression of LDL oxidation and atherosclerosis, also through their ability to regulate gene expression and functional activities of proteins that are critically involved in atherogenesis.

9.2 Vitamin E composition

Vitamin E is a term including a group of compounds called tocopherols and another one called tocotrienols. Their structures, having a chromanol unit (chroman ring with an alcoholic hydroxyl group), is similar except that tocotrienols structure has double bonds (three unsaturations in the 3', 7', 11' positions) on the lateral farnesyl tail whereas the tocopherols have a phytyl tail (Figure 9.1).

It is generally believed that there are, in nature, only four tocopherols (α-, β-, γ- and δ-T) and four tocotrienols (α-, β-, γ- and δ-T3). The α forms are trimethylic, the β and γ ones are dimethylic and, finally, those δ are monomethylic (Figure 9.2) (Brigelius-Flohé and Traber, 1999). However, there are at least 12 vitamins, known as E vitamins, counting two new tocopherols and two new tocotrienols.

Tocotrienols are antioxidants many times more potent than tocopherols but are poorly distributed to tissues in blood due to their reduced assimilation by digestion (Yoshida *et al.*, 2003). For this reason they are rapidly metabolized and eliminated from the body. On the other hand, tocotrienols are well-absorbed by the skin and thus are well suited for use as a vitamin E cream (Packer *et al.*, 2001). α-tocopherol is the vitamin E form with strongest activity although all vitamin E forms possess antioxidant activity. Specifically, d-α tocopherol, designated in this way

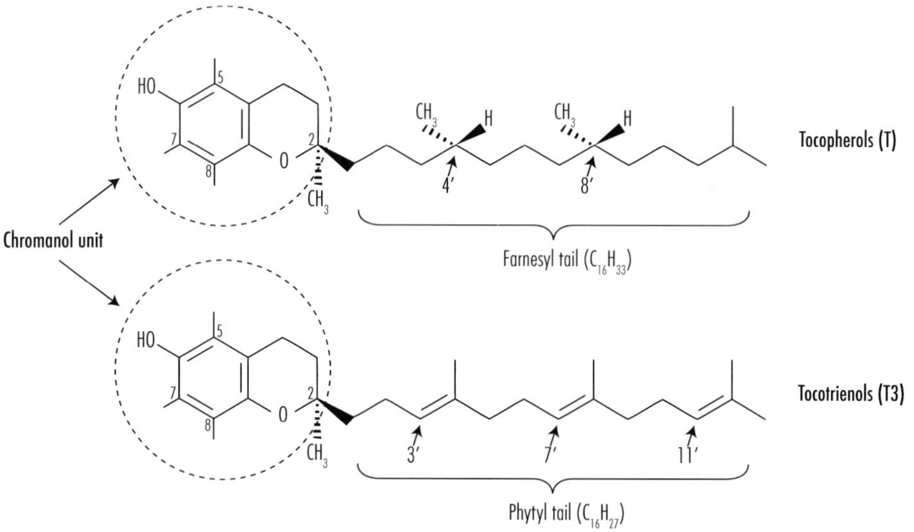

Figure 9.1. Molecular and chemical structures of vitamin E. Vitamin E is the term used for the eight related forms named tocopherols and tocotrienols. Each vitamin E form has its own biological activity but the name of the most active form is alpha-tocopherol (α-tocopherol) which natural configuration is 2R,4'R,8'R. The primary function of vitamin E is as an antioxidant. For the antioxidant activity is important the hydroxy group at C-6 of the chromanol ring which can gave its hydrogen atom to terminate the radical chain in the autoxidation reaction (see Figure 9.8).

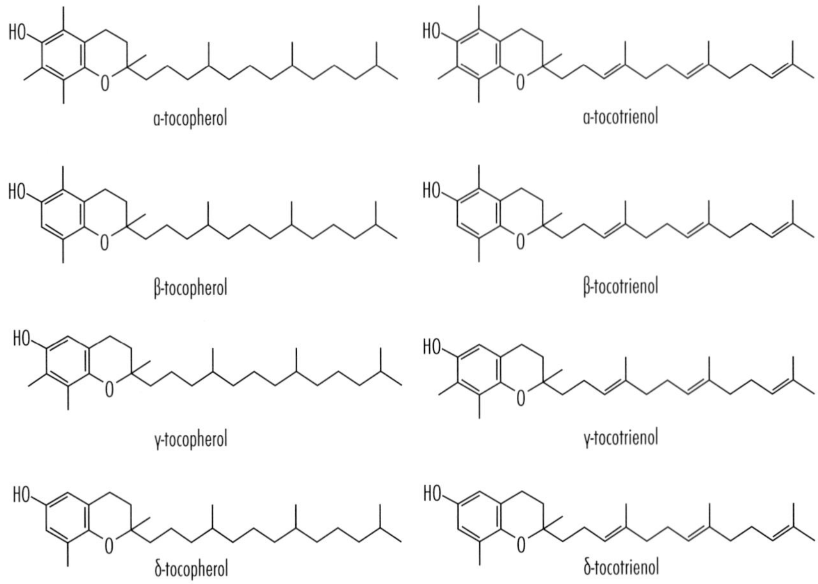

Figure 9.2. The structures of tocopherols and tocotrienols. Vitamin E refers to a family of eight molecules having a chroman ring with an alcoholic hydroxyl group and a side chain (12-carbon) containing two methyl groups in the middle and two more methyl groups at the end. For the four tocopherols the side chain is saturated, whereas for the four tocotrienols the side chain contains three double-bonds, all bringing a methyl group. The four tocopherols and the four tocotrienols have an α, β, γ and δ form. Their names depend on the number and position of the methyl groups on the chromanol ring.

on the basis of optical activity, is the most potent form. There are three asymmetric carbon atoms in tocopherol, one at the 2-position of the chromanol ring, and the other two on the aliphatic chain, at the 4' and 8' positions, all being locations of methyl groups (Figure 9.1). There is thus the possibility of eight stereoisomers (Figure 9.3).

The International Union of Pure and Applied Chemistry (IUPAC, establishing conventional names for chemicals) chooses the designation R and S for these stereoisomers rather than the 'd-' and 'l-' prefixes indicating the optical activity.

Consequently, the α-tocopherol form, usually called RRR-α-tocopherol, has the IUPAC name 2R,4'R,8'R-α-tocopherol. The 'd-α-tocopherol' term is now archaic. RRR-α-tocopherol is stable in heat and in acids; other forms are lost in heat, with storage or freezing, or oxidized by exposure to the air. All vitamin E's are weakly unstable in alkali and are readily used up when in contact with polyunsaturated oils or rancid fats and oils, which are protected from oxidative destruction by vitamin E.

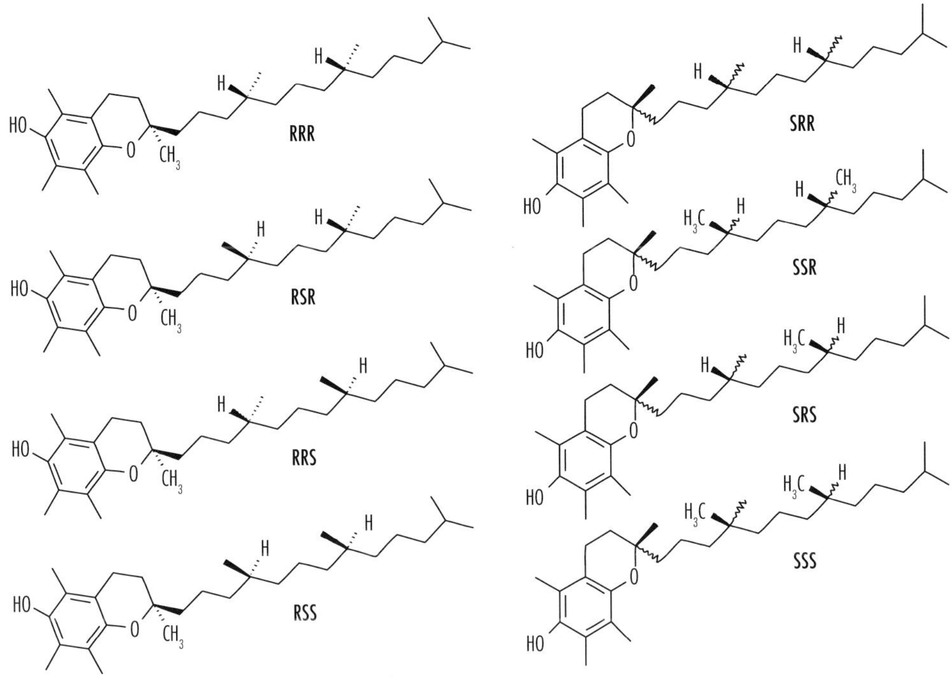

Figure 9.3. Tocopherol stereoisomers. There are three chiral centers, at positions 2′, 4′, and 8′, in the phytyl tail. There is thus the possibility of eight stereoisomers. The most abundant of the naturally-occurring forms is the *R,R,R* form.

RRR-α-tocopherol ^{1}H-NMR, ^{13}C-NMR, FT-IR and GC/MS spectra are shown in the Figure 9.4. They are acquired on Bruker VM-300 ACP, Jasco 4200, Hewlett Packard GC-MSD 5972, respectively.

9.3 Fatty acid autoxidation chemistry

To facilitate the understanding of the vitamin E antioxidant activity, it is necessary to describe the mechanism of autoxidation of polyunsaturated fatty acids. This is a process of many radical reactions. Initially, the fatty acid radical formed at once reacts with oxygen; the resulting peroxyl radical takes away a hydrogen from another fatty acid giving as product a hydroperoxide and a new fatty acid radical. In this way propagation of the chain starts. The antioxidant's rule is to interrupt this process. The starting reaction of the autoxidation process is the abstraction of a hydrogen atom from the fatty acid to yield a carbon radical. Generally, the hydrogen atom is abstracted from a carbon placed in between two double bonds (bisallylic methylene) of a polyunsaturated fatty acid such as linoleic acid (C18:2) or arachidonic acid (C20:4). In effect, its bond is weakest in the molecule, and consequently, less energy is necessary for abstraction of a bis-allylic hydrogen compared to a hydrogen that is only allylic, as, for example, in oleic acid (C18:1). For this reason polyunsaturated fatty acids are the principal targets for autoxidation.

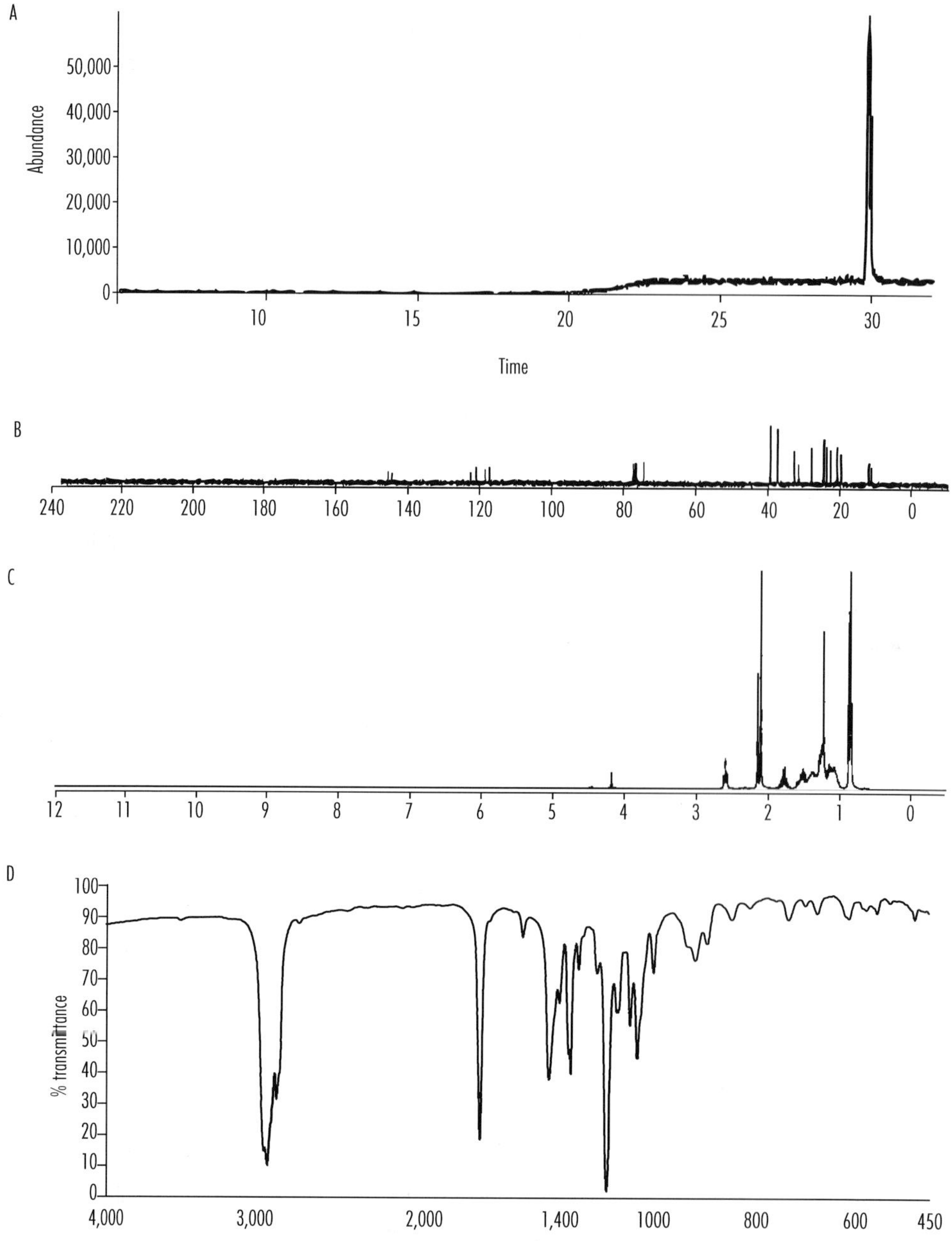

Figure 9.4. a-tocopherol instrumental characterization. The ^{1}H-NMR and ^{13}C-NMR spectra (c and d respectively) of a-tocopherol were measured for $CDCl_3$ solution at room temperature. a-Tocopherol was also analyzed by using KBr disks (D). Its estimation was carried out even by GC-MS spectrometer (m/z 537/539) (A).

They are numerous natural and synthetic 'reagents' capable of starting the autoxidation radical chain by enhancing the initial hydrogen abstraction including metal ions (Cu^{2+} and Fe^{2+}), UV and gamma-radiation, cells that form active oxygen species, and, of course, fatty acid peroxyl radicals leading to the elongation of the chain reaction (Ingold *et al.*, 1993). Briefly, the carbon radical, formed by hydrogen abstraction, immediately reacts with molecular oxygen to form a fatty acid peroxyl radical. Oxygen itself does not need to be activated because in its ground state it is already a radical ('triplet oxygen') suitable for reaction with the fatty acid radical. This reaction is extremely fast and it is only limited by the diffusion of oxygen toward the fatty acid radical (Porter *et al.*, 1995). After the first step the oxidation process can go on as follows:

- *chain propagation*: come across with a pentadiene moiety of another fatty acid. The peroxyl radical performs a hydrogen atom abstraction from a bis-allylic methylene of another fatty acid molecule. This reaction propagates the chain by transferring the radical to a new fatty acid molecule (Figure 9.5). It is the slowest reaction during autoxidation, and therefore the rate-limiting step. Consequently, an antioxidant, transferring a hydrogen atom faster to the peroxyl radical than the peroxyl radical could be terminate the propagation.
- *endoperoxide formation*: encounter with a double bond in the same fatty acid. This reaction is possible for fatty acids such as linolenic acid (C18:3) and arachidonic acid (C20:4) that have at least three double bond. The peroxyl radical, reacting in a 5-exo-cyclization with the additional double bond, furnishes a 5-membered cyclic peroxide (endoperoxide) and a carbon radical outside the endoperoxide ring ('exo') (Figure 9.6).
- *β-fragmentation*: loss of oxygen from the peroxyl radical. This is the reverse reaction to the formation of the peroxyl radical yielding molecular oxygen and the fatty acid radical and not leading to the reversal of autoxidation, because the commiting step is the initial hydrogen abstraction, and the on-rate of oxygen, as described above, is exceedingly fast (Figure 9.7).
- *chain termination*: to come up with an antioxidant. The peroxyl radical abstracting a hydrogen atom from the antioxidant forms a fatty acid hydroperoxide that is a rather stable product. This reaction is known as the 'antioxidant' reaction because it leads to the termination of the autoxidation reaction. The resulting antioxidant radical by itself refrains from starting a radical reaction (Figure 9.8).

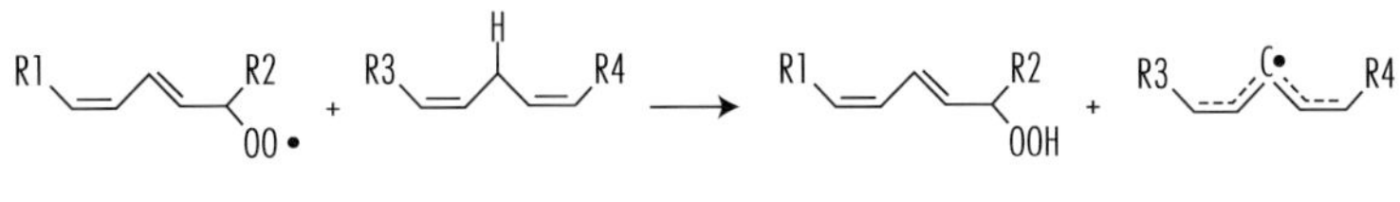

Figure 9.5. Chain propagation is a reaction which consists in the initial hydrogen abstraction of the bis-allylic methylene of a polyunsaturated fatty acid from a new fatty acid molecule.

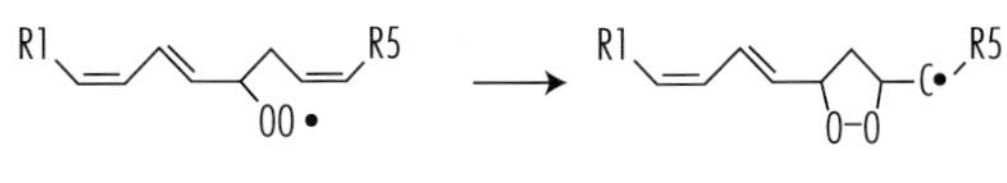

Figure 9.6. Endoperoxide formation. If the fatty acid peroxyl radical contains a β,γ-double bond, the peroxyl radical can form (5-exo cyclization) a cyclic peroxide named endoperoxide.

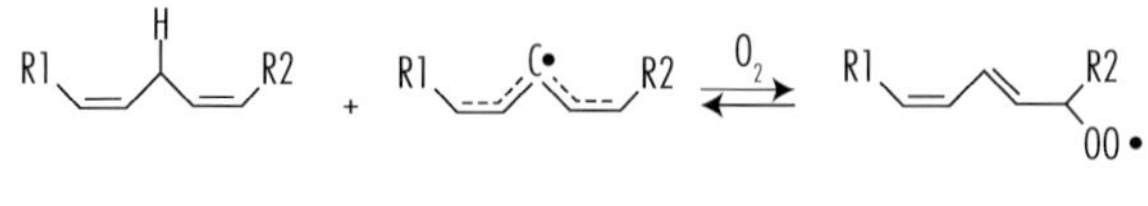

Figure 9.7. Formation of a fatty acid peroxyl radical and its β-fragmentation. The abstraction of hydrogen from a bis-allylic methylene carbon leads to a delocalized radical that immediately reacts with oxygen forming the fatty acid peroxyl radical. The loss of molecular oxygen and the conversion to the pentadienyl radical is known as β-fragmentation of the peroxyl radical.

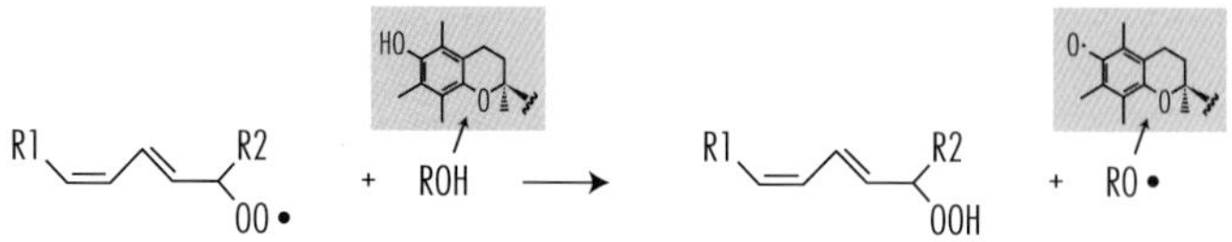

Figure 9.8. Reaction with an antioxidant (ROH, for example α-tocopherol): chain termination. Tocopherols can interrupt free radical chain reactions by capturing the free radical and for this reason it possesses antioxidant properties. The free hydroxyl group on the aromatic ring is responsible for these properties. The hydrogen from this group is given to the free radical, resulting in a relatively stable free radical form of the vitamin.

Tocopherols (vitamin E) are equipped to perform a unique function. They can interrupt free radical chain reactions by capturing the free radical; this imparts to them their antioxidant properties. The free hydroxyl group on the aromatic ring is responsible for the antioxidant properties. The hydrogen from this group is donated to the free radical, resulting in a relatively stable free radical form of the vitamin.

9.4 Antioxidant properties of vitamin E

The antioxidant action of α-tocopherol is due to its reaction with fatty acid peroxyl radicals, the primary products of lipid peroxidation to interrupt the chain propagation reaction (Burton and Ingold, 1981). The α-tocopherol reaction with the peroxyl radical is faster than that of peroxyl radicals with any other compounds. Moreover, α-tocopherol removes the radical character from the oxidizing fatty acid preventing further radical reactions and producing a stable radical (tocopheroxyl radical) that, under normal circumstances, it will react with another radical, a tocopheroxyl radical or a fatty acid peroxyl radical, to form stable non-radical products. This last reaction leads to the destruction of α-tocopherol as antioxidant. Thus, one molecule of α-tocopherol can terminate two autoxidation chains. Chemically α-tocopherol is the most active tocopherol against peroxyl radicals due to the substitution pattern of methyl groups on the chromanol ring making the hydrogen of the C-6 hydroxy group especially active, facilitating in this way the transfer of the hydrogen to a peroxyl radical and the tocopheroxyl radical formation. The tocopheroxyl radical can be reduced (restored) to tocopherol directly by CoEnzyme Q10 (Ubiquinol) or vitamin C (Shi *et al.*, 1999). For this reason, α-tocopherol is well-known as the most powerful lipid soluble antioxidant. However, recently have been developed novel synthetic antioxidants with greater antioxidant capacity than α-tocopherol's one (Wijtmans *et al.*, 2003).

9.5 Vitamin E pro-oxidant activity

α-tocopherol also shows *in vitro* pro-oxidative effects. Really, it may exert anti- and pro-oxidative effects depending on the reaction partners present. The importance of a pro-oxidant role of vitamin E *in vivo* has been demonstrated in LDL isolated from healthy volunteers (Bowry *et al.*, 1992) and patients with a specific defect in the TTP gene (Kontush *et al.*, 1996). Certainly, in the presence of co-antioxidants such as ascorbic acid, neutralizing the tocopherol radical, vitamin E does not have a pro-oxidant function. When co-antioxidants, such as vitamin C, are not available to neutralize the tocopherol radical and when oxidative stress is mild we observe a pro-oxidant activity. On the other hand, when oxidative stress is high, tocopherol radicals are more likely to neutralize (or to be neutralized by) other radicals than to mediate lipid peroxidation, also if co-antioxidants are not available (Kontush *et al.*, 1996). Smokers given a high polyunsaturated diet (safflower oil) suffer oxidative damage that is considerably worsened by the addition of α-tocopherol (Weinberg *et al.*, 2001). But for (non-smoking) rats fed salmon oil, α-tocopherol is antioxidant (Flader *et al.*, 2003). α-tocopherol showed no effect on lipid peroxidation when given alone as a supplement to healthy persons (Meagher *et al.*, 2001).

9.6 Vitamin E biological activity

Vitamin E is absorbed from the small intestine together with fat and bile salts. It is incorporated into chylomicrons and transported via the lymphatic system into the blood, which carries it to the liver to be used or stored. α-Tocopherol is the vitamin E form that predominates in blood and tissue. This is due to the action of a liver protein, α-tocopherol transfer protein, that preferentially incorporate α-tocopherol into the lipoproteins which deliver it to the different tissues. Vitamin E is found in most human body tissues and predominantly in the adipose tissue, liver and muscles. Its availability to the body is reduced from iron (this is particularly critical in the case of anaemic newborns). Iron, especially the inorganic form, depletes vitamin E absorption in the small intestine. Then, the two should not be taken together, as this causes the absorption of both to be diminished. Chlorine, ferric chloride, and rancid oils also deplete or destroy vitamin E. Furthermore, vitamin K deficiency may be aggravated by vitamin E affecting blood coagulation (Booth *et al.*, 2004). On the other hand, vitamin E deficiency results in progressive neuromuscular disease characterized by loss of reflexes, muscle weakness, loss of balance and impaired ability to coordinate voluntary movements (ataxia). In premature infants, vitamin E deficiency is associated with haemolytic anaemia, intraventricular haemorrhage and retrolental fibroplasia. In fact, there are many biological functions of vitamin E. The major biological function of vitamin E is the antioxidant one. It is protective because it helps reduce oxidation of lipid membranes and the PUFAs and prevents the breakdown of other nutrients by oxygen. These phenomena cause the free radicals production, unstable molecules that can lead to cellular and tissue injury, which leads to diseases, such as atherosclerosis, heart disease, hypertension, arthritis, senility, and even cancer. In particular, vitamin E helps to stabilize cell membranes and protect the tissues of the skin, eyes, liver, breast, and testes, which are more sensitive to oxidation. It protects the lungs from oxidative damage due to environmental substances, prevents the oxidation of some hormones, such as those released from the pituitary and adrenal glands and may help in preventing tumor growth. A review of cancer prevention by vitamin E affirmed that γ-tocopherol is the most potent form for preventing breast cancer (Jiang *et al.*, 2001). Apart from maintaining the integrity of the cell membranes in the human body, vitamin E also protects LDL from oxidation. In recent times, a role of vitamin E supplementation in the treatment of neurodegenerative diseases, such as Alzheimer´s disease and amyotrophic lateral sclerosis, is also under investigation. On the other hand, the biological activity of vitamin E which attracts more interest is the prevention of lipid peroxidation. The vitamin E polar chroman ring tends to stay near the edges of the membrane and the hydrophobic core will be buried deep into the membrane. When the tail of a phospholipid is peroxidized by a free radical it becomes more polar and migrates to the surfaces. As a result, it can meet the tocopherol chroman ring to be neutralized. After that the tocopheroxyl radical produced can be restored to tocopherol directly by vitamin C and then by glutathione or lipoic acid (via vitamin C), which are in turn reduced by NADH or NADPH (Figure 9.9).

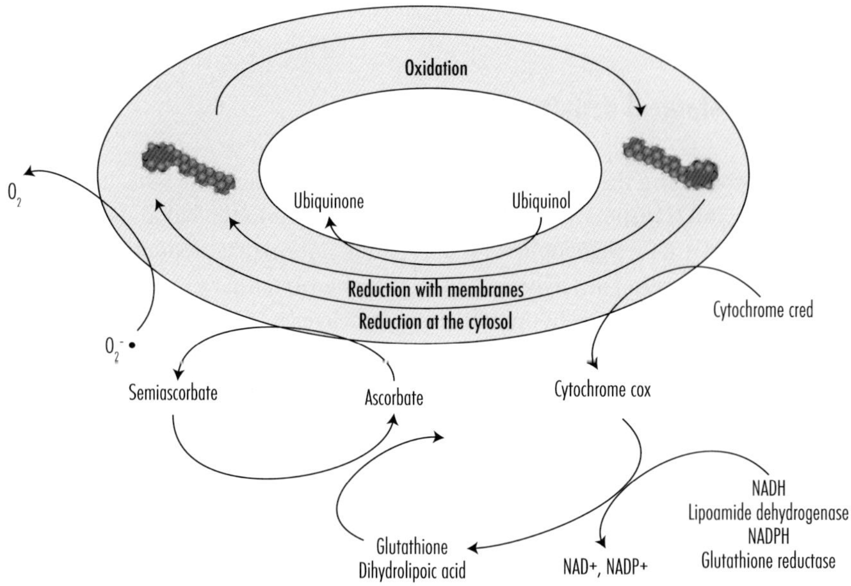

Figure 9.9. Tocopherol recycling by antioxidant network. CoEnzyme Q10 (Ubiquinol) or vitamin C regenerates vitamin E. Vitamin C is in turn regenerated by glutathione or lipoic acid. Glutathione is effective against hydrogen peroxide and hydroxyl radicals in the aqueous phase of cells. Respect to lipoic acid it is more effective than glutathione in regenerating vitamin C and can, additionally, chelate free-radical-generating metal ions. Glutathione or lipoic acid can be regenerated by NADH or NADPH (energy-containing molecules resulting from glycolysis or oxidative phosphorylation) that complete the antioxidant network for membrane lipid peroxidation.

Tocopherol can be also regenerated by CoEnzyme Q10 (Ubiquinol) that is more hydrophobic than vitamin E and is therefore less mobile in cell membranes. The CoEnzyme Q10 radical is readily reduced in mitochondria by the readily-available succinate. Like CoEnzyme Q10, vitamin K is a biological quinone antioxidant which can regenerate vitamin E from the tocopheroxyl radical (Mukai *et al.*, 1992).

Vitamin E has also been used to enhance immunity in the treatment of some viral infections of the nerves and skin. Various kidney and liver diseases and muscular dystrophy have all been treated with vitamin E. Leg cramps and circulatory problems associated with diabetes may be also helped with vitamin E treatment. For various skin rashes, including those of lupus erythematosus, vitamin E, usually along with vitamin A, may be of some help. In summary, there are several benefits of vitamin E for human body and it may be useful in treating or possibly preventing various diseases (Table 9.1).

Table 9.1. Vitamin E has numerous health benefits. It plays a role in preventing atherosclerosis and heart disease due to its effects on a number of steps in the development of atherosclerosis. Vitamin E protects against exogenous pollutants, decreases the risk of cancer and of cataracts and provides protection against premenstrual syndrome and diabetes. Vitamin E may also provide relief for muscle cramping. A role of vitamin E in the treatment of neurodegenerative diseases is also known.

Rheumatoid arthritis	Cataracts	Low sperm count
Alzheimer and Parkinson diseases	Menstrual pain	Cardiovascular disease
Diabetes	Restless leg syndrome or relief from muscle cramping	
Prostate and breast cancer	Inflammation of eye tissues	Asthma

9.7 Vitamin E and skin care

Vitamin E is well known for its skin friendly properties (Figure 9.10). There are different benefits of vitamin E on the skin (Table 9.2) and this is the main reason that it is used in nearly every skin care product. In the field of skin care there is a large body of experimental evidence pointing to vitamin E photoprotective effects.

Moreover, recent studies indicate that the use of vitamin E may provide dermatological benefits that surpass the purpose of cosmetics and may extend into an area that has been termed 'cosmeceuticals'. It is very true that vitamin E is useful in the treatment of aging. In fact, vitamin E helps skin look younger by reducing the appearance of wrinkles and fine lines whose appearance is a natural sign of aging.

There are many factors, responsible for skin damage such as sun exposure, alcohol intake and smoking. Ultraviolet rays permeate the skin and strike molecules of oxygen. This whole process

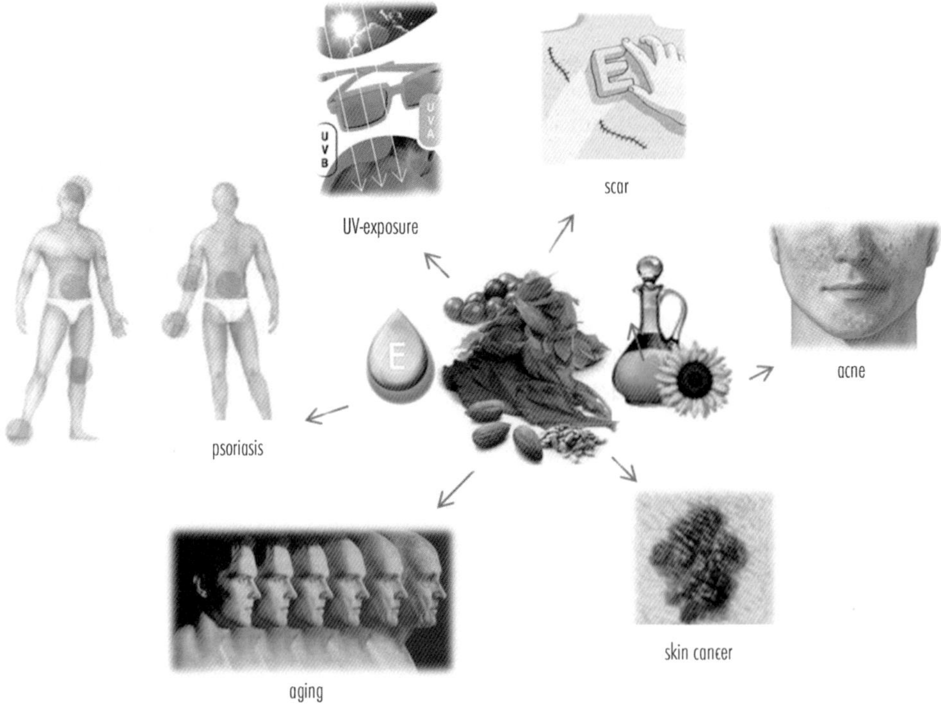

Figure 9.10. Skin friendly properties of vitamin E.

Table 9.2. In terms of skin health and skin care, vitamin E benefits are numerous.

Helps reduction of the appearance of stretch marks
Helps prevent the appearance of age spots
Reduces trans-epidermal water loss from skin
Helps maintain the skin's oil balance during the cleansing process
Strengthens the skin's barrier function
Helps regulate the level of retinol or vitamin A in the body

results in the formation of free radicals that attack lipids, proteins, and DNA. They cause proteins to link together to make skin 'leathery,' and they deplete the collagen that makes skin smooth and supple. The damage caused by UV rays results in wrinkling, age spots, and cancer. The key function of vitamin E is deactivating free radicals with its antioxidant properties. It provides one of its electrons to the electron deficient free radical and makes it more stable (Berneburg *et al.*, 1999). There are two forms of vitamin E: the natural (alcohol form d-α-tocopherol) and the synthetic form (dl-tocopherol acetate). These both possess antioxidant activity, but the natural form of vitamin E is more readily and effectively absorbed by the skin. On the other hand, the synthetic one does not penetrate the skin's surface well and provides less of the benefits of vitamin E such as antioxidant activity for the skin. Vitamin E on the skin can be applied topically in the form of vitamin E skin cream, oil, lotion and other skin care products or else taken orally. In particular, vitamin E lotions provide some benefit in preventing and treating sunburns. These formulations protect the epidermis layer of the skin from early stages of ultra violet light damage. Vitamin E preparations are also used to increase the effectiveness of sunscreens that have two purposes. Firstly they help the skin maintain moisture and its own natural oils which can be lost through exposure to the sun's radiation. Secondly, a good sunscreen protects the skin against UVA and UVB rays. It is said that the reduction in the atmosphere's ozone layer has made these rays more damaging. The body's own defense system to sunlight causes skin pigment cells, called melanocytes, to synthesize increasing amounts of melanin, which is what a tan actually is, gives our skin a protective pigmentation. It may not be quite enough these days, especially if you are fair-skinned or have a family history of skin cancer.

9.7.1 Vitamin E can aid to treat psoriasis

Psoriasis is one of the most stubborn skin diseases. It is a chronic disease, characterized by thick, red, silvery, scaled patches of skin. Generally, the skin of the person suffering from psoriasis appears red and irritated and may be covered with bright silvery scales. Sometimes, there is also a little itching. Recent studies have shown that psoriasis involves an abnormality in the mechanism in which the skin grows and replaces itself. This abnormality is related to the metabolism of amino-acids which are nature's basic building blocks for the reproduction of cell tissues. Vitamin E therapy has been found effective in the treatment of psoriasis. The patient should take this vitamin in therapeutic doses of 400 mg a day. It will help to reduce itching and scab formation.

9.7.2 The effects of topical vitamin E on the scars

If orally ingested vitamin E can help to treat erythema, a skin inflammation that results in reddish, painful, and tender lumps. Regular application of vitamin E oil is useful in the treatment and removal of many skin darkening related problems such as scars and stretch marks. Nevertheless, some studies have indicated that vitamin E creams do not help prevent surgical scarring and may actually make the scar look worse (Baumann and Spencer, 1999). In 90% of the cases in this study, topical vitamin E either had no effect on, or actually worsened, the cosmetic appearance of scars. Of the patients studied, 33% developed a contact dermatitis to the vitamin E. Therefore the use of topical vitamin E on surgical wounds should be discouraged. This is in opposition to the experience of a number of individuals that claim vitamin E creams help soften the appearance of scars.

9.7.3 Vitamin E for acne treatment

Vitamin E prevents acne via antioxidant protection. Various studies have proved the relationship between vitamin E and acne. Particularly, a study in the *Journal of Investigative Dermatology* (Thiele *et al.*, 1999) elucidated that vitamin E helps to prevent oils trapped in the pores from becoming rancid and hardened. This preventative action lowers the chances of having inflamed, painful acne lesions. However, vitamin E reaches the surface layer of the skin via oil secretion from the pores. Thus, it is important to clear trapped oils from the skin so that vitamin E can flow to the outer layers of the face and pre-empt acne lesions. Many acne sufferers report that applying vitamin E oil to affected areas reduces scars and evens the skin tone. Some people also say the vitamin E oil accelerates the healing of pimples, whiteheads, blackheads and other acne eruptions. Research has shown that topical use of vitamin E oil helps to unclog the pores, resulting in far fewer breakouts. Vitamin E also controls the retinol concentration in the body that is essential for maintaining a healthy glowing skin. Any reduction in retinol can cause cellular damage and cause severe acne. It is essential for cellular respiration, nourishment of the cells and strengthens the capillary walls thereby keeping the skin acne free.

9.7.4 Vitamin E and skin cancer prevention

One of the most important benefits of vitamin E is the prevention of skin cancer. This occurs because of its sun protection quality and of course its powerful antioxidant properties, which help reduce or prevent sun damage. There is evidence that oxidative stress is involved in the pathophysiology of melanoma and non-melanoma cancer (Sander *et al.*, 2003), and that vitamin E slows melanoma growth by promoting tumor cell apoptosis (Malafa *et al.*, 2002a,b). In reality, more specific protective effect in cancer treatment are attributed to vitamin E analogues, 'mitocans', a novel group of anticancer agents (Neuzil *et al.*, 2007). Vitamin E analogue, α-tocopheryl succinate, suppresses malignant mesotelioma in a preclinical model, alters cell cycle distribution sensitizing human osteosarcoma cells to methotrexate-induced apoptosis, promotes breast cancer tumor dormancy, and selectively induces apoptosis in neuroblastoma cells. Mechanisms of such α-tocopheryl succinate effects on tumor cells are under intensive investigation.

9.7.5 Experimental evidence on the vitamin E anti-aging activity

Vitamin E has been increasingly incorporated into anti-aging formulations as a means of upping a product's antioxidant efficacy. For this reason, and the fact that is easy to manufacture, readily available and inexpensive, it has now become the number one selling antioxidant ingredient. *Skin Pharmacology Physiology* has published a study online (Lademann, 2010) showing that using topical exogenous antioxidants on the skin may prevent or minimize free radical-induced damage. Researchers determined the antioxidative capacity of a topical skin care treatment – an oil-in-water vitamin E-containing formula – on human skin exposed to ultraviolet radiation (UVR) by using a photochemiluminescence device and biophysical methods.

Briefly, in a randomized, double blind study, either a pH-balanced vitamin E emulsion or a control lotion was applied onto the forearm skin of 10 healthy Caucasian participants. Thirty minutes after application, test sites were exposed to a UV light to induce erythema; one untreated site served as a control. Visual scoring and instrumental measurements were recorded at baseline and thereafter at 24 hours and 48 hours to determine antioxidant capacity. A few days after UV exposure, vitamin E emulsion and the vehicle control significantly suppressed visual scores when compared with the blank control. Furthermore, vitamin E emulsion significantly reduced skin blood flow volume when compared with blank control. From the test results the researchers concluded, vitamin E emulsion and its vehicle control proved effective in preventing induction of erythema and reducing inflammatory damage caused by UV exposure, and the effect of vitamin E emulsion exceeded that of an 'active control'. The results suggest that vitamin E is a valuable antioxidant providing protection to the skin and preventing visible signs of skin aging.

9.7.6 Reasons for use of vitamin E and co-antioxidants for photoprotection

Vitamin E has a greater protective effect if coupled to vitamin C. A series of studies investigating non-enzymatic *stratum corneum* antioxidants have demonstrated that vitamin E is the predominant physiological barrier antioxidant in human skin (Thiele *et al.*, 2001). When compared to nucleated epidermal layers, there is a lack of important co-antioxidants, such as vitamin C, in the *stratum corneum* as well as in the dermis. Taken together, these findings suggest that the skin barrier as well as the upper dermis reveal a lack of antioxidant protection. In fact, upon solar UV-exposure, these are the cutaneous sites exhibiting the most pronounced oxidative protein damage (Sander *et al.*, 2002). Accordingly, antioxidant supplementation with vitamin E as well as synergistically active co-antioxidants, such as vitamin C, may enhance the photoprotective strategies of sunscreens. Experimental data suggest that taking 2,000 mg of vitamin C with 1000 IU of vitamin E every day for a short time to protect skin (such as for a week during summer vacation) has the protective effect of 3,000 mg of vitamin C a day without the vitamin E. In fact, if you are getting adequate supplies of other vitamins, you can get a noticeable benefit from even very small doses of vitamin C. When vitamin C is used to complement the benefits of vitamin E on skin, as little as 60 mg of C a day can be used to reduce the processes that cause wrinkling and age spots. The combination of vitamins C and E is especially helpful in rejuvenating skin damaged by the chemicals in tobacco smoke. However, vitamin E alone protects skin in persons over 60 years of age. Taking about 1,200 IU a day for 30 days has protective effect against sunburn and seems to stop, although not reverse the wrinkling process. In particular, one year-long study of elderly volunteers found that the benefits of vitamin E on skin are greatest when it is associated with other antioxidants. The presence of vitamin C and β-carotene, supports the antioxidative, protective action of vitamin E; the same is true for mineral selenium (Longe *et al.*, 2004).

9.7.7 Allergic contact dermatitis

Although vitamin E and its derivatives are widely used in many topical cosmetic products, reports of side effects such as allergic or irritant skin reactions are rare. In clinical studies, tocopherol and tocopherol acetate were found to be safe for use in topical skin formulations since irritant or sensitizing reactions were found only in very small percentages. In case reports, however, clinical side effects have been described after topical application of vitamin E containing products, e.g. local and generalized contact dermatitis, contact urticaria, and erythema-multiforme-like eruptions (Brodkin and Bleiberg, 1965). Perrenoud *et al.* (1994) reported about 1000 cases of allergic papular and follicular contact dermatitis caused by α-tocopherol linoleate in a cosmetic line. Positive patch test reactions were also reported in several cases after application of α-tocopherol acetate (Manzano *et al.*, 1994). In general, however, positive patch test results due to α-tocopherol are rare. Some animal studies even suggest that topical vitamin E at a concentration of 20% suppressed allergic and irritant contact dermatitis, exerting a comparable effect to that of a 0.5% prednisolone ointment (Kuriyama *et al.*, 2002).

9.7.8 Practical recommendations for vitamin E dosage

The RDA for vitamin E is based on the most active and usable form called α-tocopherol. About 60% of vitamin E in the diet comes from vegetable oil or products made with vegetable oils. Therefore, good food sources of vitamin E include vegetable oils and margarines. Vitamin E is also found in fruits and vegetables, grains, nuts, seeds and fortified cereals (Table 9.3).

Based on the health promotion and prevention of diseases the US Food and Nutrition Board has set RDA values for vitamin E (α-tocopherol) (Table 9.4). One milligram of α-tocopherol equals to 1.5 International Units (IU).

Table 9.3. Vegetable oils, nuts (almonds), and green leafy vegetables are the best sources of vitamin E (α-tocopherol).

Food	Serving	α-tocopherol (mg)
Olive oil	1 tablespoon	1.9
Soybean oil	1 tablespoon	1.1
Corn oil	1 tablespoon	1.9
Canola oil	1 tablespoon	2.4
Safflower oil	1 tablespoon	4.6
Sunflower oil	1 tablespoon	5.6
Almonds	1 ounce	7.4
Hazelnuts	1 ounce	4.3
Peanuts	1 ounce	2.4
Spinach	½ cup, raw	0.3
Carrots	½ cup, raw chopped	0.4
Avocado (California)	1 fruit	2.7

Table 9.4. The RDA of vitamin E is 4-5 mg for infants, between 6 to 11 mg for children and 15 mg for adolescents and adults.

Life stage	Age	Males; mg/day (IU/day)	Females; mg/day (IU/day)
Infants	0-6 months	4 mg (6 IU)	4 mg (6 IU)
Infants	7-12 months	5 mg (7.5 IU)	5 mg (7.5 IU)
Children	1-3 years	6 mg (9 IU)	6 mg (9 IU)
Children	4-8 years	7 mg (10.5 IU)	7 mg (10.5 IU)
Children	9-13 years	11 mg (16.5 IU)	11 mg (16.5 IU)
Adolescents	14-18 years	15 mg (22.5 IU)	15 mg (22.5 IU)
Adults	19 years and older	15 mg (22.5 IU)	15 mg (22.5 IU)
Pregnancy	all ages	-	15 mg (22.5 IU)
Breast-feeding	all ages	-	19 mg (28.5 IU)

Acknowledgment

The author thanks Ms. Anna Internò, an English lecturer of Faculty of Pharmacy, for her assistance.

References

Baumann, L.S. and Spencer, J., 1999. The effects of topical vitamin E on the cosmetic appearance of scars. Dermatologic Surgery 25, 311-315.

Berneburg, M., Grether-Beck, S., Kürten, V., Ruzicka, T., Briviba, K., Sies, H. and Krutmann, J., 1999. Singlet oxygen mediates the UVA-induced generation of the photoaging-associated mitochondrial common deletion Journal of Biological Chemistry 274, 15345-15349.

Booth, S.L., Golly, I., Sacheck, J.M., Roubenoff, R., Dallal, G.E., Hamada, K. and Blumberg, J.B., 2004. Effect of vitamin E supplementation on vitamin K status in adults with normal coagulation status. American Journal of Clinical Nutrition 80, 143-148.

Bowry, V.W., Ingold, K.U. and Stocker, R., 1992. Vitamin E in human low-density lipoprotein. When and how this antioxidant becomes a pro-oxidant. Biochemical Journal 288, 341-344.

Brigelius-Flohé, R. and Traber, M.G., 1999. Vitamin E: function and metabolism. Faseb Journal 13, 1145-1155.

Brodkin, R.H. and Bleiberg, J., 1965. Sensitivity to topically applied vitamin E. Arcives of Dermatology 92, 76-77.

Burton, G.W. and Ingold, K.U., 1981. Autoxidation of biological molecules. 1. Antioxidant activity of vitamin E and related chain-breaking phenolic antioxidants *in vitro*. Journal of American Chemical Society 103, 6472-6477.

Flader, D., Brandsch, C., Hirche, F. and Eder, K., 2003. Effects of megadoses of dietary vitamin E on the antioxidant status of rats fed lard or salmon oil. International Journal for Vitamin and Nutrition Research 73, 275-283.

Ingold, K.U., Bowry, V.W., Stocker, R. and Walling, C., 1993. Autoxidation of lipids and antioxidation by alpha-tocopherol and ubiquinol in homogeneous solution and in aqueous dispersions of lipids: unrecognized consequences of lipid particle size as exemplified by oxidation of human low density lipoprotein. Proceedings of the National Academy of Sciences of the United States of America 90, pp. 45-49.

Jiang, Q., Christen, S., Shigenaga, M.K. and Ames, B.N., 2001. γ-Tocopherol, the major form of vitamin E in the US diet, deserves more attention. American Journal of Clinical Nutrition 74, 714-722.

Kontush, A., Finckh, B., Karten, B., Kohlschutter, A. and Beisiegel, U., 1996. Antioxidant and prooxidant activity of alpha-tocopherol in human plasma and low density lipoprotein. Journal of Lipid Research 37, 1436-1448.

Kuriyama, K., Shimizu, T., Horiguchi, T., Watabe, M. and Abe, Y., 2002. Vitamin E ointment at high dose levels suppresses contact dermatitis in rats by stabilizing keratinocytes. Inflammation Research 51, 483-489.

Lademann, J. (ed.), 2010. Skin Pharmacology Physiology 23, 1-335.

Longe J. (ed.), 2004. The Gale Encyclopedia of Alternative Medicine, second edition. Gale Group Publishing, Detroit, MI, USA.

Malafa, M.P., Fokum, F.D., Mowlavi, A., Abusief, M. and King, M., 2002a. Vitamin E inhibits melanoma growth in mice. Surgery 131, 85-91.

Malafa, M.P., Fokum, F.D., Smith, L. and Louis, A., 2002b. Inhibition of angiogenesis and promotion of melanoma dormancy by vitamin E succinate. Annals of Surgical Oncology 9, 1023-1032.

Manzano, D., Aguirre, A., Gardeazabal, J., Eizaguirre, X. and Diaz Perez, J.L., 1994. Allergic contact dermatitis from tocopheryl acetate (vitamin E) and retinol palmitate (vitamin A) in a moisturizing cream. Contact Dermatitis 31, 324.

Meagher, E.A., Barry, O.P., Lawson, J.A., Rokach, J. and FitzGerald, G.A., 2001. Effects of vitamin E on lipid peroxidation in healthy persons. Journal of the American Medical Association 285, 1178-1182.

Mukai, K., Itoh, S. and Morimoto, H., 1992. Stopped-flow kinetic study of vitamin E regeneration reaction with biological hydroquinones (reduced forms of ubiquinone, vitamin K, and tocopherolquinone) in solution. Journal of Biological Chemistry 267, 22277-22281.

Neuzil, J., Tomasetti, M., Zhao, Y., Dong, L.F., Birringer, M., Wang, X.F., Low, P., Wu, K., Salvatore, B.A. and Ralph, S.J., 2007. Vitamin E analogs, a novel group of mitocans as anticancer agents: the importance of being redox-silent. Molecular Pharmacology 71, 1185-1199.

Packer, L., Weber, S.U. and Rimbach, G., 2001. Molecular aspects of alpha-tocotrienol antioxidant action and cell signaling. Journal of Nutrition 131, 369S-373S.

Perrenoud, D., Homberger, H.P., Auderset, P.C., Emmenegger, R., Frenk, E., Saurt, J.H. and Hauser, C., 1994. An epidemic outbreak of papular and follicular contact dermatitis to tocopheryl linoleate in cosmetics. Dermatology (Basel) 189, 225-233.

Porter, N.A., Caldwell, S.E. and Mills, K.A., 1995. Mechanisms of free radical oxidation of unsaturated lipids. Lipids 30, 277-290.

Sander, C.S., Chang, H., Salzmann, S. and Muller, C.S., Ekanayake-Mudiyanselage, S., Elsner, P. and Thiele, J.J., 2002. Photoaging is associated with protein oxidation in human skin *in vivo*. Journal of Investigative Dermatology 118, 618-625.

Shi, H., Noguchi, N. and Niki, E., 1999. Comparative study on dynamics of antioxidative action of α-tocopheryl hydroquinone, ubiquinol, and α-tocopherol against lipid peroxidation. Free Radical Biology and Medicine 27, 334-346.

Thiele, J.J., Weber, S.U. and Packer, L., 1999. Sebaceous gland secretion is a major physiologic route of vitamin E delivery to skin. Journal of Investigative Dermatology 113, 1006-1010.

Thiele, J.J., Schroeter, C., Hsieh, S.N., Podda, M. and Packer, L., 2001. The antioxidant network of the *stratum corneum*. Current Problems in Dermatology 29, 26-42.

Weinberg, R.B., VanderWerken, B.S., Anderson, R.A., Stegner, J.E. and Thomas, M.J., 2001. Pro-Oxidant Effect of Vitamin E in Cigarette Smokers Consuming a High Polyunsaturated Fat Diet. Arteriosclerosis. Thrombosis and Vascular Biology 21, 1029-1033.

Wijtmans, M., Pratt, D.A., Valgimigli, L. and Di Labio, G.A., 2003. 6-Amino-3-pyridinols: towards diffusion-controlled chain-breaking antioxidants. Angewandte Chemie International Edition in English 42, 4370-4373.

Yoshida, Y., Niki, E. and Noguchi, N., 2003. Comparative study on the action of tocopherols and tocotrienols as antioxidant: Chemical and physical effects. Chemistry and Physics of Lipids 123, 63-75.

Key facts

- Vitamin E was discovered in 1922 as a micronutrient essential for reproduction. Vitamin E is a generic term for tocopherols and tocotrienols, and totrienols differ from the corresponding tocopherols only in their aliphatic tail.
- Foods containing fairly large amounts of tocotrienols are limited. They comprise only palm oil, rice bran oil and wheat bran, which contain much more γ-tocotrienol than α-tocotrienol.
- Ultraviolet irradiation induces serious skin problems such as sunburn and skin cancer via skin inflammation.
- Orally dosed tocotrienol is retained in vastly lower concentrations than α-tocopherol in most vital organs. After oral feeding with tocotrienols in our study, it was mostly absent in the liver, kidney or plasma but substantial amounts of α- and γ-tocotrienol were present in the skin and adipose tissues.
- In our study, dietary tocotrienols protected the skin from damage induced by ultraviolet B (UVB) irradiation in hairless mice.

Summary points

- Major sources of vitamin E are vegetable oil, vegetables and nuts. Some vegetable oils are rich in α-tocopherol, but other oils rich in γ-tocopherol. Foods containing fairly large amounts of tocotrienols are limited, comprising only palm oil, rice bran oil, and wheat bran.
- The animal body retains mainly α-tocopherol, and orally dosed tocotrienol is present in vastly lower concentrations than tocopherol in most vital organs. We observed that substantial amounts of α- and γ-tocotrienol were present in the skin of mice given palm oil extracts containing α- and γ-tocotrienols, even though the concentrations were much less than that of α-tocopherol.
- Sesame seeds and sesamin produced higher concentrations of tocotrienols in the skin of mice fed a diet containing a tocotrienol rich fraction.
- Dietary tocotrienols could protect the skin from sunburn induced by short term UVB irradiation and from tumor caused by long term irradiation in hairless mice fed diets containing a tocotrienols rich fraction extracted from palm oil or the same diet with sesamin.
- Miyazawa's research group investigated UVB-induced skin inflammation and the anti-inflammatory effect of γ-tocotrienol. They observed that oral γ-tocotrienol suppressed UVB-induced changes in skin thickness, cyclooxygenase-2 (COX-2) protein expression, and hyperplasia, but α-tocopherol suppressed them slightly or not at all.

10. Dietary tocotrienol and UVB-induced skin damage

K. Yamashita
Department of Food and Nutrition, Sugiyama Jogakuen University, 17-3 Hoshigaoka-motomachi, Chikusa-ku, Nogoya 464-8662, Japan; kanae@lime.ocn.ne.jp

Abstract

Among vitamin E forms, α-tocopherol appears predominantly in humans and animals, and tocotrienols very slightly. However, tocotrienols have been shown to have higher antioxidant, anticancer, and neuroprotective properties that different from α-tocopherol. We have observed that substantial amounts of tocotrienols are present in the skin of animals fed a diet containing a tocotrienol rich fraction together with α-tocopherol extracted from palm oil (T-mix), and that the addition of sesamin to T-mix produced much higher amounts of tocotrienols in the skin. We investigated whether dietary tocotrienols could protect the skin from damage induced by ultraviolet B (UVB) irradiation in hairless mice fed four diets: vitamin E-free, α-tocopherol alone, T-mix and T-mix with sesamin. We examined the protective effects of tocotrienol on intensity of sunburn in the short term exposure and tumor incidence in long term exposure to UVB. The strongest sunburn was observed in the VE-free group, and the next strongest in the α-tocopherol group. Weak sunburn was observed in the T-mix group and the weakest sunburn in the T-mix with sesamin group. Sesamin enhanced tocotrienol concentrations in the skin and improved the sunburn scores and incidence of tumors. These results suggest that dietary tocotrienols accumulate in the skin and protect the skin more strongly than α-tocopherol against damage induced by UVB. Miyazawa's research group also evaluated the anti-inflammatory effect of γ-tocotrienol on UVB-induced inflammatory reaction extensively using immortalized human keratinocytes and HR-1 hairless mice. They showed that orally dosed γ-tocotrienol suppressed UVB-induced changes in skin thickness, Cyclooxygenase-2 protein expression, and hyperplas in hairless mice, while α-tocopherol did not. These results suggest oral tocotrienols have potential protective power against UVB-induced skin inflammation.

Keywords: sunburn, tumor incidence, inflammation, sesamin

Abbreviations

α-TTP	α-tocopherol transfer protein
CEHC	Carboxyethylhydroxychroman
COX-2	Cyclooxygenase-2
DMBA	7,12-dimethylbenz(a)anthracene
HaCaT	Human keratinocytes
IL	Interleukin
MCP	Monocyte chemotactic protein
PGE_2	Prostaglangin E_2
ROS	Reactive oxygen species
SQ-OOH	Squalene hydroperoxide
UV	Ultraviolet
UVB	Ultraviolet B

10.1 Introduction

Vitamin E is a generic term for tocopherols and tocotrienol. It occurs naturally in eight different isoforms: α-, β-, γ-, and δ-tocopherols and α-, β-, γ-, and δ-tocotrienols (Figure 10.1). Tocotrienols differ from the corresponding tocopherols only in their aliphatic tail. Tocopherols have a phytyl side chain attached to their chromanol nucleus, whereas the tail of tocotrienols forms an isoprenoid chain with three double bonds. The various isoforms of vitamin E differ in their methyl substituents in the chromanol nucleus. The α-form contains 3 methyl groups, whereas the β- and γ-forms have two and the δ-form only one methyl group. It is well recognized that the antioxidative effects of vitamin E are very important in biological systems and foods. Since vitamin E compounds are only synthesized by plants, vitamin E is an essential nutrient for

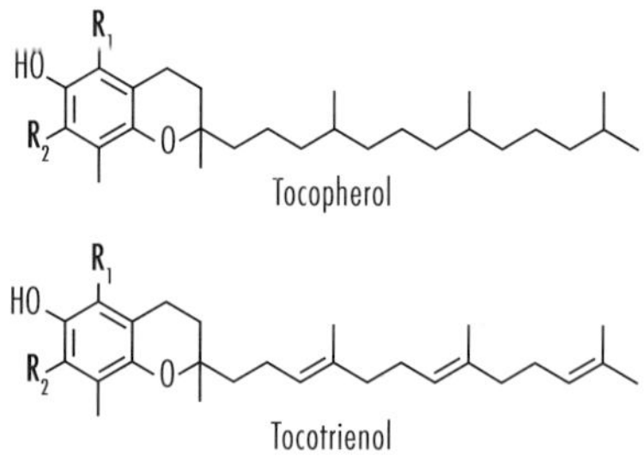

	R_1	R_2
α-	CH_3	CH_3
β-	CH_3	H
γ-	H	CH_3
δ-	H	H

Figure 10.1. Chemical structure of tocopherols and tocotrienols.

humans and animals. Major sources of vitamin E are vegetable oil, vegetables and nuts. Among foods, the content of vitamin E isoforms differs vastly. Some vegetable oils such as olive, safflower and sunflower oil are rich in α-tocopherol, but other oils such as sesame seed, soybean, corn and rapeseed oil are rich in γ-tocopherol. Foods containing fairly large amounts of tocotrienols are limited, comprising only palm oil, rice bran oil, and wheat bran (Sheppard *et al.*, 1993). The animal body retains mainly α-tocopherol because the discrimination of the eight forms of vitamin E is made by α-TTP in the liver (Hosomi *et al.*, 1997). Thus, α-tocopherol exhibits the highest bioavailability among the 8 vitamin E homologues because it has the highest affinity to the α-TTP. However, there are many reports that tocotrienols have stronger antioxidant activity than that of α-tocopherol in animal cells (Sen *et al.*, 2000; Serbinova *et al.*, 1991; Yoshida *et al.*, 2003). Orally dosed tocotrienol exists in vastly lower concentration than α-tocopherol in most vital organs but the biological activity of tocotrienols are exhibited in lower concentrations compared with α-tocopherol both *in vitro* and *in vivo* (Sen *et al.*, 2000; Shibata *et al.*, 2010). Hayes *et al.* (1993), Podda *et al.* (1996), our own studies (Ikeda *et al.*, 2000, 2001), Patel *et al.* (2006) and Shibata *et al.* (2010) have reported that substantial amounts of α- and γ-tocotrienol were present in the skin and adipose tissues after oral feeding with tocotrienols.

UV irradiation induces serious skin problems. UV light, especially UVB, produces various ROS in the skin. ROS attack lipids, proteins, and nucleic acids in the skin, and cause skin damage such as sunburn and skin cancer via skin inflammation (Matsumura and Ananthaswamy, 2004).

We speculated as to whether tocotrienols accumulated in the skin from dietary tocotrienols prevent skin damage induced by UV irradiation. We previously reported that sesame lignan inhibits vitamin E (tocopherols and tocotrienols) degradation, and produced higher vitamin E concentrations in the animal body (Yamashita, 1992; Ikeda, 2001). We examined whether or not dietary tocotrienols would protect against skin damage induced by UVB in hairless mice fed diets containing palm oil extracts rich in tocotrienols (T-mix) alone or with sesamin. There have also been studies by Miyazawa's research group concerning the anti-inflammation effects of γ-tocotrienol in which oral tocotrienol has been to suppress inflammation induced by UVB irradiation in hairless mice (Shibata *et al.*, 2010).

10.2 Tocotrienols in foods

Vitamin E occurs in nature in four tocopherols and four tocotrienols whose chroman rings, with differing numbers and positions of the methyl groups, characterize α-, β-, γ-, and δ-tocopherols and α-, β-, γ-, and δ-tocotrienols (Figure 10.1). Major sources of vitamin E from foods are vegetables and vegetable oils. Among vegetable oils, the content of vitamin E isoforms differs vastly. Some vegetable oils such as olive, safflower and sunflower oils are rich in α-tocopherol, and other oils such as sesame seed, soybean, corn and rapeseed oils are rich in γ-tocopherol. Vegetables contain mostly α-tocopherol. α-Tocopherol contents are high in dark green leafy vegetables such as spinach, Chinese chive and perilla leaves (Table 10.1). More than 90% of our consumed vitamin E is α-tocopherol and γ-tocopherol. Foods containing fairly large amounts of

Table 10.1. Tocopherol and tocotrienol content in food[1].

Product	Tocopherols		Tocotrienols	
	alpha (mg)	gamma (mg)	alpha (mg)	gamma (mg)
Vegetable oils[2]				
Corn	11.2	60.2	-	-
Cottonseed	38.9	38.7	-	-
Olive	11.9	0.7	-	-
Palm	25.6	31.6	14.6	29.7
Peanut	13	21.4	-	-
Rice bran	25.5	3.4	23.6	34.9
Safflower	34.2	-	-	-
Sesame	0.5	43.7	-	-
Soybean	7.5	79.7	0.2	-
Sunflower	48.7	5.1	-	-
Walnut	56.3	59.5	-	-
Wheat germ	133	26	2.6	-
Vegetable[3]				
Basil	3.5	-	-	-
Chinese chive	2.2	0.5	-	-
Parsley	2.5	0.8	-	-
Perilla leaves	3.7	-	-	-
Spinach	2.5	0.2	-	-
Tossa jute	6.6	-	-	-

[1] Vitamin E contents are indicated in mg per 100 g product.

[2] Sheppard *et al.* (1993).

[3] Standard tables of food composition in Japan in which tocotrienol contents have not been exhibited (2000).

Dashes (-) denote no value or trace

tocotrienols are limited; comprising only palm oil, rice bran oil and wheat bran, which contain much more γ-tocotrienol than α-tocotrienol. At any rate, it is difficult to get adequate amounts of tocotrienol from ordinary dietary habits. Therefore, it is necessary to obtain it from foods supplemented with a tocotrienol rich fraction extracted from palm oil, wheat germ or rice bran oil. Tocotrienols are safe as well as tocopherols and human studies show no adverse effects with consumption of 240 mg/day for 48 months (Tomeo *et al.*, 1995).

10.3 Bioactivity of tocotrienol

All forms of vitamin E are taken up by intestinal cells and released into the circulation with chylomicrons. At this stage there is probably no discrimination between the different forms (Kayden and Traber, 1993). The vitamins reach the liver via chylomicron remnants. In the liver, a specific protein, α-TTP, selectively sorts out α-tocopherol from all incoming tocopherols and tocotrienols for incorporation into VLDL.Vitamin E isoforms not retained are degraded to CEHC with oxidation of their side chain, a phytyl side chain for tocopherols and an isoprenoid chain for tocotrienols (Lodge *et al.*, 2001, Schultz *et al.*, 1995). Accumulations of tocopherols and tocotrienols in tissues depend on the affinity to α-TTP. α-Tocotrienol uptake and distribution after oral ingestion are less than that of α-tocopherol (about 15% of α-tocopherol). However, when animals were fed vitamin E isoforms containing α-tocopherol, the addition of α-tocopherol decreased retention of other forms (Handelman *et al.*, 1985) due to the priority α-tocopherol has in the competition to bind with α-TTP. Thus, more than 90% of vitamin E isoforms retained in the animal body is α-tocopherol and about 10% is γ-tocopherol. Other forms, including tocotrienols, are very few. Thus, among vitamin E forms, α-tocopherol primarily takes the form of vitamin E in humans and animals, and tocotrienols are very low in vitamin E activity. While the bioavalability of dietary tocotrienol as vitamin E activity is low, tocotrienol has stronger antioxidative activity than α-tocopherol. Vitamin E is a potent fat-soluble antioxidant that inhibits lipid peroxidation in the biological membrane. Tocotrienols have three double bonds in the molecules which give them rapid cellular uptake (Saito *et al.*, 2004), smooth movement in the biomembrane and more homogeneous distribution. Thus, this chemical specificity gives them stronger antioxidant capacity than the saturated side chain in tocopherols (Serbinova and Packer, 1994). Sen *et al.* (2000) showed that uptake of tocotrienols in cells from the culture medium was more efficient compared with that of α-tocopherol. These results indicate that tocotrienols may protect organ cells at low concentrations. Tocotrienols have been shown to have cholesterol-lowering, anticarcinogenic, and neuro protective properties, which may not be related to their antioxidant function. Recently, there have been many such reports, for example, Ahn *et al.* (2007) who reported that γ-tocotrienol inhibited the nuclear factor-κB (NF-κB) pathway which has a central role in tumorigenesis, Shibata *et al.* (2008) who demonstrated antiangionic effects, and Khanna *et al.* (2003) who reported neuroprotective effects. These functions are specific to tocotrienols and do not appear for α-tocopherol at the same concentration in the medium. However, these novel beneficial effects of tocotrienols have been demonstrated only in *in vitro* experiments.

10.4 Specific distribution in skin and adipose tissues of tocotrienols

More than 90% of vitamin E isoforms retained in the animal body is α-tocopherol and about 10% is γ-tocopherol. Other forms including tocotrienol, are very few. However, Hayes *et al.* (1993) observed that in hamsters fed a diet supplemented with a tocotrienol rich fraction of palm oil, tocotrienols did accumulate more than tocopherols in adipose tissue. Further, Podda *et al.* (1996) reported that the distribution of vitamin E isoforms varies from tissue to tissue. In mice fed a chow diet not specifically enriched with tocotrienols, among the tissues for which vitamin E concentrations were determined, only the skin contained tocotrienols, comprising up to 15% of the total vitamin E, and in other tissues, 99% of the vitamin E present was α- or γ-tocopherol. On the other hand, Patel *et al.* (2006) observed that with oral supplementation of tocotrienol without α-tocopherol for 3 years in nine generations of female and male rat, α-tocotrienol was delivered to all vital organs at the highest levels in the adipose, skin and gonads. Recently, Kawakami *et al.* (2007) also determined distribution of tocotrienol in rats fed a rice bran tocotrienol concentrate for 3 weeks. The rice bran tocotrienol was distributed at a significantly high level in the adipose tissue, but a substantial amount of tocotrienol also was distributed in the skin. Later, Shibata *et al.* (2010) also reported that hairless mice skin accumulated substantial amounts of γ-tocotrienol in mice fed a diet containing extracts from rice bran (~90% γ-tocotrienol).

We considered whether tocotrienols with strong antioxidative property accumulated in the skin would protect the skin from damage induced by UV irradiation. Thus, we investigated whether or not dietary tocotrienols could accumulate in the skin. The first experiment examined the concentrations of tocotrienols in several organs as well as the skin of hairless mice, nude mice and rats fed diets containing a tocotrienol rich fraction of palm oil (Ikeda *et al.*, 2000). α-tocopherol was retained abundantly in the skin, liver, kidney and plasma of mice and rats. α-tocotrienol and γ-tocotrienol were mostly not present in the liver, kidney and plasma, while substantial amounts of these tocotrienols were retained in the skin of both mice and rats. Later we reconfirmed that tocotrienols were retained in the skin and adipose tissues of rats fed the same diet for 8 weeks (Ikeda *et al.*, 2001). We reported that sesame seed lignans produced higher concentrations of α-, and γ-tocopherol in the organs by inhibiting vitamin E degradation (Ikeda *et al.*, 2002). In the tocotrienol study, sesame seed produced not only higher concentrations of α-tocopherol but also α- and γ-tocotrienols in the skin (Figure 10.2). Further, we investigated the effect of α-tocopherol on distribution of α- and γ-tocotrienols in rat organs using pure α-tocopherol, α- and γ-tocotrienols. α-tocotrienol accumulated in large amounts in the adipose tissue, skin, muscle and heart, but γ-tocotrienol was present only in the adipose tissue and skin in smaller amounts compared with α-tocotrienol. The addition of α-tocopherol causes a decrease of α-tocotrienol but not γ-tocotrienol (Ikeda *et al.*, 2003). This effect is probably produced by the accelerated degradation of α-tocotrienol by α-tocopherol because we observed an increase in metabolite of α-tocotrienol, α-CEHC, but not the γ-tocotrienol metabolite, γ-CEHC. One of the benefits of γ-tocotrienol may be that it is not affected by a large intake of α-tocopherol. While we cannot confirm this beneficial effect of γ-tocotrienol, a tissue-specific distribution of tocopherols and tocotrienols is known to exist, suggesting that these forms have unique roles in cellular function. The unique distribution of various antioxidants in tissues suggests that their distribution may be dependent upon a selective mechanism for maintaining antioxidant defenses. Thus, we examined whether the skin might accumulate dietary tocotrienols, and whether or not the tocotrienols would protect the skin from damage by UVB irradiation.

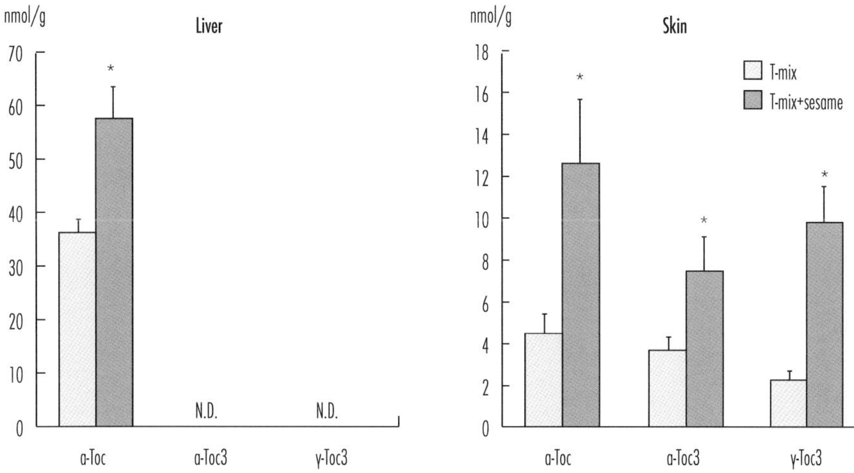

Figure 10.2. Effects of dietary sesame seed on vitamin E concentration in liver and skin of rats fed tocotrienol rich fraction extracted from palm oil (T-mix) and T-mix with sesame seed. For 8 weeks rats were fed a diet containing 220 mg/kg diet T-mix (50 mg α-tocopherol, 77.8 mg α-tocotrienol, 103 mg γ-tocotrienol), or a diet containing 220 mg/kg T-mix and 200 g/kg sesame (T-mix+sesame).
*Significant difference ($P<0.05$) from diet with only T-mix.

10.5 The protective effect of dietary tocotrienols on skin damaged by UVB irradiation

Tocotrienols topically dosed on the skin penetrate into skin tissue and protect the skin from oxidative stress by UV irradiation (Weber *et al.*, 1997). We observed that dietary tocotrienols were accumulated in the skin (Ikeda *et al.*, 2000), and further that dietary sesame seed produced higher tocotrienol concentrations in the skin (Ikeda *et al.*, 2001). We supposed that these tocotrienols taken up in the skin might protect the skin from damage induced by UVB irradiation.

The major acute effects of UV irradiation on the skin is sunburn inflammation (erythema), and chronic exposure to UV irradiation leads to photoaging, immunosuppression, and ultimately to the development of skin cancer. Thus, we investigated whether or not dietary tocotrienols would protect against skin damage induced by UVB using female hairless mice (Yamada *et al.*, 2008).

We examined the suppressive effects of tocotrienols and sesamin (sesame seed contains about 1% sesamin) on UVB-induced skin damage in two experiments, acute irradiation and chronic irradiation. In the acute effects experiment, hairless mice were given experimental diets containing palm oil extracts rich in tocotrienols (T-mix; 50 α-Toc, 70 α-Toc3 and 103 γ-Toc3 mg/kg diet) for 6 weeks and for the final 7 days mice received UVB irradiation of 180 mJ/cm^2 once a day. The skin damage was evaluated by the intensity of lesions on the final day. As shown in Figure 10.3, severe sunburn was observed in the VE-free group with the next most intense damage appearing in the

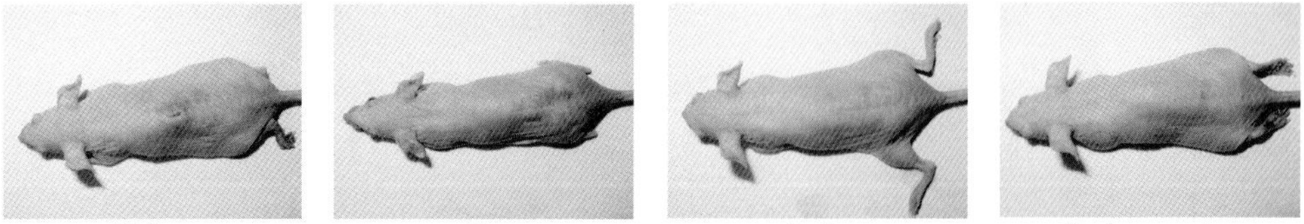

Figure 10.3. Representative photographs of skin on the backs of hairless mice in each group 7 days after UVB irradiation. The mice were fed experimental diets for 7 weeks; a diet without vitamin E (VE-free), a diet containing 50 mg/kg α-tocopherol (α-Toc), a diet containing 229 mg/kg T-mix (T-mix; contained 50 mg/kg α-tocopherol), or a diet containing 229 mg/kg T-mix and 2 g/kg sesamin (T-mix+sesamin). In the last 7 days, mice received 180 mJ/cm^2 UVB irradiation once daily. The figures given in parentheses are the mean arbitrary units of sunburn; 0, no lesion; 1, barely detectable red lesions; 2, moderate red lesions; 3, bright red lesions.

α-Toc group. T-mix alone and T-mix with sesamin alleviated sunburn. We could detect α- and γ-tocotrienols in the skin of T-mix fed groups, though the concentrations were significantly lower than those of α-tocopherol, and sesamin increased the concentration of α-tocotrienol slightly, but of γ-tocotrienol significantly (Figure 10.4). These results suggest that tocotrienols, especially γ-tocotrienol, protect the skin against damage induced by UV irradiation.

Subsequently, in chronic exposure of UVB, hairless mice fed the same experimental diets received topical treatment with DMBA on the back skin as a tumor initiator and they simultaneously received irradiation with UVB (180 mJ/cm^2) twice a week for 20 weeks. As shown in Figure 10.5, in the first 6 weeks no tumor was observed, but at the 7th week a few tumors were detected in all groups except the T-mix with sesamin group, and at the 10th week tumors appeared in all groups. α-Tocopherol suppressed the incidence of tumor in the final 4 weeks to some degree compared with vitamin E-free, and in the tocotrienol fed group incidence of tumors at 10 weeks was clearly suppressed. The tocotrienol with sesamin group had the lowest incidence of tumors among the four dietary groups.

In conclusion, our results suggested that dietary tocotrienols are taken up in the skin, and protect the skin against damage produced by UVB. Tocotrienols with sesamin enhanced tocotrienol's effect on skin photoprotection. However, α-tocopherol exhibited less protection effects against UVB-induced skin damage compared with tocotrienols.

10.6 Suppression of tocotrienol on UVB-induced inflammation

Recently, Miyazawa's research group extensively examined the mechanism of inflammation induced by UVB irradiation and the anti-inflammatory effect of tocotrienols. UVB produces various ROS in the skin. ROS attack lipids, proteins, and nucleic acids in the skin, and those oxidized materials cause skin damage. One of the most toxic factors is lipids peroxide, particularly,

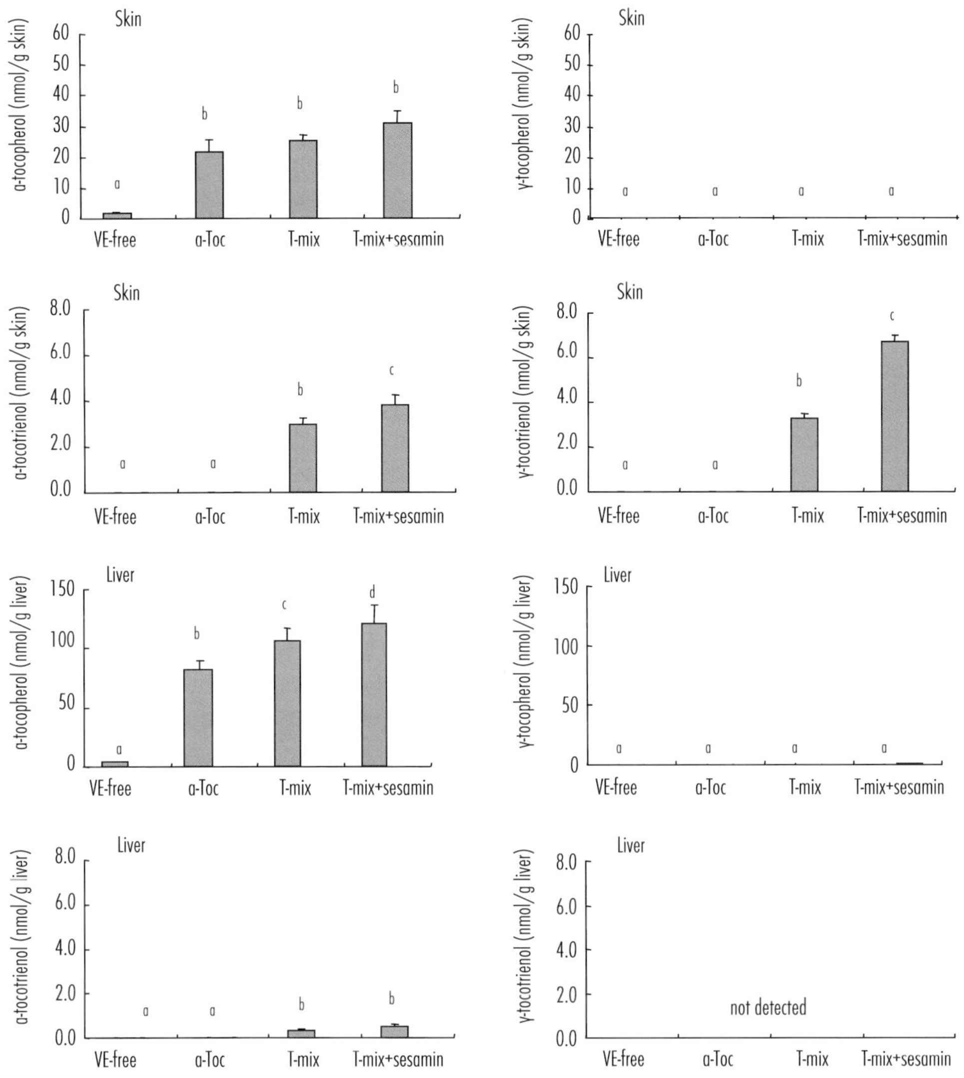

Figure 10.4. Concentrations of α-tocopherol and α- and γ-tocotrienol in the skin and liver of hairless mice. For 6 weeks, hairless mice were fed the same experimental diets shown in Figure 10.3. Values are means ±SEM. Values with different letters are significantly different, $P<0.05$.

SQ-OOH (Nakagawa *et al.*, 2010). Human skin lipids consist mainly of triacylglycerol, wax esters, squalene, and sterols. SQ-OOH produced by irradiation of UVB caused an increase in the expression of inflammatory genes such as the interleukins as well as COX-2. Miyazawa's group also extensively investigated the mechanisms of inflammation induced by UVB, using immortalized HaCaT. PGE_2 production, mRNA expression of COX-2, IL-1b, IL-6, MCP-1 and ROS generation were determined. All these inflammatory mediators increased though UVB irradiation, and γ-tocotrienol reduced the production of these inflammatory mediators. The

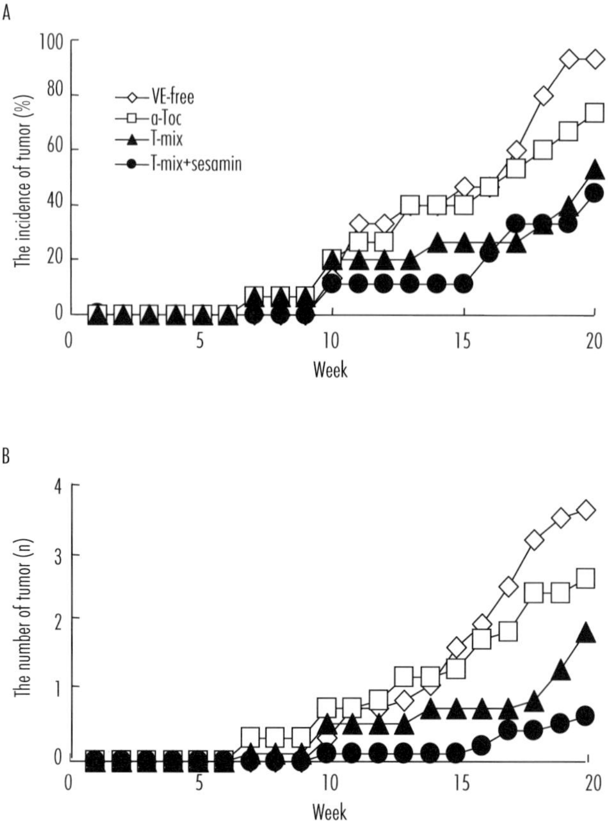

Figure 10.5. Incidence of tumor produced by UV irradiation after DMBA treatment. All hairless mice received topical treatment with DMBA on the back skin as a tumor initiator. One week later, mice were divided into four groups, and fed the same diets as in Figure 10.3. They simultaneously received irradiation of 180 mJ/cm^2 UVB light twice a week. Skin tumors more than 1 mm in size were counted once a week for 20 weeks. Upper (A) shows percentage of mice with tumor, and lower (B) is the mean number of tumors per mouse.

group further investigated the effects of orally administrated tocotrienols on UVB-induced skin damage using HR-1 hairless mice. The mice fed a vitamin E-free diet were given 2.5 mg α-tocopherol or 2.57 mg rice bran tocotrienol concentrate (RBT3 ~90% γ-tocotrienol) dissolved in vitamin E-free corn oil orally once a day for a total of 14 successive days. On days 9, 11, and 13, dorsal mouse skin was directly exposed to UVB irradiation (200 mJ/cm^2). Skin thickness was measured daily. On day 14, 24 h after the last UVB exposure, mice skins were collected and analyzed for vitamin E concentration and COX-2 protein expression. Significant incorporation of γ-tocotrienol (70.5 nmol/g) and α-tocopherol (39.6 nmol/g) in the skin was detected in groups that received tocotrienol 2.5 mg/day and α-tocopherol 2.5 mg/day respectively, compared with the vitamin E-free group (5.3 nmol/g α-tocopherol). In this experiment, γ-tocotrienol incorporated higher amounts than α-tocopherol. As shown in Figure 10.6, skin thickness and COX-2 protein expression increased from UVB irradiation. Orally dosed γ-tocotrienol suppressed these skin

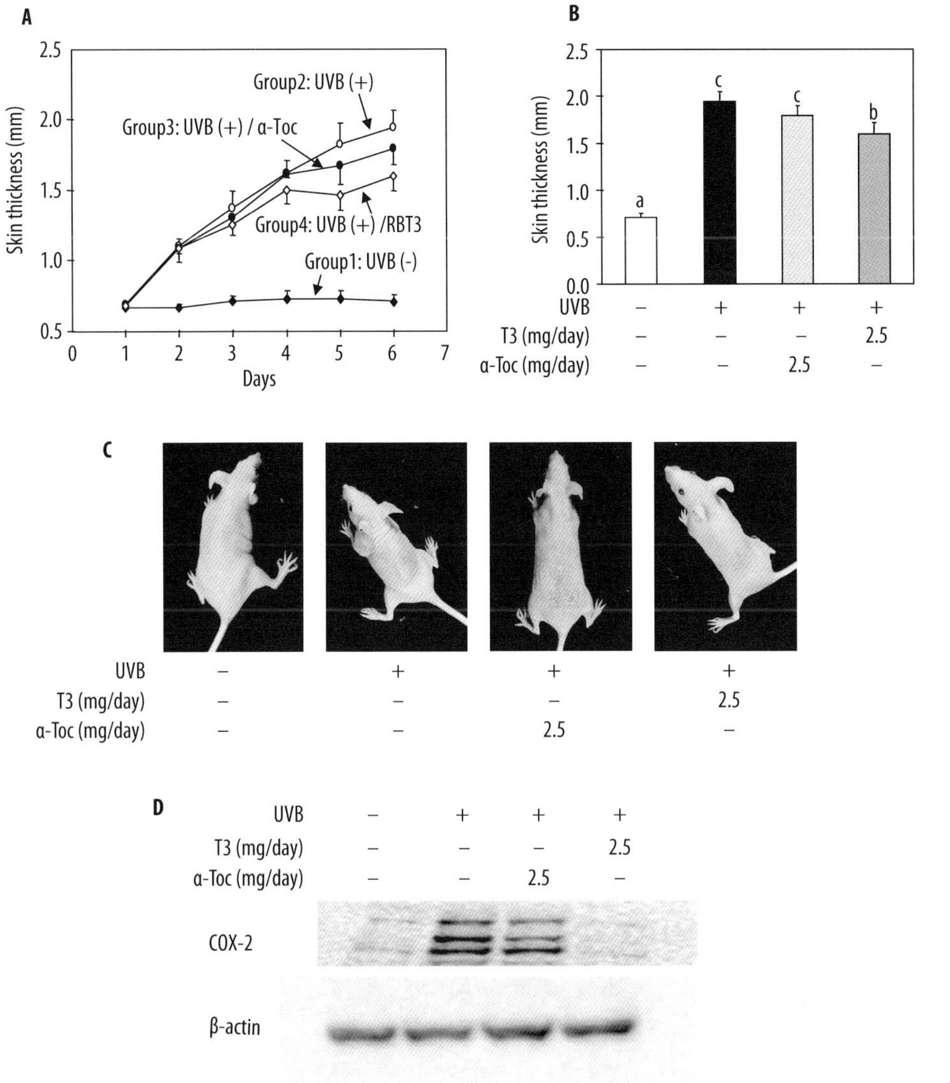

Figure 10.6. Effect of oral administrated γ-tocotrienol or α-tocopherol on the skin thickening (A,B), sunburn reaction (C), and COX-2 protein expression (D) in UVB-irradiated hairless mice. Each mouse received samples orally once a day for 14 successive days. On day 9, 11, and 13, dorsal mouse skin was exposed to UVB irradiation at a dose of 200 mJ/cm^2. Data of skin thickness on day 14 is presented (B). Values are means ±SD. Values with different letters are significantly different, $P<0.05$. On day 14, skins were collected for Western blot analysis of COX-2 protein (D). Each Western blot is a representative example of the data. (Shibata *et al.*, 2010)

disorders, but α-tocopherol had a slight or no effect. These results suggest that oral tocotrienol accumulates in the skin and may protect against UVB-induced skin inflammation.

Acknowledgments

The author gratefully acknowledges Professor Miyazawa of Tohoku University for the generous permission to use his research group data in this article.

References

Ahn, K.S., Sethi, G., Krishnan, K. and Aggarwal, B.B., 2007. Gammα-tocotrienol inhibits nuclear factor-κB signaling pathway through inhibition of receptor-interacting protein and TAK1 leading to suppression of antiapoptotic gene products and potentiation of apoptosis. The Journal of Biological Chemistry 282, 809-820.

Handelman, G.J., Machlin, L.J., Fitch, K., Weiter, J.J. and Dratz, E.A., 1985. Oral alphα-tocopherol supplements decrease plasma gammα-tocopherol levels in humans. The Journal of Nutrition 115, 807-813.

Hayes, K.C., Pronczuk, A. and Liang, J.S., 1993. Differences in the plasma transport and tissue concentrations of tocopherols and tocotrienols: observations in humans and hamsters. Proceeding Society of Experimental Biological Medicine 202, 353-359.

Hosomi, A., Arita, M., Sato, Y., Kiyose, C., Ueda, T., Igarashi, O., Arai, H. and Inoue, K., 1997. Affinity for α-tocopherol transfter protein as a determinant of the biological activities of vitamin E analogs. FEBS Letters 409, 105-108.

Ikeda, S., Niwa, T. and Yamashita, K., 2000. Selective uptake of dietary tocotrienols into rat skin. Journal of Nutritional Science and Vitaminology 46, 141-143.

Ikeda, S., Toyoshima, K. and Yamashita, K., 2001. Dietary sesame seeds elevate α- and γ-tocotrienol concentrations in skin and adipose tissue of rats fed the tocotrienol-rich fraction extracted from palm oil. The Journal of Nutrition 131, 2892-2897.

Ikeda, S. Tohyama T. and Yamashita, K., 2002. Dietary sesame seed and its lignans inhibit 2,7,8-trimethyl-2(2'-carboxyethyl)-6-hydroxychroman excretion into urine of rats fed -tocopherol. The Journal of Nutrition 132, 961-966.

Ikeda, S., Tohyama, T., Yoshimura, H., Hamamura, K., Abe, K. and Yamashita, K., 2003. Dietary α-tcopherol decreases α-tocotrienol but not γ-tocotrienol concentration in rats. The Journal of Nutrition 133, 428-434.

Kawakami, Y., Ysuzuki, T., Nakagawa, K. and Miyazawa, T., 2007. Distribution of Tocotienols in Rats Fed a Rice Bran Tocotrienol Concentrate. Bioscience Biotechnology and Biochemistry 71, 464-471.

Kayden, H.J. and Traber, M.G., 1993. Absorption, lipoprotein transport, and regulation of plasma concentrations of vitamin E in humans. The Journal of Lipid Research 34, 343-358.

Khanna, S., Roy, S., Ryu, H., Bahadduri, P., Swaan, P.W., Ratan, R.R. and Sen, C.K., 2003. Molecular basis of vitamin E action. Tocotrienol modulates 12-lipoxygenase, a key mediator of glutamate-induced neurodegeneration. The Journal of Biological Chemistry 278, 43508-43515.

Lodge, J.K., Ridlington, J., Leonard, S., Vaule, H. and Traber, M.G., 2001. α- and γ-Tocotrienols are metabolized to carboxyethyl-hydroxychroman derivatives and excreted in human urine. Lipids 36, 43-48.

Matsumura, Y. and Ananthaswarny, H.N., 2004. Toxic effects of ultraviolet radiation on the skin. Toxicology and Applied Pharmacology 195, 298-308.

Nakagawa, H., Shibata, A., Maruko, T., Sookwong, P., Tsuduki, T., Kawakami, K., Nishida, H. and Miyazawa, T., 2010. c-Tocotrienol Reduces Squalene Hydroperoxide-Induced Inflammatory Responses in HaCaT Keratinocytes. Lipids 45, 833-841.

Patel, V., Khanna, S., Roy, S., Ezziddin, O., Sen, C.K., 2006. Natural vitamin E alpha-tocotrienol: retention in vital organs in response to lonγ-term oral supplementation and withdrawal. Free Radical Reseach 40, 763-771.

Podda, M., Weber, C., Traber, M.G. and Packer, L., 1996. Simultaneous determination of tissue tocopherols, tocotrienols, ubiqunols, and ubiquinones. The Journal of Lipid Research 37, 893-901.

Saito, Y., Yoshida, Y., Nishio, K., Hayakawa, M. and Niki, E., 2004. Characterization of Cellular Uptake and Distribution of Vitamin E. Annals of the New York Academy of Sciences 1031, 368-375.

Schultz, M., Leist, M., Petrzika, M., Gassmann, B. and Brigelius-Flohe, R., 1995. Novel urinary metabolite of α-tocopherol, 2,5,7,8-tetramethyl-2 (2'-carboxyethyl)-6-hydroxychroman, as an indicator of an adequate vitamin E supply? The American Journal of Clinical Nutrition 62(suppl), 1527S-1534S.

Sen, C.K., Savita Khanna, S., Roy, S. and Packer, L., 2000. Molecular Basis of vitamin E Action. Tocotrienol Potently Inhibits Glutamate Induced pp60$^{c\text{-}Src}$ kinase Activation and Death of HT4 Neuronal Cells. The Journal of biological chemistry 275, 13049-13055.

Serbinova, E., Kagan, V., Han, D. and Packer, L., 1991. Free radical recycling and intramembrane mobility in the antioxidant properties of alpha-tocopherol and alpha-tocotrienol. Free Radical biology and Medicine 10, 263-275.

Serbinova, E.A. and Packer, L. 1994. Antioxidant properties of tocopherol and tocotrienol. Methods Enzymology, 234, 354-367

Sheppard, A.J., Pennington, J.A.T. and Weihrauch, J.L., 1993. Analysis and distribution of vitamin E in vegetable oils and foods. In: Packer, L. and Fuchs, J. (eds.) Vitamin E in Health and Disease. Marcel Dekker, Inc. New York, NY, USA, pp. 9-31.

Shibata, A., Nakagawa, K., Sookwong, P., Tsuduki, T., Tomita, S., Shirakawa, H., Komai, M. and Miyazawa, T., 2008. Tocotrienol inhibits secretion of angiogenic factors from human colorectal adenocarcinoma cells by suppressing hypoxia-inducible factor-1α. The Journal of Nutrition 138, 2136-2142.

Shibata, A., Nakagawa, K., Kawakami, Y., Tsuzuki, T. and Miyazawa, T., 2010. Suppression of γ-Tocotrienol on UVB Induced Inflammation in HaCaT Keratinocytes and HR-1 Hairless Mice via Inflammatory Mediators Multiple Signaling. Journal of Agricultural and Food Chemistry 58, 2013-2020.

Tomeo, A.C., Geller, M., Watkins, T.R., Gapor, A. and Bierenbaum, M.L., 1995. Antioxidant effects of tocotrienols in patients with hyperlipidemia and carotid stenosis. Lipids 30, 1179-1183.

Yamashita, K., Nohara, Y., Katayama, K. and Namiki, M., 1992. Sesame seed lignans and γ-tocopherol act synergistically to produce vitamin E activity in rats. The Journal of Nutrition 122, 2440-2446.

Yamada, Y., Obayashi, M., Ishikawa, T., Kiso, Y., Ono, Y. and Yamashita, K., 2008. Dietary Tocotrienol Reduces UVB-Induced Skin Damage and Sesamin Enhances Tocotrienol effects in Hairless Mice. The Journal Nutritional Science and Vitaminology 54, 117-123.

Yoshida, Y,. Niki, E. and Noguchi, N., 2003. Comparative study on the action of tocopherols and tocotrienols as antioxidant: chemical and physical effects. Chemistry and Physics of Lipids 123, 63-75.

Weber, C., Podda, M., Rallis, M., Thele, J.J. Traber, M.G. and Packer, L., 1997. Efficacy of topically applied tocopherols and tocotrienols in protection of murine skin from oxidative damage induced by UV-irradiation. Free Radical biology and Medicine 22, 761-769.

Key facts

- One third of the world's population is estimated to be deficient in dietary zinc.
- Developing countries, primarily, certain South East Asian and sub-Saharan African countries are at high risk.
- The risk in some subpopulations is as high as 73%.
- Vegetarians, alcoholics and the critically ill are especially at risk.
- Zinc body stores are estimated to be between 2-3 g.
- The musculo-skeletal reserves account for about 85% of total body reserves.
- The skin contains 6% of the total body zinc.
- Plasma zinc accounts for only 0.1% of total body zinc and is maintained between 11.5-18.5 μmol/l.
- Main losses of zinc occur in the intestine (27-90 μmol/day), urine (8-11 μmol/d), shed epithelial cells, sweat, semen, hair and menstrual bleeding.
- Red meat, fish and shellfish are excellent sources of zinc.
- Phytates and fiber present in vegetables and cereals inhibit zinc absorption.

Summary points

- Zinc has a myriad of different biological roles in the human body. These include metabolic, structural, catalytic and regulatory processes.
- Recent evidence has expanded on its immunomodulatory, antioxidant and anti-inflammatory functions as well.
- Zip4 and ZnT1 are the 2 main types of zinc transporters that are involved in zinc homeostasis.
- Zinc deficiency can be ascertained by directly measuring zinc levels, and indirectly by measuring the zinc dependent enzymes such as alkaline phosphatase. The favored method is by measuring the plasma zinc levels. These are extremely variable and fluctuate with the type of diet, nutritional status and if there is acute inflammation present.
- Skin diseases clearly linked to impaired zinc absorption or intake such as acrodermatitis enteropathica, acute and moderate zinc deficiency, have been shown to respond to oral supplementation of zinc.
- Other inflammatory diseases such as acne vulgaris, hidradentis suppurativa, folliculitis decalvans, oral ulcers, Behçets disease and rosacea have been shown to be responsive to oral supplementation with zinc. Overall, better responses were seen in patients who were clearly shown to be hypozincemic before treatment.
- Zinc has been shown to improve immune functioning against pathogens that may explain its utility in both cutaneous and disseminated leishmaniasis and leprosy.
- Zinc sulfate is usually preferred by most physicians at a dose of 110-220 mg (25-50 mg elemental zinc) up to three times a day.

11. Zinc and skin health: an overview

H.K. Bangash and A. Sethi
Section of Dermatology, Department of Medicine, The University of Chicago, 5841 South Maryland Avenue, MC 6092, Chicago, IL, 60637, USA; asethi@medicine.bsd.uchicago.edu

Abstract

Zinc is a micronutrient that plays an integral part in the normal functioning of the human body. Recent studies have elucidated the intricate physiology of zinc homeostasis. It has been shown to have immunomodulatory, anti-oxidant and anti-inflammatory roles. However, the application of this new understanding of zinc homeostasis to chronic inflammatory, infectious and auto-immune dermatological conditions is lacking. Evidence based support to the value of using oral zinc supplementation in various diseases is limited. Literature searches showed a paucity of prospective randomized controlled trials. Most literature on the use of zinc in dermatological conditions is largely anecdotal, based on small case series, or individual case reports. Diseases clearly linked to defective zinc absorption or nutritional deficiencies such as acrodermatitis enteropathica and other acquired zinc deficiency states respond rapidly to oral zinc supplementation. As a consequence, oral zinc supplementation remains a first line agent for these conditions. On the other hand, for most other conditions linked to mild to moderate deficiencies of zinc, its use remains limited as a second line agent or as a last measure in certain patients in whom conventional treatment has failed. Further research and adequate double blinded randomized control trials are needed in future. These should evaluate the efficacy of treatment with oral zinc in patients as compared to conventional treatments and would help to reveal the clinical value of using zinc supplementation.

Keywords: zinc deficiency, acrodermatitis enteropathica, acne vulgaris, rosacea, hidradentis suppurativa

Abbreviations

AE	Acrodermatitis enteropathica
ALP	Alkaline phosphatase
BD	Behçets disease
DNA	Deoxyribonucleic acid
DRI	Dietary reference intake
ENL	Erythema nodosum leprosum
IFN	Interferon
IL	Interleukin
MTF-1	Metal response element factor 1
NAE	Necrolytic acral erythema
OCMI	Oral clinical manifestation index
RDA	Recommended daily allowance
RNA	Ribonucleic acid
SLC	Solute linked carrier
ZIP	Zrt-Irt-like protein
ZnT	Zinc transporter

11.1 Introduction

Zinc is an essential trace element in the human diet. It is involved in numerous complex and important functions in the human body. These include structural, catalytic and regulatory roles. (Maverakis *et al.*, 2007) It has recently been shown to be important for normal immunological, antioxidant, metabolic and homeostatic functioning as well. (Prasad, 2009) Thus, inadequate intake of zinc can be associated with the development of various disease states.

11.1.1 Background

Hanas *et al.* (1983) were the first to show that zinc ions were required for the binding of the *Xenopus* transcription factor IIIA to DNA. Since then zinc has been shown to be an important structural component of over 2000 transcription factors, RNA and DNA polymerases as well. These transcription factors possess a zinc-finger motif that enables zinc to interact with proteins and helps to bind to them and activate them as well. Other prominent examples of zinc finger proteins include retinoic acid and vitamin D receptors. Additionally, the T-lymphocyte cell receptors also require zinc to activate tyrosine kinase (Prasad, 2009).

Zinc is required for the catalytic functioning of over 300 metalloenzymes that are involved in various metabolic reactions. Examples of zinc dependent metalloenzymes include carbonic anhydrase, alkaline phosphatase, alcohol dehydrogenase and carboxypeptidase (Lansdown *et al.*, 2007).

Zinc is stored in specific neuronal vesicles of the cerebral cortex that also contain glutamate. Along with calcium it is been shown to have an important role in synaptic signaling and normal neurophysiological functioning (Maverakis *et al.*, 2007).

Zinc plays an integral part in the functioning of the immune system, particularly in the cell-mediated immune system. It facilitates phagocytosis and intracellular killing of pathogens by macrophages, neutrophils, and natural killer cells. It assists in the generation of oxygen free radicals by the oxidative burst. Additionally, it is involved in cytokine production such as IL-2 and IFN-γ, and complement activity. IL-2 in turn activates T-lymphocytes and natural killer cells. T-lymphocytes in particular, appear to be most sensitive to zinc deficiency. The fine balance between the Th1 and Th2 lymphocytes is altered in zinc deficiency. The Th1 lymphocytic response is down-regulated whereas the Th2 lymphocyte response is up-regulated. As a consequence, the effectiveness against intracellular pathogens is diminished in zinc deficient individuals (Prasad, 2009).

Zinc has also been showed to be protective against oxidative stress. It is utilized by certain cysteine rich proteins known as thioneins, which upon binding to zinc are converted into metallothioneins. Metallothioneins aid in the cellular transport of zinc and thereby regulate the concentration of zinc in the body. The cysteine residues of metallothioneins enable them to react with oxygen free radicals. Once bound with oxygen, the cysteine residues are converted into cysteine and free zinc ions. These free zinc ions in turn up-regulate the production of more metallothioneins, aiding in its antioxidant activity. Furthermore, zinc stabilizes enzymes such as superoxide dismutase which has a marked antioxidant activity (Prasad, 2009).

11.2 Zinc transporters

Zinc homeostasis is tightly regulated by 24 different transmembrane proteins which are encoded by two SLC gene families: ZnT (SLC30) and Zip (SLC39). There are 9 ZnT and 15 Zip transporters in humans that play a critical role in zinc homeostasis (Maverakis *et al.*, 2007; Tuerk and Fazel, 2009).

The ZnT transporters function primarily to reduce intracellular zinc. They do this in one of two ways; by either facilitating the efflux of zinc from the cell or by sequestering the zinc inside the cells into intracellular vesicles known as zincosomes. ZnT1 was the first ZnT transporter to be discovered and is present in the small intestine, renal tubular epithelium and placenta (Maverakis *et al.*, 2007). It helps to transfer zinc from the enterocytes into the general circulation. It is maximally concentrated at the basolateral membrane. Deficiency of dietary zinc promotes the upregulation of ZnT1 transporters. The ZnT4 transporter is present in high concentrations in the mammary glands and is associated with milk containing vesicles. It may aid in the transfer of zinc into milk. As with ZnT1, ZnT4 receptors are also upregulated with zinc deficiency.

Conversely, the Zip transporters are involved in increasing the intracellular concentrations of zinc by facilitating the entry of zinc into cells. Zip4 is abundant in enterocytes and absorbs dietary zinc from the small intestine. Zinc deficiency increases the apical expression of Zip4 transporters enabling an increase in the intercellular concentrations of zinc (Tuerk and Fazel, 2009).

The small intestine, specifically the proximal jejunum and distal duodenum, respond to the dietary intake of zinc by altering the expression of the various zinc transporters in order to maintain a tight control of zinc levels in the body (Perafan-Riveros *et al.*, 2002).

When zinc is deficient in the diet the Zip4 transporter, that is normally concentrated at the basolateral membrane of the enterocytes, is upregulated at the apical side of the enterocytes. It thereby facilitates the entry of zinc into the cells and increases the intracellular concentration of zinc (Maverakis *et al.*, 2007).

Once inside the cell, zinc concentration can be tightly controlled by sequestering it inside zincosomes or by being bound to storage ligands which form a zinc sink. The free intracellular zinc can bind with and activate various kinases, phosphatases and transcription factors. One such transcription factor is the MTF-1, which in turn, activates the transcription of metallothioneins and other genes. Zinc can also associate with thionein and form metallothionein which is involved in buffering the levels of zinc and thus preventing toxicity from its buildup (Tuerk and Fazel, 2009). Additionally, it acts as an antioxidant and protects the cell against oxidative stress.

The ZnT1 receptors present in the basolateral membrane helps transport the zinc into the general circulation. Eighty percent of the zinc is bound to albumin in the circulation (Maverakis *et al.*, 2007).

The total amount of zinc in the human body is estimated to be between 2-3 grams. Different organs in the body contain different reserves of zinc. The musculo-skeletal accounts for most of the stored zinc, approximated to be about 85% of the total body zinc levels (Tuerk and Fazel, 2009). The skin is the second most abundant site for storage, containing 6% of all the zinc (Bae *et al.*, 2010). Plasma zinc only accounts for 0.1% of the total zinc.

Zinc is excreted in large amounts mainly from the gastrointestinal tract. Intestinal and pancreas secretions can account for the excretion of about 27-90 μmol/day. Severe losses of zinc may be seen in children with diarrhea. Minimal amounts of zinc are also lost from the urinary system as well as most the filtered zinc is reabsorbed by renal tubules. Losses in the urine can be up to 8-11 μmol/d. Zinc is also lost by shedding epithelial cells, sweat, semen, hair and by menstrual bleeding. (Tuerk and Fazel, 2009) The normal of concentration of zinc is maintained within 11.5-18.5 μmol/l.

11.3 Recommended daily allowance

The RDA of zinc is published by the Food and Nutrition Board at the national academy of sciences in 2001. The RDA is the average daily dietary intake level; sufficient to meet the nutrient requirements of nearly all (97-98%) healthy individuals in a group. This varies with age, sex and in women it also varies in pregnancy and lactation. When the RDA cannot be calculated an Adequate Intake is estimated to ensure ample zinc intake (Table 11.1, for dietary sources of zinc see Table 11.2).

Table 11.1. Recommended dietary reference intake (mg/day) for zinc with respect to age, sex, pregnancy and lactation (adapted from Trumbo *et al.*, 2001; with permission from Elsevier).

Age	Males	Females	Pregnancy	Lactation
0-6 mo	2	2	-	-
6-12 mo	3	3	-	-
1-3 y	3	-	-	-
4-8 y	5	-	-	-
9-13 y	8	8	-	-
14-18 y	11	9	12	13
19-30 y	11	8	11	12
31-50 y	11	8	11	12
51-70 y	11	8	-	-
70 + y	11	8	-	-

Table 11.2. Dietary sources of zinc (adapted from USDA, 2010).

Food	Weight (g)	Mg per measure
Oysters, cooked (fried)	85	74
Oysters, 6 medium (raw)	84	33
Fortified cereals, 1 cup	30	15
Baked beans (with pork)	253	14
Turkey	152	11
Beef	85	7-9
Crab	85	6.5
Lamb	85	6
Chicken	145	6

11.4 Zinc deficiency

11.4.1 Epidemiology

Though uncommon in the developed world, zinc deficiency is major problem in the developing world. According to the World Health Report (2002), it is estimated that one-third of the world's population is deficient in zinc. High risk populations include people from South-east Asia and sub-Saharan Africa. In certain populations the percentage may be as high as 73%. Most of these are cases of mild to moderate zinc deficiency. Inherited zinc deficiency on the other hand is very rare. It is estimated that AE occurs in 1 per 500,000 children (Maverakis *et al.*, 2007).

11.4.2 Risk factors

Populations with zinc deficient diets, vegetarians, alcoholics, and the malnourished are at risk for zinc deficiency (Table 11.3). Children and the elderly are particularly vulnerable. There have been reports of AE like lesions arising in patients afflicted with cystic fibrosis (Zedek *et al.*, 2008).

Table 11.3. Manifestations of zinc deficiency (adapted from Plum *et al.*, 2010; with permission from the authors).

Systemic manifestations	Growth retardation
	Immune dysfunction
	Infection
Skin	Skin lesions
	Acrodermatitis enteropathica
	Decreased wound healing
Neurological	Decreased nerve conduction
	Neuropsychiatric symptoms
	Neurosensory disorders
	Mental lethargy
Thymus	Thymic atrophy
Reproductive symptoms	Infertility
	Retarded genital development
	Hypogonadism

11.5 Inherited zinc deficiency

11.5.1 Acrodermatitis enteropathica

Acrodermatitis enteropathica, also known as the Danbolt-Gloss syndrome, is a rare autosomal recessive multisystem disease caused by defective intestinal zinc transportation. Danbolt and Gloss were the first to identify AE as a distinct disease in 1942. It was Barnes and Moynahan who, in 1973, established a definite link between reduced zinc levels and the development of the symptoms (Barnes and Moynahan, 1973).

This multi-system disease affects the gastro-intestinal, ophthalmologic, integumentary, immunological and endocrine systems. The classical triad associated with this condition includes dermatitis, diarrhea and alopecia. The genetic defect has been localized to a SLC) gene family, 39A4 on chromosome 8q24. This gene encodes a zinc transporter Zip4 that is present in mainly in the small intestine.

11.5.2 Epidemiology

It affects about 1 in 500,000 children. 30% of the patients have an affected sibling with the disease (Perafan-Riveros *et al.*, 2002).

11.5.3 Clinical manifestations

AE presents clinically in infancy during the weaning period when the breast-fed child is introduced to bovine milk. It can present earlier within days if the infant is bottle fed. The classical triad of dermatitis, diarrhea and alopecia are seen in only 20% of the patients.

11.5.4 Skin manifestations

The dermatological manifestations of AE show a predilection for the perioral, acral and perineal regions. It can present with a wide spectrum of morphological lesions ranging from erythematous eczematous pink scaly plaques which can become vesicular, bullous, pustular or even lead to desquamation. Angular cheilitis presents early. An overlap of psoriasis and AE has been reported recently. Chronic lesions may resemble psoriasiform lesions and may be difficult to differentiate (Maverakis *et al.*, 2007; Perafan-Riveros *et al.*, 2002).

Nail changes include nail dystrophy, onycholysis, onychodystrophy and paronychia (Maverakis *et al.*, 2007; Perafan-Riveros *et al.*, 2002). Hair changes include alopecia totalis. Hair shaft changes are seen as well. The hair shafts may show alternating dark and light bands on polarized light (Maverakis *et al.*, 2007).

11.5.5 Other findings

There can also be involvement of the eyes with blepharitis and conjunctivitis. Other systemic findings can also include hypogonadism, delayed puberty, growth failure, dysgeusia, anemia and photophobia. Zinc deficiency has been associated with impaired immune functioning and can lead to potentially deadly infections with Candida albicans, gram positives and gram negative bacteria including Pseudomonas. There has also been a single case report of septicemia seen from Klebsiella infection (Maverakis *et al.*, 2007; Perafan-Riveros *et al.*, 2002).

11.6 Acquired zinc deficiencies

11.6.1 Acute zinc deficiency

Kay *et al.* (1976) described a series of 37 very ill patients who showed marked plasma zinc depletion while they were receiving total parenteral alimentation. They presented with mental status changes and features similar to AE with a triad of dermatitis, diarrhea and alopecia. These patients responded to zinc supplementation (Kay *et al.*, 1976). Since then zinc has been added to parenteral nutritional solutions (Table 11.4).

11.6.2 Moderate systemic zinc deficiency

In 1961 Prasad *et al.* described a number of Iranian patients form Shiraz who showed a deficiency of both zinc and iron. They presented with hypochromic, microcytic anemia, hepatosplenomegaly, growth retardation, dermatitis and geophagia. Later similar individuals were also identified in Egypt. These patients were treated with zinc and iron supplementation with a marked response of their symptoms (Prasad *et al.*, 1961).

11.6.3 Acne vulgaris

In 1970, Michaelsson and Fitzherbert, reported a marked improvement of acne in patients suffering from AE who were being treated with high doses of zinc. Since then there have been a number of descriptions in the literature supporting the role of zinc in acne. Zinc has important antioxidant and anti-inflammatory roles in the body. Zinc deficiency therefore could contribute and exacerbate acne lesions. In patients with marked plasma zinc deficiency and inflammatory acne, zinc supplementation has been proven to be effective by modulating the levels of IGF-1, which are normally increased by *Propionibacterium acnes* (Isard *et al.*, 2010).

Sardana and Garc (2010). recently described 48 patients with inflammatory comedonal and papulo-pustular acne that were given oral zinc supplementation in the form of APC a novel methionine based zinc complex with additional antioxidants. They were treated thrice a day for 3 months, followed by a 4 weeks treatment free period. They reported a 79% response rate in these patients.

Table 11.4. Dermatological conditions responding to oral zinc supplementation adapted from Bae *et al.*, 2010.

Skin disease	References	Study type	Zinc formulation and dosage
Acne vulgaris	Hillstrom *et al.*, 1977	DB RCT, N=91	ZS 400 mg/d × 12 w
	Goransson *et al.*, 1978	DB RCT, N=54	ZS 600 mg/d × 12 w
	Verma *et al.*, 1980	DB RCT, N=56	ZS 600 mg/d × 12 w
	Liden *et al.*, 1980	DB RCT, N=54	ZS 600 mg/d × 12 w
	Orris *et al.*, 1978	DB RCT, N=22	ZS monohydrate, 411 mg/d
	Weimar *et al.*, 1978	DB RCT, N=52	ZS 220 mg TID × 12 w
	Dreno *et al.*, 1989	DB RCT, N=30	ZG 200 mg/d × 4 w
	Dreno *et al.*, 2001	DB RCT, N=332	ZG 30 mg/d (elemental) × 3 m
Rosacea	Sharquie *et al.*, 2006b	DB RCT, cross-over, N=25	ZS 100 mg TID × 3 m
Hidradenitis suppurativa	Brocard *et al.*, 2007	Case series, N=22	ZG 90 mg OD × 4 m
Folliculitis decalvans	Abeck *et al.*, 1992	Case series, N=3	ZS 400 mg OD × 6 m
Behçets disease	Sharquie *et al.*, 2006a	DB RCT, cross-over, N=30	ZS 100 mg TID × 3 m
Necrolytic Acral Erythema	Moneib *et al.*, 2010	Case series, N=15	ZS 440 mg OD × 1 m
Recalcitrant verrucae	Al-Gurairi *et al.*, 2002	RCT, N=80	ZS 10 mg/kg (600 mg/d) × 2 m
	Sadighha, 2009	RCT, N=26	ZS 10mg/kg (600 mg/d) × 2 m
	Yaghoobi *et al.*, 2009	RCT, N=32	ZS 10mg/kg (600 mg/d) × 2 m
Cutaneous leishmaniasis	Sharquie *et al.*, 2001	RCT, N=104	ZS 2.5, 5 or 10mg/kg TID × 45 d
Disseminated leishmaniasis	Sharquie and Najim, 2004	Single case report	ZS 10 mg/kg TID × 6 m
Leprosy	Mathur *et al.*, 1983	Case series, N=8	NA
	Mathur *et al.*, 1984	Case series, N=15	NA
	Mahajan *et al.*, 1994	Case series, N=40	ZS × 4 m
Oral ulcers	Sharquie *et al.*, 2008	DB RCT	ZS 150 mg BID × 12 w
Yellow nail syndrome	Arroyo and Cohen, 1993	Single case report	Zinc × 2 y

Abbreviations: ZS Zinc sulfate; ZG Zinc gluconate; DB Double blind; RCT Randomized controlled trial; TID Three times a day; BID Twice Daily; OD Once daily; NA Not available; d day; w week/s; m month/s; y year/s.

Zinc has a myriad of anti-inflammatory functions. These include inhibition of leukocyte migration, differentiation and recruitment of other inflammatory cells. It modulates the inflammatory response by reducing the pro-inflammatory cytokines and subsequent release of other inflammatory mediators and vasoactive amines. It inhibits sebum production and has also been shown to have a direct bacteriostatic effect on *P. acnes*. Additionally, it has an inhibitory function on *P. acnes* lipase and reduces the production of free fatty acids (Rebello *et al.*, 1986).

There is a dearth of reliable randomized controlled trials that support the role of zinc as a first line agent for acne. It has thus been used as a second line agent for the control of mild to moderate inflammatory acne.

11.6.4 Rosacea

Sharique *et al.* (2006b) conducted a randomized, controlled, double blind, crossover trial examining the effects of oral zinc supplementation on 25 patients suffering from rosacea. A disease severity score was established for each patient and they were either started oral zinc sulfate therapy or given a placebo. The group receiving the oral zinc sulfate showed a significant reduction in the disease severity score within the first month of treatment. These patients were crossed over to the placebo group, and it was observed that their mean scores rose till the fourth month of treatment however, these scores still remained lower than their pre-treatment scores. In the second group the mean scores continued to be elevated for the 3 months that they received the placebo. Once this group of patients was started on the zinc sulfate the mean scores decreased significantly to lower levels (Sharquie *et al.*, 2006b).

11.6.5 Hidradenitis suppurativa

Brocard *et al.* (2007) described 22 patients with a Hurley's grade I and II recalcitrant hidradenitis suppurativa who were treated with zinc gluconate 90 mg once per day. They reported 8 complete remissions and 14 partial remissions. Other studies are needed to confirm these effects. At the current time oral zinc salts are usually reserved for maintenance treatment once the active flare is controlled by oral antibiotics (Brocard *et al.*, 2007).

11.6.6 Folliculitis decalvans

Abeck *et al.* (1992) have published a case series in which 3 patients with folliculitis decalvans were treated with oral zinc sulfate 400 mg once per day for 6 months and 1 g of fusidic acid topically. At one year 2 of the 3 patients showed a complete response (Abeck *et al.*, 1992; James *et al.*, 2009).

11.6.7 Behçets disease

Sharique *et al.* (2006a) conducted a randomized, controlled, double-blind, crossover trial on the effects on zinc sulfate in the management of BD. Thirty age and sex matched controls and the patients were randomly assigned to two groups and either received 100 mg of oral zinc sulfate three times a day or an identical placebo. Serum zinc levels were calculated at the beginning of the trial initially and then monthly afterwards. After 3 months of treatment the patients were crossed over. At the end of the study, zinc sulfate supplementation was found to reduce the clinical symptoms of BD and could be a useful agent for controlling this disease (Sharquie *et al.*, 2006a).

11.6.8 Psoriasis

Zinc deficiency has been associated with the development of lesions that tend to mimic those seen in psoriasis. A recently described case report of a woman with psoriasiform papules and plaques in her acral and genital lesions was found to be deficient in multiple nutrients including zinc. Her lesions responded to the oral supplementation of the deficient trace minerals and lipids and were completely cleared by 2 weeks (Bolotin *et al.*, 2010). Zinc deficiency itself has not been shown to be linked with psoriasis. Furthermore, there has been no conclusive evidence to show that zinc supplementation has any role in treating psoriasis.

11.6.9 Necrolytic acral erythema

El Darouti *et al.* (1996) illustrated the link between this erythematous, hyperkeratotic rash with active hepatitis C virus infection. This rash favors the acral areas; particularly the dorsal hands and feet. NAE is an important cutaneous marker of underlying hepatitis C virus infection and is usually seen early on in the disease course. It is thought to arise due to the metabolic disturbances that occur as a consequence of hepatitis C infection and damage to the liver. Management of the rash is usually achieved by treating the underlying infection with ribavirin and interferon alpha. Oral zinc supplementation has been described for refractory lesions in the literature. This could imply an underlying systemic zinc deficiency in these patients (El Darouti and Abu el Ela, 1996; Moneib *et al.*, 2010). Recently, it has also been suggested that there could be a localized deficiency of zinc, particularly in the cutaneous acral surfaces, that could explain the preponderance of the lesions on the hands and feet (Najarian *et al.*, 2008).

11.6.10 Oral ulcers

Sharique *et al.* (2008) performed a double blind randomized control trial on 45 patients with recurrent apthous ulcers who were randomly assigned to three groups. They were either administered oral zinc sulfate 150 mg twice a day, dapsone 50 mg twice a day or a glucose placebo 250 mg. The patients were assessed for response to therapy based on an OCMI assessing the diameter of the ulcers, on day 0, 4, and then every 2 weeks till week 12.

After 12 weeks of treatment they reported a statistically significant difference in the OCMI for the patients treated with the oral zinc sulfate, and those who were treated with oral dapsone. However, zinc sulfate was reported to have a quicker therapeutic effect as compared to dapsone and was shown to be superior to dapsone in the 6th week of treatment (Sharquie *et al.*, 2008). The patients placed on placebo did not show a statistically significant change in the OCMI.

11.6.11 Recalcitrant viral warts

There is emerging evidence in the literature that zinc may be effective in treating viral warts due to its immunomodulatory functions. Al-Gurairi *et al.* (2002) described 80 patients with recalcitrant viral warts who were treated with 10 mg/kg/day of oral zinc sulfate daily (up to 600 mg/d). After

2 months of treatment he reported an overall response rate of more than 85% in the zinc sulfate treated group. None of the placebo treated patients showed any clinical response (Al-Gurairi *et al.*, 2002).

Sadighha (2009) similarly treated 26 patients with either oral zinc sulfate or a placebo. He also measured the zinc levels in the patients before the study was started, and calculated a mean zinc level of 53.3±9.7 μg/dl. There were 13 patients in each arm of the study. He treated one group with 10 mg/kg of oral zinc sulfate daily (up to 600 mg/d) and the other with a placebo. In the patients treated with oral zinc sulfate he reported a response rate of above 50% within 1 month and up to 76% after 2 months of treatment. The mean zinc levels in those who responded to the treatment was reported as 27.77 μmol/l and those who did not respond had a mean of 10.68 μmol/l (Sadighha, 2009).

11.6.12 Old world cutaneous leishmaniasis

Zinc supplementation has been shown to reduce the severity of symptoms in cutaneous leishmaniasis. Sharique *et al.* (2001) demonstrated the effect of oral zinc sulfate on 104 patients who had cutaneous leishmaniasis and in whom systemic treatment was indicated. Patients were selected if they had, multiple lesions >5, large lesions >4 cm^2 or lesions that could not be injected, or who had lesions near critical areas such as the eye.

The patients either received oral zinc sulfate or received a placebo, and were treated for 45 days. They reported a dose dependent cure rate of about 83.9%, 93.1% and 96.9% for those treated with oral zinc sulfate at a dose of 2.5 mg/kg, 5 mg/kg and 10 mg/kg respectively. They reported no response in the patients treated with the placebo (Sharquie *et al.*, 2001).

A case of disseminated cutaneous leishmaniasis in a 60 year old lady was also described by the same author, who reported complete resolution of her lesions with oral zinc sulfate at a dose 10 mg/kg/day, in 3 divided doses, administered for 6 months (Sharquie and Najim, 2004).

Zinc efficacy in cutaneous leishmaniasis may be due to a direct inhibitory effect on the leishmanial promastigotes and amastigotes (Najim *et al.*, 1998). Additionally, zinc has immune modulating functions favoring the Th1 arm; this could also enable the immune cells to eliminate the intracellular parasites more effectively (Prasad, 2009).

11.6.13 Leprosy

Zinc supplementation in patients suffering from leprosy has been shown to be effective in reducing the severity of the disease, the dosage of the anti-lepromatous therapy as well as abating the symptoms of the leprosy reactional states such as erythema nodosum leprosum. This may be attributed to the improved immune functioning and greater activity against *Mycobacterium leprae* (Prasad, 2009).

Mathur *et al.* (1983) described eight patients with chronic ENL who were treated with high doses of clofazamine and steroids. They noted a reduction in symptoms after the oral zinc sulfate was instituted. The maintenance dose of clofazamine required was reduced and the steroids were discontinued. Additionally, they reported an increased tolerance to dapsone in four out of the five patients who could not tolerate the medication previously (Mathur *et al.*, 1983).

Mahajan (1994), similarly, described 40 leprosy patients with chronic ENL on prednisolone who were supplemented with oral zinc sulfate. They also noted a reduction in both symptom severity and duration as well (Mahajan *et al.*, 1994).

Treatment with oral zinc was also shown to have an additive effect in 15 patients already on antileprosy medications such as dapsone. Those who were treated with oral zinc as well showed reduced bacterial counts, rapid resolution of clinical symptoms, including erythema and re-growth of eyebrows. Additionally, there was also a reduction in the bacterial index of granuloma and an increased neovascularization in the lesions (Mathur *et al.*, 1984).

11.6.14 Yellow nail syndrome

The yellow nail syndrome is an autosomal dominant condition that occurs due to problems with the lymphatic drainage. The clinical manifestations include lymphedema, pleural effusions and the dystrophic yellow colored nails. There is a single case report of a 53 year old woman who had complete resolution of her yellow nails and lymphedema after she was treated with oral zinc supplementation for two years (Arroyo and Cohen, 1993).

Yellow nail syndrome can spontaneously remit in approximately 30% of the patients, this could account for the apparent response to treatment with zinc (Hoque *et al.*, 2007).

11.7 Work up and evaluation of zinc deficiency

Blood zinc level measurements, both plasma and serum, are highly unreliable. Zinc levels have been shown to be dependent upon inflammation, ingestion of certain foods and the general nutritional status of the individual.

Plasma zinc measurements are generally preferred over serum zinc measurements and represents about 0.1% of the zinc in the body. Eighty percent of the plasma zinc is bound to albumin. Therefore, albumin should be measured along with the zinc levels for an accurate assessment of a patient's nutritional status (Tuerk and Fazel, 2009). The plasma zinc levels are normally maintained between 11.5-18.5 μmol/l. Zinc levels can decrease with the recent ingestion of food, particularly foods with high fiber and phytate contents. Therefore, a serum level of 7.65 μmol/l or less can be used to make a presumptive diagnosis of zinc deficiency (Maverakis *et al.*, 2007).

Zinc levels may decrease during inflammation. A murine model has shown IL-6, an acute phase reactant, to be responsible for up regulating Zip14 transporters in mouse livers and thus reducing the plasma zinc levels (Maverakis *et al.*, 2007). A similar mechanism may be responsible for the apparent hypozincemia seen in humans during acute inflammation and may be a confounding factor in patients suspected to have zinc deficiency.

Conversely, zinc levels may also be falsely elevated if the sample is hemolyzed or if the patient is already on zinc supplementation (Maverakis *et al.*, 2007). Finally, plasma zinc levels may even be found to be normal even with total body zinc deficiency due to compensatory homeostatic mechanisms that tend to elevate the levels in the blood. Other methods of diagnosing zinc deficiency that have been employed previously include measuring zinc levels in erythrocytes and hair shafts. Zinc excretion in urine as well as sweat has been used in the past. These methods are still less sensitive than measuring plasma zinc levels (Maverakis *et al.*, 2007).

Indirect methods of measuring zinc deficiency include measuring the activity of the various zinc dependent metalloenzymes. ALP is one such enzyme that is commonly measured to illustrate a dysfunction of the hepatobiliary system. Low levels of ALP may be a helpful guide to a suspected diagnosis of zinc deficiency. Levels of serum superoxide dismutase, as well as metallothionein concentrations within erythrocytes can also be measured to arrive at a diagnosis of zinc deficiency (Maverakis *et al.*, 2007). A skin biopsy may aid in the diagnosis of Acrodermatitis enteropathica (Perafan-Riveros *et al.*, 2002).

11.8 Treatment

Treatment of zinc deficiency states, both inherited and acquired, is centered upon the supplementation of zinc orally. Zinc is available in various formulations including, zinc sulfate, zinc gluconate and zinc acetate. They differ in the amount of elemental zinc that they contain and there is no clear evidence of one form being superior over the other in terms of absorption, tolerance and bioavailability. Most physicians seem to prefer the use of zinc sulfate.

For AE zinc replacement is started at 5-10 mg/kg/d of elemental zinc initially. Response to therapy is generally rapid and occurs between 24-48 hours. Complete response is seen within 2-4 weeks. Once response is achieved the patients are placed on zinc maintenance at a dose of about 1-2 mg/kg/d of elemental zinc indefinitely. Plasma zinc levels should be monitored every 3-6 months.

For all other conditions clearly linked to zinc deficiency, zinc can be supplemented in the form of oral zinc sulfate at a dose of 110-220 mg (25-50 mg elemental zinc) up to three times a day.

Side effects of zinc supplementation depend upon the formulations, but toxicity is possible and can lead to gastrointestinal symptoms, including diarrhea, nausea, vomiting, hemorrhage and renal failure.

Zinc can inhibit copper absorption and may lead to hypocupemia. This effect is used for therapeutic purpose in patients suffering from Wilson disease. However, in normal individuals reduced copper levels may eventually lead to anemia and neutropenia (Maverakis *et al.*, 2007; Perafan-Riveros *et al.*, 2002).

References

Abeck, D., Korting, H.C. and Braun-Falco, O., 1992. Folliculitis decalvans. Long-lasting response to combined therapy with fusidic acid and zinc. Acta Dermato-Venereologica 72, 143-145.

Al-Gurairi, F.T., Al-Waiz, M. and Sharquie, K.E., 2002. Oral zinc sulphate in the treatment of recalcitrant viral warts: randomized placebo-controlled clinical trial. British Journal of Dermatology 146, 423-431.

Arroyo, J.F. and Cohen, M.L., 1993. Improvement of yellow nail syndrome with oral zinc supplementation. Clinical and Experimental Dermatology 18, 62-64.

Bae, Y.S., Hill, N.D., Bibi, Y., Dreiher, J. and Cohen, A.D., 2010. Innovative uses for zinc in dermatology. Dermatologic Clinics 28, 587-597.

Barnes, P.M. and Moynahan, E.J., 1973. Zinc deficiency in acrodermatitis enteropathica: multiple dietary intolerance treated with synthetic diet. Proceedings of the Royal Society of Medicine 66, 327-329.

Bolotin, D., Racz, M.I., Petronic-Rosic, V. and Sethi, A., 2010. Acquired combined nutritional deficiency presenting as psoriasiform dermatitis. Journal of the American Academic of Dermatology 62, 715-718.

Brocard, A., Knol, A.C., Khammari, A. and Dreno, B., 2007. Hidradenitis suppurativa and zinc: a new therapeutic approach. A pilot study. Dermatology 214, 325-327.

Dreno, B., Amblard, P., Agache, P., Sirot, S. and Litoux, P., 1989. Low doses of zinc gluconate for inflammatory acne. Acta Dermato-Venereologica 69, 541-543.

Dreno, B., Moyse, D., Alirezai, M., Amblard, P., Auffret, N., Beylot, C., Bodokh, I., Chivot, M., Daniel, F., Humbert, P., Meynadier, J. and Poli, F., 2001. Multicenter randomized comparative double-blind controlled clinical trial of the safety and efficacy of zinc gluconate versus minocycline hydrochloride in the treatment of inflammatory acne vulgaris. Dermatology 203, 135-140.

El Darouti, M. and Abu el Ela, M., 1996. Necrolytic acral erythema: a cutaneous marker of viral hepatitis C. International Journal of Dermatology 35, 252-256.

Goransson, K., Liden, S. and Odsell, L., 1978. Oral zinc in acne vulgaris: a clinical and methodological study. Acta Dermato-Venereologica 58, 443-448.

Hanas, J.S., Hazuda, D.J., Bogenhagen, D.F., Wu, F.J.-H. and Wu, C.-W., 1983. Xenopus transcription factor A requires zinc for binding to the 5 S RNA gene. Journal of Biological Chemistry 258, 14120-14125.

Hillstrom, L., Pettersson, L., Hellbe, L., Kjellin, A., Leczinsky, C.G. and Nordwall, C., 1977. Comparison of oral treatment with zinc sulphate and placebo in acne vulgaris. British Journal of Dermatology 97, 681-684.

Hoque, S.R., Mansour, S. and Mortimer, P.S., 2007. Yellow nail syndrome: not a genetic disorder? Eleven new cases and a review of the literature. British Journal of Dermatology 156, 1230-1234.

Isard, O., Knol, A.C., Aries, M.F., Nguyen, J.M., Khammari, A., Castex-Rizzi, N. and Dreno, B., 2010. Propionibacterium acnes Activates the IGF-1/IGF-1R System in the Epidermis and Induces Keratinocyte Proliferation. Journal of Investigative Dermatology 131, 59-66.

James, K.A., Burkhart, C.N. and Morrell, D.S., 2009. Emerging drugs for acne. Expert Opinions on Emerging Drugs 14, 649-659.

Kay, R.G., Tasman-Jones, C., Pybus, J., Whiting, R. and Black, H., 1976. A syndrome of acute zinc deficiency during total parenteral alimentation in man. Annals of Surgery 183, 331-340.

Lansdown, A.B., Mirastschijski, U., Stubbs, N., Scanlon, E. and Agren, M.S., 2007. Zinc in wound healing: theoretical, experimental, and clinical aspects. Wound Repair and Regeneration 15, 2-16.

Liden, S., Goransson, K. and Odsell, L., 1980. Clinical evaluation in acne. Acta Dermato-Venereologica Suppl (Stockh) Suppl 89, 47-52.

Mahajan, P.M., Jadhav, V.H., Patki, A.H., Jogaikar, D.G. and Mehta, J.M., 1994. Oral zinc therapy in recurrent erythema nodosum leprosum: a clinical study. Indian Journal of Leprosy 66, 51-57.

Mathur, N.K., Bumb, R.A. and Mangal, H.N., 1983. Oral zinc in recurrent Erythema Nodosum Leprosum reaction. Lepr India 55, 547-552.

Mathur, N.K., Bumb, R.A., Mangal, H.N. and Sharma, M.L., 1984. Oral zinc as an adjunct to dapsone in lepromatous leprosy. International journal of leprosy and other mycobacterial diseases 52, 331-338.

Maverakis, E., Fung, M.A., Lynch, P.J., Draznin, M., Michael, D.J., Ruben, B. and Fazel, N., 2007. Acrodermatitis enteropathica and an overview of zinc metabolism. Journal of American Academic Dermatology 56, 116-124.

Moneib, H.A., Salem, S.A. and Darwish, M.M., 2010. Evaluation of zinc level in skin of patients with necrolytic acral erythema. British Journal of Dermatology 163, 476-480.

Najarian, D.J., Najarian, J.S., Rao, B.K. and Pappert, A.S., 2008. Hypozincemia and hyperzincuria associated with necrolytic acral erythema. International Journal of Dermatology 47, 709-711.

Najim, R.A., Sharquie, K.E. and Farjou, I.B., 1998. Zinc sulphate in the treatment of cutaneous leishmaniasis: an *in vitro* and animal study. Memórias do Instituto Oswaldo Cruz 93, 831-837.

Orris, L., Shalita, A.R., Sibulkin, D., London, S.J. and Gans, E.H., 1978. Oral zinc therapy of acne. Absorption and clinical effect. Archives of Dermatology 114, 1018-1020.

Perafan-Riveros, C., Franca, L.F., Alves, A.C. and Sanches Jr., J.A., 2002. Acrodermatitis enteropathica: case report and review of the literature. Pediatric Dermatology 19, 426-431.

Plum, L.M., Lothar, R. and Haase, H., 2010. The essential toxin: impact of zinc on human health. International Journal of Environmental Research on Public Health 7, 1342-1365.

Prasad, A.S., 2009. Zinc: role in immunity, oxidative stress and chronic inflammation. Current Opinion in Clinical Nutrition and Metabolic Care 12, 646-652.

Prasad, A.S., Halsted, J.A. and Nadimi, M., 1961. Syndrome of iron deficiency anemia, hepatosplenomegaly, hypogonadism, dwarfism and geophagia. American Journal of Medicine 31, 532-546.

Rebello, T., Atherton, D.J. and Holden, C., 1986. The effect of oral zinc administration on sebum free fatty acids in acne vulgaris. Acta Dermato-Venereologica 66, 305-310.

Sardana, K. and V.K. Garg, 2010. "An observational study of methionine-bound zinc with antioxidants for mild to moderate acne vulgaris." Dermatology and Therapy 23, 411-418.

Sadighha, A., 2009. Oral zinc sulphate in recalcitrant multiple viral warts: a pilot study. Journal of the European Academy of Dermatology and Venereology 23, 715-716.

Sharquie, K.E. and Najim, R.A., 2004. Disseminated cutaneous leishmaniasis. Saudi Medical Journal 25, 951-954.

Sharquie, K.E., Najim, R.A., Al-Dori, W.S. and Al-Hayani, R.K., 2006a. Oral zinc sulfate in the treatment of Behcet's disease: a double blind cross-over study. Journal of Dermatology 33, 541-546.

Sharquie, K.E., Najim, R.A., Al-Hayani, R.K., Al-Nuaimy, A.A. and Maroof, D.M., 2008. The therapeutic and prophylactic role of oral zinc sulfate in management of recurrent aphthous stomatitis (ras) in comparison with dapsone. Saudi Medical Journal 29, 734-738.

Sharquie, K.E., Najim, R.A. and Al-Salman, H.N., 2006b. Oral zinc sulfate in the treatment of rosacea: a double-blind, placebo-controlled study. International Journal of Dermatology 45, 857-861.

Sharquie, K.E., Najim, R.A., Farjou, I.B. and Al-Timimi, D.J., 2001. Oral zinc sulphate in the treatment of acute cutaneous leishmaniasis. Clinical and Experimental Dermatology 26, 21-26.

Trumbo, P., Yates, A.A., Schlicker, S. and Poos, M., 2001, Dietary Reference Intakes: Vitamin A, Vitamin K, Arsenic, Boron, Chromium, Copper, Iodine, Iron, Manganese, Molybdenum, Nickel, Silicon, Vanadium, and Zinc. Journal of the American Dietetic Association 101, 294-301.

Tuerk, M.J. and Fazel, N., 2009. Zinc deficiency. Current Opinion in Gastroenterology 25, 136-143.

United States Department of Agriculture (USDA), 2010. USDA Nutrient Database for Standard Reference. Release 23, available at: http://www.nal.usda.gov/fnic/foodcomp/search/.

Verma, K.C., Saini, A.S. and Dhamija, S.K., 1980. Oral zinc sulphate therapy in acne vulgaris: a double-blind trial. Acta Dermato-Venereologica 60, 337-340.

Weimar, V.M., Puhl, S.C., Smith, W.H. and tenBroeke, J.E., 1978. Zinc sulfate in acne vulgaris. Archive of Dermatology 114, 1776-1778.

Yaghoobi, R., Sadighha, A. and Baktash, D., 2009. Evaluation of oral zinc sulfate effect on recalcitrant multiple viral warts: a randomized placebo-controlled clinical trial. Journal of American Academic Dermatololgy 60, 706-708.

Zedek, D., Morrell, D.S., Graham, M., Goodman, D. and Groben, P., 2008. Acrodermatitis enteropathica-like eruption and failure to thrive as presenting signs of cystic fibrosis. Journal of American Academic Dermatology 58, S5-8.

Key facts

- Although iron is an essential trace metal, the regulatory mechanisms concerning iron uptake from the intestinal mucosa or through the skin are not known well.
- The regulatory mechanisms concerning the loss of iron in sweat and through hair and nail growth are not known well.
- Iron facilitates oxygen transport, intracellular oxidation-reduction process, and mitochondrial respiration.
- Iron deficiency elicits great damage to skin cell function.
- Excess iron leads to hyperpigmentation of the skin, hair and nails
- Ferrous ferric chloride (FFC), is important dimer of Fe (II) and Fe (III).

Summary points

- Iron and its special form such as FFC are very important for normal structure and function of the skin, hair and nails.
- The skin is involved in regulating the body iron content.
- Excess iron is lost in sweat and through hair and nail growth.
- Low concentration of FFC can stimulate the function of mammalian skin.
- FFC can stimulate the proliferation of human fibroblasts synergistically with low-molecular weight collagen, and also stimulate the differentiation of mouse melanocytes additionally with Chinese wolfberry and Siberian ginseng.
- FFC can stimulate the proliferation and differentiation of mouse and human skin cells from a distance.

12. Iron and skin health: iron stimulates skin function

T. Hirobe
Radiation Effect Mechanisms Research Group, National Institute of Radiological Sciences, Anagawa, Inage-ku, Chiba 263-8555, Japan; thirobe@nirs.go.jp

Abstract

Iron (Fe) and the Fe (II) and Fe (III) dimer are important for normal development of the skin and its appendages such as hair and nails. Although iron is an essential trace metal, the mechanisms regulating iron uptake from the intestinal mucosa and the skin are not well understood. The skin is involved in regulating bodily iron content. Excess iron is lost through perspiration as well as hair and nail growth, but the mechanisms controlling these actions are also not well known. The human body contains 3-5 g of iron, of which up to 75% may be bound in haemoglobin, with lesser amounts in ferritin, myoglobin and transferrin. However, the minimal levels necessary for the structure and function of the skin seem to be quite low. A normal iron concentration is required for maintaining healthy epidermis, dermis, hair and nails. Fe (II) and Fe (III) dimers such as ferrous ferric chloride are very important for regulating proliferation and differentiation of mouse and human skin cells. Low ferrous ferric chloride concentrations can stimulate mammalian skin functions. Moreover, ferrous ferric chloride can stimulate fibroblast proliferation synergistically with low-molecular-weight collagen and melanocyte differentiation in combination with herbal medicines. Most importantly, ferrous ferric chloride can stimulate proliferation and differentiation of mouse and human skin cells from a distance without being added to the culture medium. These results suggest that ferrous ferric chloride is involved in regulating skin homeostasis through the regulation of the skin-cell turnover.

Keywords: keratinocyte, fibroblast, melanocyte, hair, epidermis

Abbreviations

bFGF	Basic fibroblast growth factor
DBcAMP	Dibutyryl adenosine 3':5'-cyclic monophosphate
DDW	Deionised and distilled water
FFC	Ferrous ferric chloride
MDM	Melanoblast-defined medium
MDMD	Melanocyte-proliferation medium
MDMDF	Melanoblast-proliferation medium
MDMF	Fibroblast-proliferation medium
MDMM	Melanocyte-differentiation medium
MSH	Melanocyte-stimulating hormone
NMF	Natural moisturizing factor
PKA	Protein kinase A
PKC	Protein kinase C
ROS	Reactive oxygen species
SOD	Superoxide dismutase

12.1 Introduction

The skin is the largest organ of the human body. It possesses many functions such as protection of the human body from the invasion of chemical substances, bacteria and viruses by keratinization (barrier function); control of body temperature by contraction and expansion of blood vessels or capillaries; a role as a sensory organ (detecting external stimuli and sending the related information to the brain); absorption of oily and other substances from the epidermis; secretion (sweat or lipids are secreted from sweat glands or sebaceous glands); cellular immunity (protect the skin from invasion of alien substances); respiration (oxygen is incorporated into many capillaries distributed in the skin) and covering the internal organs (the integrity of the skin is supported by collagen and elastin fibers in the dermis). Thus, the skin is a very complex organ of the human body. These skin functions are supported by various cells, including epithelial tissue cells such as keratinocytes and sebaceous gland cells, connective tissue cells such as fibroblasts and endothelial cells, immune system cells such as T cells, leucocytes and Langerhans cells and cells of ectodermal origin such as melanocytes and Merkel cells (Hirobe, 2005).

Iron is the most abundant trace metal in the human body. Approximately, 3-5 g of iron is present in most healthy adults. Some ferrous iron (Fe^{2+}) is absorbed from the skin, but the major portion is obtained from diet. Iron absorption in the small intestine is not an efficient process, but adult men are believed to absorb about 1 mg of iron into the plasma daily from diet, while women absorb about 2 mg (Scrimshaw and Young, 1976). An adult man with an average weight of 70 kg requires about 10 mg/day of iron to maintain a healthy condition (Consolazio *et al.*, 1963). The amount of iron absorbed from diet varies greatly with the form in which the iron is present in the intestinal mucosa (Fe^{2+} is more easily absorbed than ferric iron, Fe^{3+}), concentrations of other metallic elements or chelating substances and the requirement of the body at the time. Soluble complexes of iron and lipid are more easily absorbed (Taylor *et al.*, 1974). The mechanisms of iron uptake from the intestine and skin are largely unknown. Vitamin C is an important factor involved in the absorption and metabolism of iron in the human body; hypo-vitaminosis has been identified as a cause of follicular keratitis, skin hemorrhages and abnormal hair and epidermal dysgenesis (Green *et al.*, 1968). Iron plays an important role in the development, maintenance and function of the skin, in addition to its major function in the metabolism and mobilization of oxygen radicals (Sato, 1991). The human skin surface may permit the absorption of some inorganic and organic substances even in very low concentrations. In the case of iron, where optimal concentrations required for skin function are very low, any absorption may influence the appearance of the skin and the balance of trace metals present (Lansdown, 2001). There is evidence to show that relative concentrations of iron, zinc, calcium, copper and magnesium present in the skin are specific for age, sex and race as well as body region (Lansdown, 2001).

This review focuses on the importance of iron in the skin homeostasis and includes a detailed discussion of how the appearance of the skin, hair and nails can indicate iron excess or deficiency. A recent study undertaken by our laboratory on the role of a special aqueous form of iron, a dimer of ferrous chloride and ferric chloride named FFC, in the regulation of proliferation and differentiation of mouse and human skin cells is also reported.

12.2 Iron is an important factor for skin cell function

The transfer of iron into skin cells is believed to occur from the plasma pool in the human body. Some iron is probably re-absorbed into the plasma from epidermal cells for storage. However, 6-10 mg/l of iron may be lost daily in sweat secreted from the apocrine gland (Adams *et al.*, 1980). The iron content in the human skin varies with age, sex, race, skin region and health condition. Studies using dogs have shown that skin regions with the greatest iron concentrations are those with the highest vascularity, suggesting that a large proportion of iron is present within blood vessels and red blood cells (Lansdown, 2001). Keratinocytes, melanocytes, dermal papillae fibroblasts, nail plate epithelial cells, sebaceous gland cells, apocrine gland cells and dermal macrophages seem to be particularly sensitive to iron deficiencies. Severe iron deficiency, such as anaemia can be recognized by fatigue, eczema in the corners of the mouth, skin irritation, skin itching, hair loss and nail malformation. Thus, iron is a very important trace metal for the structure and function of the skin.

12.3 Iron and the epidermis

The epidermis is a stratified squamous epithelium consisting mainly of cells with two different origins, keratinocytes and melanocytes. Keratinocytes comprise the bulk of the epthelium, undergo keratinization and form the dead superficial layers (mostly keratin) of the skin. These superficial keratinized cells continuously desquamate from the surface and are replaced by cells derived from mitotic cells in the lowest layer (basal layer). The cells are displaced to successively higher levels by the new cell populations below them. As they move upward, they produce keratin and accumulate it in the cytoplasm (spinous layer and granular layer), and finally, almost all cells are occupied by keratin (keratinized layer or cornified layer). For a long time the keratinized layer was believed to possess no important biological role in the function of the epidermis. However, the layer is now known to play a very important role in retaining water by the action of seramid or NMF such as caspase-14 (Denecker *et al.*, 2007). Melanocytes are cells located mainly in the basal layer that do not keratinize, but produce melanin pigment (Hirobe, 2005). Most of these melanocytes migrate from the epidermis to the hair follicles and colonize hair bulb melanocytes in hairy skin (Hirobe, 2005). In the glabrous skin of the ear, nose, foot and tail, numerous differentiated melanocytes are found even in the epidermis of adult mice. However, in the epidermis of hairy skin, epidermal melanocytes are found only during the early weeks after birth (Hirobe, 2005). On the other hand, in human skin, an epidermal melanin unit consisting of keratinocytes and basal layer melanocytes is important for the protection of skin against the harmful effects of ultraviolet light (Hirobe, 2005).

Iron serves a primary function in mammalian cells by facilitating oxygen transport, intracellular oxidation-reduction and mitochondrial respiration (Sato, 1991). This is related to the regulation of mitochondrial DNA synthesis in metabolically active cells, such as basal keratinocytes, dermal papillae fibroblasts and nail bed epithelial cells, which continuously undergo a sequence of proliferation and senescence (Lansdown, 2001). Transferrin-bound iron is absorbed into keratinocytes by receptor-mediated endocytosis. The iron is then released for incorporation in ribonucleotide reductase and other enzymes (Lasky *et al.*, 1988). Other *in vitro* studies have suggested that ferrienzymes are important for maintaining living cells and transferrin inactivation through competitive binding of toxic metals inhibits cell proliferation in iron deficiency (Chitambar and Seligman, 1986).

12.4 Iron and skin appendages

Growth and differentiation of specialized skin appendages such as hair and nails is affected by iron content (Comaish, 1971). The texture and color of hair are good indicators of the nutritional state, especially iron concentration. Iron deficiency seems to induce hair loss or premature greying as well as hair follicle atrophy. Although blood ferritin levels of 40 mg/l or greater may be necessary for hair growth, the amount of iron required for hair growth seems to vary with age, sex and race (Rushton and Ramsey, 1992). Iron deficiency seems to induce hair loss along with splitting, narrowing and drying of the hair shaft. Twenty percent of 96 patients showing diffuse hair loss were reported to exhibit iron deficiency without anaemia (Hard, 1963). The changes were transitory and repletion of iron levels restored normal hair growth (Hard, 1963). Rushton (1990) reported that 72% of women with iron deficiency showed an increased number of hair follicles in telogen (resting stage). Hair color is also influenced by iron concentration. Higher iron concentrations have been reported in humans with red or dark brown hair (Dutcher and Rothman, 1951). However, in patients with haematochromatosis, 75% were found to possess hair loss and epidermal atrophy, suggesting that very high iron concentrations are harmful to hair (Blankenship, 1971) and upset the balance necessary for hair growth (Lansdown, 2001). Iron deficiency leads to transitory hypochromasia or loss of color due to reduced or no melanin pigments in the hair shaft (Sato, 1991). The iron content of the hair of a woman who regularly washed her hair with water with high iron content increased and her hair became darker (Platchek and Lubach, 1989). Occupational exposure to iron may lead to hyperpigmentation of the skin, hair and nails. Although short-term exposure seems to possess low toxicity, chronic exposure may possess more lasting effects (Hare, 1951).

Nail differs from hair in that it is composed of hard keratin. Nails are a sensitive indicator of mineral deficiency. Nail malformation, inflammation of the nail plate and reduced growth are associated with iron deficiency (Beveridge *et al.*, 1965; Sobolewski *et al.*, 1978). On the other hand, exposure of human nails to excess iron such as 10% ferric chloride leads to rapid oxidation of tyrosine and darkening of the nail through increased melanogenesis (Harrison and Cremona, 1972). Although wide variations exist, levels of iron and other minerals present in the human nail measured by X-ray diffraction and atomic absorption spectrophotometry may reflect the nutritional state of the human body (Sobolewski *et al.*, 1978). Leuconychia (white nail) and pale nail coloration in anaemic patients seem to be due to reduced blood flow and heme deficiency in the nail bed, while koilonychia (nail flattening) may be related to defects in cystine metabolism in the nail plate (Jalili and Al-Kassab, 1959). Patients with haemochromatosis show abnormally high iron concentrations in nails and hyperpigmentation of the skin. The patients often show a loss of transparency in their nails and deformities of the nail bed attributable to iron toxicity (Fletcher *et al.*, 1999). Moreover, exposure to high iron concentrations may cause a discoloration of the nails without obvious signs of toxicity (Beveridge *et al.*, 1965; Sobolewski, *et al.*, 1978).

12.5 The Fe (II) and Fe (III) dimer and skin function

12.5.1 Effects of FFC on proliferation and differentiation of mouse skin cells *in vitro*

FFC® (Akatsuka Corporation, Mie, Japan) is an aqueous solutions of the dimer. FFC is an aqueous complex of ferrous chloride and ferric chloride (Sugi and Yamashita, 1991), which functions in both oxidation and reduction (Figure 12.1). FFC is thought to stimulate the cellular function of living organisms by regulating the oxidation and reduction status of cells. Water containing FFC can be obtained by soaking FFC ceramic beads (Akatsuka) in water. These FFC ceramic beads are prepared by heating hardened soil with a FFC solution to 800~1,200 °C. FFC ceramic beads are made of porous ceramic treated with FFC solution. FFC ceramic-treated water (FFC water) contains various minerals, including Ca, Mg, K, Na, Fe and Cu, in addition to FFC.

Brain (2006), using neutron activation analysis, reported that about 10 mg of Fe was released from 1 g of FFC ceramic beads after soaking the ceramic beads in water for 1 h. He also reported that haemoglobin content and red blood cell numbers in rats were increased by daily administration

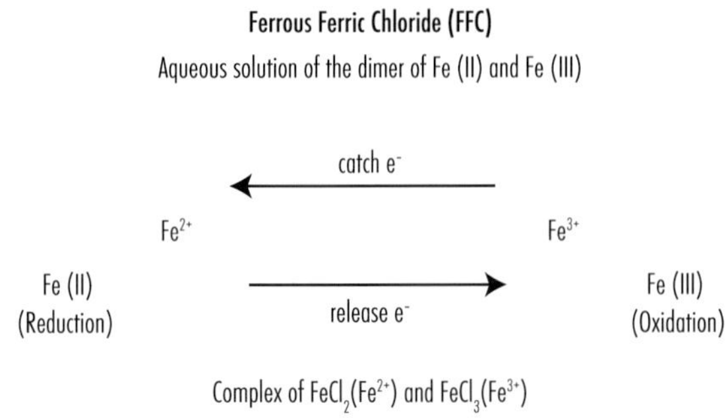

Figure 12.1. Diagram showing the action of FFC. FFC contains Fe^{2+} and Fe $^{3+}$. Fe^{2+} releases electron and changes to Fe^{3+} (reduction), and Fe^{3+} catches electron and changes to Fe^{2+} (oxidation).

of FFC water. His report suggests that FFC is released from FFC ceramic beads and that FFC stimulates biological activities of red blood cells.

FFC was first manufactured as follows; ferric chloride ($FeCl_3 \cdot 6H_2O$) was dissolved in a mixed solution of ammonium formate (2 M), hydroxylamine (1 M) and formamide (1 M) to a final concentration of 1 M and then diluted with DDW to 10^{-8}-10^{-14} mM. Another $FeCl_3 \cdot 6H_2O$ (1 g/10 ml solution) was dissolved in the solution and then gradually heated at 100 °C until FFC could be crystallized (Sugi and Yamashita, 1991). The proportion of Fe (II) to Fe (III) was 4:6 (for a 10^{-8} mM solution), 6:4 (for a 10^{-12} mM solution) and 7:3 (for a 10^{-14} mM solution). An analysis using ion-exchange chromatography and X-ray diffraction suggested that Fe (II) and Fe (III) formed a dimer.

In our study, we investigated the effects of FFC on mouse and human skin cells in culture. One day before the initiation of culture experiments, 40 g of FFC ceramic beads was soaked in 2 l of DDW for 24 h in a styrene foam box at room temperature (this DDW is henceforth referred to as FFC-DDW (20 g/l)). In FFC-DDW, FFC was present in addition to various minerals such as Ca, Mg, K, Na, Fe and Cu (Brain, 2006). The culture medium (Ham's F-10) was prepared using FFC-DDW. For culturing mouse skin cells, five serum-free primary culture media were prepared (Table 12.1).

When epidermal cell suspensions derived from newborn B10 mice were cultured in MDM, MDMM, MDMD and MDMDF with FFC, large keratinocyte colonies with numerous enlarged keratinocytes were observed. The number of keratinocytes per colony at 2 days was doubled in all cases, suggesting that FFC stimulates keratinocyte proliferation and differentiation.

Since MDM does not contain melanogens, it can be used to determine whether FFC stimulates differentiation of mouse melanoblasts in primary cultures of epidermal cell suspensions. We found that although melanoblasts did not increase in number, the percentage of differentiated melanocytes in the melanoblast-melanocyte population increased (>40%) (Hirobe, 2007).

Table 12.1. The five kinds of culture media used in this study.

Culture media	Name	Component[1]	Melanoblasts or melanocytes at 14 days
MDM	Melanoblast-defined medium	Ins, BSA, EA, PEA, SE	Pure melanoblasts
MDMM	Melanocyte- differentiation medium	MDM + α-MSH (100 nM)	Pure melanocytes
MDMF	Fibroblast-proliferation medium	MDM + bFGF (2.5 ng/ml)	Pure fibroblasts
MDMD	Melanocyte- proliferation medium	MDM + DBcAMP (0.5 mM)	Pure melanocytes
MDMDF	Melanoblast-proliferation medium	MDMD + bFGF (2.5 ng/ml)	Pure melanoblasts (90%) and melanocytes (10%)

Abbreviations: Ins: insulin, BSA: bovine serum albumin. EA: ethanolamine, PEA: phosphoethanolamine, SE: sodium selenite, α-MSH: melanocyte-stimulating hormone, DBcAMP: dibutyryl adenosine 3′:5′-cyclic monophosphate, bFGF: basic fibroblast growth factor.

[1]All these components were added to Ham's F-10 medium.

α-MSH stimulates melanocyte differentiation *in vivo* and *in vitro* (Hirobe, 2005). MDMM supplemented with α-MSH can determine whether FFC stimulates melanocyte differentiation in cooperation with α-MSH. When epidermal cell suspensions derived from neonatal mouse skin were cultured in MDMM with FFC, melanoblasts and melanocytes failed to increase in number, but melanocyte differentiation was stimulated. In addition, dendritogenesis, cell expansion and pigmentation were increased (Hirobe, 2007).

Basic FGF stimulates fibroblast proliferation (Hirobe, 2009b), and MDMF supplemented with bFGF can be used to determine whether FFC stimulates proliferation of dermal fibroblasts in cooperation with bFGF. When dermal cell suspensions derived from neonatal mouse skin were cultured in MDMF with FFC, fibroblasts increased in number. In addition, projection formation and cell expansion were increased.

DBcAMP stimulates both proliferation and differentiation of cultured melanocytes in the presence of keratinocyte-derived factors (Hirobe, 2005). DBcAMP-containing MDMD can determine whether FFC stimulates melanocyte proliferation and differentiation in cooperation with DBcAMP. When epidermal cell suspensions were cultured in MDMD with FFC, the number of mouse melanocytes doubled. Dendritogenesis, cell expansion and pigmentation were also increased by FFC (Hirobe, 2007).

In the presence of DBcAMP, bFGF stimulates melanoblast proliferation in cooperation with keratinocyte-derived factors (Hirobe, 2005). Use of MDMDF can determine whether FFC stimulates melanoblast proliferation in cooperation with bFGF and DBcAMP. When epidermal cell suspensions were cultured in MDMDF with FFC, the number of mouse melanoblasts doubled (Hirobe, 2007).

After depletion of keratinocytes (14 days in primary culture), FFC induced differentiation of melanocytes cultured in MDM and, in addition, FFC induced the proliferation of melanocytes cultured in MDMD as well as the proliferation of melanoblasts and melanocytes cultured in MDMDF (Hirobe, 2007). These results suggest that FFC can directly stimulate melanocyte proliferation and differentiation.

Melanocyte proliferation and differentiation can be induced by several growth factors, cytokines and hormones (Hirobe, 2005). It should be emphasized that FFC can stimulate proliferation of melanoblasts and melanocytes and, in addition, induce the differentiation of melanocytes in the absence of these growth factors or cytokines. Proliferation and differentiation of mammalian melanocytes is stimulated through PKC as well as MAP kinase and PKA signaling pathways (Hirobe, 2005). FFC is assumed to activate these signaling pathways even in the absence of growth factors and cytokines.

The extent to which FFC stimulates proliferation of keratinocytes is similar to that of melanoblasts/melanocytes and fibroblasts (approximately 2-fold increase), suggesting that proliferation of the three types of cells constituting the skin is equally stimulated by FFC and that skin homeostasis is controlled by FFC. Moreover, FFC stimulated differentiation of keratinocytes, melanocytes and fibroblasts, suggesting that renewal of cells constituting the skin is accelerated by FFC. Thus, FFC is assumed to ensure healthy skin by replacing old keratinocytes, melanoblasts/melanocytes and fibroblasts with new ones.

12.5.2 Effects of FFC-containing drinks on proliferation and differentiation of mouse melanocytes

Pairogen® (Akatsuka) is a refreshing drink that consists of 90% FFC water in combination with other purified ingredients, such as rice vinegar, Japanese apricot vinegar, apple vinegar, persimmon vinegar, vitamins B2, B6 and C, glucose, fructose, citric acid and malic acid. Pairogen Gold® (Akatsuka) is another refreshing drink consisting of 90% FFC water; the remaining 10% consists of other purified ingredients, such as apple vinegar, vitamins B1, B2, B6 and C, glucose, fructose, citric acid, malic acid, polydextrose and extracts of Korean ginseng (*Panax ginseng*), Chinese wolfberry (*Lycium chinense*) and Siberian ginseng (*Eleutherococcus senticosus*).

We studied the effects of these drinks on skin cells and found that keratinocyte proliferation and differentiation as well as melanoblast proliferation in cells cultured in MDM supplemented with Pairogen and Pairogen Gold were similar to those observed with FFC alone. However, melanocyte differentiation was greatly stimulated by Pairogen Gold (Hirobe, 2009a). To clarify which herbal medicines in Pairogen Gold were responsible for this synergistic effect, extracts of Korean ginseng, Chinese wolfberry and Siberian ginseng were added to MDM with FFC. Addition of any one of the three extracts to MDM with FFC, or a combination of Korean ginseng plus Chinese wolfberry or Korean ginseng plus Siberian ginseng with FFC failed to increase the degree of melanocyte differentiation. In contrast, the combination of Chinese wolfberry plus Siberian ginseng with FFC markedly increased the percentage of melanocytes to a level similar to that obtained with Pairogen Gold (10^{-2}). Three combinations with FFC showed a melanocyte differentiation level similar to that induced by the two combinations of Chinese wolfberry plus Siberian ginseng with FFC. The percentages of melanocytes obtained with Pairogen Gold, Chinese wolfberry plus Siberian ginseng, and Korean ginseng plus Chinese wolfberry plus Siberian ginseng were significantly greater than those obtained with FFC alone. In contrast, single treatments and other combinations without FFC failed to exceed melanocyte differentiation level induced by FFC alone (Hirobe, 2009a).

When MDMD and MDMDF were supplemented with Pairogen and Pairogen Gold, proliferation and differentiation of mouse melanoblasts/melanocytes were stimulated. The optimal concentration of Pairogen and Pairogen Gold for stimulating the proliferation and differentiation of melanoblasts/melanocytes was 10^{-3}-10^{-4}.

12.5.3 Effects of FFC on mouse skin cells *in vivo*

FFC Super Essence Plain Type® (FFC Plain; Akatsuka) contains FFC water, various plant-derived moisturizing factors, and various plant extracts containing skin-activating factors. The other type of FFC lotion, FFC Super Essence Moisture Type® (FFC Moisture; Akatsuka) contains 3 oils (rose oil, grape seed oil and jojoba oil) as well as emulsifiers in addition to the ingredients contained in FFC Plain. The moisturizers present in the FFC lotions prevent evaporation of water, thereby allowing FFC to remain functional for a long period. To study the effects of these lotions, we applied FFC Plain and FFC Moisture (0.25 ml) to the dorsal skin of B10 mice daily from 0.5 days after birth. The epidermis of 2.5- and 5.5-day-old B10 mice treated with FFC Plain and FFC Moisture became thicker than that of control mice and contained more keratinocytes. Moreover, the number of mitotic keratinocytes in the epidermis treated with FFC Plain and FFC Moisture was greater than in control epidermis. The same tendency was observed in keratinocytes from developing hair follicles. At 5.5 days, hair follicles in the skin treated with FFC Plain and FFC Moisture were larger than those in the control skin. The number of mitotic keratinocytes per epidermis in the control skin gradually decreased after birth, whereas those in the skin treated with FFC Plain and FFC Moisture markedly increased. No difference was observed between FFC Plain and FFC Moisture. These results suggest that FFC Plain and FFC Moisture stimulate keratinocyte proliferation in the epidermis and hair follicles. Moreover, FFC Plain and FFC Moisture increased the thickness and density of the keratinized layer, suggesting that they also stimulate keratinocyte differentiation (Hirobe, 2009b).

In addition, the dorsal skin became darker than the control skin following the application of FFC Plain and FFC Moisture. The number of melanocytes in the epidermis, developing hair follicles, hair bulbs and dermis in FFC-treated skin was increased by both FFC Plain and FFC Moisture. Moreover, in melanocytes treated with FFC Plain and FFC Moisture, dendritogenesis, tyrosinase activity and cell size were greater than those in control melanocytes, suggesting that FFC Plain and FFC Moisture stimulate melanocyte differentiation through stimulation of tyrosinase activity, melanosome formation and melanosome transport. The number of melanoblasts in the skin treated with FFC Plain and FFC Moisture was similarly increased (Hirobe, 2009b).

The number of fibroblasts in the dermis gradually decreased after birth in the control skin, whereas the number of fibroblasts treated with FFC Plain and FFC Moisture increased after birth. FFC Plain and FFC Moisture almost doubled the number of fibroblasts. The mitotic index of fibroblasts similarly decreased after birth in the control skin, whereas the mitotic index of fibroblasts treated with FFC Plain and FFC Moisture increased after birth. FFC Plain and Moisture almost doubled the mitotic index of fibroblasts. In addition to stimulating fibroblast proliferation, FFC Plain and FFC Moisture increased the density of fibres around fibroblasts, suggesting that FFC Plain and FFC Moisture stimulate the production of fibres, namely fibroblast differentiation (Hirobe, 2009b).

Application of FFC Plain or FFC Moisture to the skin of B10 mice after birth stimulated the development of hair. B10 mice generally lose their hair at telogen (Hirobe, 2009b). This hair loss is caused by the expression of the alopecia gene. In most B10 mice, some or all cutaneous hair except for those on the head are lost from 2-3 weeks after birth. Regrowth of hair begins after 3-4 weeks, and complete cutaneous hair recovery is observed. In mice applied FFC Plain and FFC Moisture, almost no hair loss was observed, suggesting that the FFC lotions inhibited the alopecia-induced hair loss (Hirobe, 2009b).

12.5.4 Effects of FFC-containing drinks on proliferation and differentiation of human skin cells *in vitro*

The drink Non-calorie Pairogen® (Akatsuka) also contains FFC water as its main component (90%) in combination with other purified ingredients (10%), such as rice vinegar, Japanese apricot vinegar, apple vinegar, persimmon vinegar, lemon vinegar, vitamins B2, B6, B12 and C, citric acid, malic acid as well as low-molecular-weight collagen and hyaluronic acid. When we tested the effects of this drink on skin cells in culture, we found that Non-Calorie Pairogen doubled the number of cultured mouse primary keratinocytes and melanoblasts/melanocytes, and tripled the number of pure mouse fibroblasts observed in culture. Non-calorie Pairogen also stimulated the differentiation of mouse keratinocytes, melanocytes and fibroblasts (Hirobe, unpublished).

FFC, Pairogen, Pairogen Gold and Non-calorie Pairogen also stimulated proliferation of cultured human keratinocytes (almost double) (Hirobe, 2009c). In addition, all these drinks and FFC almost doubled the number of human melanocytes (Hirobe, 2009c). All the drinks and FFC also stimulated proliferation of human fibroblasts. Pairogen, Pairogen Gold and FFC almost doubled the number of fibroblasts. However, Non-calorie Pairogen almost tripled the number of fibroblasts. Non-calorie Pairogen was the most effective in stimulating human fibroblasts, suggesting that low-molecular-weight collagen or hyaluronic acid, which are present only in Non-calorie Pairogen, are involved in stimulating fibroblast proliferation in cooperation with FFC. To clarify which of these factors was responsible for the observed effects, low-molecular-weight collagen and hyaluronic acid were added to the culture medium with FFC. The combination of low-molecular-weight collagen with FFC dramatically increased the number of fibroblasts to a level similar to that observed using Non-calorie Pairogen. Three combinations of FFC, low-molecular-weight collagen and hyaluronic acid failed to exceed the proliferation level induced by FFC plus low-molecular-weight collagen. Low-molecular-weight collagen alone, hyaluronic acid alone and collagen plus hyaluronic acid also failed to exceed the control proliferation level. These results suggest that FFC synergistically stimulates fibroblast proliferation in combination with low-molecular-weight collagen (Hirobe, 2009c).

12.5.5 Effects of FFC on proliferation and differentiation of cultured human and mouse skin cells from a distance

FFC Plain or FFC Moisture was applied under the culture dish or on the covers of culture dishes, and the effect of FFC on proliferation and differentiation of skin cells was tested from this distance. When FFC Plain or FFC Moisture was applied under the culture dishes (1 mm away) or on the top of the covers of culture dishes (1 cm away), large keratinocyte colonies containing numerous (almost double) mouse and human (Figure 12.2) keratinocytes were observed. When the covers of the culture dishes were applied with FFC Plain or FFC Moisture, proliferation of mouse and human (Figure 12.3) melanoblasts and melanocytes were also stimulated. FFC lotions also stimulated melanocyte differentiation as well as melanogenesis and dendritogenesis. When the covers of the culture dishes were applied with FFC Plain and Moisture, proliferation of mouse and human (Figure 12.4) fibroblasts also increased. FFC Plain and FFC Moisture almost doubled the number of fibroblasts. In fibroblasts treated with FFC Plain and FFC Moisture from a distance, larger fibroblast colonies with more enlarged fibroblasts were observed, suggesting that the FFC lotions stimulated differentiation as well as proliferation of fibroblasts (Hirobe, 2011).

Human melanoblasts and melanocytes were then treated with FFC Plain and FFC Moisture from distances of 2 to 5 cm away. To perform this experiment, 1 (2 cm away), 2 (3 cm away), 3 (4 cm away) and 4 (5 cm away) dishes were placed between the dish of cultured cells and the covers of the 100 mm bacterial dishes under which FFC Plain or FFC Moisture was applied. No increase in the number of cells was observed from a distance of 2 to 5 cm away. Similarly, stimulation of melanogenesis and dendritogenesis of melanocytes by the FFC lotions was not observed from a distance of 2 to 5 cm away (Hirobe, 2011).

FFC Plain and FFC Moisture were then applied on Teflon covers (Teflon is known to inhibit the diffusion of substances). Polystyrene dishes used for culturing mouse keratinocytes and melanoblasts/melanocytes were placed between the Teflon covers and the 100 mm bacterial dishes. FFC Plain and FFC Moisture applied on the Teflon covers stimulated proliferation and differentiation of mouse keratinocytes and melanoblasts/melanocytes. Therefore, it is conceivable that FFC can stimulate proliferation and differentiation of these cells from 1 cm away through polystyrene, Teflon, air and culture medium. Thus, FFC can stimulate proliferation and differentiation of human and mouse skin cells.

All the data reported are summarized in Table 12.2. These results suggest that FFC can act on skin cells through physical factors rather than chemical factors. FFC is considered to be important for maintaining skin health and skin homeostasis. Thus, FFC-containing drinks, such as Pairogen, Pairogen Gold and Non-Calorie Pairogen, may be important for maintaining

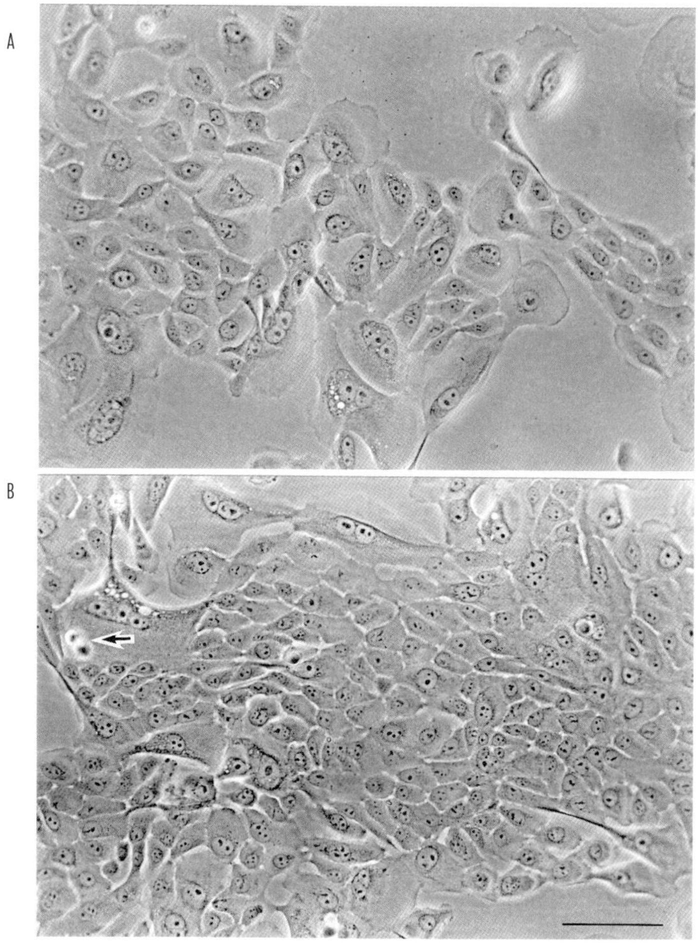

Figure 12.2. Effects of FFC Plain on the proliferation of human keratinocytes. Keratinocytes were cultured for 5 days in KG2 medium (A) or in KG2 under the influence of FFC Plain (B) that was painted on the top of the cover of the culture dish (1 cm away). FFC induced larger keratinocyte colonies with more numerous keratinocytes. Many mitotic keratinocytes (arrow) are observed. Phase-contrast microscopy. Scale bar, 100 μm.

the healthy condition of the human body through activation of keratinocyte, melanocyte and fibroblast function (Figure 12.5). Moreover, FFC may be increasingly used in cosmetics and skin care products, because it can stimulate proliferation and differentiation of human skin cells.

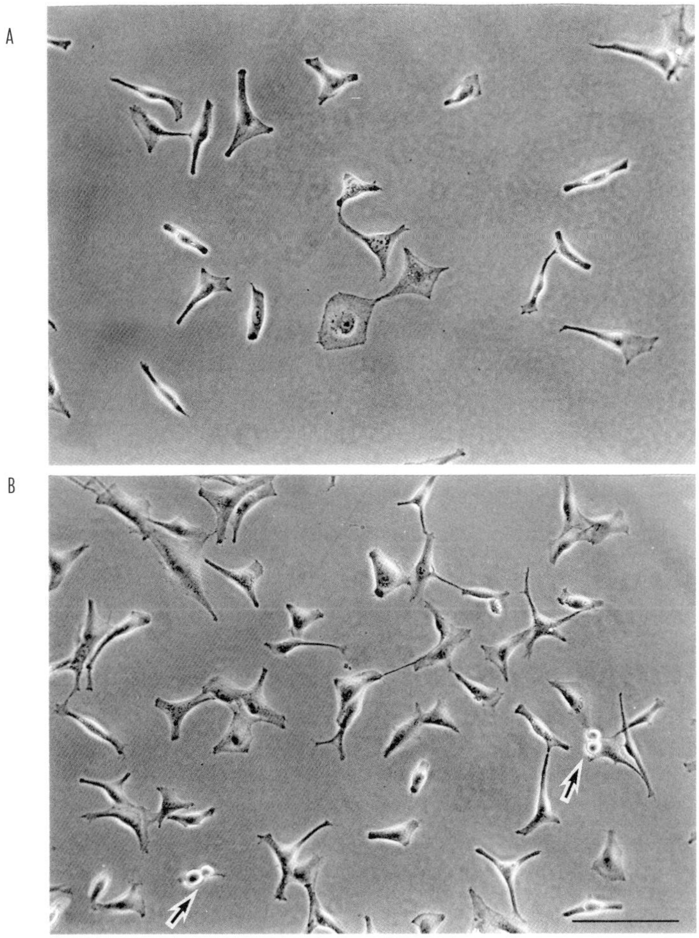

Figure 12.3. Effects of FFC on the proliferation of human melanocytes. Human melanocytes were cultured for 8 days in MDMD supplemented with transferrin (Tf) and endothelin-1 (ET-1; A) or MDMD + Tf + ET-1 under the influence of FFC Plain (B). FFC Plain stimulated the proliferation of melanocytes from 1 cm away. Many mitotic melanocytes (arrows) are observed. Phase-contrast microscopy. Scale bar, 100 μm.

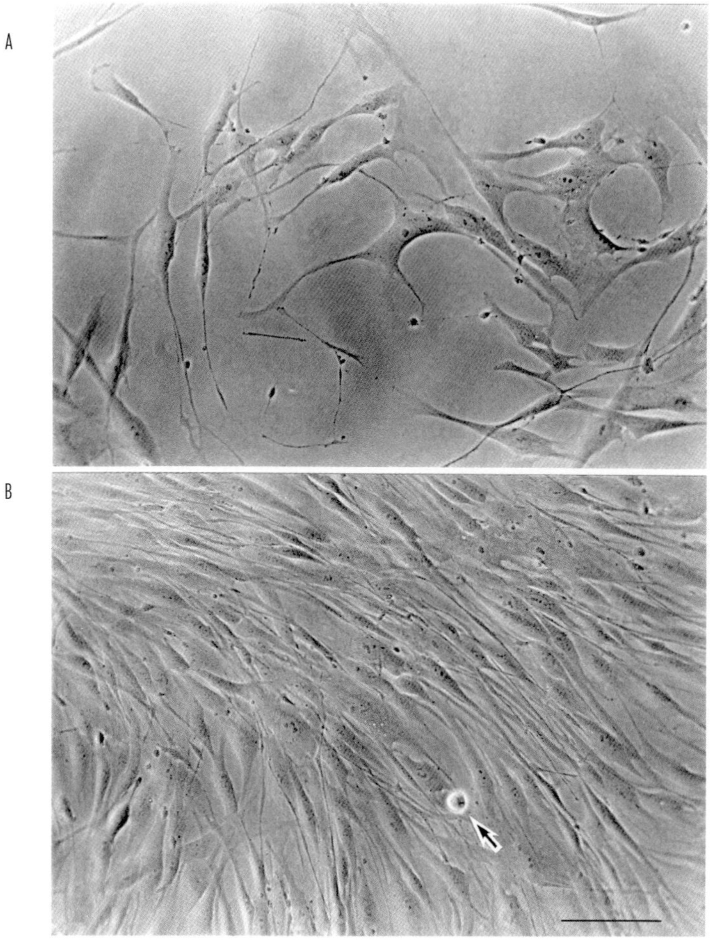

Figure 12.4. Effects of FFC on the proliferation of human fibroblasts. Fibroblasts were cultured for 1 day in 106S+ medium (A) or in 106S+ under the influence (1 cm away) of FFC Plain (B). FFC Plain dramatically increased the number of fibroblasts compared with control culture. Many mitotic fibroblasts (arrow) are observed. Phase-contrast microscopy. Scale bar, 100 μm.

Table 12.2. Effects of FFC, FFC-containing drinks and FFC-containing skin lotions on the mouse and human skin cells.

Cells	Species	FFC (culture)	Pairogen (culture)	Gold (culture)	Non-calorie (culture)	FFC lotion	
						(painted)	(1 cm away)
(A) Proliferation							
K	Mouse	+	+	+	+	+	+
	Human	+	+	+	+		+
M	Mouse	+	+	+	+	+	+
	Human	+	+	+	+		+
F	Mouse	+	+	+	++*	+	+
	Human	+	+	+	++*		+
(B) Differentiation							
K	Mouse	+	+	+	+	+	+
	Human	+	+	+	+		+
M	Mouse	+	+	++**	+	+	+
	Human	+	+	+	+		+
F	Mouse	+	+	+	+	+	+
	Human	+	+	+	+		+

FFC, Pairogen and Pairogen Gold stimulated the proliferation of keratinocytes, melanoblasts/melanocytes and fibroblasts to two-fold. Non-calorie Pairogen stimulated the proliferation of keratinocytes and melanoblasts/melanocytes to two-fold, while the drink stimulated the proliferation of fibroblasts to three-fold. FFC stimulated the proliferation of mouse and human fibroblasts synergistically with low molecular weight collagen FFC (++*). FFC, Pairogen, Pairogen Gold and Non-calorie Pairogen stimulated the differentiation of keratinocytes, melanocytes and fibroblasts. Pairogen Gold greatly stimulated the differentiation of mouse melanocytes. FFC stimulated the differentiation of melanocytes additionally with Chinese wolfberry and Siberian ginseng included in Pairogen Gold (++**).

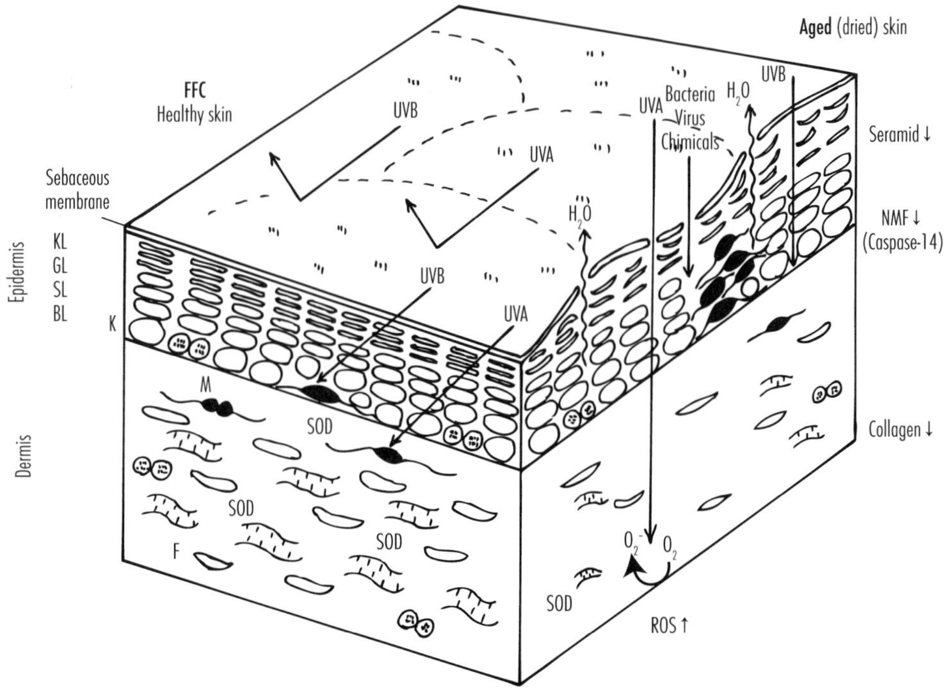

Figure 12.5. Diagram showing the effects of FFC on the human skin. In aged skin (right), the skin activity is greatly diminished owing to the reduction in the production and release of seramid, NMF such as caspase-14 and collagen synthesis. Ultraviolet radiation A (UVA), UVB, bacteria, virus and chemicals easily enter into the epidermis and dermis due to the destruction of sebaceous membrane. Water is also lost from the epidermis. Radical scavengers such as SOD greatly decrease in the skin, and consequently ROS increase follows by abnormal collagen fibers. However, in FFC-treated skin (left), proliferation and differentiation of skin cells are activated. Since sebaceous membrane is healthy, water loss is limited and the entrance of UVA and UVB is prohibited. Even if the UVA and UVB penetrate into the epidermis and dermis, melanocytes in the epidermis and dermis can inhibit the harmful effect of UVA and UVB. Healthy skin cells may produce plenty of SOD. Thus, ROS production is very limited, consequently normal collagen fibers are retained. Thus, the normal healthy condition of the skin is maintained by FFC.
KL: keratinized layer, GL: granular layer, SL: spinous layer, BL: basal layer.

References

Adams, W.S., Leslie, A. and Levin M.H., 1950. The dermal loss of iron. Proceedings of the Society of Experimental Biology and Medicine 74, 46-48.

Beveridge, B.R., Bannerman, R.M., Evanson, J.M. and Witts, L.J., 1965. Hypochromic anemia: a retrospective study and follow-up of 378 in-patients. Quarterly Journal of Medicine 34, 145-161.

Blankenship, M.L., 1971. Dysplastic hairs in iron deficiency anemia. Cutis 7, 467-468.

Brain, J., 2006. Healthy aging: environmental factors, the role of drinking water and iron. In: Proceedings of Anti-aging International Symposium and Exposition in Tokyo 2006, June 15-17, 2006, Tokyo, Japan, p. 14.

Chitambar, C.R. and Seligman, P.A., 1986. Effects of different transferrin forms on transferrin receptor expression, iron uptake, and cellular proliferation of human leukemic HL60 cells. Mechanisms responsible for the specific cytotoxicity of transferrin-gallium. Journal of Clinical Investigation 78, 1538-1546.

Comaish, S., 1971. Metabolic disorders of hair growth. British Journal of Dermatology 84, 83-86.

Consolazio, C.F., Matoush, L.O. Nelson, R.A., Harding, R.S. and Canham, J.E., 1963. Excretion of sodium, potassium, magnesium and iron in human sweat and the relation of each to balance and requirements. Journal of Nutrition 78, 407-415.

Denecker, G., Hoste, E., Gilbert, B., Hochepied, T., Ovaere, P., Lippens, S., Van den Broecke, C., Van Damme, P., D'Herde, K., Hachem, J.P., Borgonie, G., Presland, R.B., Schoonjans, L., Libert, C., Vandekerckhove, J., Gevaert, K., Vandenabeele, P. and Declercq, W., 2007. Caspase-14 protects against epidermal UVB photodamage and water loss. Nature Cell Biology 9, 666-674.

Dutcher, T.F. and Rothman, S., 1951. Iron, copper and ash content of human hair of different colors. Journal of Investigative Dermatology 17, 65-68.

Fletcher, L.M., Halliday, J.W. and Powell, L.W., 1999. Interrelationships of alcohol and iron in liver disease with particular reference to the iron-binding proteins, ferritin and transferrin. Journal of Gastroenterology and Hepatology 14, 202-214.

Green, R., Charlton, R. and Seftel, H., 1968. Body iron excretion in man. A collaborative study. American Journal of Medicine 45, 336-353.

Hare, P.J., 1951. A case of occupational iron pigmentation of the skin. British Journal of Dermatology 63, 63-66.

Hard, S., 1963. Non-anemic iron deficiency as an etiologic factor in diffuse loss of hair of the scalp of women. Acta Dermato-Venereologica 43, 562-569.

Harrison, W.W. and Clemena, G.G., 1972. Survey analysis of trace elements in human fingernails by spark source mass spectrometry. Clinica Chimica Acta 36, 485-492.

Hirobe, T., 2005. Role of keratinocyte-derived factors involved in regulating the proliferation and differentiation of mammalian epidermal melanocytes. Pigment Cell Research 18, 2-12.

Hirobe, T., 2007. Ferrous ferric chloride stimulates the proliferation and differentiation of cultured keratinocytes and melanocytes in the epidermis of neonatal mouse skin. Journal of Health Science 53, 576-584.

Hirobe, T., 2009a. Ferrous ferric chloride induces the differentiation of cultured mouse epidermal melanocytes additionally with herbal medicines. Journal of Health Science 55, 86-94.

Hirobe, T., 2009b. Ferrous ferric chloride stimulates the skin cell function and hair growth in mice. Biological and Pharmaceutical Bulletin 32, 1347-1353.

Hirobe, T., 2009c. Ferrous ferric chloride stimulates the proliferation of human skin keratinocytes, melanocytes, and fibroblasts in culture. Journal of Health Science 55, 447-455.

Hirobe, T., 2011. Stimulation of the proliferation and differentiation of skin cells by ferrous ferric chloride from a distance. Biological and Pharmaceutical Bulletin 34, 987-995.

Jalili, M.A. and Al-Kassab, S., 1959. Koilonychia and cystine content of nails. Lancet 274, 108-110.

Lansdown, A.B.G., 2001. Iron: a cosmetic constituent but an essential nutrient for healthy skin. International Journal of Cosmetic Science 23, 129-137.

Laskey, J., Webb, I. and Schulman, H.M., 1988. Evidence that transferrin supports cell proliferation by supplying iron for DNA synthesis. Experimental Cell Research 176, 87-95.

Platschek, H. and Lubach, D., 1989. Brown hair and nail discoloration by water containing iron. Hautarzt 40, 441-442.

Rushton, D.H., 1990. Biochemical and trichological characterization of diffuse alopecia in women. British Journal of Dermatology 123, 187-197.

Rushton, D.H. and Ramsey, I.D., 1992. The importance of adequate serum ferritin levels during oral cyproterone acetate and ethinyl oestradiol treatment of diffuse androgen-dependent alopecia in women. Clinical Endocrinology 36, 421-427.

Sato S., 1991. Iron deficiency: structural and microchemical changes in hair, nails, and skin. Seminars in Dermatology 10, 313-319.

Scrimshaw, N.S. and Young, V.R., 1976. The requirements of human nutrition. Scientific American 235, 51-64.

Sobolewski, S., Lawrence, A.C.K. and Bagshaw, P., 1978. Human nails and body iron. Journal of Clinical Pathology 31, 1068-1072.

Sugi, J. and Yamashita, S., 1991. The method for obtaining ferrous ferric chloride. Bulletin of the Society of Sea Water Science Japan 42, 11.

Taylor, T.V., Rimmer, S., Day, B., Butcher, J. and Dymock, I.W., 1974. Ascorbic acid supplementation in the treatment of pressure-sores. Lancet 304, 544-546.

Nutraceuticals and skin

Key facts

- Excessive sun light exposure is associated with an increased oxidative stress and skin premature aging.
- Oxidative stress, generated by free radicals, leads to oxidation of skin lipids, proteins and DNA, while topical and systemic photoprotectants reduce these damages when properly used.
- The global personal care (GPC) market increasing by 5 to 6% year was evaluated around US$ 328 billions in 2009; about 10% is represented by nutricosmetics growing faster increasing by 15 to 20% year.
- The GPC market has registered four mega trends impact: globalization, low carbon economy, health & wellness, functionality and performance.
- The use of solaria 10 times per year adds a negligible risk for developing skin cancer, but use 3 time weekly from age 20 to 50 could double the likelihood of this development
- Genetics and especially childhood sun exposure play an important role in the development of skin cancers.
- Currently, skin cancer prevention focuses on sun avoidance and use of sunscreens, photoprotective nutrition supplements and protective clothing.
- ROS deplete and damage non-enzymatic and enzymatic antioxidant defence systems of the skin and, activating cytoplasmatic signal transduction pathways in resident fibroblasts, induce senescence and connective tissue degradation.
- p53, called *guardian of the genome*, is a tumor suppressor gene inactivated by UV rays and activated by extra virgin olive oil.
- The tumor suppressor gene p53 is the most frequent target for mutations in human cancer, as actinic keratoses.
- Nutritional skin care and photoprotectants will be driven by the increased number of young and older health-conscious consumers. In anyway the photoprotectant nutricosmetics and cosmeceuticals will capture part of the future multibillion global market of skin care produces.

Summary points

- Photoaging occurs as result of cumulative UV radiation, while chronological aging is the result of passage of time.
- The sun-exposed skin damages are superimposed on chronological aging, while UV radiation is the most important factor in premature photoaging.
- The increased life-time sun exposure has increased aging processes and skin malignancy. Thus, the necessity of a topical and systemic photoprotection by the use of antioxidant compounds as vitamin E, carotenoids and PUFA.
- The skin antioxidant defence is influenced, in fact, by nutritive factors and the photoprotective effects of nutrient supplementation should prevent UV damages and premature skin aging.
- In conclusion nutricosmetics aim to create radiant beauty from the inside out, while nutraceuticals provide medical or health benefits.
- However the mechanism of action of the phoprotective compounds has to be understood in a better way by further research studies.

13. Skin photoprotection and nutraceuticals: an overview

P. Morganti
International Society of Cosmetics Dermatology (I.S.C.D), Via Innocenzo XI, 41, 00165 Rome, Italy; morganti@iscd.it

Abstract

The skin is increasingly exposed to environmental pollutants and UV rays thus increasing its risk for photo-oxidative damage. Sunscreens protect the skin against these risks and, to maintain the protective level effective during prolonged outdoor activity, these topical sunscreens are administered together with nutriceuticals and nutricosmetics. Human dietary intervention studies using carotenoids combined with vitamins, polyunsaturated fatty acid, minerals and natural phytoextract ingredients show promise in protecting against pollutant and solar-induced effects. Testing of new cosmeceuticals and nutricosmetics is at an early stage of development and further work is required to achieve suitable protocols. It is in this direction that future research is likely to be most beneficial with regard to discovering all around systemic protectants.

Keywords: carotenoids, vitamin E, vitamin C, nutricosmetics, photoprotection

Abbreviations

CN	Chitin-nanofibrils
EPA	Eicosapentaenoic acid
DHA	Docosahexaenoic acid
GTP	Green tea polyphenols
PUFA	Polyunsaturated fatty acid
ROS	Reactive oxygen species

13.1 Introduction

Sunlight, as part of our everyday life, can be both tonic or toxic to human skin. It promotes the synthesis of vitamin D and stimulates the production of the photo-protective melanin, under the form of a healthy looking tan; but as the critical major oxidizing and hazardous agent to the human biological system, it induces photoaging and skin cancer (Holick, 2001; Tannini *et al.*, 2002; Urbach, 2001) Aging is associated, in fact, with progressive skin changes that are strongly influenced by environmental factors, as life-time UV exposure, smoking, life-style, and genetic factors (Kaminer, 1995; Kennedy *et al.*, 2003; Makratonaki, 2010: Reylye *et al.*, 2006; Yaar *et al.*, 2002; Yaar and Gilchrest, 2003; Youn *et al.*, 2003).

In most cases, sun damage is indeed responsible for the majority of clinically evident age-associated cutaneous changes (Lavker, 1995; Urbach, 2001; Yaar and Gilchrest, 2007). In sun-exposed skin, the cumulative oxidative damage is superimposed on naturally intrinsic aging occurring in all the skin types, which manifest as dry, wrinkled, lax, unevenly pigmented, and brown spotted skin. UV rays cause a large number of deleterious effects on human skin, including sunburn, immuno-suppression, and skin cancer, in addition to premature aging.

As a consequence collagen fibrils become disorganized, their synthesis decreases, while skin levels of type I and III collagen precursors and cross-links are reduced, and there is an abnormal accumulation of elastin containing material (Talwar *et al.*, 1995; Varani *et al.*, 2004). This in effect determines the main differences between chronological aging and skin photoaging (Figure 13.1). The unifying mechanism inducing these changes is the oxidative stress generated by the ROS, such as the superoxide anion, the hydroxyl radical, and hydrogen peroxide. These free radicals deplete and damage the skin's enzymatic and non-enzymatic antioxidant defence systems causing age related cancers, cardiovascular complications, acute and chronic inflammations, and neurodegenerative lesions (Breimer, 1990; Jaswal, 2004). An imbalance between ROS and the skin's antioxidant protection mechanisms leads to oxidation of macromolecules, including DNA, lipids, and proteins, resulting in loss of structural and/or functional integrity of key components of the epidermal barrier. It has been shown, in fact, that photoaged skin exhibits high levels of different markers of oxidative stress, including the accumulation of lipid peroxides, glycation end-products, and oxidized proteins (Jeanmaire *et al.*, 2001; Sander *et al.*, 2002a,b; Tanaka *et al.*, 2001). Whether generated as by-products of cellular metabolism or whether they are of environmental origin as UV rays, ROS can modify amino-acid structure, fragment polypeptide chains, and the crosslinks of peptides and proteins, leading to the formation of DNA lesions (De Gruijl, 2000; Kvam and Tyrrel, 1997; Stadtman, 2001). Thus the cumulative DNA damages, accompanying the chronological genetic aging (intrinsic) and the environmental component of photoaging (extrinsic) result in large part from modified DNA status during an individual's lifespan (Krutmann and Gilchrest, 2006; Vij, 2000; Yaar and Gilchrest, 2007). Human studies to date have underlined the possibility of preventing or at least minimizing ROS-induced photoaging by the external use of efficient sunscreens or by the internal use of nutraceuticals (Heinrich *et al.*, 2003; Morganti, 2009a,b,c; Morganti *et al.*, 2004a,b). The mainstay of skin protective strategy appears to be a correct photo-protection obtained by UV filters used not only in sunscreens, but increasingly present at relevant concentrations in cosmetic products for daily use, such as foundations, make-ups, creams and lotions. However the more promising strategy for safeguarding the skin at 360° from dangerous UV rays, seems to be represented by

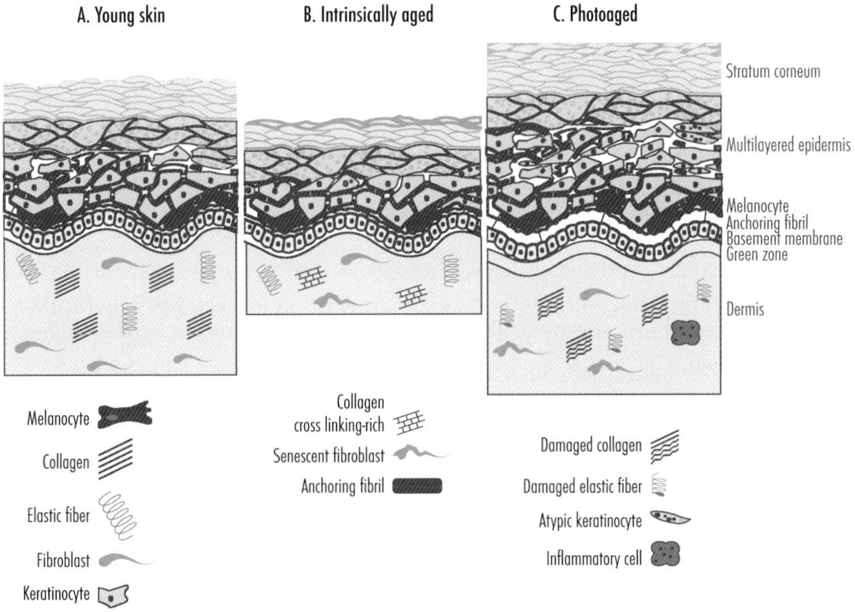

Figure 13.1. Young skin (A) reveals a balanced composition and distribution of keratinocytes of the multilayered epidermis and a normal *stratum corneum*, together with a normal anchoring of basement membrane, and distinct extracellular matrix components, such as fibroblasts, elastic fibers and collagen fibers. The distribution of melanocytes is normal also. Intrinsically aged skin (B) is atrophic with a decrease in epidermal and dermal thickness. The content of cross-links in collagen is increased, meanwhile both collagen and elastic fibers are reduced in quantity together with fibroblasts that become senescent. Photoaged skin (C) presents hyperplasia with increase thickness of the *stratum corneum*, the epidermis, and the dermal compartment. The distribution of the melanocytes becomes inhomogeneous resulting in pigmentary changes, anchoring fibrils are reduced in number and microfibrillar components results severely damaged; long-term sun exposure also results in a state of inflammation.

the simultaneous use of both internal and external products containing topical and systemic photo-protective agents (Krutmann and Yaroshi, 2006 Morganti *et al.*, 2001).

13.2 Topical photoprotecion

Sunscreens are the active ingredients used in topical photoprotection. They are divided into physical and chemical filters but are generally used in combination, and should protect against UVB and UVA rays. Moreover they have to be photo-stable and water resistant (Krutmann, 2001; Morganti *et al.*, 2008a, 2009). Chemical filters have the capacity to absorb the UVA and/or UVB photons releasing them as heat and lower-energy photons, while physical filters contemporary reflect and scatter all the UV rays. However, the photoprotective activity of topical sunscreens can be ineffective when their filters are poorly bonded at the *stratum corneum* level, are photo-unstable, and are easily degraded by the skin enzymes, thus generating ROS and subsequent photo-toxic and photo-allergic reactions. To obtain a more stable and broader photoprotective activity it seems necessary to use both chemical and physical filters in the same cosmetic formulation, that combines topical and endogenous photo-protection as seen in nutricosmetics or nutraceuticals (Morganti, 2007a,b, 2009a; Morganti and Morganti, 2007a; Morganti *et al.*, 2001, 2002, 2009).

13.3 Systemic photoprotection

While topical sunscreen products provide photoprotection through mechanisms which are based on absorption or reflection of UV rays, systemic photoprotectants, rely on their ability to prevent some of the biochemical and molecular consequences occurring in the skin, after UV irradiation has been absorbed. Such agents would be expected to act by using one or more of the following mechanisms, i.e. (1) as a biochemical barrier to UV rays; (2) as a modulating ingredient of the antioxidant cell network; or (3) as a booster to the immune system. Whereas some agents have shown in protecting against certain solar-induced effects by one or more of these mechanisms, others have demonstrated limitations as universal photo-protectants (Black and Rhodes, 2001; Morganti and Morganti, 2007b; Morganti *et al.*, 2004a,b, 2008; Strickland, 2001; Stahl and Krutmann, 2006).

13.3.1 The market

Today the nutricosmetic ingredient market includes four principal segments: carotenoids (23.3%), vitamins (23.1%), PUFA (4%) and others (49.6%). According to Frost and Sullivan (Dokos, 2010), the global Personal Care Market (Figure 13.2), has been valuated around US$ 328 billion in 2009, exhibiting a healthy 5 to 6 percent annual growth rate. Into this market, while cosmetic and toiletry products represent US$ 250 bn, nutraceuticals have a turnover of US$ 65 bn, cosmeceuticals US$ 11 bn, and nutricosmetics US$ 2.1 bn only (Figure 13.3). However, as opposed to the other sector, both cosmeceuticals and nutricosmetics differently within the other segments are expected to grow by 2015 to an annual growth rate of at least 17%. The market driver

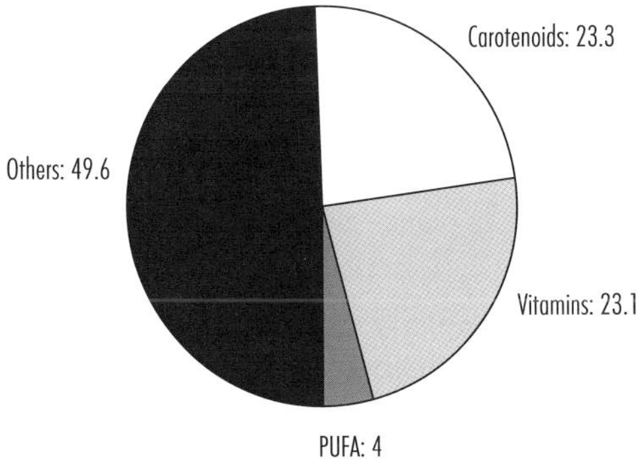

Figure 13.2. Nutricosmetics ingredients market.
PUFA: polyunsaturated fatty acid

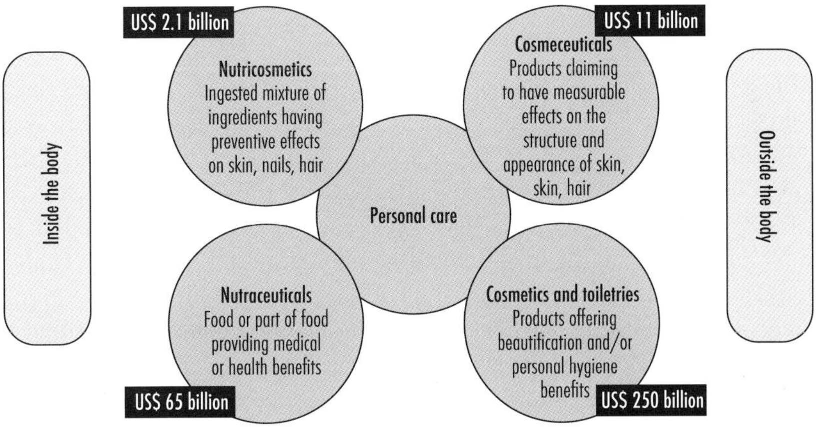

Figure 13.3. Personal care market.

of this increase is "the aging population with health concerns and interest in their expected quality of life" (Dokos, 2010). As a result cosmeceutical and nutricosmetic markets could realize their potential if they will harnessed the opportunity within the baby boomer generation. But the aging population could be persuaded to switch from cosmetics and nutraceuticals to cosmeceuticals and nutricosmetics if scientists and manufacturers could provide them with more scientific data and proofs about their properties and benefits. However, personal care products that offer photoprotective and anti-aging efficacy are expected to continue exhibiting a healthy growth over the next years.

13.3.2 Photoprotective ingredients

Carotenoids

They are a class of molecules found in plants, algae and in some fungi and bacteria that protect against excess light. Chemically, typical carotenoid pigments are linear polyenes, consisting of eight isoprenoid residues, while carotenoids containing oxygen, such as lutein and astaxantin, form a subclass called the xanthophylls (Figure 13.4) (Szaboles, 1990). Highly effective antioxidant carotenoids are capable of neutralizing singlet oxygen and peroxyradicals, frequently formed during photo-oxidative processes (oxidative stress). Probably for these reasons recent human studies have shown that oral intake of carotenoids has interesting photoprotective effects (Biagini *et al.*, 2006; Morganti, 2009b,c; Morganti *et al.*, 2004a,b; Roberts *et al.*, 2009; Scarmo *et al.*, 2010).

Thus, in normal healthy volunteers a significant reduction in solar simulator-induced erythema formation was seen after 12 weeks of oral administration of β-carotene and lycopene (Stahl and Sies, 2007). The same results were also obtained by oral intake of 16 mg/day of lycopene for a period of 10 weeks (Stahl and Sies, 2002), and by a combination of 8 mg lutein given by oral route together with 8 mg lycopene and 8 mg of β-carotene (Heinrich *et al.*, 2003). This study showed a modest protective effect by supplementation with a natural carotenoid mixture against UVA and UVB-induced erythema, while in another study the ingestion of a lycopene-rich product for 10-12 week demonstrated photoprotective effects against UV-induced erythema (Stahl *et al.*, 2006a,b). The carotenoid mix containing mainly β-carotene, astaxanthin, zeaxanthin, and lutein, was given for 24 weeks resulting in an interesting photoprotective effect (Lee *et al.*, 2000). A further study showed that an oral administration of a mixture of lutein and zeaxanthin increases skin hydration, while decreasing skin lipid peroxidation after two months as compared to placebo (Hamilton *et al.*, 2004). Moreover the effects on human skin were further analyzed in a double blind, placebo-controlled study showing that oral lutein intake increased skin hydration and

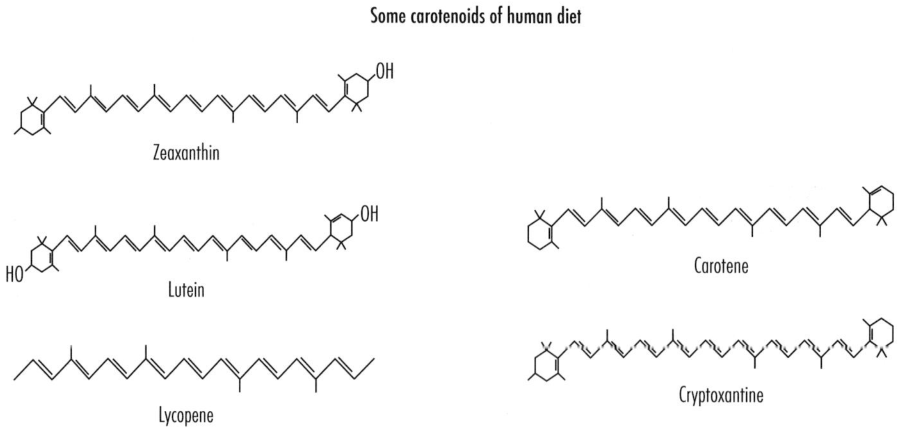

Figure 13.4. Carotenoids of the human diet.

elasticity, while contemporarily increased photo-protection which was greater with the combined oral and topical administration as compared with oral or topical treatment alone (Morganti, 2009a,b,c; Palombo *et al.*, 2007). Oxo-carotenoids seems, in fact, to have a superior preventive effects against photo-oxidative changes in cell culture also (Camera *et al.*, 2009). However, it is important to underline how the photo-protective effectiveness of these molecules depends on the type of carotenoid used, and the intake of a balanced and controlled diet (Burri *et al.*, 1999; Palombo *et al.*, 2007). However, for each molecule used a critical dose range has to be defined at which photo-protection can be achieved with the best result without causing side effects. In conclusion, the most recent studies show that protection against UV induced erythema can be achieved modulating the diet and formulating well studied nutraceuticals or nutricosmetics based on the right mix of carotenoids, vitamins and minerals (Morganti, 2009a,b,c).

13.3.3 Vitamins E and C

Vitamin E (tocopherols and tocotrienols) as an endogenous and lipo-soluble antioxidant compound is widely present in human skin, acting as scavenger of ROS generated in primary or secondary reactions following UV irradiation (Traber and Sies, 1996). To have a photo-protection activity at the cell membrane level vitamin E or its biologically active metabolite has to be available in sufficient amounts at the target site (Stahl *et al.*, 2002). Therefore, concentration of both nutraceutical photo-protectant compounds and the vehicle used must be well defined in order to guarantee the maximum antioxidant action at the skin level. Other important considerations are the safety of the recommended dose range and its effects over a long period of time (Stahl *et al.*, 2006). Differently from vitamin E, vitamin C (ascorbate) is an enzymatic water soluble cofactor present in high concentrations in many tissues. It plays a fundamental role in regenerating vitamin E from the tocopheryl radical, and is necessary for the production of collagen fibers to maintain connective tissue in its normal state, as well as for wound healing, as scavenger of the superoxide radical anion, hydrogen peroxide, hydroxyl radical and the singlet oxygen (Frei *et al.*, 1989; McCall and Frei, 1999; Weber *et al.*, 1996). For all these reasons vitamins E and C are included in nutraceutical supplements to reduce skin photo-damages, rebalance free radical-mediated pathologic processes, and prevent photoaging. Using a combined treatment of 2 g/day of α-tocopherol and 3 g/day of ascorbate over a period of 50 days, the minimal erythemal producing dose of irradiation was significantly increased from about 100 mj/cm^2 at the beginning to about 180 mj/cm^2 after supplementation (Eberlei-Konig, 1998). There is evidence, in fact, that vitamins E and C improve, to some extent, the efficacy of carotenoids as photo-protectant agents.

Thus it has been shown that supplementation with both vitamin E and β-carotene was more effective than β-carotene treatment alone (Stahl *et al.*, 2000), and also that carotenoids (β-carotene and lycopene) combined with vitamin C and selenium demonstrated a higher protective activity for subjects affected by photo-allergy and photo-sensitivity in comparison with the vehicle used alone (Morganti *et al.*, 2004a,b). A similar carotenoid mixture was used in a study where the protective effect of lutein alone or in combination with tocopherol, vitamin C and melatonin were investigated (Biagini *et al.*, 2006). These and other data show that antioxidants provide protection against erythema formation in humans and may be useful in diminishing sensitivity towards UV-light. In conclusion, vitamins E and C seem to have an interesting activity against photoaging, especially when used in combination with carotenoids. However, it is important to underline that photoprotection can be achieved only when an optimal dose range is reached in the skin.

13.4 Polyunsaturated fatty acid

PUFAs, depending on the position of the first unsaturated double bond from the methyl end of the molecule are categorized into omega-3, omega-6 and omega-9 families (Anonymous, 1992) (Figure 13.5). The omega-6 family, largely derived from vegetable oils, includes the pro-inflammatory mediator arachidonic acid, while the omega-3 family, including EPA and DHA are derived mainly from fish oil. PUFA, the structural components in human bio-membranes, are also used for the synthesis of eicosanoids-hormone-like substances which modulate immunological function (Rhodes *et al.*, 1995, 2000). Recently, scientific studies have drawn attention to the anti aging properties of α-linolenic acid showing a link between inflammation and skin-aging (Giacomoni, 2005; Giacomoni and Rein, 2001, 2004) and the protective activity of EPA (Kim *et al.*, 2006). Therefore, a balanced supplement of omega-3 and omega-6 fatty acids are recommended for patients with inflammatory skin conditions, and/or as skin photoprotectants. It has been shown, in fact, that supplementation with 10 g of fish oil containing the omega-3 fatty acids EPA and DHA for three months reduced susceptibility to erythema formation. It seems

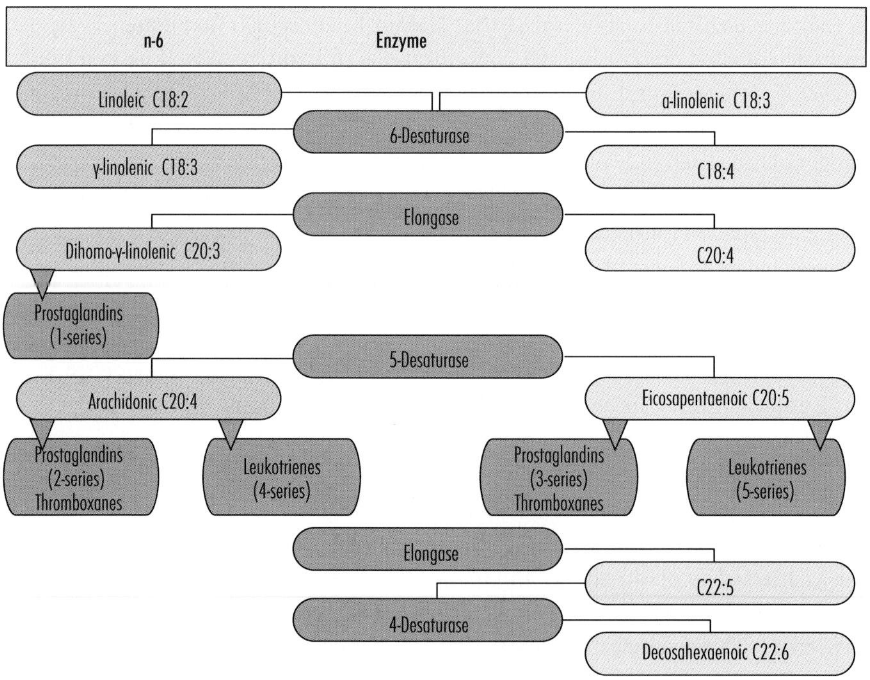

Figure 13.5. Polysunsaturated fatty acids.

to have, in fact, a pronounced protection against UVB-induced p53 production (Rhodes *et al.*, 2000). Thus, PUFAs have the potential of a therapeutic effect on inflammatory skin disorders because of their effects on the fluidity and function of immune cell membranes, and as precursors for the immune modulators eicosanoids. In any way, many scientific controversies still exist with regard to adequate omega-3 intake levels and their role in specific physiological functions and pathologies, but consumer awareness can only benefit from public sector involvement and an open debate. The scientific outlook for this ingredient remains positive while remains a lack of consumer knowledge to be overcome (Hudson, 2010). However, a collaboration EU program is currently under way to examine the nature and mechanism of the photo-protective properties of purified EPA, also because it has been recently reported that the omega-3 fatty acid, one of the most popular functional ingredients used today, will have a global volume consumption increase of 24% during the 2009-2014 forecast period (Hudson, 2010).

13.4.1 Skin photoprotection by innovative bio-textiles

Skin photoprotection may be obtained by the use of innovative bio-textiles capable to screen not only UVB, but UVA 1, UVA 2 and blue light also. For this purpose, CN-complexed with lutein-UVA-filter were melted with other natural or chemical raw materials (cellulose, rayon, etc.) to obtain innovative photoprotective fibers by electro-spinning (Morganti, 2008). The CN-lutein complex is added as suspension in one of the process steps preceding the spinning of the cellulose/rayon solution to form fibers and new anti-aging bio-textiles. In the same way it should be possible to use other CN-complexes to obtain antibacterial and/or wound healing bio-fibres for producing sport/outdoor and medical/health care textiles.

In conclusion, nanotechnology-related textiles can play an important role in the medical sector. Currently, woven and non-woven anti-bacterial fabrics are the most used applications of nanotechnology in the medical textile segment, being used to prevent infection or deodorize medical clothing and wound dressing, or protect the skin from the UV damages (Mantovani *et al.*, 2010). It is important to underline how the annual turnover of the textile industry in 2008 was over 203 billion euro and the sector employs 2-3 millions of workers in more than 145,000 companies, while the global market for medical textile was about 8 billion euro with an increase of more than 15% year (Mantovani *et al.*, 2010).

13.5 Conclusion

Although carotenoids, E, C vitamins, and PUFAs are used in different cosmeceutical and nutricosmetic products, many other emerging ingredients are gaining in popularity (Chaunday, 2010). The specific role shown in skin health from each of these ingredients, classified as *others* (Figure 13.2), suggests that they may facilitate the development of new products useful to obtain a radiant beauty from the inside out.

These ingredients with a 49.6% market share, represent, in fact, the largest market segment among the many ingredients used to formulate the so called nutricosmetics This category includes botanical antioxidants (Afaq and Mukhtar, 2002), ectoin (Buenger and Driller, 2004), proanthocyanidins (Rapport *et al.*, 2003), phytosterols, (Widyarini *et al.*, 2001), aloe vera, (Vogler and Ernst, 1999), resveratol (Reagan-Shaw *et al.*, 2004), melatonin (Biagini *et al.*, 2006), coenzyme Q10 (Puizina-Ivic *et al.*, 2010), glucosamine and pomegranate (Chaudary, 2010), vitamin K2 (Geleijnse *et al.*, 2004), polyglucosides, such as chitin and chitosan (Janak *et al.*, 2011) etc. All are supposed to have different kinds of benefits, ranging from the protection of the human body against environmental aggressions, to combating the cause of osteoporosis, to the possibility of increasing the immune response that can protect the skin against aging and photoaging effects. However, antioxidant compounds as GTP, given orally or topically, prevent single-strand breaks in DNA as guardian of genoma by scavenging the ROS generated by UV radiation (Katiyar *et al.*, 2000; Lane, 1992). Topical and/or oral administration of GTP within 30 min prior UV irradiation resulted in reduced production of cyclobutyl pyrimidine dimers in the epidermis and dermis of human volunteers, reducing UVB-induced erythema, postaglandin synthesis, and the influx of inflammatory cells into human skin (Strickland, 2001).

The growth trend in the sector of nutraceuticals and nutricosmetics is expected to continue as "health-conscious consumers become increasingly aware and confident about selecting the right products for their nutrition and personal care" (Dokos, 2010). But solid scientific research studies have to be done on the ingredients used and on the marketed products to demonstrate their real efficacy on the human body. Moreover, a more effective technical service and formulation assistance is necessary at all levels for ascertaining the product's suitability and stability, and communicating the obtained benefits to clients and consumers. We are living much longer with the consequences of a higher UV exposure. Thus, there is a greater necessity to screen or filter solar rays by the use of photoprotectant products applied externally and taken internally, to neutralize or modulate ROS activity, repairing all damages before they can provoke disease, as skin aging and cancer. Prevention is, in fact, preferable to treatment, both economically and socially. The multiplicity of sun-protection and anti-aging strategies, and the different products employed (sunscreens, antioxidant and immuno-regulatory compounds (external and internal) show that just as *all roads lead to Rome,* many are the ways to protect the body from premature aging. The future challenge will be to better understand the cell's vital machinery and the mechanism of action all the active ingredients have, when used for nutraceutical and nutricosmetic formulations. These products must prevent, in fact, the damaging effects of UV rays, chemical and pollutants, on the many biological processes involved in aging, photoaging and cancer.

References

Afaq, F. and Mukhtar, H., 2002. Photochemoprevention by botanical antioxidants. Skin Pharmacology and Applied Skin Physiology 15, 297-306.

Anonymous, 1992. British Foundation's Task Force- Report Unsaturated Fatty acids. Chapman & Hall, London, UK.

Biagini, G., Gabbanelli, F., Lucarini, G., Kyrikidou, K., Mattioli-Belmonte, M. and Morganti, P., 2006. The green chemistry. Activity of an antioxidant network melatonin rich. SOFW-Journal 132(3), 16-26.

Black, H.S. and Rhodes, L.E., 2001. Systemic photo-protection dietary intervention and therapy. In: Giacomoni, P.U. (ed.), Sun Protection in man. Elsevier, New York, NY, USA, pp. 573-591.

Breimer, L.H.,1990. Molecular mechanism of oxygen radical carcinogenesis and mutagenesis: the role of DNA base damage. Molecular Carcinogenesis 3, 188-197.

Buenger, J. and Driller, H., 2004. Ectoin: an effective natural substance to prevent UVA-induced premature photoaging. Skin Pharmacology and Physiology 17, 232-237.

Burri, B.J., Clifford, A. and Dixon, Z., 1999. Beta-carotene normalizes oxidative damage in carotenoid-depleted women. In: Kumpulainen, J.T. and Salonen, J.T. (eds.), Natural antioxidants and anticarcinogens in nutrition, health and diseases. Royal Society of Chemistry, Cambridge, UK, pp. 231-234.

Camera, E., Mastrofrancesco, A., Fabbri, C., Daubrawa, F., Picardo, M., Sies, H. and Stahl, W., 2009. Astaxanthin, cant haxanthin and beta-carotene differently affect UVA-induced oxidative damage and expression of oxidative stress-responsive enzymes. Experimental Dermatology 18, 222-231.

Chaudary, R., 2010. Good nutrition for healthy skin. Nutraceutical Business & Technology 6, 12-17.

De Gruijl, F.R., 2000. Photocarcinogenesis: UVA vs UVB. Methods in Enzymology 319, 359-366.

Dokos, L. 2010. Nutricosmetics ingredients markets challenge and growth opportunity in the EU and USA. Cosmetics Paris, France, April 13.

Eberlein-König, B., Placzek, M. and Przybilla, B., 1998. Protective effect against sunburn of combined systemic ascorbic acid (vitamin C) and D-alpha-tocopherol (vitamin E). Journal of the American Academy of Dermatology 38, 45-48.

Frei, B., England, L. and Ames, B.N., 1989. Ascorbate is an outstanding antioxidant in human blood plasma. Proceedings of the National Academy of Science of the United States of America 86, 9748-9753.

Geleijnse, J.M., Vermeer, C., Grobbee, D.E., Schurgers, L.J., Knapen, M.H.J., Van der Meer, I.M., Hofman, A., Witteman and J.C.M., 2004. Dietary intake of menaquinone is associated with a reduced risk of coronary heart disease: The Rotterdam study. The American Society for Nutritional Sciences 134, 3100-3108.

Giacomoni, P.U. and Rein, G., 2001. Factor of skin aging share common mechanisms. Biogerontology 2, 219-229.

Giacomoni, P.U. and Rein, G., 2004. A mechanistic model for the aging of human skin. Micron 35, 179-184.

Giacomoni, P.U., 2005. Aging science and the cosmetics industry. The micro-inflammatory model serves as a basis for developing effective anti-aging products for the skin. EMBU report 6, 545-548.

Hamilton, L., Brown, M. and Long, S., 2004. Improving resistance of the skin barrier to UV damage through oral supplementation. Proceedings International Federation of Societies Cosmetic Chemists 23rd Congress Orlando, FL, USA, pp. 221-225.

Heinrich, H., Gärtner, C., Weil, M., Eichler, O., Sies, H., Tronnier, H. and Stahl, W., 2003. Supplementation with β-carotene or a similar amount of mixed carotenoids protects humans from UV-induced erythema. Journal of Nutrition 133, 98-101.

Holick, M.F., 2001. A perspective on the beneficial effects of moderate exposure to sunlight: bone health, cancer prevention, mental health and wellbeing. In: P. Giacomoni (ed.), Sun Protection in man. Elsevier, New York, NY, USA, pp. 11-38.

Hudson, E., 2010. Omega-3 fatty acids: the whole package. Nutraceuticals, Business and Technology 6, 30-32.

Jaswal, A.K., 2004. Signalling in coordinated activation of antioxidant gene expression. Free Radicical Biology and Medicine 36, 1199-1207

Jeanmaire, C., Danolx, L. and Pauly, G., 2001. Glycation during human dermal intrinsic and actinic ageing: an *in vivo* and *in vitro* model study. British Journal of Dermatology 145, 10-18.

Janak, K., Vidanarachchi, J.K., Maheshika, S., Kurukulasuriya, S. and Se-Kwon, K., 2011. Chitin, chitosan and their oligosaccharides in food industry. In: Se-Kwon, K. (ed.), Chitin, chitosan and their oligosaccharides and their derivatives. CRC Press, New York, NY, USA, pp, 543-560.

Kaminer, M., 1995. Photodamage: magnitude of the problem. In: Gilchrest, B.A. (ed.), Photodamage. Blackwell, Cambridge, MA, USA, pp. 1-11.

Katiyar, S.K., Ahmad, N. and Mukhtar, H., 2000. Green tea and skin. Archives of Dermatology 136, 989-994.

Kennedy, C., Bastiaens, M.T., Bajdik, C.D., Willemze, R., Westendorp, R.G. and Bouwes Bavinck, J.N., 2003. Effects of smoking and sun on aging skin. Journal of Investigative Dermatology 120, 548-554.

Kim, H.H., Cho, S., Lee, S., Kim, K.H., Cho, K.H., Eun, H.C. and Chung, J.H., 2006. Photoprotective and anti-skin-aging effects of eicosapentaenoic acid in human skin *in vivo*. Journal of Lipid Research 47, 921-930.

Krutmann, J., 2001. New developments in photo-protection of human skin. Skin Pharmacology and Applied Skin Physiology 14, 401-407.

Krutmann, J. and Gilchrest, B., 2006. Photoaging of skin. In: Gilchrest, B.A. and Krutmann, J. (eds.). Skin Aging, Springer-Verlag, Berlin, Germany, pp. 33-43.

Krutmann, J. and Yaroshi, D., 2006. Modern photo-protection of human skin. In: BA Gilchrest, J Krutmann (eds.), Skin aging, Springer-Verlag, Berlin, Germany, pp. 103-112.

Kvam, E. and Tyrrell, R.M., 1997. Induction of oxidative DNA base damage in human skin cell by UV and near visible radiation. Carcinogenesis 18, 2379-2384.

Lane, D., 1992. p53, guardian of genome. Nature 358, 15-16.

Lavker, R., 1995. Cutaneous aging: chronological versus photoaging. In: Gilchrest, D. (ed.), Photodamage vol.1: Blackwell, Cambridge, UK, pp. 123-135.

Lee, J., Jiang, S., Levine, N. and Watson, R.R., 2000. carotenoid supplementation reduces erythema in human skin after simulated solar radiation exposure. Proceedings of Society for Experimental Biology and Medicine 223, 170-174.

Makratonaki, E., 2010. Skin ageing and hormones. EADV News 35 (summer), 10-11.

Mantovani, E., Piergiorgio, Z., Conde, J., Sitja, R. and Periales, F., 2010. ObservatoryNANO: Report on Nanotechnology & textiles Medical Textiles. Sport/Outdoor Textiles. Available at: http://www.observatorynano.eu/project/document/3144/.

McCall, M.R. and Frei, B., 1999. Can oxidant vitamins materially reduce oxidative damage in humans? Free Radical Biology and Medicine 26, 1034-1053.

Morganti, P., 2007a. Where nutriceuticals meet cosmecuticals. Journal of Applied Cosmetology 25, 111-120.

Morganti, P., 2007b. Beauty from inside out. Nutracos, VI (4), 5-8.

Morganti, P., 2008. Leather and textile chemicals: chitin nanofibrils in textiles. Specialty Chemicals Magazine 28, 26.

Morganti, P., 2009a. Natural products work in multiple ways. In: Tabor, A. and Blair, R. (eds.), Nutritional Cosmetic: Beauty from Whitin. William Andrew Publishing, New York, NY, USA, pp. 185-198.

Morganti, P., 2009b. The antioxidant benefits of oral carotenoids for protecting the skin against photoaging. In: Tabor, A. and Blair, R. (eds.), Nutritional Cosmetic: Beauty from Whitin. William Andrew Publishing, New York, NY, USA, pp. 185-198.

Morganti, P., 2009c. The photoprotective activity of nutraceuticals. Clinics in Dermatology 27, 166-174.

Morganti, P. and Morganti, G., 2007a. Cosmeceutical and Nutraceuticals to age well. Eurocosmetics 15, 6-10.

Morganti, P. and Morganti, G., 2007b. Carotenoids and vitamins to prevent photodamage and skin aging. Eurocosmetics 15, 10-17.

Morganti, P., Fabrizi, G. and Morganti, G., 2001. Topical and Systemic Photoprotectants to Prevent Light- induced Reactions. Eurocosmetics 9, 18-21.

Morganti, P., Bruno, C., Guarneri, F., Cardillo, A., Del Ciotto, P. and Valenzano, F., 2002. Role of Topical Nutritional Supplement to Modify the Oxidative Stress. International Journal of Cosmetic Science 24, 331-339.

Morganti, P., Bruno, C., Fabrizi, G., Valenzano, F. and Morganti, G., 2004a. The Antioxidant Network of skin and eye: efficacy of carotenoids, Journal of Applied Cosmetology 22, 133-142.

Morganti, P., Fabrizi, G. and Bruno, C., 2004b. Protective effects of Oral Antioxidants on Skin and Eye Function, Skin MED: Dermatology for the Clinician 3, 310-316.

Morganti, P., Ruocco, E., Guarneri, F., Cardillo, A., Pagliarello, C. and Fabrizi, G., 2008a. A Sunscreen formulation for acne –prone skin. C&T USA 123, 73-79.

Morganti, P., Fabrizi, G., Ruocco, E., Del Ciotto, P., Palombo, P., Palombo, M. and Cardillo, A., 2009. Chitin nanofibrils to improve photo-protection. C&T USA 124, 66-73.

Morganti, P., Martin-Sousa, D. and Morganti, G., 2008b. Skin activity of lutein. Söfw Journal 134, 2-11.

Palombo, P., Fabrizi, G., Ruocco, V., Flühr, J., Roberts, R. and Morganti, P., 2007. Beneficial long-term effects of combined oral/topical antioxidant treatment with carotenoids lutein and zeaxanthin on human skin: A double-blinded, placebo-controlled study in humans. Skin Pharmacology and Physiology 20, 199-210.

Puizina-Ivic, N., Miric, L., Carija, A., Karlica, D. and Marasovic, D., 2010. Modern approach to topical treatment of aging skin. Collegium Antropologicum 34, 1145-1153.

Rapport, L., Lockwood, B. and Berra, B., 2003. Nutraceutici. MDM Medical Media, Milan, Italy [in Italian].

Reagan-Shaw, S., Afagi, F., Aziz, M.H. and Amhad, D.N., 2004. Modulations of critical cell cycle regulatory events during chemioprevention of ultraviolet B-mediated responses by resveratrol in SKH-1 hairless mouse skin. Oncogene 23, 5151-5160.

Reylye, H., Peterson, I., Johansen, M., Klitkou, L., Jenne, B. and Christensen, K., 2006. Influence of environmental factors on facial ageing. Age Aging 35, 110-115.

Rhodes, L.E., Azurdia, R.M., Dean, M.P., Moison, R., Steenwinkel, N.J., Bejrsbergen and G.M.J., 2000. Systemic eicosapentaenoic acid reduces UVB-induced erythema and p53 induction in skin, while increasing oxidative stress, in a double-blind randomized study. British Journal of Dermatology 142, 601-602.

Rhodes, L.E., Durhan, B.H., Fraser, W.D. and Friedman, P.S., 1995. Dietary fish oil reduces basal and ultraviolet B-generated PGE_2 levels in skin and increases the threshold to provocation of polymorphic lighteruption. Journal of Investigative Dermatology 105, 532-535.

Roberts, R.L., Green, J. and Lewis, B., 2009. Lutein and zeaxanthin in eye and skin healthy. Clinics in Dermatology 27, 195-201.

Sander, C.S., Hansel, A., Heisemann, S.H., Elsner, P. and Thiele, J.J., 2002a. *In vivo* evidence for a link between photoaging and oxidative stress in human skin. Journal of Investigative Dermatology 119, 331A.

Sander, C.S., Chang, H., Salzimann, S., Muller, C.S., Ekanayake-Mudiyanselage, S., Elsner, P. and Thiele, J.J., 2002b. Photoaging is associated with protein oxidation in human skin *in vivo*. Journal of Investigative Dermatology 118, 618-625.

Scarmo, S., Carmel, B., Lin, H., Leffel, D.J., Welch, E., Bhosale, P., Bernstein, P.S. and Maine, S.T., 2010. Significant correlations of dermal total carotenoids and dermal lycopene with their respective plasma levels in healthy adults. Archives of Biochemical and Biophysics 504, 34-39.

Stadtman, E.R., 2001. Protein oxidation in aging and age-related diseases. Annals of the New York Academy of Sciences 928, 22-38.

Stahl, W. and Sies, H., 2002. Carotenoids and protection against solar UV radiation. Skin Pharmacology and Applied Skin Physiology 15, 291-296.

Stahl, W. and Krutmann, J., 2006. Systemic photoprotection through carotenoids. Hautarzt 57, 281-285.

Stahl, W. and Sies, H., 2007. Carotenoids and flavonoids contribute to nutritional protection against skin damage from sunlight. Molecular Biotechnology 37, 26-30.

Stahl, W., Vanderberg, H., Arthur, J., Bast, A., Dainty, I. and Faulks, R.M., 2002. Bioavailability and metabolism. Molecular Aspects of Medicine 23, 39-100.

Stahl, W., Heinrich, V., Jungmann, H., Sies, H. and Tronnier, H., 2000. Carotenoids and carotenoids plus vitamin E protect against ultraviolet light-induced erythema in humans. American Journal of Clinical Nutrition 71, 795-798.

Stahl, W., Heinrich, V., Aust, O., Tronnier, H. and Sies, H. 2006. Lycopene-rich products and dietary photoprotection. Photochemical and Photobiological Sciences 5, 238-242.

Stahl, W., Mukhtar, H., Afaq F. and Sies, H., 2006. Vitamins and polyphenols in systemic photoprotection In: In: Gilchrest, B.A. and Krutmann, J. (eds.), Skin aging. Springer-Verlag, Berlin, Germany, pp. 113-121.

Strickland, F.M., 2001. Boosting the immune system. In: Giacomoni, P.U. (ed.), Sun protection in man. Elsevier, New York, NY, USA, pp. 613-636.

Szaboles, J., 1990. Plant carotenoids in carotenoids: chemistry and biology. In: Krinsky, N.I., Mathews-Roth, M.M. and Taylor, S.F. (eds.), Plenum Press, New York, USA, pp. 39-58.

Talwar, H.S., Griffiths, C.E., Fisher, G.S., Hamilton, T.A. and Voorhees, J.J., 1995. reduced type I and type III procollagens in photodamage adult human skin. Journal of Investigative Dermatology 122, 285-290.

Tanaka, N., Tajima, S., Ishibashe, A., Uchida, K. and Shigematsu, T., 2001. Immunohistochemical detection of lipid peroxidation products, protein-bound acrolein and 4-hydroxynonenal protein adducts, in actinic elastosis of photodamaged skin. Archives of Dermatology Research 293, 363-367.

Tannini, R.M., Langiou, P., Adami, H.O. and Trichopulos, D., 2002. Comments on "evidence supporting the role of vitamin D insufficiency among free-living healthy young adults. American Journal of Medicine 112, 659-662.

Traber, M.G. and Sies, H., 1996. Vitamin E in humans: demand and delivery. Annual Review of Nutrition 16, 321-347.

Urbach, F., 2001. The negative effects of solar radiation: a clinical overview. In: Giacomoni, P. (ed.), Sun Protection in man. Elsevier, New York, NY, USA, pp. 39-67.

Varani, J., Schuger, L., Dame, M.K., Leanhard, C.H., Fligiel, S.E.G., Kang, S., Fisaer, G.J. and Voorhees, J.J., 2004. Reduced fibroblast interaction with intact collagen as a mechanism for depressed collagen synthesis in photodamaged skin. Journal of Investigative Dermatology 122, 1471-1479.

Vij, G., 2000. Somatic mutations and aging: a re-valuation. Mutat. Research 447, 117-135.

Vogler, B.K. and Ernst, E., 1999. Aloe vera: a system review of its clinical effectiveness. British Journal of General Practitioners 49, 823-829.

Weber, P., Pendich, A. and Schalch, W., 1996. Vitamin C and human health-a review of recent data relevant to human requirements. International Journal of Vitamin and Nutrition Research 66, 19-30.

Widyarini, S., Spinks, N., Husband, H.S. and Reele, V.E., 2001. Isoflavonoid compounds from red clover (trifolium pratense) protect from inflammation and immune suppression induced by UV radiation. Photochemical and Photobiology Sciences 74, 465-470.

Yaar, M., Eller, M.S. and Gilchrest, B.A., 2002. Fifty years of skin aging. Journal of Investigative Dermatology Symposium Proceedings 7, 51-58.

Yaar, M. and Gilchrest, B.A., 2003. Aging of skin. In: Freedberg, I.M., Eisen, A.Z., Walff, K., Austen, K.F., Goldsmith, L.A. and Katz, S. (eds.), Fitzpatrick's Dermatology in general medicine, Vol. 2 Mc Graw-Hill, New York, NY, USA, pp. 1386-1398.

Yaar, M. and Gilchrest, B.A., 2007. Photoaging: mechanism, prevention and theraphy. British Journal of Dermatology 157, 874-887.

Youn, C.S., Kwon, O.S., Won, C.H., Hwang, E.J., Park, B.J., Eun, H.C. and Chung, H., 2003. Effects of pregnancy and menopause on facial wrinkling in women. Acta Dermato-Venereologica 83, 419-424.

Key facts

- Flaxseed oil, also called linseed oil, is oil rich in α-linolenic acid (ALA).
- Borage oil is made of the seeds of borage, and the oil is rich in γ-linolenic acid (GLA). Borage oil is discussed as an additional treatment for patients with atopic dermatitis.
- Both ALA and GLA are fatty acids with 18 carbon atoms; ALA is a n-3 fatty acid and GLA is a n-6 fatty acid. Another n-6 fatty acid with 18 carbon atoms is linoleic acid, which is present in most plant oils.
- n-3 Fatty acids and n-6 fatty acids are essential nutrients for humans. Examples for n-3 fatty acids are eicosapentaenoic acid (EPA) and docosahexaenoic acid and for n-6 fatty acids, di-homo- γ-linolenic acid (DHGLA) and arachidonic acid (AA). EPA and docosahexaenoic acid can be obtained from fish oil, AA from meat fat. DHGLA is the elongation product of GLA in the pathway of human fatty acid synthesis.
- The difference between ALA and GLA and the other essential fatty acids is the chain length. ALA and GLA are fatty acids with 18 carbon atoms, whereas EPA, DHGLA and AA consist of 20 carbon atoms.
- For people, avoiding animal food, plant oils rich in essential fatty acids are an alternative source.
- EPA, DHGLA and AA are precursors of different eicosanoids, including several prostaglandins and leukotriens. Another biological function is related to their influence on membrane fluidity.
- Essential fatty acids cannot be synthesized by humans, however, ALA and GLA can be converted to the longer chain analogous, the efficacy of this reaction is limited.

Summary points

- Flaxseed- and borage oil are rich in α-linolenic acid or γ-linolenic acid, respectively, and may serve as alternative sources for essential fatty acids.
- The application of borage oil is discussed as an additional therapy in inflammatory skin disorders like atopic dermatitis
- Sensitive skin is a condition with increasing incidence; however, the cause is not completely understood.
- Human intervention studies showed that the dietary intake of flaxseed- or borage oil ameliorated symptoms related to sensitive skin.
- Flaxseed- and borage oil ingestion reduced immune response to nicotinate exposure and increased barrier function of human skin.

14. Effect of flaxseed- and borage oil ingestion on skin conditions

S. De Spirt[1], W. Stahl[1] and U. Heinrich[2]
[1]Institute of Biochemistry and Molecular Biology I, Faculty of Medicine, Heinrich-Heine-University Dusseldorf, P.O. Box 101007, 40001 Dusseldorf, Germany; [2]DermaTronnier, Institute for Experimental Dermatology, University of Witten-Herdecke, Alfred-Herrhausen-Str. 44, 58455 Witten, Germany; silke.spirt@uni-duesseldorf.de

Abstract

Skin lipids play an essential role in barrier function and skin homeostasis. An imbalance in skin lipid composition is associated with disorders like atopic dermatitis. Sensitive skin is a phenomenon with increasing incidence but the ultimate cause is not known yet. Sensitive skin usually reacts with an inflammatory response to non pathogenic stimuli. Dietary intake of oils rich in essential fatty acids are discussed as a supportive therapy in inflammatory skin disorders and thus might be also be suitable for the treatment of sensitive skin. Data on the effects of oils rich in essential fatty acids on such conditions are scarce. The following article describes responses of sensitive skin in women following the application of oils rich in essential fatty acids (flaxseed oil, borage oil and safflower seed oil). Flaxseed oil and borage oil both improved selected parameters related to the condition of sensitive skin.

Keywords: flaxseed oil, borage oil, sensitive skin, barrier function

Abbreviations

AA	Arachidonic acid
ALA	α-linolenic acid
DHA	Docosahexaenoic acid
DHGLA	Di-homo-γ-linolenic acid
EFA	Essential fatty acid
EPA	Eicosapentaenoic acid
GLA	γ-linolenic acid
LA	Linoleic acid
PUFA	Polyunsaturated fatty acid
SC	*Stratum corneum*
TEWL	Transepidermal water loss

14.1 Introduction

Lipids are biologically active compounds important for the physiological functions of human skin. Different classes of lipids are present in this tissue, some of them with unusual structures compared to lipid patterns in other organs. Topical treatment of the skin with lipids provided by cosmetics or pharmaceutical products is frequently used to maintain skin health or treat disorders. Additionally, nutritional supply with suitable lipids attracts increasing attention. Lipids are a heterogeneous group of hydrophobic substances, including fats, free fatty acids, sterols, waxes, ceramides, triglyceride, fat soluble vitamins, phospholipids and others. Biological functions are related to energy supply, cell signalling, and the formation of cell membranes. In nutritional science the importance of specific lipids in particular fatty acids including EFA is recognized. One of the first publications, which mentioned the importance of regularly lipid intake was written by Burr and Burr in 1929. After this initial work the importance of fatty acids which are not synthesized by higher animals (EFAs) was revealed and is still in the focus of present research. EFAs are needed for proper function of all tissues including the skin, e.g. LA deficiency results in impaired skin barrier function and scaling (Ziboh *et al.*, 2000).

EFAs are PUFAs and separated into two groups, n-6 and n-3 fatty acids. General dietary advice, not only related to skin health, recommends to increase the amount of PUFAs in the diet. n-6 fatty acids are abundant in the Western diet, with an increasing proportion (Simopoulos, 2008). Several studies have shown that higher intake of n-3 fatty acids is related to the prevention of cardiovascular diseases, diabetes and other chronic disorders, mainly those related to inflammation. Thus, it is recommended to increase the ratio of n-3 to n-6 fatty acids in the diet (Simopoulos, 2008). n-6 Fatty acids are present in meat fat and various plant oils. Major sources of n-3 fatty acids are marine fish oils and a few selected plant oils.

14.2 Essential fatty acids

Characteristic for each EFA is the position and number of double bonds. Among the n-6 fatty acids are linoleic acid (18:2), γ-linolenic acid (18:3), di-homo-γ-linolenic acid (20:3) and arachidonic acid (20:4). Important n-3 fatty acids are α-linolenic acid (18:3), eicosapentaenoic acid (20:5) and docosahexaenoic acid (22:6). Examples of C18 fatty acids are given in Figure 14.1. The human organism is able to synthesize longer chain and higher desaturated fatty acids starting with ingested C18 EFAs in a multiple step reaction catalyzed by a set of desaturases and elongases, e.g. LA can be converted to GLA, DHGLA and AA (Figure 14.2). The same set of enzymes converts ALA into stearidonic acid and EPA. Desaturation by the Δ6-desaturase is the rate limiting step. The activity of this enzyme is different in various tissues, and low in the epidermis (Ziboh *et al.*, 2002). Generally, the affinity of the Δ6-desaturase is higher for ALA, than for LA (Simopoulos, 2008).

C20 fatty acids are important precursors of lipids which play a major role in the immune response of the skin. AA is the parent compound of series 2 prostaglandins and thromboxanes as well as series 4 leukotriens and other oxidized derivates, like hydroxy-, or hydroxyperoxieicosatetraenoic acids, with pro-inflammatory properties. It is also known that AA derived compounds mediate anti-inflammatory properties and are a part of the self regulation system of an inflammatory response. An example is prostaglandin E_2 which has in addition to its pro-inflammatory properties an inhibitory effect on some pro-inflammatory cytokines production (Calder, 2009; Simopoulos, 2008).

EPA is often assigned as the anti-inflammatory counterpart of AA. It replaces AA in the phospholipids of biological membranes of inflammatory cells and is an alternative substrate for eicosanoid synthesis, sharing the same enzyme with AA as substrate (Simopoulos, 2008). EPA is the parent compound of series 3 prostaglandins and thromboxanes and series 5 leukotriens when

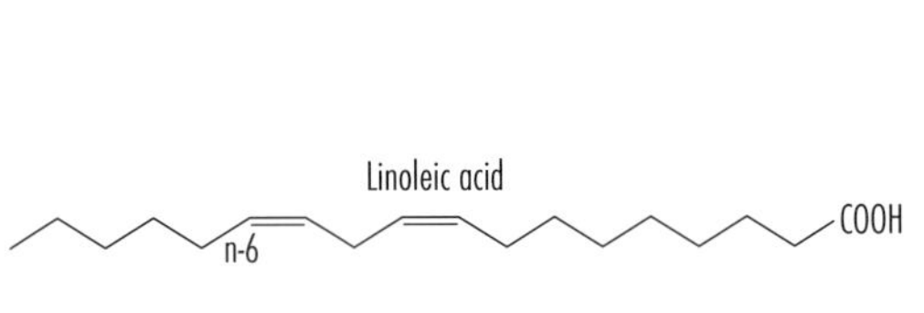

Figure 14.1. Examples of unsaturated essential fatty acids delivered via the diet. Plant oils are sources of essential fatty acids with mainly 18 carbon atoms. Essential fatty acids are divided into n-3 and n-6 fatty, which describes the position of the first double bond in relation to the methyl endgroup.

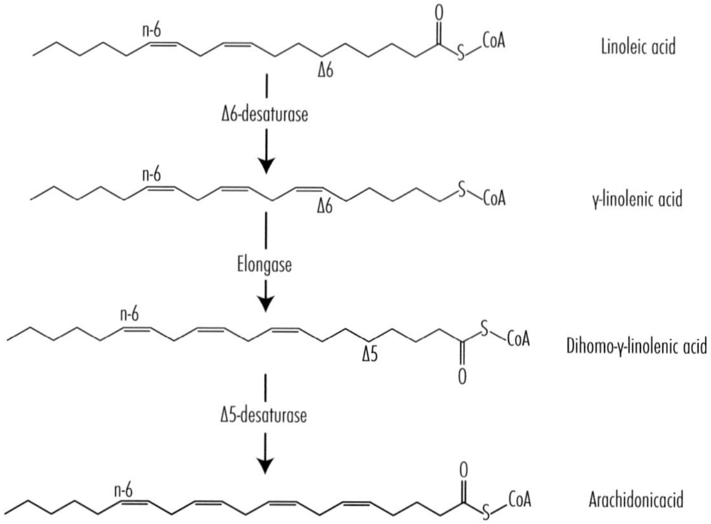

Figure 14.2. Essential fatty acids - metabolism of n-6 fatty acids. Essential fatty acids can be converted into longer chained fatty acid derivatives. Desaturase activity is low in the epidermis (Ziboh *et al.*, 2000, 2002). n-3 fatty acid, share the same set of enzymes.

the amount of EPA in the diet is increased, the production of AA-derived eicosanoids is inhibited. Some of these eicosanoids, like leukotrienB_5, have a lower inflammatory potential. Additionally, recent data show that EPA and also DHA, a further reaction product of the n-3 metabolism, are precursors of resolvins, lipid mediators, with anti-inflammatory properties (Calder, 2009). Besides AA and EPA, DHGLA is the third C20 FA with an important role in the inflammation process, especially in the skin (Ziboh *et al.*, 2002).

14.3 Selected sources of essential PUFAs

Oils rich in PUFAs are an important component of the diet and are relevant to maintain human health. Plant oils provide among other fatty acids EFA composed of 18 carbon atoms (C18), whereas animal fats additionally contain EFAs with 20 carbon atoms (C20) and more.

LA is present in most of the dietary plant-based oils and deficiency in humans is rare. Another dietary n-6 FA found in plant oils is GLA. A suitable source of this fatty acid, e.g. for the preparation of supplements or cosmetic products, is a plant-based oil derived from the seeds of borage. In this oil average contribution of GLA to total fatty acids varies between 16-26%. Evening primrose oil is another source of GLA with about 7-14% of total fatty acid (Foster *et al.*, 2010). However, the normal daily consumption of GLA is low.

Plant sources of the n-3 fatty acid ALA are flaxseed oil (also called linseed oil), and seed oils of canola, soybean or walnut. Additionally, chloroplasts contain ALA (Simopoulos, 2008). The use of the oils mentioned above in dermatology is not as common as the application of borage oil. More often n-3 fatty acid rich fish oil supplements, mainly produced from salmon, are applied. However, the use of flaxseed oil in supplements or as culinary oil is increasing and the oil is often recommended for people who avoid eating fish or taken fish oil supplements.

14.4 Lipids and skin

Little is known on the biological effects of EFAs in skin, beside the role of LA and the C20 fatty acids as eicosanoid precursors. The human skin consists of epidermis, dermis and a subcutaneous fat layer. The outer layer is the *stratum corneum*, which plays an essential role in the formation of the skin barrier and maintenance of the barrier function, protecting our body against chemicals, microorganisms and physical injuries (outside-inside barrier). Further the SC is involved in the regulating of TEWL and electrolyte transport (inside-outside barrier) (Feingold, 2009; Proksch *et al.*, 2008).

The SC is composed of corneocytes, protein enriched cells embedded in a complex extracellular lipid matrix (Proksch *et al.*, 2008). Both components built a unique system which was formerly described as a 'brick and mortar' like structure, whereby the corneocytes are compared to 'bricks' and the lipid matrix to the 'mortar' (Elias, 1983). Today it is known that the SC lipid matrix is not such a simple homogenous system. Due to the composition and the complex physicochemical interaction between the different components crystalline or more fluid domains are formed which is crucial for skin barrier function (Jungersted *et al.*, 2008; Rawlings, 2003).

Corneocytes are differentiated keratinocytes, the main cell type of the human epidermis. The lipid pattern of the SC is unique compared to other tissues. It evolves during the differentiation process in the epidermis when corneocytes are formed. Precursors are polar lipids like glycosphingolipids, free sterols, or phospholipids stored in keratinocytes. They form organelles, so called lamellar bodies which are unique in the epidermis. Specific lipid metabolizing enzymes are active in these organelles. At the end of the differentiation process the lipids are secreted to the extracellular matrix, and the polar lipids are enzymatically converted to the non polar lipids. Main lipids are ceramides, free fatty acids and cholesterol (Feingold, 2009; Proksch *et al.*, 2008; Rawlings, 2003).

In skin different ceramides are identified, one of the ceramides (ceramide 1) is conjugated with an ester bond to C18 fatty acids, mainly LA, but also oleic acid or stearic acid have been identified. All of them have important roles related in the stabilization of skin barrier (De Sousa Neto *et al.*, 2011; Masukawa *et al.*, 2009; Rawlings, 2003).

Further, free fatty acids with chain length from mainly 12 to 24 carbon atoms, are important for microbial defence, lamellar body secretion and maintenance of the epidermal pH of 4-5.5 (Feingold, 2009; Jungersted *et al.*, 2008). Cholesterol is important for the stability and fluidity of the SC and also plays a role in the process of desquamation (Jungersted *et al.*, 2008; Proksch *et al.*, 2008).

14.5 Skin effects related to the intake of flaxseed- or borage oil.

Two recent studies evaluated the influence of the intake of flaxseed oil on skin conditions. In the first study the effect of flaxseed oil was compared to that of borage oil, in the second study with safflower oil, which is rich in LA (De Spirt *et al.*, 2009; Maurette, 2008; Neukam *et al.*, 2011). Aim of both studies was to investigate whether the intake of the oils modulates skin conditions, like barrier function and affects skin sensitivity. Study participants were healthy women with sensitive skin. They received ~2 g of the oils per day over a period of 12 weeks. Within this time different skin parameters were modulated following the intake of flaxseed- or borage oil. Changes were documented with objective and valid methods. The parameters are related to skin sensitivity, barrier function and surface structure of the skin. In the first study the effect was more pronounced in the flaxseed oil group (De Spirt *et al.*, 2009). In the second study the flaxseed oil effects were comparable to those found in the first study. However, intake of safflower oil resulted in only moderate effects regarding skin roughness (one parameter describing skin surface) and hydration, which is related to the barrier function (see below) (Neukam *et al.*, 2011). Bioavaibility and exposure to fatty acids was demonstrated by measuring plasma fatty acid composition. Upon intervention with flaxseed oil, ALA increased. With borage oil GLA and with safflower oil LA plasma levels increased. No increase was observed in the longer chained PUFAs, except for a small increase in DHGLA in the borage oil group observed in the first study. Also a slight increase in EPA in the flaxseed oil group was observed, but this effect was statistically not significant (De Spirt *et al.*, 2009; Neukam *et al.*, 2011).

14.5.1 Skin sensitivity

Skin sensitivity is a common problem in the Western population. It has been published that more than 50% of the people consider themselves to have 'sensitive skin' (Farage and Maibach, 2010). The underlying pathophysiological and pathobiochemical mechanisms are widely unknown and a generally accepted definition of the disorder is not available yet. Summarizing various definitions the phenomenon may be defined as skin that reacts with erythema and/or subjective symptoms to not pathogenic stimuli (Saint-Martory *et al.*, 2008). Reported subjective symptoms are e.g. pricking, burning, pain, pruritus, external causes are wind, heat, cold, water, exposure to cosmetics or stress (Saint-Martory *et al.*, 2008). People with chronic dermatological disorders as well as healthy people can be affected (Kligman *et al.*, 2006). Due to the different and unspecific signs of sensitive skin, diagnoses is complicated, and more than one test correlating to inflammation/immune response, barrier function and hydration has to be performed for verification (Primavera and Berardesca, 2005).

Individual skin sensitivity can be tested with the nicotinate test, which is influenced by the individual immune response and barrier function of the SC. In this test topical application of methyl nicotinate induces a vasodilation of skin blood vessels and an increased redness of the skin (erythema) (Primavera and Berardesca, 2005). Vasodilation is part of the inflammation process which is usually more pronounced in people with sensitive skin. Dilation of blood vessel is among others factors regulated by eicosanoids. Under treatment with flaxseed oil and borage oil a reduced nicotinate response was determined, whereas the reduction with flaxseed oil was more pronounced. Safflower oil had no effect (De Spirt *et al.*, 2009; Neukam *et al.*, 2011). Data on direct effects of plant derived essential C18 fatty acid on immune response are scarce, effects are reported for AA, EPA, DHGLA or DHA (Calder, 2009; Ziboh *et al.*, 2002).

Conversion of C18 fatty acids into C20 fatty acids which are precursors of eicosanoids may provide an explanation for the observed effect following plant oil ingestion. Enzymatic conversion must occur in other organs than the skin, because the epidermis is lacking Δ6- desaturase activity. So only the enzyme catalyzed reaction of GLA to DHGLA is possible (Ziboh *et al.*, 2002). A positive effect of borage oil or evening primrose oil on skin conditions of patients suffering from atopic dermatitis is discussed, because eicosanoids formed by DHGLA have less inflammatory potential than AA derived eicosanoids and DHGLA replaces AA in the biological membrane of immune cells (Simopoulos, 2008). DHGLA which reaches the skin is not further metabolized to AA, because human epidermis is also low in Δ5-desaturase activity (Ziboh *et al.*, 2002).

It is likely, that the reduced inflammatory reaction to nicotinate exposure in women with sensitive skin after borage oil supplementation is due to the same mechanism, which is discussed in the context with atopic dermatitis. However, the effect of flaxseed oil is difficult to explain. ALA is not converted to EPA in the epidermis, because Δ6 desaturase is required for this reaction. The conversion may occur in other organs, and EPA is delivered to the epidermis with the blood. Intervention with flaxseed oil led in addition to increased ALA plasma levels and also to slightly increased (statistically not significant) EPA levels (De Spirt *et al.*, 2009). Further intervention studies have shown that higher amounts of ALA must be supplied to obtain measurable responses of EPA in the blood cells (Brenna *et al.*, 2009). Thus, the implication of EPA on the inflammatory reaction of the skin can not be ruled out, but a direct proof is not provided by these studies.

It was also discussed that LA, which is abundant in all plant oils, has a direct influence on the immune response of the skin. LA present in the skin is not converted into GLA. Studies with guinea pigs have shown that epidermal lipoxygenases are very active, and that 13-S-hydroxyeicosatrienoic acid can be formed from LA catalyzed by 15-lipoxygenase, a probably unique reaction of the epidermis (Ziboh *et al.*, 2000, 2002). This Mono-hydroxy fatty acid is suggested to have anti-inflammatory/antiproliferative effects, modulating signalling pathways (Ziboh *et al.*, 2002). However, intervention with the LA-rich safflower oil had no effect in the nicotinate test (Neukam *et al.*, 2011). Thus, the pathway discussed above is likely of minor importance and further mechanisms are operative.

To initiate the vasodilatory effect of nicotinate in the skin, the compound must penetrate the different skin layers to reach the microcirculation system of the dermis. Penetration is affected by the barrier efficacy of the skin. The effect of plant oils on the barrier function is discussed in the next paragraph.

14.5.2 Skin barrier function

As mentioned above, the inflammatory response in the nicotinate test was reduced after supplementation with flaxseed or borage oil, which may be a result of an improved outside- inside barrier function. The inside-outside barrier function was tested in the studies measuring TEWL and hydration (De Spirt *et al.*, 2009; Neukam *et al.*, 2011). TEWL is defined as passive penetration of water through the skin, and is mainly affected by properties of the lipid bilayers. An increased TEWL is often but not always correlated with a diminished skin hydration (Fluhr *et al.*, 2008; Proksch *et al.*, 2008). Hydration is further dependent on natural moisture factors, which are built during the formation of the SC (Fluhr *et al.*, 2008).

Intake of both, flaxseed oil and borage oil reduced TEWL and increased hydration of the skin during the studies (De Spirt *et al.*, 2009; Neukam *et al.*, 2011), suggesting that here the effects are directly correlated. The effect of flaxseed oil intake was more pronounced than that of borage oil. Also safflower oil led to a slight significant increase of hydration.

The condition of the skin surface reflects the morphological status of the entire skin, which undergoes a permanent alteration. Ageing, changes in barrier function or skin diseases alter the surface. Skin aging is a complex process inducing changes to the skin structure which are hardly reversible. Wrinkles are formed, the skin turns rougher, color and pigmentation change, a loss of firmness and elasticity is observed, hair-loss occurs and repair of lesions is less efficient (Callaghan and Wilhelm, 2008).

A reduction of the barrier function during aging increases roughness of the skin surface and reduces smoothness (Callaghan and Wilhelm, 2008). Supplementation with flaxseed oil improved roughness, scaling and smoothness, borage oil treatment improved roughness and scaling, whereas safflower oil only affected roughness of the skin. Wrinkling was not modulated with any of these oils (De Spirt *et al.*, 2009; Neukam *et al.*, 2011).

Similar effects were also observed in other studies. The effect of borage oil (1.5-3 g) consumption on skin parameters of elderly healthy people was investigated by Brosche and Platt (2000). TEWL was diminished, hydration was slightly improved (Brosche and Platt, 2000). Supplementation with 3 g of evening primrose oil led to a decreased TEWL and an increased hydration, additionally skin structure was improved (Muggli, 2005).

The biochemical mechanisms underlying the effects of the oils on barrier function are not understood. LA deficiency results in impaired barrier function and scaling (Ziboh *et al.*, 2000) which might be due to the fact that LA is a constituent of ceramide 1, an important structural element of the skin barrier (Rawlings, 2003). Borage oil, evening primrose oil and flaxseed oil all provide LA. However, with safflower oil, which had the highest amount of LA (72%), only hydration was moderately affected, whereas with flaxseed which provided only 16% LA, effects were more pronounced (De Spirt *et al.*, 2009; Neukam *et al.*, 2011). Thus, LA content is not the only factor responsible for the observed changes. Many other factors and compounds are likely involved in water homeostasis and barrier function (Rawlings, 2003).

14.6 Conclusion

Plant oils rich in ALA or GLA are nutrients with beneficial effect on skin homeostasis as shown in human intervention studies. Typical skin parameters related to sensitivity and barrier function were improved; however, the underlying mechanisms are not completely understood yet.

References

Brenna, J.T., Salem, N., Sinclair, A.J. and Cunnane, S.C., 2009. Alpha-Linolenic acid supplementation and conversion to n-3 long-chain polyunsaturated fatty acids in humans. Prostaglandins, Leukotrienes, and Essential Fatty Acids 80, 85-91.

Brosche, T. and Platt, D., 2000. Effect of borage oil consumption on fatty acid metabolism, transepidermal water loss and skin parameters in elderly people. Archives of Gerontology and Geriatrics 30, 139-150.

Burr, G.O. and Burr, M.M., 1929. A new deficiency disease produced by the rigid exclusion of fat from the diet. Journal of Biological Chemistry 82, 345-367.

Calder, P.C., 2009. Polyunsaturated fatty acids and inflammatory processes: New twists in an old tale. Biochimie 91, 791-795.

Callaghan, T.M. and Wilhelm, K.P., 2008. Clinical perspectives and clinical methods in the evaluation of ageing skin. International Journal of Cosmetic Science 30, 323-332.

De Sousa Neto, D., Gooris, G. and Bouwstra, J., 2011. Effect of the ω-acylceramides on the lipid organization of *stratum corneum* model membranes evaluated by X-ray diffraction and FTIR studies (Part I). Chemistry and Physics of Lipids 164, 184-195.

De Spirt, S., Stahl, W., Tronnier, H., Sies, H., Bejot, M., Maurette, J.-M. and Heinrich, U., 2009. Intervention with flaxseed and borage oil supplements modulates skin condition in women. The British Journal of Nutrition 101, 440-445.

Elias, P.M., 1983. Epidermal lipids, barrier function, and desquamation. The Journal of Investigative Dermatology 80, 44-49.

Farage, M.A. and Maibach, H.I., 2010. Sensitive skin: closing in on a physiological cause. Contact Dermatitis 62, 137-149.

Feingold, K.R., 2009. The outer frontier: the importance of lipid metabolism in the skin. Journal of Lipid Research 50, 417-422.

Fluhr, J.W., Darlenski, R., Angelova-Fischer, I., Tsankov, N. and Basketter, D., 2008. Skin irritation and sensitization: mechanisms and new approaches for risk assessment. 1. Skin irritation. Skin Pharmacology and Physiology 21, 124-135.

Foster, R.H., Hardy, G. and Alany, R.G., 2010. Borage oil in the treatment of atopic dermatitis. Nutrition 26, 708-718.

Jungersted, J.M., Hellgren, L.I., Jemec, G.B.E. and Agner, T., 2008. Lipids and skin barrier function--a clinical perspective. Contact Dermatitis 58, 255-262.

Kligman, A.M., Sadiq, I., Zhen, Y. and Crosby, M., 2006. Experimental studies on the nature of sensitive skin. Skin Research and Technology 12, 217-222.

Masukawa, Y., Narita, H., Sato, H., Naoe, A., Kondo, N., Sugai, Y., Oba, T., Homma, R., Ishikawa, J., Takagi, Y. and Kitaha T., 2009. Comprehensive quantification of ceramide species in human *stratum corneum*. Journal of Lipid Research 50, 1708-1719.

Maurette, J.-M., 2008. Flaxseed oil, a fish oil challenger? Oléagineux, Corps Gras, Lipides 15, 257-261.

Muggli, R., 2005. Systemic evening primrose oil improves the biophysical skin parameters of healthy adults. International Journal of Cosmetic Science 27, 243-249.

Neukam, K., De Spirt, S., Stahl, W., Bejot, M., Maurette, J.-M., Tronnier, H. and Heinrich, U., 2011. Supplementation of flaxseed oil diminishes skin sensitivity and improves skin barrier function and condition. Skin Pharmacology and Physiology 24, 67-74.

Primavera, G. and Berardesca, E., 2005. Sensitive skin: mechanisms and diagnosis. International Journal of Cosmetic Science 27, 1-10.

Proksch, E., Brandner, J.M., Jensen and J.-M., 2008. The skin: an indispensable barrier. Experimental Dermatology 17, 1063-1072.

Rawlings, A.V., 2003. Trends in *stratum corneum* research and the management of dry skin conditions. International Journal of Cosmetic Science 25, 63-95.

Saint-Martory, C., Roguedas-Contios, A.M., Sibaud, V., Degouy, A., Schmitt, A.M. and Misery, L., 2008. Sensitive skin is not limited to the face. The British Journal of Dermatology 158, 130-133.

Simopoulos, A.P., 2008. The importance of the omega-6/omega-3 fatty acid ratio in cardiovascular disease and other chronic diseases. Experimental Biology and Medicine 233, 674-688.

Ziboh, V.A., Cho, Y., Mani, I. and Xi, S., 2002. Biological significance of essential fatty acids/prostanoids/ lipoxygenase- derived monohydroxy fatty acids in the skin. Archives of Pharmacal Research 25, 747-758.

Ziboh, V.A., Miller, C.C. and Cho, Y., 2000. Metabolism of polyunsaturated fatty acids by skin epidermal enzymes: generation of antiinflammatory and antiproliferative metabolites. The American Journal of Clinical Nutrition 71, 361-366.

Key facts

- A wide variety of skin-care products are on the market.
- In recent years, new insights into the relationship between food, nutrition and skin has led to the development of dietary supplements from specific food components and plant sources for beneficial effects on the skin.
- Among them, red ginseng (RG) has been reported to have skin protection effect.
- Specifically, dietary RG prevents wrinkle formation, pigmentation and hyperprolfieration as well as improving skin hydration.
- However, the skin protection effects of dietary RG are varied depending on the amount. It is important to utilize an optimal amount of dietary RG for skin protection.

Summary points

- RG is one of the most traditional medicinal herbs, and contains active components such as ginsenosides.
- Ingested RG is influenced by intestinal bacteria, and further metabolized prior to reaching target tissues; therefore, the components themselves or their metabolites are able to exert beneficial effects on skin protection.
- Specifically, dietary RG inhibits wrinkle formation in both photo aged and intrinsically aged dermis with increased expression of procollagen I and decreased expression of MMPs due to altered expression or activity of AP-1 transcriptional factors.
- In addition, dietary RG protects the epidermis from photo aging induced pigmentation and hyperproliferation with inhibited expression of TGF-β1.
- Although epidermal hydration is caused by photo- and intrinsic aging, 0.5% dietary RG only improves hydration in photo-aged epidermis with increased content of ceramides, the major lipids of epidermal barrier. 1% dietary RG has no beneficial effect on it. The optimal amount of dietary RG should be defined, and mechanistic research on dietary RG for its beneficial effects on skin protection must be further elucidated in depth to provide solid scientific values.

15. Dietary red ginseng and skin protection

J.J. Wee[1] and Y. Cho[2]
[1]Ginseng Research Institute, Korea Ginseng Corporation, 22 Gajeong-ro, Shinseong-dong, Yuseong-gu, Daejeon, 305-805, Korea; [2]Department of Medical Nutrition, Graduate School of East-West Medical Science, Kyung Hee University, Yongin-si, Gyeonggi-do, 446-701, Korea; choyunhi@khu.ac.kr

Abstract

Red ginseng (RG) (the steamed root of *Panax ginseng* C.A. Meyer) has been widely used in the Orient and West as a tonic agent, health food, and/or alternative herbal medicine for a wide range of therapeutic applications. RG has also been reported to have therapeutic applications for burn wound healings, inflammation, allergies and aging of skin, but most of these beneficial effects of RG in skin have been based on *in vitro* research or topical application. In particular, the controversy with respect to scientific evidence on the dietary effects of RG, evaluations of its clinical efficacy on skin protection, and its solid scientific mechanism, remains to be elucidated in depth. However, coupled with developing new insights into the relationship between food, nutrition and skin in recent years, substantial advances have been made in studies of dietary RG and skin protection. In addition, studies on the absorption and metabolism of ginsenosides, the major active components of RG, have been receiving attention. This review paper is concerned with recent developments in our understanding of the metabolism of RG in the human body, as well as the dietary effects of RG on skin protection. Special attention is paid to the dietary effects of RG on wrinkle formation, skin pigmentation, epidermal hyperproliferation, and hydration.

Keywords: wrinkle, pigmentation, hyperproliferation, hydration

Abbreviations

AQP-3	Aquaporin-3
aSMase	acid sphingomyelinase
β-GlcCer'ase	β-Glucocerebrosidase
CER	Ceramide
C-K	Compound K
EGF	Epidermal growth factor
GC	Glucosylceramide
IGF-1	Insulin-like growth factor-1
IL	Interleukin
MAP	Mitogen-activated protein
MMP	Matrix metalloproteinase
NMF	Natural moisturizing factor
PPD	Protopanaxadiol
PPT	Protopanaxatriol
RG	Red ginseng
ROS	Reactive oxygen species
SC	*Stratum corneum*
SM	Sphingomyelin
SMase	Sphingomyelinase
SPT	Serine palmitoyltransferase
TGF-β1	Transforming growth factor-β1
TNF-α	Tumor necrosis factor-α
UV	Ultraviolet

15.1 Introduction

Ginseng radix, the dried root of *Panax ginseng* C.A. Meyer is widely used as a traditional oriental medicine as well as a supplemental medicine in Western countries (Radad *et al.*, 2006). Two distinct varieties of *Panax ginseng* are commercially available, namely red and white ginseng. The difference between them is their method of processing that results in different components; white ginseng (Ginseng radix alba) is produced by air-drying the roots under the sun, while RG (Ginseng radix rubra) is produced by steaming the roots followed by drying. The steaming processing for RG preparation has been found to exhibit inactivation of catabolic enzymes, thereby preventing deterioration of ginseng quality and lengthening its storage period over 10 years (Nam, 2005). In addition, the steaming processing for RG causes further degradation or transformation of ginsenosides (saponins), the major bioactive components of ginseng, which results in the occurrence of more ginsenosides in RG than in white ginseng.

Ingested RG is influenced by the gastrointestinal fluid in the digestive tracts, and is further metabolized prior to reaching target tissues (Hasegawa, 2004; Wang *et al.*, 2001). Each component itself or metabolized components of the ingested RG exert beneficial effects for various diseases including hypertension, diabetes, atherosclerosis, cancer, and post-menopausal disorder, and also counteracts weakness (Radad *et al.*, 2006). Recently, RG extracts have been used in the treatment of ultraviolet (UV)-irradiated tumors, burn wound healings, inflammation, and allergies of skin (Bae *et al.*, 2005; Kimura *et al.*, 2006; Lee *et al.*, 2009a) but most of these beneficial effects of RG in skin have been based on *in vitro* research or topical application. In particular, the controversy with respect to scientific evidence on the dietary effects of RG, evaluations of its clinical efficacy on skin protection, and its solid scientific mechanism, remains to be elucidated in depth. However, substantial advances have been made in studies of dietary RG and skin protection for the prevention of wrinkle formation, pigmentation, epidermal hyperproliferation and dryness. In addition, studies on the absorption and metabolism of ginsenosides have been receiving attention. This review paper is concerned with recent developments in our understanding of the major active components of RG, the metabolism of ingested RG in the human body, and dietary effects of RG in skin protection.

15.2 Ginsenosides and polysaccharides in red ginseng

Studies on the chemical components of ginseng have been performed largely, leading to the identification of ginsenosides, polysaccharides, polyacetylenic alcohols, antioxidants, and peptides. Among them, ginsenosides have been the target of numerous researches in chemistry, biochemistry, and pharmacology as the main active components of ginseng. Ginsenosides are classified into three groups by type of aglycone, namely protopanaxadiol, protopanaxatriol and oleanolic acid. The genuine structures of the PPD and PPT ginseng sapogenins are dammar-24-ene-3β,12β,20(S)-triol (PPD) and dammar-24-ene-3β,6α,12β,20(S)-tetrol (PPT), respectively (Huang, 1999). Ginsenosides have been named as ginsenoside Rx (x=o, a1, a2, b1, b2, b3, c,d,e, f, g1, g2, h1, h2) according to their moving distance on a thin layered chromatography plate (Shibata *et al.*, 1995). Their chemical structures differ from one another by the type of aglycone, number of sugars and their configuration, as well as their site of attachment (Figure 15.1). Sugars bind to the C-3 and C-20 positions of PPD, and the C-6 and C-20 positions of PPT (Figure 15.1A and B). Among the ginsenosides, ginsenoside Ro belongs to the oleanane-type saponins, which are commonly encountered in plants (Figure 15.1C). Another oleanane-type ginsenoside is polyacetylene ginsenoside Ro, which contains a polyacetylenyl ester at the C-6' position of a glucosyl moiety (Zhang *et al.*, 2002).

Naturally occurring ginsenosides undergo structural conversion during the steaming process such as in the manufacturing of red ginseng, in acidic conditions of the stomach, or by bacterial metabolism in the gastrointestinal tract. Some partially deglycosylated saponins such as ginsenosides Rh1, Rg2, and Rg3 are obtained from red ginseng as byproducts produced during steaming via deglycosylation at C-20 (Figure 15.1A and B). Ginsenoside Rg3 is generated from ginsenosides Rb1, Rb2, Rc, and Rd via deglycosylation at C-20. Ginsenoside Rg3 is further converted to ginsenoside Rh2 via deglycosylation of a terminal glucose at C-3. Ginsenoside Rh1 is generated from ginsenoside Rg1, and ginsenoside Rg2 is generated from ginsenoside Re both

A

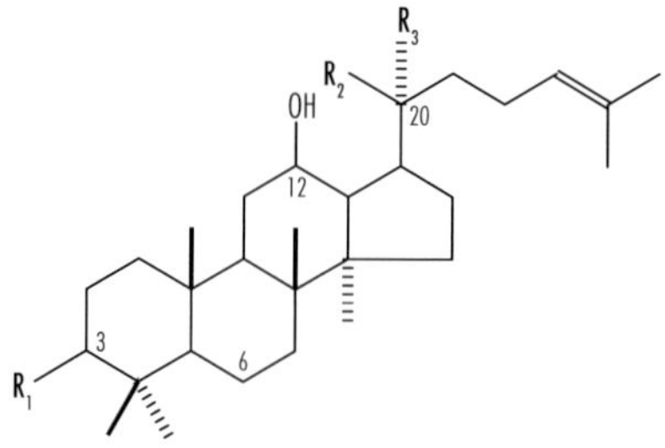

	R_1	R_2	R_3
Ginsenoside Rb_1	O-Glc(2→1)Glc	O-Glc(6→1)Glc	CH_3
Ginsenoside Rb_2	O-Glc(2→1)Glc	O-Glc(6→1)Arap	CH_3
Ginsenoside Rc	O-Glc(2→1)Glc	O-Glc(6→1)Araf	CH_3
Ginsenoside Rd	O-Glc(2→1)Glc	O-Glc	CH_3
20(S)-ginsenoside Rg_3	O-Glc(2→1)Glc	OH	CH_3
20(R)-ginsenoside Rg_3	O-Glc(2→1)Glc	CH_3	OH
Ginsenoside Rh_2	O-Glc	OH	CH_3
Ginsenoside Rs_1	O-Glc(2→1)Glc(6)Ac	O-Glc(6→1)Arap	CH_3
Malonyl ginsenoside Rb_1	O-Glc(2→1)Glc(6)Ma	O-Glc(6→1)Glc	CH_3
Compound K	OH	O-Glc	CH_3
20(S)-protopanaxadiol	OH	OH	CH_3

B

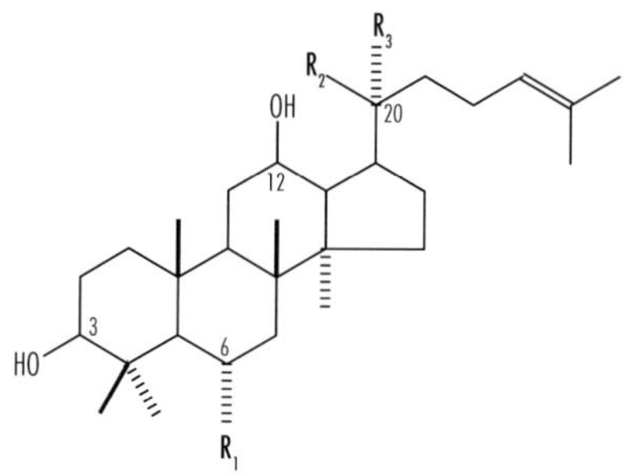

	R_1	R_2	R_3
Ginsenoside Re	O-Glc(2→1)Rha	O-Glc	CH_3
Ginsenoside Rf	O-Glc(2→1)Glc	OH	CH_3
Ginsenoside Rg_1	O-Glc	O-Glc	CH_3
20(S)-ginsenoside Rg_2	O-Glc(2→1)Rha	OH	CH_3
20(R)-ginsenoside Rg_2	O-Glc(2→1)Rha	CH_3	OH
20(R)-ginsenoside Rh_1	O-Glc	CH_3	OH
20(S)-protopanaxatriol	OH	OH	CH_3

C

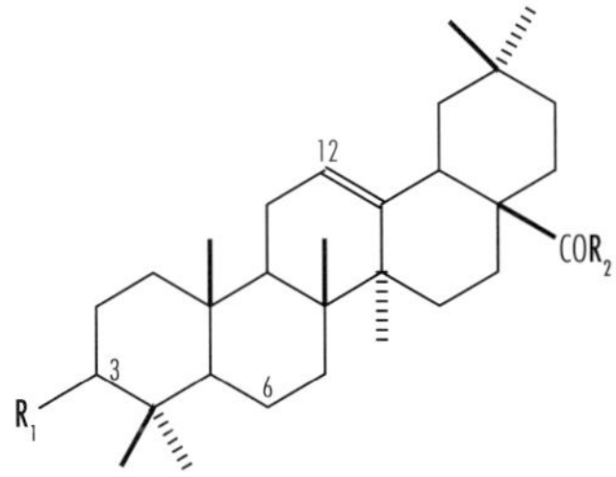

	R_1	R_2
Ginsenoside Ro	O-GlcUA(2→1)Glc	O-Glc

Figure 15.1. Chemical structures of protopanaxadiol (PPD) (A), protopanaxatriol (PPT) (B) and oleanane-type (C) ginsenosides.
Ac: Acetyl, Araf: α-L-arabinofuranosyl, Arap: α-L-arabinopyranosyl, Glc: β-D-glucopyranosyl, GlcUA: β-D-glucuronic acid, Ma: Malonyl, Rha: α-L-rhamnopyranosyl.

via deglycosylation at C-20. Elimination of the sugar moiety and subsequent epimerization of the hydroxyl group at the C-20 position yields the 20(R)-forms of ginsenosides Rh1, ginsenosides Rg2 and ginsenosides Rg3 as epimers. Ginsenoside Rs1-Rs7, containing acetyl group in the glucosyl moieties at C-3 or C-6 positions of their aglycones, are found only in red ginseng, thus it appears that the deacetylating enzymes are inactivated during the steaming process (Kasai *et al.*, 1983). The steaming or decoction of ginseng roots can also modify the chemical structure of side chains at the C-20 position by dehydration or hydration, leading to the generation of ginsenosides Rh4, Rg5, Rg6, and Rf2 (Park *et al.*, 2005). Christensen's review is helpful to better understand the chemistry of ginseng (Christensen, 2009).

Ginseng polysaccharides have also received attention in immunomodulating effects and acidic polysaccharides, such as ginsenan S-IA and ginsenan S-IIA are isolated from *Panax ginseng*. Ginsenan S-IIA is found to be composed of L-arabinose, D-galactose, D-glucose, and D-galacturonic acid, in the molar ratio of 15:10:2:5 (Tomoda *et al.*, 1993).

15.3 Absorption, metabolism and excretion of red ginseng in human body

Chemical components of red ginseng may be absorbed, in part, as each component itself or undergo metabolic changes both in the gastrointestinal tract and in the liver after absorption. In this respect, numerous studies have focused on the degradation or biotransformation of ginsenosides by bacterial metabolism in the gastrointestinal tract, in particular, compound K (20-O-β-D-glucopyranosyl-20(S)-protopanaxadiol, C-K), which is generated by the stepwise cleavage of sugars attached to the C-3 or C-20 hydroxyl groups of the aglycone.

The main human intestinal bacteria involved in this metabolism are species of the genera *Prevotella, Eubacterium, Streptococcus, Bacteroides,* and *Bifidobacterium*. The potential of intestinal bacteria to hydrolyze naturally occurring ginsenoside Rb1 to C-K (or termed M1) is found in 79% of fecal specimens from 58 human subjects. The major bacterial species possessing Rb1-hydrolyzing potential is *Prevotella oris* strains (Hasegawa *et al.*, 1997). *Bacteroides* spp., *Eubacterium* spp., and *Bifidobacterium* spp. potently transform ginsenoside Rc to C-K. In particular, *Bifidobacterium* K-103 and *Eubacterium* A-44 transform Rc to C-K via Rd (Bae *et al.*, 2002). Rg1 is converted into three metabolites, namely Rh1, F1, and 20(S)-PPT (or termed M4) through rat intestinal bacteria metabolism *in vitro* and *in vivo* (Wang *et al.*, 2001). The metabolic pathways of naturally occurring ginsenosides in the gastrointestinal tract can be summarized as follows: PPD-type, Rb1 → [M10 (Rd) → M5 (F2) or M9 → M13] → C-K (M1) → PPD (M12); Rb2 → M6 → M2 → C-K → PPD; Rc → M7 → M3 → C-K → PPD; PPT-type, Re → Rg1 → M11 (F1) or M8 (Rh1) → PPT (M4); Rg1 → F1 or Rh1 → PPT (Figure 15.2) (Hasegawa, 2004). Further, C-K is esterified with fatty acids in the liver at C-3 of the aglycone moiety or C-6 of the glucose moiety (Hasegawa *et al.*, 2000).

Since metabolic pathways of ginsenosides have been elucidated in the gastrointestinal tract, much attention has been placed on the pharmacokinetics of the metabolites, including C-K, PPD, and PPT. After the oral administration of ginseng saponins to rats (1 g/kg/day), C-K is detected in both blood and urine, whereas PPD and PPT are detected only in the blood. C-K is also detected in human urine (about 0.2 ug/ml) at 16-24 h after oral administration of ginseng extract (150 mg/kg/day) (Hasegawa *et al.*, 1996). After the oral administration of standardized Ginsana extract G115 (Pharmaton S.A., Lugano, Switzerland), three hydrolysis products of the PPT ginsenosides, namely Rh1, F1, and C-K, which are not originally present in the extract, are identified in both plasma and urine (Tawab *et al.*, 2003). Another pharmacokinetic study reveals that intact Rb1 is not detectable in serum at 24 h following the oral administration of Rb1, whereas the level of C-K in the serum reaches a maximum at 8 h (8.5 ug/ml) (Wakabayashi *et al.*, 1997). In contrast, when ginsenoside Rb1 (200 mg/kg) is orally administered to germ-free rats, neither C-K nor any other metabolites are detected in the plasma, intestinal tract, or cumulative feces 7 or 15 h after administration, whereas most of the ginsenoside Rb1 administered is recovered from the intestinal tract, indicating poor absorption of Rb1 (Akao *et al.*, 1998). Recently, a human study demonstrates that after oral administration of 12 g of powdered *Panax ginseng* to healthy subjects, C-K is absorbed into the blood until 24 h after dosing, with a Tmax of 10.76±2.07 h (Lee *et al.*, 2009b). However, it is unfortunate that no data on intact ginsenosides are included in the study. The studies above demonstrate that after administration of total saponins or ginseng extracts, original ginsenosides such as Rb1 and Rg1 are hardly detected in the plasma, urine, or feces, suggesting their poor bioavailability. Nonetheless, a pharmacokinetic study demonstrates that after oral administration of *Panax notoginseng* extract, Rb1 and Rg1 are absorbed through the digestive tract, with oral bioavailabilities of 4.35 and 18.40%, respectively. Also, for Rb1, the half-life of α-phase is found to be 23.40 min, and that of β-phase is 17.96 h, and for Rg1 the half- life of α-phase is 24.23 min and that of β-phase is 14.13 h, indicating that they both make two-compartment models (Xu *et al.*, 2003). Furthermore, another study demonstrates that after oral administration of *Panax notoginseng* extract, in which the major saponins are ginsenosides Ra3, Rb1, Rd, Re, Rg1, and notoginsenoside R1, ginsenosides Ra3, Rb1, Rd are measurable in rat

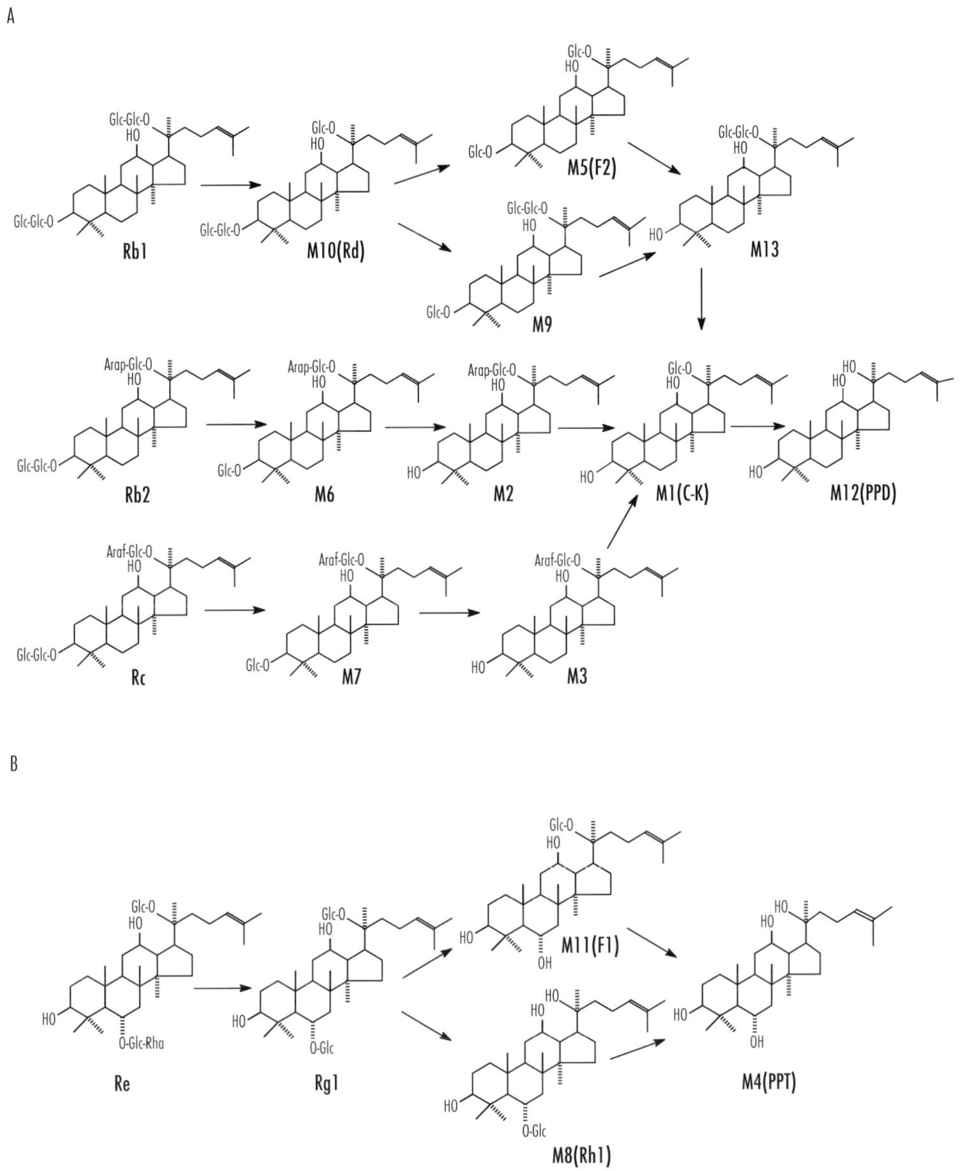

Figure 15.2. Metabolic pathways of (PPD) (A) and protopanaxatriol (PPT))-type (B) ginsenosides by intestinal flora.
Arap: α-L-arabinopyranosyl, Araf: α-L-arabinofuranosyl, Glc: β-D-glucopyranosyl, Rha: α-L-rhamnopyranosyl.

plasma up to 48 h and their maximal plasma concentration occurs 6-10 h after dosing, supporting the absorption of naturally occurring ginsenosides into the blood as intact forms (Liu *et al.*, 2009). The oral bioavailabilities of Rh2 are assessed to be 17.6% and 24.8% in male and female dogs, respectively (Xie *et al.*, 2005a), whereas in another study Rh2 bioavailability is about 5%

in rats and 16% in dogs (Gu *et al.*, 2009). Ginsenoside Rg3 and its metabolites Rh2 and PPD are simultaneously determined in rat plasma after oral administration of ginsenoside Rg3 (Xie *et al.*, 2005b). In contrast, ginsenoside Rg3 is not detected in rat plasma after oral administration, while 0.97-1.15% of ginsenoside Rg3, and hydrolysis and oxygenated metabolites are detected and identified in the rat feces (Qian *et al.*, 2005). In a human study that used intramuscular injection instead of oral administration, ginsenoside Rg3 surprisingly persists in the blood for 9 days (Zhao *et al.*, 2010). As described, the metabolic transformations of ginsenosides in the gastrointestinal tract play a role in the efficacy of ginseng. However, more studies are needed to conclude that the conversion of naturally occurring ginsenosides to C-K is essential for ginsenosides to be biologically active, even though it is obvious that C-K is generated following ginseng ingestion.

15.4 Dietary red ginseng and dermal protection

15.4.1 Dietary red ginseng and wrinkle formation

Wrinkles are major clinical manifestation of skin aging. Although intrinsic factors (the passage of time) also attribute somehow, wrinkle formation is mainly affected by extrinsic factors such as UV irradiation, free radicals and mechanical damages. Compared to fine wrinkle formation with time passage, extrinsic factors cause deep, coarse wrinkle formation in exposed skin (Helfrich *et al.*, 2008). Physiologically, wrinkle formation is associated with disorganization and the reductions of type I collagen, the principal component of the dermal layer. Inversely, wrinkle formation is also related to the increased levels of MMPs, endoproteases with a broad range of substrate specificities that are capable of degrading all extracellular matrixes including collagen and thus contribute to wrinkle formation (Helfrich *et al.*, 2008). Although both intrinsic and extrinsic factors of skin aging induce the expression of MMPs, UV radiation is one of the most important extrinsic factors of premature skin aging (Fisher *et al.*, 1997), thereby most of the effects of dietary RG on wrinkle formation have been reported in UV irradiated animals.

The pre-oral administration of RG extract powders (20 or 60 mg/kg BW, twice daily) for 16 days prevents reductions of skin elasticity induced by acute UVB-irradiation in male C57BL/6J mice (Kim *et al.*, 2008). When RG extract containing 242.5 mg/g saponins and ginsenoside Rb1 (43.5mg/g) as standardized major active components, is supplemented at either 0.5% (group H0.5+UV) or 2.5% (group H2.5+UV) of the diet along with chronic UV irradiation for 12 weeks in male SKH-1 hairless mice, visiometer analysis of skin replicas reveals that wrinkle formation in groups H0.5+UV and H2.5+UV is significantly less than that in UV irradiated group (group +UV) in 5 weeks (Kang *et al.*, 2009). However, wrinkles are also formed in group UV-non irradiated group (group -UV) after 12 weeks by intrinsic aging. There are no significant differences in wrinkle formation between -UV and +UV groups in 12 weeks. In addition, groups H0.5+UV and H2.5+UV do not exert inhibition effects of wrinkle formation, implying that although RG extract can inhibit wrinkle formation in photoaging, it is not effective for inhibiting the wrinkle formation of intrinsic aging. However, the protein expression of procollagen type I in the dermis of group H2.5+UV is significantly higher than that of group +UV in 12 weeks. In contrast, protein expression of MMP-1, one of the major MMPs in skin, is significantly decreased in group H2.5+UV while it is highly increased in group +UV. Although visiometer analysis of skin replicas is not enough to support, these data remains the possibility, in which the wrinkle formation induced by intrinsic aging may also be protected by dietary supplementation of RG extract.

15.4.2 Dietary red ginseng mix with *Torilis fructus* and *Corni fructus*, and wrinkle formation

When orally supplemented, RG extract powder is often prepared as a mixture with *Torilis fructus* and *Corni fructus* extracts (Cho *et al.*, 2009; So *et al.*, 2008). Based on the long history of its combination with RG as an effective treatment for itching, eczema, and edema (Anonymous, 1991; Jung and Sin, 1990), *Torilis fructus,* the fruit of *Torilis japonica,* is an efficacious anti-inflammatory (Jung and Sin, 1990). Similar to *Torilis fructus, Corni fructus*, the fruit of *Cornus officinalis* Sieb et Zucc, also has anti-inflammatory effects, and its antioxidant, anti-microbial, and anti-allergy activities have also been reported (Jung and Sin, 1990). When RG mix with these two herbs (300~1000 mg/kg BW/day) is supplemented with chronic UV-irradiation for 8 weeks in male SKH-1 hairless mice, mRNA and protein expressions of MMP-3 in dermis are decreased in a dose-dependent manner (So *et al.*, 2008). Of several possible mechanisms to alter the expression of MMPs or procollagen by UV irradiation, the photochemical generation of ROS, and subsequently increased expression or activated signaling of cytokines and growth factors in epidermal keratinocytes and dermal fibroblasts, have been most accepted (Callaghan and Wilhelm, 2008). Activated receptors of these cytokines and growth factors stimulate signal transduction cascades that induce MAP (mitogen-activated protein) kinase (p38) and transcription factor, AP-1, which stimulates the transcription of MMP genes (Rittie and Fisher, 2002). In fibroblasts, AP-1 is also reported to inhibit procollagen gene expression (Rittie and Fisher, 2002). Besides inhibited expression of MMP-3, RG mix inhibits the protein expression of TNF-α, as well as the protein expression and activation of MAP kinase and c-Jun, a major component of AP-1. These inhibited effects are most significant in group receiving a high dose (1000 mg/kg BW/day).

The efficacy of RG mix with *Torilis fructus* and *Corni fructus* in inhibiting wrinkle formation is further confirmed in a human study (Cho *et al.*, 2009). When healthy female volunteers over 40 years of age receive capsules contained RG extract mixed with the two herbs (3 g/day; RG: *Torilis fructus*: *Corni fructus* = 45.3:36.4:18.2 w/w/w) for 24 weeks, facial wrinkle formation is improved by 14.7~19.0% in 12 weeks and these improvement is further magnified (14.1~23.5%) at 24 weeks. In addition, mRNA and protein expressions of procollagen type I are increased while MMP-9 expression is not altered at 24 weeks. Furthermore, the length of fibrillin-1 fiber, another wrinkle-related biochemical marker, is elongated, suggesting that dietary RG mix does protect the wrinkle formation caused by intrinsic aging. The mechanism of intrinsic aging is far less clear than that of photoaging. Many theories have been proposed to explain intrinsic aging. As one example, the free radical theory states that intrinsic aging results from excess ROS generated as a consequence of oxidative metabolism in the human body (Hensley and Floyd, 2002). As a result, accumulated ROS induces cellular damages including oxidation of DNA resulting in mutations, oxidation of protein resulting in reduced function, and oxidation of membrane lipids resulting in reduced transport efficiency and possibly altered transmembrane signaling. Given the central role of ROS in both photoaging and intrinsic aging, it is possible that these two processes have common molecular mediators. In photoaging, AP-1 transcriptional factor is a critical mediator involved in both overexpression of MMPs and the reduction of type I procollagen (Rittie and Fisher, 2002). In intrinsically aged human skin, mRNA and protein expressions of c-Jun, a major component of AP-1 factor, are also increased, and an increase of AP-1 regulated MMP-1 and MMP-9 activities are further reported in aged human skin *in vivo* (Chung *et al.*, 2000). AP-1 factors are a common mediator of both photoaging and intrinsic aging, and a possible mechanism for the beneficial efficacy of dietary RG itself or when mixed with *Torilis fructus* and *Corni fructus* extracts is a decrease in AP-1 activity with a significant reduction of c-Jun protein expression, thereby causing increased expression of type I procollagen or the reduction of MMPs in aged skin (Figure 15.3).

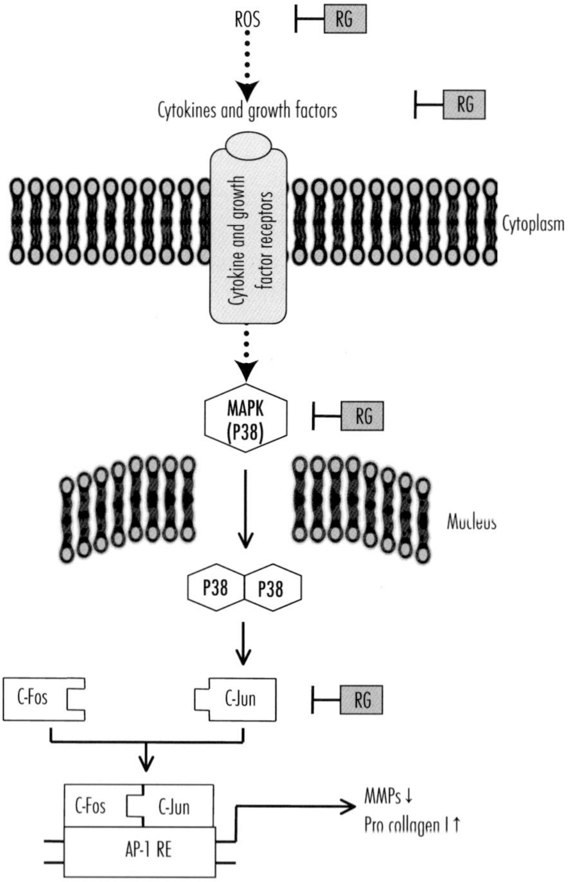

Figure 15.3. Schematic diagram of anti-wrinkle mechanism of dietary red ginseng in aged skin. MMP: Matrix metalloproteinases, MAPK: Mitogen-activated protein kinase, RG: Red ginseng.

15.5 Dietary red ginseng and epidermal protection

15.5.1 Dietary red ginseng and skin pigmentation

Photoaging also includes the clinical manifestation of irregular, dark pigmentation, which is caused by melanin production. Melanocytes in the epidermis are responsible for melanin production, in which melanosomes, lysosome like organelles filled with melanin, are produced. Repeated UV irradiation facilitates the distribution of melanosomes to the supranuclear areas of keratinocytes thereby protecting them against DNA damage, and it also further stimulates melanin production with melanocyte proliferation (Yamaguchi and Hearing, 2006). The pre-oral administration of RG extract (20 or 60 mg/kg, twice daily) prior to acute UVB-irradiation prevents skin pigmentation as well as reductions of skin elasticity induced by acute UVB-irradiation, which are parallel with reduced mRNA and protein expressions of TGF-β1 in the skin of male C57BL/6J mice (Kim *et al.*, 2008). Of several cytokines and growth factors in which mRNA and protein expressions are increased by UV-irradiation (Callaghan and Wilhelm, 2008), TGF-β1 can activate tumor progression, and furthermore, TGF-β1 is reported to promote melanocyte proliferation (Kawakami *et al.*, 2002). Therefore, it seems likely that the preventive effect of dietary RG extract on UV-induced pigmentation may be due to a reduction of TGF-β1 mediated melanocyte proliferation, thereby resulting in decreased melanin production.

15.5.2 Dietary red ginseng and epidermal hyperproliferation

In intrinsically aged skin with time passage, the epidermis is thinner with flattening of the dermal-epidermal junction (Helfrich *et al.*, 2008). The dermis also becomes atrophic, with decreased numbers of fibroblasts and decreased levels of collagen (Helfrich *et al.*, 2008). Photoaged skin, in contrast, is associated with epidermal hyperproliferation and increased epidermal thickness (Kim *et al.*, 2009; Kim *et al.*, 2008). The pre-oral administration of RG extracts (20 or 60 mg/kg, twice daily) prior to the acute UVB-irradiation prevents the increase of skin thickness as well as epidermal and corium thickness (Kim *et al.*, 2008). The dietary supplementation of RG along with chronic UV irradiation for 5 weeks also significantly decreases epidermal hyperproliferation to a similar level of the non UV-irradiated, normal control group (Kim *et al.*, 2009). In non UV-irradiated normal epidermis, EGF and IGF-1 enhance keratinocyte proliferation and TGF-β1 functions as a proliferation inhibitor of keratinocytes (Gniadecki, 1998). Coupled with the photochemical induction of cytokines and growth factor, and the subsequent signaling activation of these receptors, UV irradiation induces abnormal skin hyperproliferation, in which IL-1 is likely to initiate an excess of TGF- β1, activating the proliferation of keratinocytes (Gniadecki, 1998). As with the prevention of UV irradiated skin damage (increase of skin pigmentation and reduction of skin elasticity), the reduction of epidermal thickness with dietary RG seems to be due to decreased levels of TGF-β1 (Kim *et al.*, 2008), thereby preventing the hyperproliferation of keratinocytes in UV irradiated skin.

15.5.3 Dietary red ginseng and epidermal hydration

Maintenance of epidermal hydration is essential for skin flexibility, functioning, and maturation processes against arid environments (Harding *et al.*, 2000). Although other factors such as NMF, glycerol content, and the epidermal water-glycerol transporter, AQP-3, are also related, the maintenance of an optimal level of epidermal hydration largely depends upon the epidermal barrier, which functions as an efficient barrier against water loss through the skin (Elias and Feingold, 2006). The epidermal barrier is composed of intracellular lipid lamellae with Cers, cholesterol, and fatty acids in the SC, the outermost layer of the epidermis (Elias and Feingold, 2006). In UV-irradiated skin, the decrease in epidermal hydration correlates with the degree of epidermal barrier disruption (Holleran *et al.*, 1997). Moreover, it has been suggested that the disruption of the epidermal barrier in UV-irradiated skin is attributable to altered Cer levels in the epidermis. When RG extract is supplemented at either 0.5% (group H0.5) or 1.0% (group H1.0) of the diet, along with chronic UV irradiation for 5 weeks in male SKH-1 hairless mice, epidermal levels of hydration and total Cers in group H0.5 are higher than those of the UV irradiated group (group UV) and increased to a similar level of the non UV-irradiated, normal control (group C) (Kim *et al.*, 2009). Levels of total Cers are regulated by a balance between the activity and/or expression of ceramide-generating enzymes; SPT in the *de novo* synthesis pathway, SMase, and β-GlcCer'ase, as well as degradative enzymes such as ceramidase (Elias and Feingold, 2006). Although protein expression of SPT in group UV does not differ from that in group C, the decreased level of total Cers in group UV is paralleled with the highly increased protein expression of ceramidase. In group H0.5, protein expression of SPT is significantly higher than that of group C, but protein expression of ceramidase is higher than that of group C and comparable to that of group UV. However, epidermal levels of hydration and total Cers in group H1.0 do not differ from those in group UV. In addition, the protein expressions of SPT and ceramidase in group H 1.0 do not differ from those in group UV.

Cers are composed of a heterogenous family of various species (Cer 1-9) (Elias and Feingold, 2006). Once synthesized in *de novo* pathway, ceramides are glycosylated to GCs, and choline is phosphorylated to SMs, which become major precursors of each Cer species in the epidermal maturation process. As a general precursor, GCs are hydrolyzed by β-GlcCer'ase, which catalyses the hydrolysis of the glucose moiety linkage in GC to yield Cer 1 through Cer 9. As another precursor, SMs are hydrolyzed by aSMase, which catalyses the hydrolysis of the phosphodiester linkage in SM to yield Cer 2 and Cer 5 specifically. When altered levels of each Cer were further determined in group H0.5, the levels of Cer 2, a major ceramide in the epidermis of hairless mice (Elias and Feingold, 2006), and other Cer species such as Cer 5, and Cer 6/7 were significantly higher than those of group UV and were similar to those of group C (Figure 15.4). In addition, levels of acylGC, GC2, 3, and 4 in the epidermis of group H0.5 were similar to those of group C. However, SM did not significantly differ between groups. Similarly, protein expressions of β-GlcCer'ase and aSMase in group H0.5 were significantly increased to similar levels as group C (Figure 15.5). Taken together, these data suggest that dietary supplementation of 0.5% RG improves epidermal hydration in chronically UV-irradiated mice in parallel with ceramide accumulation that is coupled to elevated expression of SPT as well β-GlcCer'ase and aSMase

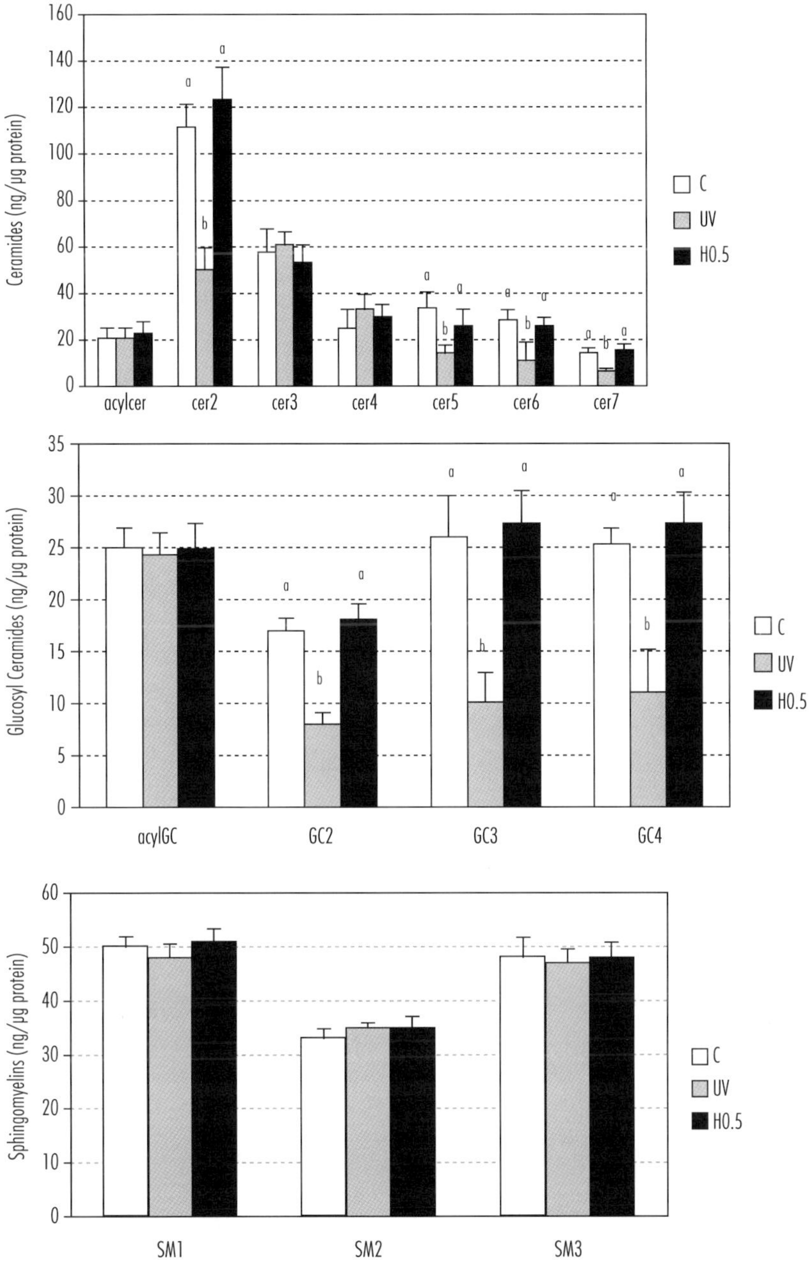

Figure 15.4. Levels of ceramides (Cer), glucosylceramides (GC) and sphingomyelins (SM) in the epidermis of hairless mice fed control diet without UV irradiation for 5 wks (group C), and UV radiated hairless mice fed control diet (group UV) or diet supplemented with 0.5 % (H0.5) powdered extracts of red ginseng in parallel with UV irradiation for 5 wks. Total lipids were fractionated into Cer, GC and SM fractions, which were further separated by high performance thin layered chromatography (HPTLC) into (A) cer 1-7, (B) acyl GC, GC B-D and (C) SM1, 2, and 3. Levels of each fraction were expressed as ng/μg protein. Values are mean ± SD (n=10). Values without a common letter are significantly different (*P*<0.05) using one-way ANOVA and Tukey's multiple comparison.

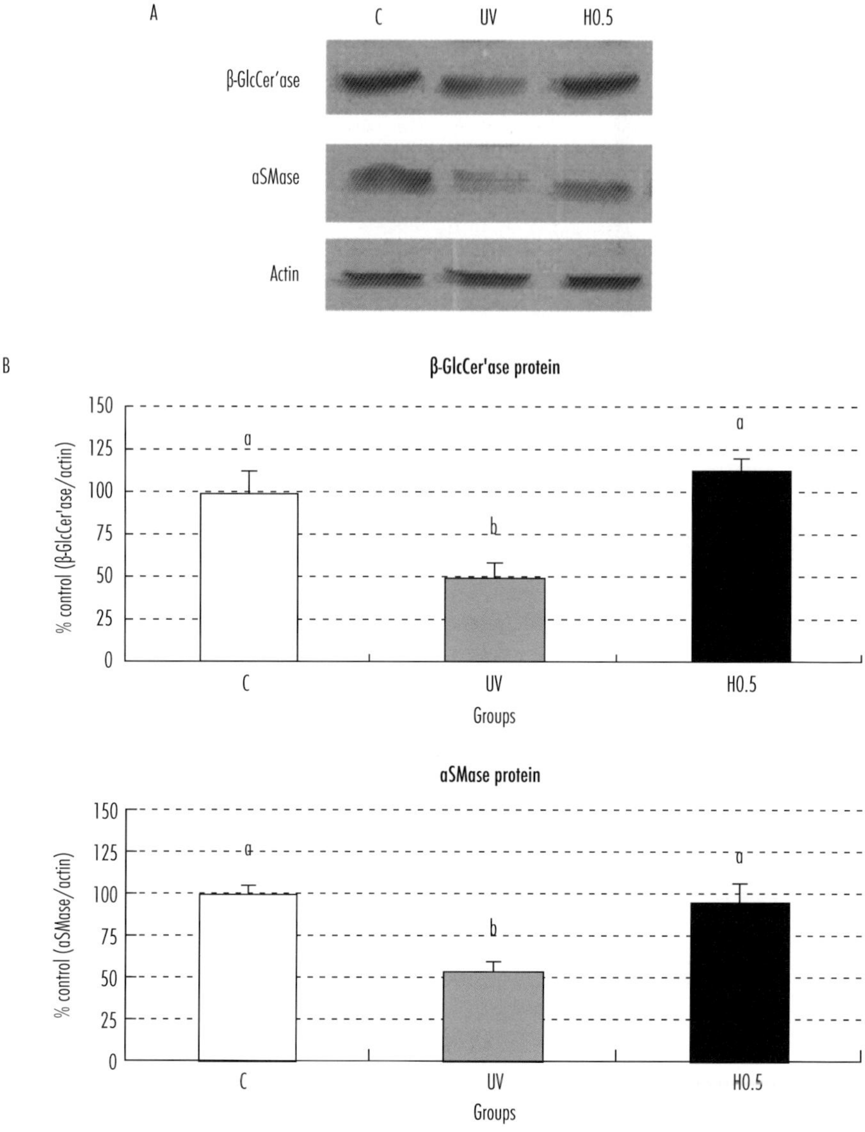

Figure 15.5. Altered protein levels of glucocerebrosidase (β-GlcCer'ase) and acid sphingomyelinase (aSMase) in the epidermis of hairless mice fed control diet without UV irradiation for 5 wks (group C), and UV radiated hairless mice fed control diet (group UV) or diet supplemented with 0.5 % (H0.5) powdered extracts of red ginseng in parallel with UV irradiation for 5 wks. (A) Representative expression of β-GlcCer'ase and aSMase proteins in epidermis of mice. Protein extracts (15 μg/each) from groups C, UV and H0.5 were subjected to 10% SDS-PAGE, and immunoblotted with anti-rabbit monoclonal antibody against β-GlcCer'ase (60 KDa), anti-goat polyclonal antibody against aSMase (55-90 KDa) or with actin specific antibody. (B) The signal intensities from multiple experiments of (A) were quantified and the integrated areas were normalized, first to the corresponding value of actin and then to the signal observed in normal control group (group C). All values are mean ± SD (n=10). Values without a common letter are significantly different (*P*<0.05) using one-way ANOVA and Tukey's multiple comparison.

in the metabolic pathway. However, the beneficial effects of dietary RG are not observed in the epidermis of H1.0 (Kim *et al.*, 2009). Since RG extract is comprised of multiple components, individual constituents may have additive or opposing effects on epidermal hydration. Therefore, it appears to be important to utilize an optimal amount of RG for dietary supplementation.

In the case of intrinsic aging, in which healthy female volunteers over 40 yrs receive capsules contained RG extract mixed with *Torilis fructus* and *Corni fructus* extracts for 24 weeks, epidermal hydration is not improved (Cho *et al.*, 2009). Although supplementation of dietary RG at the level of 0.5% in the diet enhances Cer content in UV-aged epidermis (Kim *et al.*, 2009), and saponins and phenolic components of RG such as maltol, salicylic acid, vanillic acid and *p*-coumaric acid exhibited free radical scavenging activity for skin protection (Kang *et al.*, 2006; Kim *et al.*, 1997), the inhibition of lipid peroxidation of the epidermal barrier and/or enhanced Cer content may not be enough to improve epidermal hydration in the case of intrinsic aging (Cho *et al.*, 2009). However, dietary supplementation of 0.5% dietary RG does improve epidermal hydration in photoaged epidermis with increased generation of ceramides, specifically Cer 2, 5, and 6 from GC and SM due to elevated expression of SPT protein as well as of β-GlcCer'ase and aSMase proteins.

References

Akao, T., Kida, H., Kanaoka, M., Hattori, M. and Kobashi, K., 1998. Intestinal bacterial hydrolysis is required for the appearance of compound K in rat plasma after oral administration of ginsenoside Rb1 from Panax ginseng. The Journal of Pharmacy and Pharmacology 50, 1155-1160.

Anonymous, 1991. Use of herbal ingredients. Il-wolseogag, Seoul, South-Korea, 535 pp.

Bae, E.A., Choo, M.K., Park, E.K., Park, S.Y., Shin, H.Y. and Kim, D.H., 2002. Metabolism of ginsenoside R(c) by human intestinal bacteria and its related antiallergic activity. Biological & Pharmaceutical Bulletin 25, 743-747.

Bae, E.A., Han, M.J., Shin, Y.W. and Kim, D.H., 2005. Antiallergic and antipsoriatic effects of Korean red ginseng. The Journal of Ginseng Research 29, 80-85.

Callaghan, T.M. and Wilhelm, K.P., 2008. A review of ageing and an examination of clinical methods in the assessment of ageing skin. Part I: Cellular and molecular perspectives of skin ageing. International Journal of Cosmetic Science 30, 313-322.

Cho, S., Won, C.H., Lee, D.H., Lee, M.J., Lee, S., So, S.H., Lee, S.K., Koo, B.S., Kim, N.M. and Chung, J.H., 2009. Red ginseng root extract mixed with *Torilus fructus* and *Corni fructus* improves facial wrinkles and increases type I procollagen synthesis in human skin: a randomized, double-blind, placebo-controlled study. The Journal of Medicinal Food 12, 1252-1259.

Christensen, L.P., 2009. Ginsenosides chemistry, biosynthesis, analysis, and potential health effects. Advances in Food and Nutrition Research 55, 1-99.

Chung, J.H., Kang, S., Varani, J., Lin, J., Fisher, G.J. and Voorhees, J.J., 2000. Decreased extracellular-signal-regulated kinase and increased stress-activated MAP kinase activities in aged human skin *in vivo*. Journal of Investigative Dermatology 115, 177-182.

Elias, P.M. and Feingold, K.R. (eds.), 2006. Skin barrier. *Stratum corneum* lipid processing: the final steps in barrier formation. Taylor & Francis, New York, NY, USA, 231 pp.

Fisher, G.J., Wang, Z.Q., Datta, S.C., Varani, J., Kang, S. and Voorhees, J.J., 1997. Pathophysiology of premature skin aging induced by ultraviolet light. New England Journal of Medicine 337, 1419-1428.

Gniadecki, R., 1998. Regulation of keratinocyte proliferation. General Pharmacology 30, 619-622.

Gu, Y., Wang, G.J., Sun, J.G., Jia, Y.W., Wang, W., Xu, M.J., Lv, T., Zheng, Y.T. and Sai, Y., 2009. Pharmacokinetic characterization of ginsenoside Rh2, an anticancer nutrient from ginseng, in rats and dogs. Food and Chemical Toxicology 47, 2257-2268.

Harding, C.R., Watkinson, A., Rawlings, A.V. and Scott, I.R., 2000. Dry skin, moisturization and corneodesmolysis. International Journal of Cosmetic Science 22, 21-52.

Hasegawa, H., 2004. Proof of the mysterious efficacy of ginseng: basic and clinical trials: metabolic activation of ginsenoside: deglycosylation by intestinal bacteria and esterification with fatty acid. The Journal of Pharmacological Sciences 95, 153-157.

Hasegawa, H., Lee, K.S., Nagaoka, T., Tezuka, Y., Uchiyama, M., Kadota, S. and Saiki, I., 2000. Pharmacokinetics of ginsenoside deglycosylated by intestinal bacteria and its transformation to biologically active fatty acid esters. Biological & Pharmaceutical Bulletin 23, 298-304.

Hasegawa, H., Sung, J.H. and Benno, Y., 1997. Role of human intestinal *Prevotella oris* in hydrolyzing ginseng saponins. Planta Medica 63, 436-440.

Hasegawa, H., Sung, J.H., Matsumiya, S. and Uchiyama, M., 1996. Main ginseng saponin metabolites formed by intestinal bacteria. Planta Medica 62, 453-457.

Helfrich, Y.R., Sachs, D.L. and Voorhees, J.J., 2008. Overview of skin aging and photoaging. Dermatology Nursing 20, 177-183.

Hensley, K. and Floyd, R.A., 2002. Reactive oxygen species and protein oxidation in aging: a look back, a look ahead. Archives of Biochemistry and Biophysics 397, 377-383.

Holleran, W.M., Uchida, Y., Halkier-Sorensen, L., Haratake, A., Hara, M., Epstein, J.H. and Elias, P.M., 1997. Structural and biochemical basis for the UVB-induced alterations in epidermal barrier function. Photodermatol Photoimmunol Photomed 13, 117-128.

Huang, C.C., 1999. The Pharmacology of Chinese Herbs. CRC Press, Boca Raton, FL, USA.

Jung, B.S. and Sin, M.G., 1990. Herbal Dictionary. Youngrimsa, Seoul, South-Korea, 416 pp.

Kang, K.S., Kim, H.Y., Pyo, J.S. and Yokozawa, T., 2006. Increase in the free radical scavenging activity of ginseng by heat-processing. Biological and Pharmaceutical Bulletin 29, 750-754.

Kang, T.H., Park, H.M., Kim, Y.B., Kim, H., Kim, N., Do, J.H., Kang, C., Cho, Y. and Kim, S.Y., 2009. Effects of red ginseng extract on UVB irradiation-induced skin aging in hairless mice. The Journal of Ethnopharmacology 123, 446-451.

Kasai, R., Besso, H., Tanaka, O., Saruwatari, Y.I. and Mizutare, T., 1983. Saponins of red ginseng. Chemical & Pharmaceutical Bulletin 31, 2120-2125.

Kawakami, T., Soma, Y., Kawa, Y., Ito, M., Yamasaki, E., Watabe, H., Hosaka, E., Yajima, K., Ohsumi, K. and Mizoguchi, M., 2002. Transforming growth factor beta1 regulates melanocyte proliferation and differentiation in mouse neural crest cells via stem cell factor/KIT signaling. The Journal of Investigative Dermatology 118, 471-478.

Kim, H., Oh, I., Park, K.H., Kim, N.M., Do, J.H. and Cho, Y., 2009. Stimulatory effect of dietary red ginseng on epidermal hydration and ceramide levels in ultraviolet-irradiated hairless mice. Journal of Medicinal Food 12, 746-754.

Kim, H.J., Chun, Y.J., Park, J.D., Kim, S.I., Roh, J.K. and Jeong, T.C., 1997. Protection of rat liver microsomes against carbon tetrachloride-induced lipid peroxidation by red ginseng saponin through cytochrome P450 inhibition. Planta Medica 63, 415-418.

Kim, Y.G., Sumiyoshi, M., Kawahira, K., Sakanaka, M. and Kimura, Y., 2008. Effects of Red Ginseng extract on ultraviolet B-irradiated skin change in C57BL mice. Phytotherapy Research 22, 1423-1427.

Kimura, Y., Sumiyoshi, M., Kawahira, K. and Sakanaka, M., 2006. Effects of ginseng saponins isolated from Red Ginseng roots on burn wound healing in mice. British Journal of Pharmacology 148, 860-870.

Lee, H.J., Kim, J.S., Song, M.S., Seo, H.S., Moon, C., Kim, J.C., Jo, S.K., Jang, J.S. and Kim, S.H., 2009a. Photoprotective effect of red ginseng against ultraviolet radiation-induced chronic skin damage in the hairless mouse. Phytotherapy Research 23, 399-403.

Lee, J., Lee, E., Kim, D., Yoo, J. and Koh, B., 2009b. Studies on absorption, distribution and metabolism of ginseng in humans after oral administration. The Journal of Ethnopharmacology 122, 143-148.

Liu, H., Yang, J., Du, F., Gao, X., Ma, X., Huang, Y., Xu, F., Niu, W., Wang, F., Mao, Y., Sun, Y., Lu, T., Liu, C., Zhang, B. and Li, C., 2009. Absorption and disposition of ginsenosides after oral administration of *Panax notoginseng* extract to rats. Drug Metabolism and Disposition 37, 2290-2298.

Nam, K.Y., 2005. The comparative understanding between red ginseng and white ginsengs, processed ginsengs (*Panax ginseng* C.A. Meyer). The Journal of Ginseng Research 29, 1-18.

Park, J., Rhee, D.K. and Lee, Y.C., 2005. Biological activities and chemistry of saponins from *Panax ginseng* C.A. Meyer. Phytochemistry Reviews 4, 159-175.

Qian, T., Cai, Z., Wong, R.N., Mak, N.K. and Jiang, Z.H., 2005. *In vivo* rat metabolism and pharmacokinetic studies of ginsenoside Rg3. Journal of chromatography. B, Analytical Technologies in the Biomedical and Life Sciences 816, 223-232.

Radad, K., Gille, G., Liu, L. and Rausch, W.D., 2006. Use of ginseng in medicine with emphasis on neurodegenerative disorders. The Journal of Pharmacological Sciences 100, 175-186.

Rittie, L. and Fisher, G.J., 2002. UV-light-induced signal cascades and skin aging. Ageing Research Reviews 1, 705-720.

Shibata, S., Tanaka, O., Shoji, J. and Saito, H., 1995. Chemistry and pharmacology of *Panax ginseng*. In: Wagner, H., Hikino H. and Farnsworth, N.R. (eds.), Economical and Medicinal Plant Research, vol 1. Academic Press, New York, NY, USA, pp. 217-284.

So, S.H., Lee, S.K., Hwang, E.I., Koo, B.S., Han, G.H., Chung, J.H., Lee, M.J. and Kim, N.M., 2008. Mechanisms of Korean red ginseng and herb extracts(KTNG0345) for anti-wrinkle activity. The Journal of Ginseng Research 32, 39-47.

Tawab, M.A., Bahr, U., Karas, M., Wurglics, M. and Schubert-Zsilavecz, M., 2003. Degradation of ginsenosides in humans after oral administration. Drug Metabolism & Disposition 31, 1065-1071.

Tomoda, M., Hirabayashi, K., Shimizu, N., Gonda, R., Ohara, N. and Takada, K., 1993. Characterization of two novel polysaccharides having immunological activities from the root of *Panax ginseng*. Biological & Pharmaceutical Bulletin 16, 1087-1090.

Wakabayashi, C., Hasegawa, H., Murata, J. and Saiki, I., 1997. *In vivo* antimetastatic action of ginseng protopanaxadiol saponins is based on their intestinal bacterial metabolites after oral administration. Oncology Research 9, 411-417.

Wang, Y., Liu, T.H., Wang, W. and Wang, B.X., 2001. Research on the transformation of ginsenoside Rg1 by intestinal flora. Zhongguo Zhong Yao Za Zhi 26, 188-190.

Xie, H.T., Wang, G.J., Lv, H., Sun, R.W., Jiang, X.L., Li, H., Wang, W., Huang, C.R. and Xu, M.J., 2005a. Development of a HPLC-MS assay for ginsenoside Rh2, a new anti-tumor substance from natural product and its pharacokinetic study in dogs. European Journal of Drug Metabolism and Pharmacokinetics 30, 63-67.

Xie, H.T., Wang, G.J., Sun, J.G., Tucker, I., Zhao, X.C., Xie, Y.Y., Li, H., Jiang, X.L., Wang, R., Xu, M.J. and Wang, W., 2005b. High performance liquid chromatographic-mass spectrometric determination of ginsenoside Rg3 and its metabolites in rat plasma using solid-phase extraction for pharmacokinetic studies. Journal of Chromatography. B, Analytical Technologies in the Biomedical and Life Sciences 818, 167-173.

Xu, Q.F., Fang, X.L. and Chen, D.F., 2003. Pharmacokinetics and bioavailability of ginsenoside Rb1 and Rg1 from *Panax notoginseng* in rats. The Journal of Ethnopharmacology 84, 187-192.

Yamaguchi, Y. and Hearing, V.J., 2006. Chapter 6: Melanocyte distribution and function in human skin: effects of UV radiation. In: Hearing, V.J. and Leong, S. (eds.), From melanocytes to malignant melanoma: the progression to malignancy. Humana Press, Totowa, NJ, USA, pp. 101-115.

Zhang, H., Lu, Z., Tan, G.T., Qiu, S., Farnsworth, N.R., Pezzuto, J.M. and Fong, H.S., 2002. Polyacetyleneginsenoside-Ro, a novel triterpene saponin from *Panax ginseng*. Tetrahedron Letters 43, 973-977.

Zhao, Q., Zheng, X., Jiang, J., Zhou, H. and Hu, P., 2010. Determination of ginsenoside Rg3 in human plasma and urine by high performance liquid chromatography-tandem mass spectrometry. Journal of Chromatography. B, Analytical Technologies in the Biomedical and Life Sciences 878, 2266-2273.

Key facts

- Solar ultraviolet (UV) radiation acts as a tumor initiator, tumor promoter as well as a complete carcinogen for induction of skin cancer.
- Exposure of the skin to solar UV radiation results in induction of oxidative stress, inflammation and inflammatory mediators, DNA damage and suppression of immune system.
- Altogether, UV radiation-induced adverse effects may result in various skin diseases including the development of skin cancers, which include melanoma and non-melanoma skin cancers.
- According to current projections, individuals with fair skin, particularly Caucasians, will develop at least one non-melanoma skin cancer during their life-time.
- Chemopreventive agents, such as phytochemicals, which possess anti-inflammatory, antioxidant properties, DNA repair abilities and can stimulate immune system, may be considered as ideal agents for the prevention of UV radiation-induced adverse effects.

Summary

- Grapes are one of the most widely consumed fruits all over the world.
- Seeds of the grapes are rich source of proanthocyanidins. Proanthocyanidins are found in the form of dimers, trimers, and polymerized oligomers of monomeric catechins.
- Grape seed proanthocyanidins (GSPs) possess strong anti-inflammatory and antioxidant properties.
- Supplementation of control diet with GSPs inhibits UV radiation-induced inflammatory responses, oxidative stress and inhibits UV radiation-induced immunosuppression in laboratory animals.
- GSPs inhibit UV radiation-induced immunosuppression through the induction of interleukin-12 in mice. Interleukin-12 has been shown to have immunostimulatory effect and has the ability to repair UVB-induced DNA damage.
- Dietary GSPs stimulate the development of $CD8^+$ effector T cells and enhance the secretion of Th1 cytokines. Th1 cytokines stimulate or activate the immune reactions or immune system.
- Dietary GSPs inactivate or diminish the development of $CD4^+$ regulatory T cells in UVB-exposed mice. GSPs also inhibit the secretion of Th2 cytokines which are mostly immunosuppressive in nature.
- Based on laboratory studies conducted in animal model, it is suggested that routine consumption of proanthocyanidins may provide efficient photoprotection against harmful effects of solar ultraviolet radiation.
- Proanthocyanidins in combination with sunscreens or skin care lotions may provide an effective strategy for reducing (non-)melanoma skin cancers and other skin disorders caused by excessive UV exposures.
- As grape seed proanthocyanidins are not genotoxic and posses low toxicity as indicated by some *in vitro* tests and *in vivo* animal toxicity studies reviewed elsewhere, the grape seed proanthocyanidins may be of interest for attenuation of the adverse UV radiation-induced effects on human skin.

16. Dietary grape seed proanthocyanidins and skin cancer

S.K. Katiyar
Department of Dermatology, School of Medicine, and Department of Environmental Health Sciences, School of Public Health, Veterans Affairs Medical Center, 1670 University Boulevard, Volker Hall 557, University of Alabama at Birmingham, Birmingham, AL 35294, USA; skatiyar@uab.edu

Abstract

Grapes (*Vitis vinifera*) are one of the most widely consumed fruits in the world. Grape seeds are the rich source of proanthocyanidins, and have been shown to have antioxidant, anti-inflammatory and anti-carcinogenic properties. Dietary grape seed proanthocyanidins (GSPs) inhibit ultraviolet radiation-induced skin cancer in mice in terms of tumor incidence and tumor multiplicity. Supplementation of diet with GSPs inhibited ultraviolet radiation-induced depletion of antioxidant defense enzymes, and ultraviolet radiation-induced oxidative stress-mediated phosphorylation of proteins of mitogen-activated protein kinase family and NF-κB signaling pathways in mouse skin. Dietary GSPs inhibit ultraviolet radiation-induced inflammation in the mouse skin, which includes the inhibition of infiltration of proinflammatory leukocytes and the levels of myeloperoxidase, cyclooxygenase-2, prostaglandin E_2 and proliferating cell nuclear antigen in the skin and skin tumors compared to non-GSPs-treated ultraviolet irradiated mouse skin and skin tumors. As ultraviolet radiation-induced immune suppression has been implicated in the development of skin cancer, the effect of dietary GSPs on ultraviolet radiation-induced immunosuppression was also investigated. Treatment of mice with dietary GSPs inhibits ultraviolet radiation-induced suppression of immune system, which was associated with the enhancement of the levels of interleukin-12 cytokine. The administration of GSPs did not prevent ultraviolet radiation-induced immunosuppression in interleukin-12 knockout mice but prevent it in their wild-types. Cell populations responsible for the GSPs-mediated inhibition of ultraviolet radiation-induced immunosuppression have been characterized. Adoptive transfer experiments revealed that naïve mice that received either $CD8^+$ effector T cells or $CD4^+$ regulatory T cells from GSPs-treated, ultraviolet radiation-exposed mice were able to show immunological responses after sensitization and subsequent challenge with 2,4-dinitrofluorobenzene. Studies suggest that GSPs inhibit ultraviolet-induced immunosuppression through interleukin-12-dependent stimulation of $CD8^+$ effector T cells and to diminish the development of regulatory $CD4^+$ T cells in mice.

Keywords: inflammation, oxidative stress, photocarcinogenesis, T cells, ultraviolet radiation

Abbreviations

CHS	Contact hypersensitivity
COX-2	Cyclooxygenase-2
GSPs	Grape seed proanthocyanidins
IL	Interleukin
MAPK	Mitogen-activated protein kinase
NF-κB	Nuclear factor-kappa B
PG	Prostaglandin
UV	Ultraviolet

16.1 Introduction

Proanthocyanidins are naturally occurring compounds that are widely found in fruits, vegetables, nuts, seeds, flowers and bark. They are a class of phenolic compounds that take the form of oligomers or polymers of polyhydroxy flavan-3-ol units, such as (+)-catechin and (-)-epicatechin (Yamakoshi *et al.*, 2002). These compounds are mostly found in pine bark, grape seed and red wines. However, bilberry, cranberry, black currant, green tea, black tea, and other plants also contain these flavonoids. The seeds of the grape (*Vitis vinifera*) are particularly rich source of proanthocyanidins. The GSPs are mainly found in the form of dimers, trimers and highly polymerized oligomers of monomeric catechins (Prieur *et al.*, 1994; Silva *et al.*, 1991). In comparison to other sources of proanthocyanidins, GSPs have been studied and explored in detail for their skin photoprotective potential. Therefore, emphasis has been placed to highlight and discuss the skin photoprotective effects of GSPs in this chapter. GSPs have been shown to be potent antioxidants and free radical scavengers (Bagchi *et al.*, 1997). In addition to have antioxidant activity, GSPs have been shown to have anti-inflammatory and anti-carcinogenic activity in cutaneous system and also in different tumor models (Mittal *et al.*, 2003; Nandakumar *et al.*, 2008). GSPs were subjected to a limited toxicity testing which included acute and subchronic toxicity in rats, and genotoxicity testing, comprising test for induction of gene mutation in bacteria, test for induction of chromosomal aberrations in mammalian cell *in vitro*, and mouse micronucleus test *in vivo*, the results of which indicate that these compounds are of a low toxicity and have no genotoxic potential (Yamakoshi *et al.*, 2002). As there has been considerable interest in the use of botanicals or phytochemicals for the prevention of various diseases, phytochemicals might be of interest as photoprotective agents against solar ultraviolet radiation, which is a major environmental factor affecting the skin and also responsible for the risk for skin cancers. Research on proanthocyanidins is however limited and many questions still remain to be answered. Although, dietary GSPs inhibit both chemical carcinogen- and UV radiation-induced skin tumor development, the present book chapter will highlight the investigations, developments and knowledge on the photoprotective effects of proanthocyanidins including molecular targets or mechanism *in vitro* cell culture and *in vivo* animal models. The bioavailability and metabolism of GSPs also have been discussed.

16.2 Chemistry of proanthocyanidins

Proanthocyanidins are synonymous with condensed tannins, and also known as oligomeric proanthocyanidins, pycno-genols or leukocyanidins, oligomers or polymers of flavan-3-ols and these units are linked mainly through C4→C8 bond, but the C4→C6 linkage also exists (Figure 16.1). These linkages are called B-type linkages. An additional ether bond between C2→C7 resulting in doubly linkage of the flavan-3-ol units is called an A-type linkage. The most common types of compounds and linkages are shown in Figure 16.1. The proanthocyanidins that exclusively consist of epicatechin units are designated procyanidins, the most abundant type of proanthocyanidins in plants. The less common proanthocyanidins containing epigallocatechin subunits are called prodelphinidin. The flavan-3-ol subunits may carry acyl substituents like gallic acid or glycosyl substituents like the sugars both of which may be linked at the C3 or C5 position of the oligomers (Santos-Buelga and Scalbert, 2000). The knowledge on the distribution and nature of proanthocyanidins in foods has until recently very limited; however the reported content of proanthocyanidins in various food items varies due to different analytical methods or to the

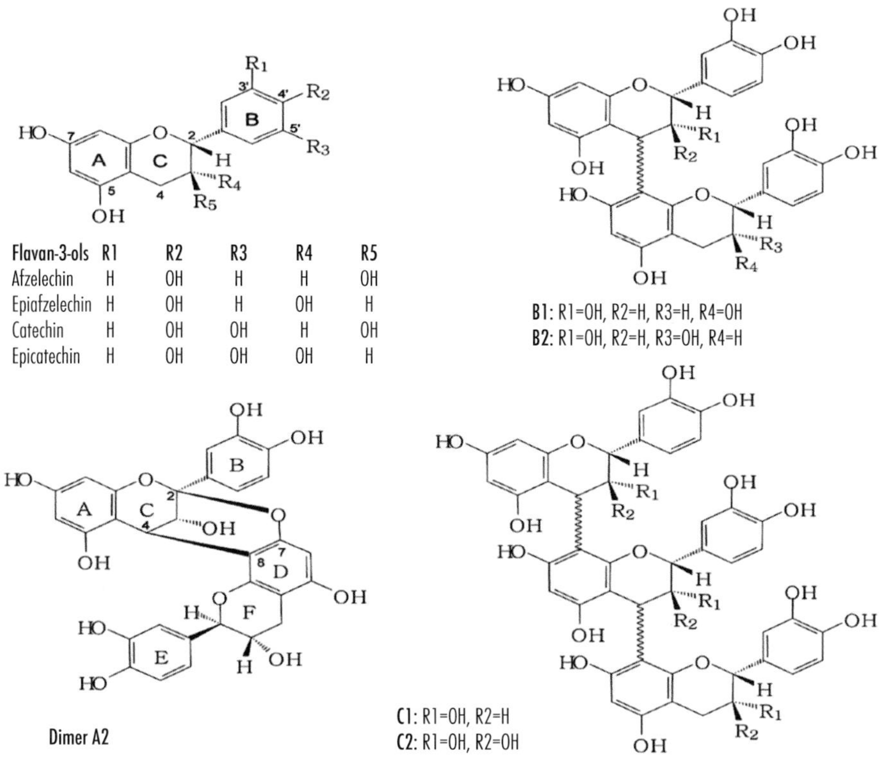

Flavan-3-ols	R1	R2	R3	R4	R5
Afzelechin	H	OH	H	H	OH
Epiafzelechin	H	OH	H	OH	H
Catechin	H	OH	OH	H	OH
Epicatechin	H	OH	OH	OH	H

Figure 16.1. Chemical structures of the monomer (flavan-3-ols), dimers (B1 and B2) and trimers (C1 and C2) are shown, which are normally found in grape seed proanthocyanidins. An example of the A-type double linkage is shown as dimer A2.

nature of the samples analyzed, variety, stage of ripeness, part of the food, level of processing, etc. (Santos-Buelga and Scalbert, 2000). Most of the plant-based foods, like fruits and berries, but also nuts, beans, some cereals foods, such as barley and sorghum, spices curry and cinnamon, wine and beers were found to contain exclusively the homogeneous B-type procyanidins.

16.3 Solar ultraviolet radiation

Solar UV radiation constitutes approximately 5% of the electromagnetic spectrum that reaches the earth's surface. The UV radiation which reaches on the earth's surface consists of 5-6% UVB and 90-95% UVA. Negligible amounts of UVC reach the earth's surface due to the filtering capacity of the ozone layer. Solar UV spectrum is mainly divided in to three main categories based on their wavelengths and adverse biological effects (Baliga and Katiyar, 2006; deGruijl and Van der Leun, 1994), as briefly discussed below:

1. Short-wave UVC (200-290 nm) spectrum: UVC radiation largely absorbed by the atmospheric ozone layer and does not reach to the surface of the earth. These wavelengths have enormous energy and are mutagenic in nature, and therefore are much more harmful than UVA or UVB radiation.
2. Mid-wave UVB (290-320 nm) spectrum: UVB radiation constitutes approximately 5% of the total solar UV radiation and responsible for variety of skin diseases including the nonmelanoma and melanoma skin cancers. UVB radiation can penetrate inside epidermis of the skin and can induce oxidative stress, DNA damage, cancer and premature aging of the skin. Excessive exposure of UVB radiation decreases the levels of antioxidant defense enzymes in the skin, impairing their ability to protect from harmful effects (deGruijl and Van der Leun, 1994; Katiyar *et al.*, 2007).
3. Long-wave UVA (320-400 nm) spectrum: UVA comprises the largest spectrum of solar UV radiation (90-95%). UVA penetrates deeper into the epidermis and dermis of the skin and it has recently been shown that extensive UVA exposure can lead to benign tumor formation as well as malignant cancers (Baliga and Katiyar, 2006). Its exposure induces the generation of singlet oxygen and hydroxyl free radicals, which can cause damage to cellular macromolecules, like proteins, lipids and DNA (Mukhtar and Elmets, 1996). UVA is a significant source of oxidative stress in human skin and causes photoaging in the form of skin sagging rather than wrinkling and can suppress some immunological functions (Katiyar, 2006; Ullrich, 1995).

UV radiation, particularly UVB, can act as a tumor initiator, tumor promoter and co-carcinogen (Baliga and Katiyar, 2006). Exposure of the skin to UVB radiation induces a variety of biological effects, including inflammation, sunburn cell formation, immunologic alterations, induction of oxidative stress and DNA damage, which play important roles in the generation and maintenance of UV-induced neoplasms (Hruza and Pentland, 1993; Mukhtar and Elmets, 1996). Although skin possesses an elaborate defense system consisting of enzymatic and non-enzymatic components to protect the skin from adverse biological effects of UV radiation; the excessive exposure to UV radiation overwhelms and depletes the cutaneous defense system and its ability, leading to the development of various skin diseases including the risk of cutaneous malignancies (Katiyar, 2006, 2007b; Mukhtar and Elmets, 1996).

16.4 Phytochemicals and photoprotection

There has been a considerable interest among human population for the use of naturally occurring botanicals or phytochemicals for the prevention of UV-induced photodamage, including the risk of skin cancer. Botanical supplements, specifically dietary botanicals, possessing anti-inflammatory, immunomodulatory and antioxidant properties are among the most promising group of compounds that can be exploited as ideal chemopreventive agents for variety of skin disorders in general and skin photoprotection in particular. Recent advances in our understanding at the cellular and molecular levels of carcinogenesis have led to the development of promising strategies for the prevention of cancer or so called 'chemoprevention'. Chemoprevention is a means of cancer control by the use of specific natural or synthetic chemical substances which can suppress, retard or reverse the process of carcinogenesis. In this respect, chemoprevention offers a realistic promise or strategy for controlling the risk of cancer. Chemopreventive approach appears to have practical implications in reducing skin cancer risk because unlike the carcinogenic environmental factors that are difficult to control, individuals can modify their dietary habits and lifestyle in combination with a careful use of skin care products to prevent photodamage and photocarcinogenesis. Studies from Dr. Katiyar's laboratory have shown the efficacy of various phytochemicals, such as green tea polyphenols, silymarin, honokiol and GSPs, against UV radiation-induced inflammation, oxidative stress and photocarcinogenesis (Katiyar, 2008; Nandakumar *et al.*, 2008). Here we will summarize and discuss the photoprotective potential of proanthocyanidins and particularly GSPs.

16.5 Prevention of experimental photocarcinogenesis by dietary GSPs

Photocarcinogenesis is a complex, multi-step process. UV-induced tumor development generally is considered to consist of three distinct stages: (1) tumor initiation, which is associated with the genotoxic effects of UV light on normal cells; (2) tumor promotion, which consists of clonal expansion of initiated cells, is a reversible stage; and (3) tumor progression, which consists of malignant transformation of papillomas to carcinomas through a process that appears to require additional genotoxic stimuli (Figure 16.2). As purified GSPs are dark brown in color, and may not be appreciated for direct topical application on the skin; however, their inclusion in diet is well tolerated. Therefore, the animal studies in author's laboratory were conducted after supplementing the GSPs with control AIN76A diet.

The chemical composition of the grape seed proanthocyanidins used in the authors laboratory (Gravinol, Kikkoman Corporation, Noda, Japan) is presented in Table 16.1. To evaluate the efficacy of GSPs in the prevention of photocarcinogenesis, GSPs were supplemented with AIN76A control diet of SKH-1 hairless mice at the levels of 0.2 and 0.5% (w/w). Administration

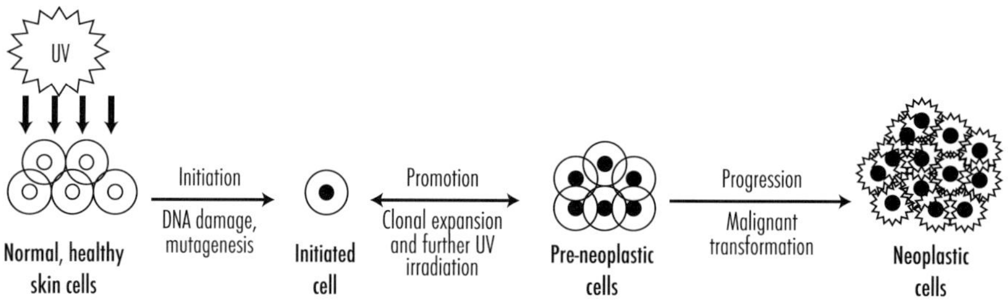

Figure 16.2. Schematic representation of UV radiation-induced multi-stage skin carcinogenesis. Multiple UV exposures are required for tumor growth and progression. Tumor promotion stage is reversible and suitable for controlling the risk of skin cancer by adopting chemopreventive strategy. UV radiation acts as a tumor initiator, promoter and complete carcinogen.

Table 16.1. Chemical composition of grape seed proanthocyanidins used in author's laboratory.

Components	% of total GSPs.
Total proanthocyanidins	89.3
Dimers	6.6
Trimers	5.0
Tetramers	2.9
Oligomers	74.8
Total monomeric flavanols	6.6
(+)-Catechin	2.5
(–)-Epicatechin	2.2
(–)-Epigallocatechin	1.4
(–)-Epigallocatechin-3-gallate	0.5
Moisture	2.2
Protein	1.1
Ash	0.8

Grape seed proanthocyanidins (GSPs) are provided regularly by Kikkoman Corporation (Tokyo, Japan) for the research work conducted in Dr. Katiyar's research laboratory. No financial conflict of interest.

of dietary GSPs inhibit UV radiation-induced skin tumor development in a dose-dependent manner in terms of percent of mice with tumors, tumor multiplicity, or tumor size as compared to non-GSPs-fed control group of mice (Mittal *et al.*, 2003). In the same study the control group (non-GSPs-fed group) exhibited 100% tumor incidence at week 15th of tumor promotion whereas the tumor incidence in mice which were given GSPs-supplemented diet (0.5%, w/w) exhibited only a 60% tumor incidence at this time point.

The photoprotective efficacy of GSPs also was determined in terms of their ability to prevent the malignant transformation of papillomas to carcinomas (Mittal *et al.*, 2003). Histological observations indicated that papillomas had started to transform into carcinomas during the 21st week of the tumor promotion protocol. A total of 70% mice in the non-GSPs-fed control group developed carcinomas, whereas only 25% of the mice in the GSPs-fed group developed carcinomas. These observations indicate that dietary GSPs may have the ability to inhibit UVB-induced transformation of benign papillomas to carcinomas.

16.6 Molecular mechanisms in prevention of photocarcinogenesis by GSPs

16.6.1 Inhibition of UV-induced oxidative stress and oxidative stress-mediated signaling

Wavelengths in the UVA and UVB region of the solar spectrum are absorbed by the skin and produce oxidative stress that contributes to the development of skin cancer (Katiyar, 2006). Acute UVB exposure as well as repeated multiple UVB exposures resulted in a reduction in the levels of antioxidant defense enzymes. The levels of reduced glutathione, glutathione peroxidase, and catalase in the UV-exposed skin as compared to the levels in the skin of control mice that were not exposed to UVB irradiation were decreased (Sharma *et al.*, 2007). The depletion in the levels of glutathione, glutathione peroxidase, and catalase in response to acute and repeated UV irradiation was reduced significantly on provision of GSPs (0.2% and 0.5%, w/w) in the diet. Catalase has a role in the catalytic conversion of H_2O_2 to oxygen and water and thus contributes to reductions in the levels of oxidative stress. It would be expected that the prevention of the UV-induced depletion of the antioxidant defense system would result in suppression of oxidative stress and the oxidative stress-mediated adverse effects in the skin. Oxidative stress may cause damage at the cellular level, as well as at the molecular level, and this can result in cutaneous inflammation, lipid and protein oxidation, DNA damage, and activation or inactivation of certain enzymes (Girotti, 1990; Sharma *et al.*, 2007). All these events potentially contribute to UVB-induced photodamage of the skin.

Under the above mentioned experimental conditions, acute or repeated exposure of the mouse skin to UVB resulted in an enhancement of the intracellular release of reactive oxygen species, including H_2O_2, in the skin compared to the non-UVB-exposed control mice. Administration of dietary GSPs significantly decreased UVB-induced intracellular release of reactive oxygen species in both acutely or repeatedly UVB exposed mouse skin (Sharma *et al.*, 2007). Oxidation of some amino acid residues of proteins leads to the formation of carbonyl derivatives that affects the nature and function of the proteins (Levine, 2002); thus, the analysis of carbonyl groups has been used as a measure of oxidative damage of proteins under conditions of oxidative stress. It was observed that acute and repeated UV exposure of the mouse skin resulted in a multi-fold increase in the levels of protein carbonyls as compared to the levels observed in the skin of non-UV-exposed control mice. The enhancement of protein carbonyl formation or protein oxidation after UV irradiation of the skin was reduced on provision of GSPs in the diet. Similarly, dietary GSPs significantly decreased UVB-induced production of nitric oxide in the skin under conditions of both acute and repeated UVB exposures (Sharma *et al.*, 2007).

MAPKs are important upstream regulators of transcription factor activities that control cellular proliferation, differentiation, and apoptosis in response to external signals or stimuli. UV-induced oxidative stress has been implicated in the activation of MAPK proteins. It was observed that UVB-induced phosphorylation of proteins of the MAPK family, such as ERK1/2, JNK and p38, in mouse skin was decreased by the dietary GSPs. These *in vivo* observations suggest that dietary GSPs may prove helpful in ameliorating the harmful effects caused by UV exposure through decreased reactive oxygen species production. Similar results were obtained when the photoprotective effects of GSPs were examined in terms of UV-induced oxidative stress in normal human epidermal keratinocytes *in vitro* (Mantena and Katiyar, 2006). Treatment of human epidermal keratinocytes with GSPs decreased UV-induced oxidative stress-mediated phosphorylation of the proteins of the MAPK family and activation of the transcription factor, NF-κB. There is evidence that NF-κB regulates a wide variety of genes that encode proteins that are involved in inflammation and carcinogenesis (Luque and Gelinas, 1997). Activation of NF-κB can up-regulate the expression of proinflammatory cytokines and inflammatory gene products, such as COX-2 and inducible nitric oxide synthase (Luque and Gelinas, 1997). Most of these genes have been shown to be up-regulated in human cancers, suggesting that inhibition of NF-κB, and subsequently NF-κB-targeted genes, might decrease the development of cancers including skin cancer. The decreasing expression of these NF-κB-targeted genes in UV-exposed mice by GSPs may explain the anti-proliferative, anti-oxidative, and anti-carcinogenic effects of GSPs (Sharma *et al.*, 2007).

16.6.2 Anti-inflammatory effect of grape seed proanthocyanidins

In addition to several other adverse biologic effects of UV radiation, UV-induced chronic and sustained inflammation has been implicated in skin tumor development. UV-induced inflammatory responses, which are characterized by increased blood flow and vascular permeability, result in the development of edema, erythema, hyperplastic responses, and increases in the levels of COX-2 and PG metabolites (Mukhtar and Elmets, 1996). UV-induced inflammation is considered as an early and important event in tumor promotion or the growth of skin tumors. Chronic inflammation plays an important role in all three stages of tumor development, i.e. initiation, promotion and progression (Mukhtar and Elmets, 1996; DiGiovanni, 1992). Therefore the control on UVB-induced inflammatory responses is considered as an important strategy to prevent skin cancer risk.

UVB radiation-induced infiltration and accumulation of activated macrophages and neutrophils in the mouse skin is a characteristic feature of skin inflammation, and the quantification of infiltrating leukocytes in skin is used routinely as a measure for the intensity of inflammation (Katiyar, 2006; Mukhtar and Elmets, 1996; Stanley *et al.*, 1991). The levels of myeloperoxidase are commonly used as a quantitative marker of inflammatory infiltrates. Normal skin exhibits low background levels of myeloperoxidase whereas skin that is inflamed by an infection, wounding, or exposure to UV radiation exhibits higher levels of myeloperoxidase. The analysis of the effects of GSPs showed that dietary GSPs inhibited UVB-induced infiltration of leukocytes in the mouse skin as well as reduced the levels of myeloperoxidase, and that suggest the lower levels of inflammation in UVB-irradiated mouse skin by dietary GSPs (Sharma and Katiyar, 2010). The reduction in myeloperoxidase activity in GSPs-fed group of mice after UVB exposure indicates the inhibition of influx of leukocytes to the inflamed skin. In the same study, dietary GSPs inhibited UVB-induced hyperplastic response in the skin. To compare the hyperplastic responses among the treatment groups, the epidermal thickness as well as cellular layers was measured along the entire length of the skin sections from the dermo-epidermal junction to the top of *stratum corneum*. In mice given the GSPs-supplemented diet, there was a significant reduction (>60%, $P<0.01$) in UVB-induced increase in epidermal thickness and vertical thickness of epidermal cell layers ($P<0.05$).

One of the most important enzymes in the process of inflammation and tumor development in UV-carcinogenesis is inducible COX-2. COX-2 is a rate-limiting enzyme for the generation of PG metabolites from arachidonic acid (Langenbach *et al.*, 1999). COX-2 overexpression has been linked to the pathophysiology of inflammation and cancer due to enhanced synthesis of PG metabolites which have been shown to be a potential contributing factor in the development of non-melanoma skin cancers (Mukhtar and Elmets, 1996; Williams *et al.*, 1999). A number of studies have demonstrated overexpression of COX-2 in chronically UVB-irradiated skin, as well as in UVB induced premalignant lesions and squamous cell carcinomas (Buckman *et al.*, 1998). A role for COX-2 in photocarcinogenesis is also supported by several studies that demonstrate that inhibition of COX-2 activity by specific inhibitors can partially block carcinogenesis induced by long-term UVB exposures. Administration of GSPs in diet reduced the expression of COX-2 in UVB-exposed skin. The levels of PGE_2 in the skin of the UVB-exposed mouse skin were also significantly decreased in GSPs-fed mice ($P<0.001$) compared with non-GSPs-fed UVB-exposed mouse skin. Exposure of the skin to UVB radiation also resulted in the induction of proinflammatory cytokines, and these responses are further enhanced by UVB-induced infiltrating leukocytes at the UVB-irradiated site of the skin. It has been shown that chronic exposure of the skin to UVB radiation resulted in significantly higher accumulation of TNF-α, IL-1β and IL-6 in the skin of mice as compared to non-UVB-exposed control mice. However, the levels of these proinflammatory cytokines were significantly inhibited by dietary GSPs in the UVB-irradiated skin of mice compared to non-GSPs-treated UVB exposed mice (Sharma and Katiyar, 2010)

In addition to the analysis of the effect of dietary GSPs on the status of UVB-induced inflammation and their mediators in the skin of mice, studies were also conducted to check the effect of GSPs on the status of these inflammatory mediators in the UVB-induced skin tumors. The UVB-induced skin tumors were subjected to the analysis of COX-2, proliferating cell nuclear antigen, PGE_2 and proinflammatory cytokines using various laboratory techniques. Resultant data revealed that GSPs inhibited the levels COX-2, PGE_2, and proliferating cell nuclear antigen in the skin tumors compared to non-GSPs-treated skin tumors. The levels of cyclin D1, another marker of cellular proliferation, were also inhibited by the administration of GSPs in the skin tumors than in the tumors of non-GSPs-treated mice. Analysis of the levels of proinflammatory cytokines, such as TNF-α, IL-1β and IL-6, in tumor samples was also performed. Dietary GSPs inhibited the levels of TNF-α ($P<0.001$), IL-6 ($P<0.01$) and IL-1β ($P<0.01$) in the skin tumors compared to non-GSPs-treated skin tumors (Sharma and Katiyar, 2010). Collectively, overall data suggest that anti-photocarcinogenic activity of GSPs is associated with the inhibition of UVB-induced inflammation and inhibition of inflammatory mediators in mouse skin.

16.6.3 Dietary GSPs inhibit UV-induced immunosuppression: role of interleukin-12

As UV-induced immunosuppression is considered to be a risk factor for the development of skin cancer (Katiyar, 2007a; Yoshikawa *et al.*, 1990; Meunier *et al.*, 1998), prevention of UV-induced immunosuppression represents a potential strategy for the management of skin cancer. When the effect of dietary GSPs on UV-induced immunosuppression was examined using a mouse contact hypersensitivity model, in which the contact sensitizer, 2,4-dinitrofluorbenzene is applied topically at the UV-exposed skin, the exposure of C3H/HeN mice resulted in suppression of the CHS response in terms of swelling of ear skin. The provision of GSPs in the diet of control mice that were not UVB irradiated did not affect their ability to generate a local CHS response to 2,4-dinitrofluorobenzene (Sharma and Katiyar, 2006). UVB irradiation of mice that did not receive GSPs resulted in a significantly lower (75% suppression, $P<0.001$) local CHS response than that observed in the mice that did not receive GSPs but were UV irradiated, confirming the immunosuppressive effect of the UVB irradiation. In contrast, the mice that received GSPs in the diet exhibited a significant reduction in UVB-induced suppression of the local CHS response. Similar effects of GSPs were also observed in systemic model of CHS where mice were sensitized with 2,4-dinitrofluorobenzene on a skin site distant from the site that is exposed to UV radiation (Sharma and Katiyar, 2006). These observations indicate that dietary GSPs are capable of protecting mice from UVB-induced immunosuppression in a local as well as systemic CHS model of immunosuppression. Several mechanisms have been proposed in UV-induced immune suppression. There are studies that implicate the immunoregulatory cytokine, IL-12, in the induction and elicitation of CHS, and CHS are considered to be a Th1-mediated immune response. Langerhans cells, which are critical antigen presenting cells in the induction phase of CHS (Toews *et al.*, 1980) have been described as an additional source of IL-12. On the other hand, IL-10 possesses immunosuppressive activity and inhibits antigen presentation in *in vitro* and *in vivo* systems (Fiorentino *et al.*, Mukhtar and Elmets, 1996). Dietary administration of GSPs to C3H/HeN mice resulted in a reduced level of IL-10 in the UV-irradiated skin, as well as in the draining lymph nodes, as compared to the controls that were not treated with GSPs. These *in vivo* effects of GSPs suggest a possible mechanism by which dietary GSPs decrease UVB-induced immune suppression in mice (Sharma and Katiyar, 2006).

IL-12 regulates the growth and functions of T-cells and especially augments the development of Th1 type cells by stimulating the production of IFN-γ (Katiyar, 2007b; Manetti *et al.*, 1993; Kobayashi *et al.*, 1989). Intraperitoneal injection of recombinant IL-12 in mice prevents UV-induced immune suppression (Schmitt *et al.*, 1995). The provision of dietary GSPs resulted in higher levels of IL-12 in the skin and draining lymph nodes of UVB-exposed C3H/HeN mice than those observed in UVB-exposed mice that did not receive GSPs (Sharma and Katiyar, 2006). These higher levels of IL-12 may contribute to stimulation of the immune responses. To further confirm the role of IL-12 in GSPs-mediated prevention of UV-induced immunosuppression, C3H/HeN mice were treated *i.p.* with an anti-IL-12 monoclonal antibody. In GSPs-treated mice, the *i.p.* injection of the anti-IL-12 antibody significantly reversed or blocked the preventive effect of GSPs on UV-induced suppression of CHS (Sharma and Katiyar, 2006). These studies provide convincing evidence that prevention of UV-induced suppression of CHS by GSPs is mediated, at least in part, through IL-12 induction, and that the protection from UVB-induced immunosuppression afforded by dietary GSPs may be associated with the protection from UVB-induced photocarcinogenesis in mice. Possible molecular targets of dietary GSPs on skin photoprotection in animal models are summarized in Figure 16.3.

16.6.4 Dietary GSPs inhibit UV-induced immunosuppression: role of T cells

To further elucidate the mechanisms by which GSPs counteract UVB-induced immunosuppression, Vaid *et al.* (2011) conducted some *in vivo* experiments and examined the role of T cells in this system. *In vivo* experiments suggested that the GSPs-induced prevention of immunosuppression is transferable to naïve mice. Further, adoptive transfer approaches were used to characterize

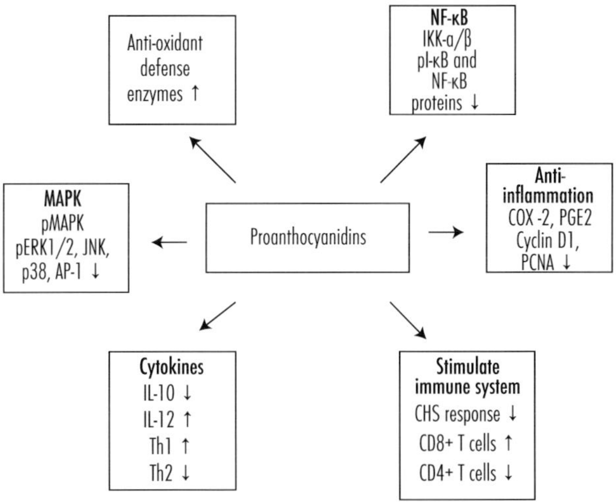

Figure 16.3. Molecular targets of grape seed proanthocyanidins (GSPs) in skin photoprotection. Dietary GSPs inhibit UVB-induced oxidative stress, inflammation, and immunosuppression. GSPs stimulate the development of CD8+ effector T cells while suppress the growth and development of CD4+ regulatory T cells, and their respective cytokines.
CHS: Contact hypersensitivity, COX-2: Cyclooxygenase-2, IL: Interleukin, MAPK: Mitogen-activated protein kinase, NF-κB: Nuclear factor-kappa.

the cells that mediate the chemopreventive effects of GSPs on immune system *in vivo*. Studies revealed that dietary GSPs prevent UVB-induced immunosuppression through stimulation and/or enhanced development of $CD8^+$ effector T cells and that the dietary GSPs enhance the ability of the $CD8^+$ T cells to secrete Th1 cytokines. Thus, these results suggest that GSPs stimulate the development and the activity of $CD8^+$ effector T cells. The studies conducted by these authors also suggested that inhibition of the development of regulatory T-cells and/or inactivation of $CD4^+$ suppressor T cells also plays a role in the prevention of UVB-induced suppression of the CHS response by GSPs and this was borne out by the significant inhibitory effects of dietary GSPs on the ability of the $CD4^+$ T cells to produce Th2 cytokines (IL-4 and IL-10) (Vaid *et al.*, 2011). Thus, $CD8^+$ T cells are the critical effector cells, a finding that is in accordance with the findings of other investigators (Gocinski and Tigelaar, 1990; Xu *et al.*, 1996). Xu *et al.* (1996) reported that CHS is mediated through $CD8^+$ T cells, whereas $CD4^+$ Th2 cells exhibit an inhibitory effect on CHS, and this observation was supported by the findings of Gocinski and Tigelaar (1990) and Anderson *et al.* (1995) who reported that depletion of $CD4^+$ T cells before sensitization results in an enhanced ear swelling response. Moreover, the transfer of $CD4^+$ regulatory T cells from those donor mice that were treated with IL-12 into naïve mice resulted in an enhanced CHS response further suggests that IL-12 plays a role in inhibiting $CD4^+$ T cells with this concept being supported by the inhibitory effects of IL-12 on the secretion of Th2 type cytokines, and this effect was found to be similar in magnitude to the effects of treatment of donor mice with GSPs+ UVB treatment.

As cytokines play a critical role in immune system, the cytokine profiles of $CD8^+$ and $CD4^+$ T cells obtained from mice that were exposed to UVB but not treated with GSPs and those that were GSPs treated and UVB exposed were determined, and to delineate the relationship of these profiles with the inhibitory effect of dietary GSPs on the UVB-induced immunosuppression (Vaid *et al.*, 2011). It was observed that the levels of Th1 cytokines (IFNγ, IL-2) were much higher in $CD8^+$ T cells from GSPs-treated mice, whereas the Th2 cytokines (IL-4 and IL-10) were hardly detectable. These alterations in cytokine profile of $CD8^+$ T cells under the influence of GSPs may have a role in the enhancement of immune reactions. IFNγ-producing T cells are important effector cells in the CHS response and also are involved in reducing the development of UVB-induced skin tumors (Vaid *et al.*, 2011). These observations also suggest that the immunopreventive effect of GSPs against UVB-induced immunosuppression are mediated, at least in part, through the inhibition of the development and/or inactivation of $CD4^+$ regulatory T cells, which was evident in the significant reduction of Th2 cytokine (IL-4 and IL-10) levels in the cell supernatants obtained from bone marrow derived dendritic cells-stimulated $CD4^+$ T cells as compared to the $CD4^+$ T cells obtained from UVB irradiated mice that were not treated with GSPs. The Th2 cytokines, IL-4 and IL-10, have been implicated in the immunosuppressive effects and the development of Th2 or $CD4^+$ cells. This property of GSPs can be used as an alternative strategy to augment the induction of $CD8^+$ effector T cells and to diminish the development of $CD4^+$ regulatory T cells and that may lead to the prevention of photocarcinogenesis in humans.

16.7 Bioavailability and metabolism of proanthocyanidins

The study of bioavailability and metabolism of any medicinal or edible phytochemical is an important part of all investigations. Some studies have been performed to examine the bioavailability and metabolism of proanthocyanidins. One-half of 88 tested foods derived from plants were found to be dietary sources of proanthocyanidins, which suggests that these are among the most abundant polyphenols in our diet (Gu *et al.*, 2003). Polymeric proanthocyanidins are not absorbed as such in the gut (Manach *et al.*, 2005). The detection of proanthocyanidin dimers B1 and B2 in human plasma was reported in studies (Holt *et al.*, 2002). The absorption of these dimers was 100-fold lower than that of the monomeric flavanols. However, these compounds were found to have direct effects on the intestinal mucosa and protect it against oxidative stress or the actions of carcinogens. In addition, the consumption of proanthocyanidin-rich foods, such as cocoa, red wine, or grape seed extracts, has been shown to increase the plasma antioxidant capacity, to have positive effects on vascular function, and to reduce platelet activity in humans. These procyanidin-rich sources always contain 5-25% monomers or other polyphenols with which the proanthocyanidins may have effects through interactions with other components, such as lipids or iron, in the gut (Manach *et al.*, 2005). Rios *et al.* (2002) demonstrated the prevention of the polymer-degradation during the stomach transit due to the buffering effect exhibited by the food bolus, making the acidic conditions milder than required for proanthocyanidin hydrolysis.

The incubation of purified, ^{14}C-labeled, proanthocyanidin oligomers with human colonic microflora led to the formation of *m*-hydroxyphenylpropionic acid, *m*-hydroxyphenylacetic acid, and their *p*-hydroxy isomers, *m*-hydroxyphenylvaleric acid, phenylpropionic acid, phenylacetic acid, and benzoic acid (Déprez *et al.*, 2000). Gonthier *et al.* (2003) showed that the extent of degradation into aromatic acids decreased by 21-times in the polymer feed as compared to the catechin monomers fed to rats, probably because of the antimicrobial properties and protein-binding capacity frequently described for proanthocyanidins. But the degradation of proanthocyanidins into microbial metabolites needs more investigations.

Acknowledgments

The work reported from Dr. Katiyar's laboratory was supported by the funds from National Cancer Institute/NIH (CA104428) and Department of Veterans Administration Merit Review Award (S.K.K.). The content of this chapter does not necessarily reflect the views or policies of the funding sources. Grateful thanks are due to our former and current postdoctoral fellows for their outstanding contributions.

References

Anderson, C., Hehr, A., Robbins, R., Hasan, R., Athar, M., Mukhtar, H. and Elmets, C.A., 1995. Metabolic requirements for induction of contact hypersensitivity to immunotoxic polyaromatic hydrocarbons. Journal of Immunology 155, 3530-3537.

Bagchi, D., Garg, A., Krohn, R.L., Bagchi, M., Tran, M.X. and Stohs, S.J., 1997. Oxygen free radical scavenging abilities of vitamins C and E, and a grape seed proanthocyanidin extract *in vitro*. Research Communication Molecular Pathology and Pharmacology 95, 179-189.

Baliga, M.S. and Katiyar, S.K., 2006. Chemoprevention of photocarcinogenesis by selected dietary botanicals. Photochemistry Photobiology Science 5, 243-253.

Buckman, S.Y., Gresham, A., Hale, P., Hruza, G., Anast, J., Masferrer, J. and Pentland, A.P., 1998. COX-2 expression is induced by UVB exposure in human skin: implications for the development of skin cancer. Carcinogenesis 19, 723-729.

DeGruijl, F.R. and Van der Leun, J.C., 1994. Estimate of the wavelength dependency of ultraviolet carcinogenesis in humans and its relevance to the risk assessment of stratospheric ozone depletion. Health Physiology 67, 319-325.

Déprez, S., Brezillon, C., Rabot, S., Philippe, C., Mila, I., Lapierre, C. and Scalbert, A., 2000. Polymeric proanthocyanidins are catabolized by a human colonic microflora into low molecular weight phenolic acids. Journal of Nutrition 130, 2733-2738.

DiGiovanni, J., 1992. Multistage carcinogenesis in mouse skin. Pharmacology Therapy 54, 63-128.

Fiorentino, D.F., Zlotnik, A., Vieira, P., Mosmann, T.R., Howard, M., Moore, K.W. and O'Garra, A., 1991. IL-10 acts on the antigen-presenting cell to inhibit cytokine production by Th1 cells. Journal of Immunology 146, 3444-3451.

Girotti, A.W., 1990. Photodynamic lipid peroxidation in biological systems. Photochemistry Photobiology 51, 497-509.

Gocinski, B.L. and Tigelaar, R.E., 1990. Roles of CD4+ and CD8+ T cells in murine contact sensitivity revealed by *in vivo* monoclonal antibody depletion. Journal of Immunology 144, 4121-4128.

Gonthier, M.P., Donovan, J.L., Texier, O., Felgines, C., Remesy, C. and Scalbert, A., 2003. Metabolism of dietary procyanidins in rats. Free Radical Biology and Medicine 35, 837-844.

Gu, L., Gu, L., Kelm, M.A., Hammerstone, J.F., Beecher, G., Holden, J., Haytowitz, D. and Prior, R.L., 2003. Screening of foods containing proanthocyanidins and their structural characterization using LC-MS/MS and thiolytic degradation. Journal of Agriculture and Food Chemistry 51, 7513-7521.

Holt, R.R., Lazarus, S.A., Sullards, M.C., Zhu, Q.Y., Schramm, D.D., Hammerstone, J.F., Fraga, C.G., Schmitz, H.H. and Keen, C.L., 2002. Procyanidin dimer B2 [epicatechin-(4β-8)-epicatechin] in human plasma after the consumption of a flavanol-rich cocoa. American Journal of Clinical Nutrition 76, 798-804.

Hruza, L.L. and Pentland, A.P., 1993. Mechanisms of UV-induced inflammation. Journal of Investigative Dermatology 100, 35S-41S.

Katiyar, S., Elmets, C.A. and Katiyar, S.K., 2007. Green tea and skin cancer: Photoimmunology, angiogenesis and DNA repair. Journal of Nutritional Biochemistry 18, 287-296.

Katiyar, S.K., 2006. Oxidative stress and photocarcinogenesis: Strategies for prevention. In: Singh, K.K. (ed.), Oxidative Stress, Disease and Cancer. Imperial College Press, London, UK, pp. 933-964.

Katiyar, S.K., 2007a. UV-induced immune suppression and photocarcinogenesis: Chemoprevention by dietary botanical agents. Cancer Letters 255, 1-11.

Katiyar, S.K., 2007b. Interleukin-12 and photocarcinogenesis. Toxicology and Applied Pharmacology 224, 220-227.

Katiyar, S.K., 2008. Grape seed proanthocyanidines and skin cancer prevention: Inhibition of oxidative stress and protection of immune system. Molecular Nutrition and Food Research 52, S71-S76.

Kobayashi, M., Fitz, L., Ryan, M., Hewick, R.M., Clark, S.C., Chan, S., Loudon, R., Sherman, F., Perussia, B. and Trinchieri, G., 1989. Identification and purification of natural killer cell stimulatory factor (NKSF), a cytokine with multiple biologic effects on human lymphocytes. Journal of Experimental Medicine 170, 827-845.

Langenbach, R., Loftin, C.D., Lee, C. and Tiano, H., 1999. Cyclooxygenase-deficient mice. A summary of their characteristics and susceptibilities to inflammation and carcinogenesis. Annals of New York Academy of Sciences 889, 52-61.

Levine, R.L., 2002. Carbonyl modified proteins in cellular regulation, aging, and disease. Free Radical Biology and Medicine 32, 790-796.

Luque, I. and Gelinas, C., 1997. Rel/NF-κB and IκB factors in oncogenesis. Seminal Cancer Biology 8, 103-111.

Manach, C., Williamson, G., Morand, C., Scalbert, A. and Rémésy, C., 2005. Bioavailability and bioefficacy of polyphenols in humans. I. Review of 97 bioavailability studies. American Journal of Clinical Nutrition 81, 230S-242S.

Manetti, R., Parronchi, P., Giudizi, M.G., Piccinni, M.P., Maggi, E., Trinchieri, G. and Romagnani, S., 1993. Natural killer cell stimulatory factor (interleukin-12 [IL-12]) induces T helper type 1 (Th1)-specific immune responses and inhibits the development of IL-4-producing Th cells. Journal of Experimental Medicine 177, 1199-1204.

Mantena, S.K. and Katiyar, S.K., 2006. Grape seed proanthocyanidins inhibit UV radiation-induced oxidative stress and activation of MAPK and NF-κB signaling in human epidermal keratinocytes. Free Radical Biology and Medicine 40, 1603-1614.

Meunier, L., Raison-Peyron, N. and Meynadier, J., 1998. UV-induced immunosuppression and skin cancers. Review Medicine Interne 19, 247-254.

Mittal, A., Elmets, C.A. and Katiyar, S.K., 2003. Dietary feeding of proanthocyanidins from grape seeds prevents photocarcinogenesis in SKH-1 hairless mice: Relationship to decreased fat and lipid peroxidation. Carcinogenesis 24, 1379-1388.

Mukhtar, H. and Elmets, C.A., 1996. Photocarcinogenesis: Mechanisms, models and human health implications. Photochemistry and Photobiology 63, 355-447.

Nandakumar, V., Singh, T. and Katiyar, S.K., 2008. Multi-targeted prevention and therapy of cancer by proanthocyanidins. Cancer Letters 375, 162-167.

Prieur, C., 1994. Oligomeric and polymeric procyanidins from grape seeds. Phytochemistry 36, 781-790.

Rios, L.Y., Bennett, R.N., Lazarus, S.A., Rémésy, C., Scalbert, A. and Williamson, G., 2002. Cocoa procyanidins are stable during gastric transit in humans. American Journal of Clinical Nutrition 76, 1106-1110.

Santos-Buelga, C. and Scalbert, A., 2000. Proanthocyanidins and tannin-like compounds- nature, occurrence, dietary intake and effects on nutrition and health. Journal of Science Food and Agriculture 80, 1094-1104.

Schmitt, D.A., Owen-Schaub, L. and Ullrich, S.E., 1995. Effect of IL-12 on immune suppression and suppressor cell induction by ultraviolet radiation. Journal of Immunology 154, 5114-5120.

Sharma, S.D. and Katiyar, S.K., 2006. Dietary grape-seed proanthocyanidins inhibition of ultraviolet B-induced immune suppression is associated with induction of IL-12. Carcinogenesis 27, 95-102.

Sharma, S.D. and Katiyar, S.K., 2010. Dietary grape seed proanthocyanidins inhibit UVB-induced cyclooxygenase-2 expression and other inflammatory mediators in UVB-exposed skin and skin tumors of SKH-1 hairless mice. Pharmaceutical Research 27, 1092-1102.

Sharma, S.D., Meeran, S.M. and Katiyar, S.K., 2007. Dietary grape seed proanthocyanidins inhibit UVB-induced oxidative stress and activation of mitogen-activated protein kinases and nuclear factor-κB signaling in *in vivo* SKH-1 hairless mice. Molecular Cancer Therapeutics 6, 995-1005.

Silva, R.C., Rigaud, J., Cheynier, V. and Chemina, A., 1991. Procyanidin dimers and trimers from grape seeds. Phytochemistry 30, 1259-1266.

Stanley, P.L., Steiner, S., Havens, M. and Tramposch, K.M., 1991. Mouse skin inflammation induced by multiple topical applications of 12-O-tetradecanoylphorbol-13-acetate. Skin Pharmacology 4, 262-271.

Toews, G.B., Bergstresser, P.R. and Streilein, J.W., 1980. Epidermal Langerhans cell density determines whether contact hypersensitivity or unresponsiveness follows skin painting with DNFB. Journal of Immunology 124, 445-453.

Ullrich, S.E., 1995. Potential for immunotoxicity due to environmental exposure to ultraviolet radiation. Human Experimental Toxicology 14, 89-91.

Vaid, M., Singh, T., Li, A., Katiyar, N., Sharma, S., Elmets, C.A., Xu, H. and Katiyar, S.K., 2011. Proanthocyanidins inhibit UV-induced immunosuppression through IL-12-dependent stimulation of $CD8^+$ effector T cells and inactivation of $CD4^+$ T cells. Cancer Prevention Research 4, 238-247.

Williams, C.S., Mann, M. and DuBois, R.N., 1999. The role of cyclooxygenases in inflammation, cancer, and development. Oncogene 18, 7908-7916.

Xu, H., DiIulio, N.A. and Fairchild, R.L., 1996. T cell populations primed by hapten sensitization in contact sensitivity are distinguished by polarized patterns of cytokine production: interferon gamma-producing (Tc1) effector CD8+ T cells and interleukin (IL) 4/IL-10-producing (Th2) negative regulatory CD4+ T cells. Journal of Experimental Medicine 183, 1001-1012.

Yamakoshi, J., Saito, M., Kataoka, S. and Kikuchi, M., 2002. Safety evaluation of proanthocyanidins-rich extract from grape seeds. Food and Chemical Toxicology 40, 599-607.

Yoshikawa, T., Rae, V., Bruins-Slot, W., Van den Berg, J.W., Taylor, J.R. and Streilein, J.W., 1990. Susceptibility to effects of UVB radiation on induction of contact hypersensitivity as a risk factor for skin cancer in humans. Journal of Investigative Dermatology 95, 530-536.

Key facts

- Olive oil has been used to protect the skin since ancient times.
- In the skin, UV radiation causes the formation of oxygen free radicals (ROS).
- To protect the skin, besides reducing exposure to sun rays, it's necessary to have a balanced diet of polyunsaturates and rich in antioxidant agents.
- The correct ω-6/ω-3 ratio offers the skin immune protective, anti-inflammatory and anti tumoral activity.
- Olive oil supplies a balanced ratio of fatty acids.
- Extravirgin olive oil contains some minor components of high biological activity with antioxidant action (α-tocopherol, carotenoids, polyphenols).
- The polyphenols present in olive oil are potent antioxidants with lipophilic and hydrophilic activity.
- The market offers extravirgin olive oil, simple olive oil, and sansa olive oil.
- Simple olive oil and sansa oil, being refined oils, lose almost all their minor components so have less biological value.
- Olive oil's characteristics are not altered with non prolonged cooking at normal cooking temperatures.
- Olive oil undergoes rigid controls by the European Union Authority to certify it is genuine and unadulterated.

Summary points

- UVA and UVB rays damage the skin by initiating a peroxidative process that induces aging and tumor risk.
- The skin's congenital antioxidant defenses are not sufficient to protect against prolonged exposure to UV rays.
- 30 minutes of solar ray exposure reduces by 50-60% the skin α-tocopherol content.
- It is necessary to integrate the skin's defenses, not only topically, but also through the intake of foods rich in antioxidant agents.
- Hydroxytirosol and oleoeuropein, polyphenols present in extravirgin olive oil, are powerful antioxidants.
- ω-3 protects the skin and combats oxidative stress.
- When taken in excess, ω-6 favors oxidative stress and tumor formation.
- A balanced diet must have a ω-3/ω-6 ratio equal to 10:1, even better if 5:1.
- The topical use of olive oil inhibits neoplastic risk activating onco-suppressor protein p53.
- Due to olive oil's balanced fatty acid composition, and the presence of antioxidant agents, it acts against skin aging and tumor risk.
- In order to maintain extravirgin olive oil's antioxidant properties, it should not be exposed to light or air.

17. Olive oil as a skin protector

P. Viola[1], F. Nobili[2] and M. Viola[3]
[1]Nutritional Section of Olive and Olive Oil National Academy, Accademia Nazionale dell'Olivo e dell'Olio, piazza della Libertà 12, 06049 Spoleto, Italy; [2]National Research Institute on Food and Nutrition, via Ardeatina 546, 00178 Rome, Italy; [3]St. Lucia Foundation, Institute of Cure and Scientific Research, via Ardeatina 306, 00179 Rome, Italy; andulivo@virgilio.it

Abstract

Olive oil, in its extra virgin form has protective action towards the skin, as well as a preventive activity towards aging and chronic degenerative diseases. This is due to its balanced fatty acid composition which has an ideal ratio between the ω-6 and the ω-3 series, and its low saturate and high monounsaturate content. Its greatest health action however is due to the numerous minor components, including some antioxidants that combat peroxidative risk caused by age, sun rays and dietary errors. Among the antioxidants are tocopherol in its alpha form (the real vitamin E), some carotenoids (β-carotene and lutein) and, in particular, numerous phenol compunds with highly interesting biological action such as hydroxytirosol and oleoeuropein. Also present is a significant quantity of triterpene hydrocarbons (squalene), which filters singlet oxygen at the skin level. Besides the extra virgin type, the market also offers simple "olive oil" and "sansa of olive oil", which have the same fatty acid composition as "extra virgin", but have a notably lower quantity of antioxidant substances and therefore are of minor biological value. In the European Union rigorous controls are enacted by the authorities to verify the quality and genuineness of commercial products and prohibit the sale of olive oil that is not officially labeled. Olive oil is a delicate product that should not be exposed to light and air, so it must be served directly from its original bottle, which should be recapped after every use.

Keywords: skin, fatty acids, olive oil, oxidative stress, antioxidants

Abbreviations

AA	Arachidonic acid
DHA	Docosahexaenoic acid
DNA	Deoxyribonucleic acid
EFA	Essential fatty acid
EPA	Eicosapentaenoic acid
DHGLA	Di-homo-γ-linolenic acid
LC-PUFA	Long chain polyunsaturated fatty acid
LT	Leukotriene
MUFA	Monounsaturated fatty acid
PG	Prostaglandin
PUFA	Polyunsaturated fatty acid
ROS	Reactive oxygen species (oxygen free radicals)
SAFA	Saturated fatty acid
TX	Thromboxane
UVA	Ultraviolet A
UVB	Ultraviolet B

17.1 Introduction

Since ancient times Mediterranean populations have used olive oil to keep their skin young and healthy. Today scientific research has not only confirmed this oil's effectiveness when used topically, but has also shown that olive oil has a protective action on skin even when it is taken in the diet.

Not all dietary fats are the same. Depending on their origin, they differ notably in their fatty acid composition and in the presence of minor components. These different compositions influence the biological value of the various fats, which can be evaluated using the following parameters:

- Fatty acid composition (saturated, monounsaturated, polyunsaturated ω-6, polyunsaturated ω-3)
- Ratio polyunsaturates/saturates
- Ratio ω-6/ω-3
- Ratio polyunsaturates/antioxidants
- Presence of minor components (vitamins, sterols, carotenoids, polyphenols, etc.)

17.2 Olive oil

Olive oil contains a high percentage of monounsaturated oleic acid (18:1 ω-9), a moderate amount of SAFAs and a sufficient quantity of PUFAs with a balanced ratio between linoleic acid (18:2 ω-6) and α-linolenic acid (18:3 ω-3), averaging 8:1 (Table 17.1).

Table 17.1. The percentage of fatty acids composition of olive oil.

Palmitic acid (16:0)	7.5-20
Palmitoleic acid (16;1 ω-7)	0.3-5.5
Stearic acid (18:0)	0.5-5.0
Oleic acid (18:1 ω-9)	55.0-83.0
Linoleic acid (18:2 ω-6)	3.5-21.0
a-linolenic acid (18:3 ω-3)	0.2-1.5

Maximum limits fixed by the International Oleic Council – Madrid.

European Union regulation no. 2568/91 defines "virgin" as an oil obtained from the mechanical processing of whole olives that are pressed immediately after harvesting. The further distinction of "extra virgin" is used if the acidity (expressed in oleic acid) is <0.8%, and "virgin" if the acidity is <2%. The simple term "olive oil" is given to a refined oil (for a high acidic content and/or containing impurities) to which 5-10% virgin oil has been added, having an acidity of <1%. The term "sansa olive oil" denotes an oil from sansa (the residue left after the olives are pressed), which is refined and to which a moderate amount of virgin olive oil is added; it has an acidity of <1%.

Extra virgin olive oil has remarkable health properties, not only for its balanced composition of fatty acids, but also for its high content of minor components (1.5-2%) of notable biological value, such as vitamin E, polyphenols, some carotenoids, triterpene hydrocarbons, phospholipids, phytosterols, chlorophyll and numerous aromatic compounds.

In refined oils, though the acidic composition is unaltered, the content of the minor components is notably reduced. Therefore, it is understandable that simple "olive oil" and "sansa olive oil" have a biological value inferior to extra virgin olive oil.

Today, thanks to the advances in agronomic technology, the majority of olive oils sold in Europe belongs to the extra virgin category. Besides their nutritional qualities, these oils contain pleasant organoleptic qualities, that range from sweet to fruity.

In some countries outside Europe, the market offers a "light olive oil" usually composed of refined olive oil associated with a small quantity of virgin olive oil and an undefined quantity of seed oil (sunflower, or soybean, or grape-stone). Obviously this oil has modest biological value.

The European Commission has established precise regulations to verify the quality and purity of the product. There is a rigorous control on the part of the Authorities established by the European Union, which is constantly updated. The whole olive oil production is well controlled in compliance with European Regulation 1019/20002 that prohibits the sale of olive oil not officially labeled.

Extra virgin olive oil offers a variation of sensorial effects according to the differences in cultivar, the geographical area of origin, the specific year of harvesting and the period in which it was harvested (early or late in the season). These variations affect its organoleptic qualities, but not its biological value. The oil which offers health protection activity is virgin oil, since this activity is influenced by a balanced acidic composition, as well as its content of minor components. Especially important are those components possessing antioxidant power that contrast oxidative stress provoked by ROS. The minor components which most interest researchers today are polyphenols (Table 17.2) that, besides preventing peroxidative risk, offer numerous other useful protections for the body (Del Rio *et al.*, 2010; Servili *et al.*, 2005) (Table17.3).

Antioxidants can be liposoluble or hydrosoluble and therefore act both in lipophilic compartments (biologic membrane and lipoproteins) and in hydrophilic compartments (blood and tissues). Olive oil is the only dietary fat that has both liposoluble and hydrosoluble polyphenols (Servili *et al.*, 2005).

Table 17.2. Phenolic compounds in olive oil.

Phenolic acids and derivatives	**Secoiridoids**
cinnamic acid	oleoeuropein aglycon (3,4 DHPEA-EA) (hydroxytyrosol)
caffeic acid	dyaldehydic form of decarboxymethyl elenoic acid
p-aminobenzoic acid	bound to 3,4 DHPEA (3,4 DHPEA-EDA)
protocatechuic acid	bound to p-HPEA (p-HPEA-EDA)
syringic acid	p-HPEA derivatives
p-coumaric acid	oleoeuropein
o-coumaric acid	ligustroside aglycon (p-HPEA-EA)
ferulic acid	dyaldehydic form of oleoeuropein aglycon
gallic acid	dyaldehydic form of ligustroside aglycon
4-(acetossietil-)-1-2-dihydroxybenzene	
benzoic acid	
Lignans	**Phenolic alcohols**
(+)-1-acetoxypinoresinol	(3,4 dihydroxyphenol) ethanol (3,4-DHPEA)
(+)-pinoresinol	(p-hydroxyphenol) ethanol (p-HPEA)
	(3,4 dihydroxyphenyl) ethanol-glucoside
Flavones	**Hydroxy-isocromans**
apigenin	-
luteolin	-

Table 17.3. Biological activity of polyphenols.

- Act directly as antioxidants
- Protect and reload α-tocopherol
- Increase retinol (vitamin A) and β-carotene activity
- Bind metal ions that catalyze free radical formation
- Lower cholesterol blood levels
- Increase the body's immune defences
- Slow tumor growth
- Inhibit some carcinogenic chemicals
- Inhibit phospholipase A_2 enzymes therefore arachidonic acid release
- Inhibit cyclo-oxygenase enzymes so reducing thromboxane formation and thereby platelet aggregation that leads to thrombosis
- Inhibit lipo-oxygenase enzymes thereby reducing leukotriene formation which is responsible for inflammatory phenomena
- Exercise an anti-allergic activity
- Protect the skin
- Give the oil a pleasant taste

17.3 Composition of skin lipids

Skin lipids are different from those of other tissues in that they are primarily synthesized *in situ* from glucose. These fatty acids can have an uneven number of carbon atoms and are characterized by chains that can be branched and quite long (even more than 24 carbon atoms), with different unsaturations from those found in other tissues. Besides these fatty acids the skin contains EFAs, squalene, wax and wax esters (Table 17.4).

As for the EFA, linoleic acid (18:2 ω-6) is present in quite notable amounts, while α-linolenic acid (18:3 ω-3) is found only in trace amounts. Important however is the presence of their long chain derivatives of 20-22 carbon atoms, or LC-PUFA, as dihomo-γ-linolenic, or DHGLA (20:3 ω-6), arachidonic, or AA (20:4 ω-6), eicosapentaenoic , or EPA (20:5 ω-3) and docosaesaenoic, or DHA (22:6 ω-3). However, though it needs long chain derivatives (LC-PUFA), the skin does not

Table 17.4. Sebum skin composition (in %).

Free fatty acids	6.3-56.0	Wax	12.3-25.0
Squalene	6.5-18.0	Triglycerides	5.5-37.5
Other hydrocarbons	0.5-10.0	Mono and di-glycerides	3.1-13.5
Unidentified minor components	5.0-12.0		

possess elongase and desaturase enzymes and therefore it cannot form them from composts of 18 carbon atoms, as linoleic and α-linolenic. Therefore LC-PUFA have to arrive in the skin already preformed. The skin, however, has cyclo-oxygenase and lipo-oxygenase, enzymes that form prostaglandins, thromboxanes and leukotrienes from LC-PUFA (Christiansen *et al.*, 1991). This is illustrated in Figure 17.1 which shows the formation of LC-PUFA and subsequent eicosanoids (PG, TX and LT).

It is especially important that the two series (ω-6 and ω-3) are present in the diet in a correct ratio, since they perform in various, and in many respects, contrasting actions. A correct ratio between the two series is therefore fundamental since EPA and DHA have an anti-inflammatory and immunoprotective action on the skin, as well as an anti-tumoral and cardioprotective one, in contrast to AA that has pro-inflammatory and immune suppressive effects (Soiland and Drevon, 2000)

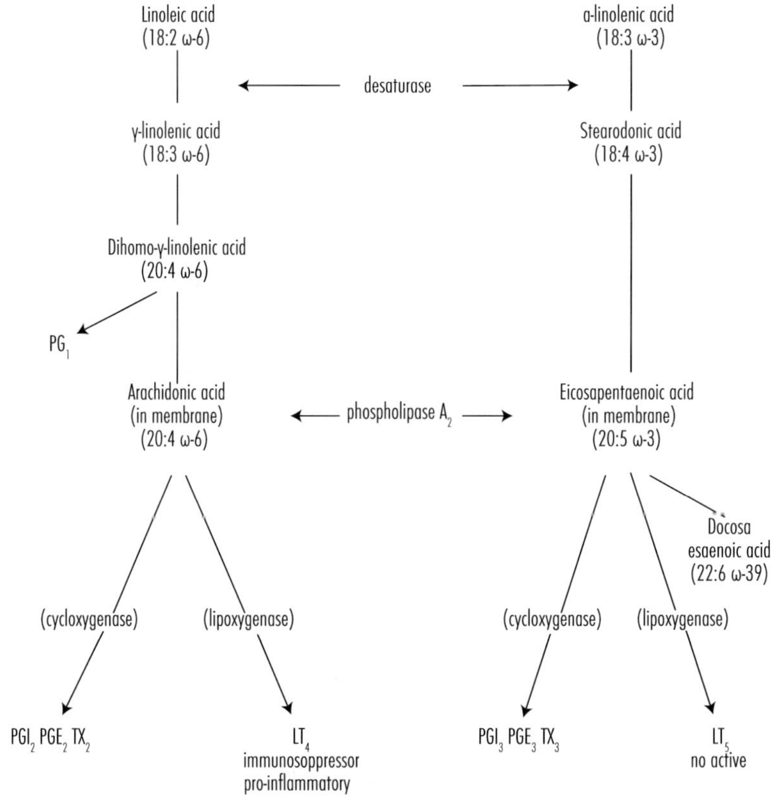

Figure 17.1. Formation of the LC-PUFA and the subsequent formation of prostaglandins (PG), thromboxanes (TX) and leukotrienes (LT).

17.4 The skin's requirement of essential fatty acids

A sufficient dietary intake of EFA maintains skin homeostasis since these fatty acids play an important role as a barrier for the skin, promoting its hydration and preventing eventual atopic alterations, as well as psoriasis, acne, and eczema. In the case of prolonged deficiencies (less than 1% of total calories), there are marked alterations, such as hemorrhagic spots, erythema and an increase in *perspiratio insensibilis* with hypo-elastic, dry and dehydrated skin. There is also an increase in DNA synthesis with epidermis proliferation and consequent scaling.

As for ω-6, the main protective activity at the skin level seems to be due to a long chain derivative of linoleic acid (that of course must arrive already preformed), and DHGLA (20:3 ω-6) (Kawashima *et al.*, 2008), which improves the cutaneous barrier as well as promoting hydration and retarding aging. DHGLA also has a specific therapeutic effect in the treatment of atopic dermatitis which is not the case with its precursor, linoleic acid. In fact, the skin concentration of linoleic acid of subjects suffering from this malady, appears increased. This denotes a reduced capacity of the liver to convert linoleic acid into DHGLA.

DHGLA's therapeutic activity stems from an inhibition enacted by one of its derivatives PG_1 (prostaglandin with only one double bond) on leukotrienes LT_4 with inflammatory action derived from AA, but also from a reduction in the formation of immunoglobulin IgE, so important in allergy formation. The body's production of DHGLA is however modest because the body rapidly converts it into its higher derivative, AA (C20:4 ω-6). To increase its presence it is necessary to take in foods that are rich in DHGLA such as borage oil, enoterra oil and blackcurrant oil.

It should be noted that DHGLA formation can be inhibited by the peroxidative action of ROS that limits Δ-6-desaturase (which forms LC-PUFA), which ultimately confirms the usefulness of including antioxidants in the diet to protect the skin. However, it is important that there is also an adequate amount of AA (20:4 ω-6), which is the successive step after DHGLA formation in the elongation of the ω-6 series (Figure 17.1). In fact, AA, by modifying the sebum, is capable of diminishing excessive skin scaling.

In any case LC-PUFA ω-3 (EPA and DHA) are noteworthy, not only for their anti-inflammatory and immunoprotective activity, but also because they regulate skin proliferation and mediate keratinogenesis by way of prostaglandins PG_3 (with 3 double bonds) derived from EPA.

Recognizing the importance of an adequate dietary intake of EFA for the skin's physiological maintenance, and their usefulness in topical treatments for the above-mentioned pathologies, it should not be forgotten that an excess of PUFA ω-6 in the diet can pose problems because they can constitute a target for ROS, which promotes skin aging, photo-exposure damage and tumor growth (Black *et al.*,1992, Homer *et al.*, 2006). Furthermore, it should be noted that the competitive conflict in the body between the linoleic (ω-6) and the α-linolenic (ω-3) series also depends on chain elongation, even in the higher derivatives. The presence in the diet of a high percentage of linoleic acid (18:2 ω-6), as seen with seed oil consumption, tends to subtract desaturase enzyme from α-linolenic acid (18:3 ω-3) the formation of AA (20:4 ω-6) instead of EPA (20:5 ω-3) and DHA (22:6 ω-3).

It is important to point out that ω-6 and ω-3 series must be present in the diet in correct amounts in the treatment of skin inflammatory diseases. In fact, in these disturbances there is a marked cutaneous increase in AA and its pro-inflammatory derivatives (PGE_2 and LT_4). The intake of AA (as well as of its precursor linoleic acid) can exacerbate the malady. On the contrary, the intake of EPA and DHA is capable of reducing lesions, by means of the formation of prostaglandins PG_3 (with three double bonds), also because leukotrienic derivatives of EPA (LT_5) have practically no pro-inflammatory activity (Soiland *et al.*, 2000).

Besides competing with ω-3, PUFA ω-6 constitute a peroxidative target for ROS, promoted by UV rays (Kielbassa *et al.*, 1997), which produce photo-aging, inflammation and immune-suppression by reducing the number of dermal Langerhans cells that elaborate the antigenic molecules that initiate the immune response. In epidermal cells, ROS formation can come about also for normal metabolic oxy-redox reactions that take place inside the cell in the respiratory chain. ROS formation determines phospholipid peroxidation within the membrane and mitochondria which undergo a destructuralization with functional modifications that can even lead to cell death.

In conclusion, to protect dermal cell integrity, it seems necessary to ensure an "optimum" dietary intake of PUFA ω-6; neither too little nor too much, since, as we have seen, an excess can worsen solar ray effects with ROS formation. The intake of PUFA ω-3 is recommended for their resistance to peroxidation and, especially as LC-PUFA, for their protective action on the body in general and on the skin in particular (Black *et al.*, 1992, Soiland *et al.*, 2000). It is very important both for the body and the skin that a balanced ratio between the ω-6 and ω-3 series (not greater than 10:1) is maintained. Today the optimum ratio recommended by many researchers is around 5:1, especially in infants and the elderly. This balanced ratio can be maintained not only by increasing the ω-3, but also by reducing the ω-6, which can be obtained by preferring MUFA (oleic acid) highly resistant to peroxidative processes. Currently the ω-6/ω-3 ratio in Western populations is 20:1, and can sometimes reach 30:1.

The seed oils commonly used have a ω-6/ω-3 ratio which is not adequate. This can be seen in Table 17.5, with the exception of soybean oil, which has a good ratio of the two series, but also contains a high level of linoleic acid and lacks a sufficient protection by vitamin E (tocopherol in α form).

Therefore, we recommend as a main energy source, the use of MUFA and not PUFA ω-6, remembering that MUFA reduce peroxidative risk and exert protective activity on skin, vascular risk and neoplastic risk (Table 17.6; Visioli and Galli, 1998).

It seems necessary to point out that ω-3, though they seem resistant to peroxidative risk, should always be combined with an adequate intake of antioxidants (tocopherols, carotenoids and polyphenols).

Table 17.5. ω-6/ω-3 ratio in various dietary fats.

Olive oil	9:1	Sesame seed oil	80:2
Soybean oil	7:1	Grape-stone oil	130:1
Peanut oil	27:1	Sunflower oil	150:1
Corn oil	80:1		

Table 17.6. Recommended amount of fat in the diet.

Overall needs	30% of total calories
Saturated fatty acids	6-7%
Monounsaturated fatty acids	14%
Polyunsaturated ω-6	5-6%
Polyunsaturated ω-3	1.5-3%

17.5 ROS peroxidation and skin damage

Oxygen represents a biological paradox in that, while it is indispensable for the life of aerobic organisms, it constitutes a potential toxin in constant search of electrons to compensate its two unpaired orbits circling its periphery. Starting out as O_2 the final aim is to become H_2O, which can happen by various pathways, such as the subtraction of a hydrogen atom from a macromolecule, like an enzyme, a protein, DNA or a PUFA (Figure 17.2) forming as a result, however, a new free radical, will lead to radical intermediate compounds with high toxicity.

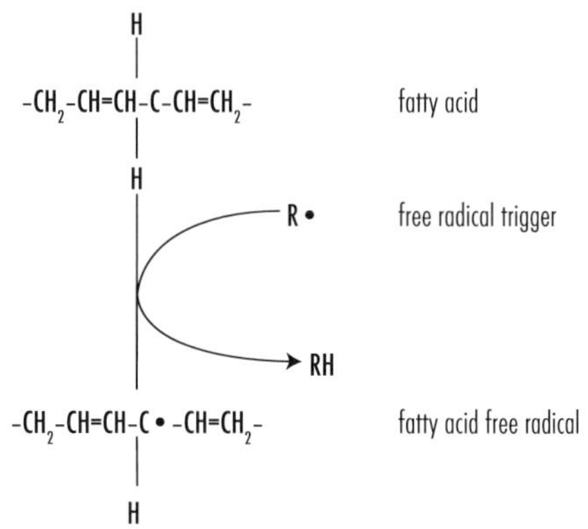

Figure 17.2. Free radical action on polyunsaturated fatty acids. Subtraction of a hydrogen atom from a polyunsaturated fatty acid by an oxygen free radical and the formation of a new free radical.

Despite the severe action against this process of cell integrity, we are protected (within certain limits) by antioxidants, which can be present congenitally (first line of defence), or taken in with the diet (second line of defence). It is understandable that since it is not possible to increase congenital antioxidants to prevent ROS damage, we should try to increase our dietary antioxidant intake, as well as avoiding all the known factors that favor ROS formation and oxidative stress

The skin is the body's largest organ, situated as an interface between the body and the external environment. For this reason, it is highly exposed to adverse environmental factors. Therefore it is a tissue with high peroxidative risk, determined by external disturbances, but also by internal disturbances, all of which overlap to a certain degree. In particular, the three reasons for its high peroxidative risk are:

- it is in direct contact with both atmospheric and endogenous molecular oxygen;
- it is exposed to the sun's UV rays;
- its lipid content is susceptible to peroxidative initiation.

External disturbances, besides those due to exposure to atmospheric oxygen and air pollution, are caused mainly by solar radiation, which induces an energy transference in the skin that causes a photosensibilized molecule to absorb photons and puts the molecule in a state of excitement. Consequently, ROS are formed with the ability to penetrate the epithelial tissue disrupting its structure and functionality, creating cellular damage (with intra-cutaneous formation of lipofuscin, a sign of aging) and tumor risk increase.

All the components that make up the solar spectrum that reach the earth's surface can induce free radical formation in human skin. This phenomenon manifests itself at all wavelengths on all the skin's layers, from the *stratum corneum* to the hypodermis. The most intense free radical formation takes place, however, in a relatively small band between 320 and 380 nm, peaking at 360 nm. This corresponds well with UVA's role in photo-aging and neoplastic induction (Kielbassa *et al.*, 1997).

The internal disturbances are represented by hormonal modifications, the effects of smoking, and especially from poor diet (obesity, alcohol abuse, polyunsaturates excess, protein deficiency, and antioxidant deficiency). In particular, during energetic utilization of foods in the respiratory chain, some oxygen (2-4%) is not usually used for normal combustion of nutrients to produce ATP, but sidetracks this process forming unstable and very reactive composts (ROS). If these composts are not neutralized by antioxidants, they have a tendency to stabilize themselves at the expense of other molecules from which they subtract an electron, producing cellular structure modifications sometimes with serious functionality repercussions. Therefore, internal disturbances represent a risk factor that is aggravated by the negative effects of external disturbances and which can be further aggravated when there is an incorrect fatty acid intake such as a prevalence of PUFA ω-6. All of this can, however, be counteracted if MUFA (oleic acid) and PUFA ω-3 are prevalent in the diet, along with a sufficient quantity of antioxidants.

In summary, the skin is an organ with high metabolic activity that is highly exposed to oxidative stress risk due to ROS and UV action. Such stress determines modifications of collagen and elastin and pigment accumulation typical of aging, as lipofuscin and ceroids that are formed thanks to the oxidative polymerization of skin lipids.

The dietary make-up is therefore very important for ensuring a correct intake of antioxidant agents and fatty acid composition. Lipid peroxidation reflects an ROS attack against PUFA ω-6 with consequent free radical formation of fatty acids, that in the presence of oxygen, leads to the formation of lipid peroxy-radicals. These radicals then extract a hydrogen atom and form unsaturated hydroperoxides, plus a free radical which can propagate a chain reaction that can be stopped by an antioxidant or by the formation of a polymer which is often toxic:

R• + PUFA = PUFA• + RH
PUFA• + O_2 = PUFA-O-O•
PUFA-O-O• + PUFA• = PUFA-O-O

The peroxidative risk of fatty acids is tied to the number of methylenes, therefore the oxidation rate increases from saturated to monoenoic, dienoic, trienoic, tetraenoic acids in the proportion of 0 - 0.025 - 1 - 2 - 4. Therefore, AA (20:4 ω-6) is 160 times more susceptible to oxidation than oleic acid (18:1 ω-9) (Sevanian, 1988).

It is understandable therefore that an unbalanced acidic composition of dietary fats can determine skin damage and influences carcinogenesis, as was shown for the first time in 1930 (Watson and Mellanby, 1930), even though for a long time this data was given little consideration. In 1973, however, another researcher (Pinkney, 1973) observed that 78% of the subjects that habitually consumed more than 10% PUFA in their diet presented with notable precocious aging signs of their skin, and looked older than their chronological age. In this group furthermore, 6% had undergone removal of one or more malignantly suspicious skin lesions.

Despite these warnings, only recently has it been demonstrated that a high level of ω-6 fatty acids can exacerbate tumor formation in the skin particularly during the post-initiation phase of carcinogenesis caused by ultraviolet rays. Experimental studies have shown, in fact, that the degree of carcinogenesis exacerbation increases with the levels of ω-6 intake, with lipid peroxidation levels that progress linearly with the amount of ω-6 consumed (Homer and Black, 2006). On the contrary, ω-3 inhibit UV induced carcinogenesis, since they are resistant to peroxidative phenomena.

To confirm what has already been stated, we would also like to recall the studies carried out in 1986 (Harman, 1986) that treated two groups of rats with two diets rich in PUFA (charthamus) and MUFA (olive oil) respectively. The rats were further radiated, and on examination it was seen that the rats given olive oil clearly presented with less serious lesions than the rats taking chartamus oil. The author maintains that since ultraviolet light induces skin modifications that mime those of normal aging, the results observed should indicate that subjects who usually consume olive oil may have younger looking skin than those who usually consume oils rich in PUFA such as charthamus and corn oils.

17.6 Antioxidants present in olive oil

As stated above, olive oil's acidic composition is balanced with a prevalence of MUFA (oleic acid), very little SAFA and the necessary and adequate presence of EFA with the correct ω-6/ω-3 ratio. This composition is not seen in the majority of other dietary fats (Table 17.7).

Above all virgin olive oil contains many antioxidant agents:

- Vitamin E (α-tocopherol)
  Tocopherol contained in olive oil is of the α form, the only one used by the body. The ratio of vitamin E (α-tocopherol) to PUFA should never be inferior to 0.5; in olive oil it is 1.5, while this ratio is almost never respected in seed oils. In these oils in fact, the tocopherols present are mainly made up of γ and δ forms (Bauernfeind, 1981; McLaughlin and Weihrauch, 1979), as can be seen in Table 17.8, which are active *in vitro*, but scarcely used by the body (Table 17.9). In fact when the body recognizes their presence, it eliminates them by way of the biliary tract, and only accepts those with the α form, which is therefore the only active one (Raber and Kajden, 1989).
  Virgin olive oil contains 150-200 mg/l of α-tocopherol, with an optimum relationship with the PUFA present.
- Carotenoids
  Extra virgin olive oil contains (even if in limited amounts) some carotenoids, (β-carotene and lutein) that gives the oil a yellow color. Among these are in particular β-carotene that protects the skin and which must be taken in daily, but not in high quantities because this could cause lung tumors (in smokers). Furthermore, olive oil promotes intestinal absorption of lycopene (highly active against skin photo-aging), and lutein, which seems to act in synergy with lycopene.

Table 17.7. Average fatty acid composition in various dietary fats (%).

	Saturated	Monounsaturated	Polyunsaturated ω-6	Polyunsaturated ω-3
Butter	45-55	35-55	1.5-2.5	traces
Lard	40-46	42-44	6-8	0.5-0.9
Olive oil	8-12	65-83	6.5-12	0.2-1.5
Peanut oil	17-21	40-70	13-28	-
Sunflower oil	5-13	21-55	55-58	-
Corn oil	12-18	32-35	34-62	0.1-2.5
Soybean oil	8-18	20-26	46-52	6.5-10.5
Grape-stone oil	8.9-9.3	15-18	65-68	0.15-0.3

Table 17.8. Average percentage of α-tocopherol in various dietary fats.

Dietary fat	α-tocopherol	γ-tocopherol	δ-tocopherol
Olive oil	21.40	-	-
Corn oil	6.00	43.70	2.00
Peanut oil	13.00	21.60	2.10
Sunflower oil	48.70	5.10	0.80
Soybean oil	10.10	59.30	26.40
Margarine (medium)	6.80	18.30	3.80
Butter	1.68	0.14	-
Lard	1.20	0.70	-

Table 17.9. Biological activity of tocopherols.

RRR - α-tocopherol	1
RRR - β-tocopherol	0.5
RRR - γ-tocopherol	0.1
RRR - δ-tocopherol	0.03

- Polyphenols
  Polyphenols (Table 17.2) are also present in high quantities. The most biologically important are hydroxytyrosol and oleoeuropein, both with an anti-inflammatory action that inhibits phospholipase A_2 (that releases AA), cyclo-oxygenase and lipo-oxygenase (reducing formation of thromboxane TX_2 and leukotrienes LT_4). Other polyphenols present in a lesser quantity are caffeic acid, ferulic acid, vanillic acid, verbascoside and lignans. It is important to mention ferulic acid that inhibits xanthine-oxidase, responsible for the formation of superoxide anions and therefore lipid peroxidation (Choquenet *et al.*, 2008; Saija *et al.*, 1999), vanillic acid, important for its action on vanilloid receptors present in skin and mucosa (receptors that activate substances capable of inhibiting the secretion of pro-inflammatory substances as interleukine-8 and PGE_2) and verbascoside, able to inhibit 5α-reductase, the enzyme correlated to acne's etiopathogenesis (De Luca and Korkina, 2008).

On the whole it can be said that the antioxidants contained in olive oil may be present in relatively low concentrations, but their activity is notably effective due to the synergistic action among the single components that increase the antioxidative potential with respect to that observed when the composts are tested singularly.

17.7 Squalene

Squalene is a component of sebum that has activity against solar radiation by filtering singlet oxygen (Kelly, 1995; Desai *et al.*, 1996). It is present in a high quantity in extra virgin oil (400-450 mg/100 g) and in lesser amounts in refined olive oil and seed oils. Once consumed, it is distributed ubiquitously in all tissues where it acts as an anti-tumor agent, but the greatest concentration is found in the skin where it integrates into the *stratum corneum*, reinforcing the structural equilibrium of the lipidic film necessary for the skin barrier function. It also exerts a notable anti-tumoral activity, so much so that some researchers feel olive oil's anti-tumoral effect is related mainly to its high squalene content (Desai *et al.*, 1996; Smith *et al.*, 1999).

It should also be noted that squalene is susceptible to peroxidative phenomena with the formation of a monohydroperoxide that can promote the appearance of seborrheic dermatitis. In order to perform the above-mentioned protective activity against oxygen singlets it therefore needs the simultaneous presence of antioxidants such as α-tocopherol and polyphenols; exactly the situation that is present in extra virgin olive.

17.8 Olive oil's skin protection activity

As previously stated, to prevent skin damage, the body needs essential PUFA in adequate doses that should not exceed physiological requirements, which for linoleic acid is 2%, and for α-linolenic is 0.2-0.5% of the total calorie count. These quantities are contained in a daily dose of 50 grams of olive oil. To satisfy energetic nutritional requirements, MUFA should supply the majority of fatty acids consumed. As for ω-3, many authors prefer to increase this amount, especially in the form of LC-PUFA directly through the consumption of fish, while also encouraging MUFA (olive oil) use.

The most important aspect of extra virgin oil is the presence of antioxidant agents. Exposure to solar radiation determines, in fact, a serious loss of antioxidant protection in the skin. In particular it has been observed that after 30 minutes of UV exposure, the skin's α-tocopherol content is reduced by 50-60%. Topical application of α-tocopherol clearly reduces the damage. This positive effect is seen, even if only to a minor extent, if α-tocopherol is taken orally (Tavakkol *et al.*, 2004).

It should be remembered that UV rays exert a negative effect on all antioxidants, especially on carotenoids, which are diminished not only at the skin level but also in the blood. An adequate antioxidant intake, especially through vegetables, but even via extra virgin oil (that contains α-tocopherol, β-carotene, lutein, and polyphenols) seems to be highly indicated for skin protection (Arief *et al.*, 2000; Lee *et al.*, 2000). Lutein seems to work synergetically with lycopene to reduce photo-aging and neoplastic risk in the skin and other tissues. Furthermore, olive oil promotes the intestinal absorption of carotenoids and lutein in particular (Lasskshimirayana *et al.*, 2005).

Recently it has been demonstrated that topical oleoeuropein has a potent antioxidant activity by acting directly in the skin as a free radical scavenger (Ancora *et al.*, 2000). From this study we can hypothesize that the same oleoeuropein skin effects can be had with oral use of olive oil. Another skin protector present in olive oil is squalene, which acts as an oxygen singlet scavenger reducing photo exposure damage and tumor risk (Desai *et al.*, 1996).

Lastly, it is interesting to highlight that β-sitosterol, one of the phytosterols in olive oil, has a cosmetic action which seems to inhibit the transformation of testosterone into dihydrotestosterone regulating sebum and thereby improving oily skin (Castellani and Zumiani, 2000).

17.9 Skin aging

The skin is the mirror of the aging process, but its appearance is affected not only by age but also by exposure to the sun and diet. Physiological aging is due to free radical activity. Besides modifying DNA replication, ROS determine progressive damage to phospholipids of the biological membrane causing structural and functional changes. At the same time, the membrane undergoes increased AA release which with the action of cycloxygenase and lipoxygenase leads to the formation of PGE_2, TX_2 and LT_4 determining excess platelet aggregation and diminished immunity which promotes vascular damage, neoplastic risk and aging.

To retard the aging process and consequent skin damage, it is necessary to ensure an adequate dietary supply of antioxidants in the form of extra virgin olive oil together with fruit and a preventive and reparative action against UVA damage (protein p53, an anti-tumor protein, is produced by gene p53, the partial or total privation of this gene plays an important role in the genesis of tumors).

17.10 Conclusions

Olive oil's protective action is linked to the antioxidant action of its minor components (α-tocopherol, carotenoids, polyphenols) and to its balanced acidic composition which, besides an adequate amount of linoleic and α-linolenic acids, has oleic acid resistant to peroxidative risk. It is advisable, however, to consume fruits and vegetables as well, because this promotes the combination of various antioxidants, which then act synergetically with the antioxidants present in extra virgin oil, strengthening their effects. The presence of only one antioxidant would be much less effective, and in high doses could even cause contrary effects of pro-oxidation. (Nobili *et al.*, 2009). Lastly, also useful for protecting the skin is the regular intake of blue fish, thereby increasing the intake of LC-PUFA ω-3, while it is recommended to avoid oils too rich in linoleic acid, to keep a ratio of ω-6/ω-3 close to 10:1, or even better 5:1.

Olive oil is consumed raw and cooked. In the latter case, it does not really undergo any serious modifications, as long as the cooking process does not last too long and the temperature is not too high. The oils subject to greater modifications are those that contain high quantities of PUFA ω-6, in contrast to olive oil which contains primarily oleic acid (MUFA) and is rich in antioxidants. However, even if olive oil undergoes only modest alterations when fried, it is always preferable to use it uncooked, adding it to foods after they are cooked.

Lastly, it should not be forgotten that olive oil is a perishable item that changes with time, and air and light exposure. To maintain its health characteristics (along with its gastronomic ones) it is necessary to consume the product within a relatively short time period (one year) and eliminate the old and traditional cruets that tend to expose the oil to air. It would therefore be better to use the original bottle the oil comes in, close it after every use and avoid light exposure, even though most producers use dark UV-resistant bottles.

References

Ancora, C., Roma, C. and Vettor, M., 2004. Evaluation of cosmetic efficacy of oleoeuropein. The new frontiers of Dermo-Cosmetology. Efficacy, Stability and Safety. 7 ISCO Congr. Rome, Italy.

Arief, B., Nazim, U.A., Wu, A., Toshinori, B., Nikaido, O., Toshihiko, O., Masato, U. and Masamitso, U., 2000. Protective effect of topically applied olive oil against photocarcinogenesis following UVB exposure of mice. Carcinogenesis 21, 2085-2090.

Bauernfeind, J., 1981. Tocopherols in food. In: Machlin L.J. (ed.), Vitamin E: a comprehensive treatise. Marcel Dekker, New York, NY, USA, pp. 99-167.

Black, H.S., Thornby, J.L. and Lenger, W., 1992. Influence of dietary omega-6, omega-3 fatty acids sources and promotion stages of photocarcinogenesis. Photochemicals and Photobiology 56, 195-201.

Castellani, L. and Zumiani, G., 2000. L'oleosità cutanea. La Pelle 7, 41-43.

Choquenet, B., Couteau, C. Paparis, E. and Coiffard, L.J., 2008. Interest of ferulic acid ethyl ester in photoprotective creams: measure of efficacy by *in vitro* method. Natural Product Research 22, 1467-1471.

Christiansen, E.N., Lund, J.S. Rotveit, T. and Rustan, A.C., 1991. Effect of dietary n-3 and n-6 fatty acids in fatty acids desaturation in rat liver. Biochemica et Biophysica Acta, 1082, 57-62.

Del Rio, D., La Costa, L.G., Lean, M.E. and Crozier, A., 2010. Polyphenols and health. What compounds are involved ? Nutrition, Metabolics & Cardiovascular Diseases 20, 1-6.

De Luca, C. and Korkina, L.G., 2008. Marcatori di qualità alimentare e sensoriale su oli e olive. In: Progetto Strategico "Qualità Alimentare" INRAN, 25-27 febr. 2008, Rome, Italy, pp. 461-488.

Desai, K.N., Wie, H. and Lamartiniere, C.A., 1996. The preventive and therapeutic potential of squalene containing compounds on tumor promotion and regression. Cancer Letters 101, 93-96.

Harman, D., 1986. Swiss male mice: Effect of dietary fats in skin damage by ultraviolet light. Olive Oil in Medicine, COI Publisher, Madrid, Spain.

Homer, S. and Black, S:, 2006. Carcinogenesis and antioxidant action of some nutritional lipids. Journal of Plastic Dermatology 2, 2-6.

Kawashima, H., Tateishi, N., Shisishi, A., Teraoka, N., Tanaka, T., Tanaka, A., Matsuda, H. and Kiso, Y., 2008. Oral administration of diomo-γ-linolenic acid prevents development of atopic dermatitis in NC/Nga mice. Lipids 43, 37-43.

Kelly, G.S., 1999. Squalene and its potential clinical uses. Alternative Medicine Review 4, 29-36

Kielbassa C., Roza L. and Epe B., 1997. Wavelength dependence of oxidative DNA damage induced by UV and visible light. Carcinogenesis 18, 811-816.

Lasskshimirayana, R., Raju, M., Keshava Prakash, M.N. and Bakatan, V., 2009. Phospholipid, oleic acid micelles and dietary olive oil influence the lutein absorption and activity of antioxidant enzymes. Lipids 44, 799-806.

Lee, E.H., Faulhber, D., Hanson, K.M, Ding, W., Peters, S., Kodali, S. and Grandstein, R.D., 2000. Dietary lutein reduces ultraviolet radiation-induced inflammation and immunosoppression. Journal of Investigative Dermatology 122, 510-517.

McLaughlin, P.J and Weihrauch, J.L., 1979. Vitamin E content in foods. Journal of American Diet Association 75, 647-665.

Nobili, F., Audisio, M., Finotti, E. and Garaguso, I., 2009. Morphological and chemical/pyhysical. (2amidinopropane) dihydrocloride. Journal of Food Science and Nutrition 38, 35-48

Pinkney, E.R., 1973. The potential toxicity of excessive polyunsaturates. Do not let the patient harm himself. American Heart Journal 25, 723-728.

Raber, M.J. and Kajden, H.I., 1989. Preferential incorporation of α-tocopherol vs γ-tocopherol in human lipoproteins. American Journal of Clinical Nutrition 49, 517-526.

Saija, A., Tomaino, A., Lo Cascio, R., Trombetta, D., Proteggente, A., De Pasquale, A., Uccella, N. and Bonina, F., 1999. Ferulic and caffeic acids as potential protective agents against photooxidative skin damage. Journal of the Science of food and Agriculture 79, 476-480.

Servili, M., Selvaggini, R., Esposto, S., Taticchi, A., Urbani, S. and Montedoro, G.F., 2005. La composizione fenolica dell'oliva e dell'olio vergine. In: Gli antiossidanti nell'olio di oliva. 10 June 2005 Spoleto, Italy, pp. 7-26.

Sevanian, A., 1988. The impact of lipid unsaturation ad autoxidation on membrane structure. In: The Molecular Mechanism of Aging. The Role of Dietary Lipids. Medical Scientific Meeting, Lucca, Italy, 27-28 may 1988, pp. 45-54.

Soiland, E. and Drevon, C.A., 2000. The effect of very long-chain n-3 fatty acids on immune function-related skin diseases. European Journal of Cancer 36, 1235-1247.

Smith, T.J., Yang, G.Y., Seril, D.N., Liao, J. and Kim, S., 1998. Inhibition of 4-(methilnitrosamino)-1-3-pyridyl-1-butanone-induced lung tumorigenesis by dietary olive oil and squalene. Carcinogenesis 19, 703-706.

Tavakkol, A., Nabi, Z., Soliman, N. and Polefka, T.G., 2004. Delivery of vitamin E to the skin by a novel liquid cleaner: composition of topical versus oral supplementation. Journal of Cosmetic Science 55, 177-187.

Visioli F. and Galli, C., 1998. Natural antioxidants and prevention of coronary heart disease: the potential role of olive oil and its minor constituents. Journal of Nutrition, Metabolism and Cardiovascular Diseases 5, 306-314.

Watson, A.F. and Mellanby, E., 1930. Tar cancer in mice II. The condition of the skin when modified by external treatment or diet, as a factor in influencing this cancerous reaction. British Journal of Experimental Pathology 11, 311-316.

Key facts

- Garlic acts as a mild antibiotic and can kill certain antibiotic resistant strains of bacteria.
- Garlic helps to prevent several cancers, especially of the digestive system and on the skin.
- Skin carcinogenesis is a multistep process and garlic inhibits the first two steps of the development of skin cancer
- Garlic consumption as diet delays the onset of skin papilloma formation and subsequently reduces both the number and size of the skin papillomas in mice.
- Garlic boosts the cellular defense system by upregulating several phase II detoxification enzymes thus act as a potential chemopreventive agent against skin cancer.
- Garlic extracts downregulate lipid peroxidation in the skin of the papillomas and maintains protein homeostasis of the skin cells.
- Diet rich in garlic downregulates cyclooxygenase 2 (COX-2), which is a marker of carcinogenesis, thus delays the occurrences of skin cancer in mice.
- Organosulfur components of garlic are mostly responsible for its beneficial action.
- P53 and Caspase3 are also regulated by garlic in order to prevent skin carcinogenesis.

Summary points

- Garlic (*Allium sativum*) belongs to the species in the onion family *Alliaceae*.
- Use of garlic in diet helps prevent various infection and acts against inflammation and infection, including colds, cough and several other medical ailments.
- Whole garlic and several organosulfur compounds such as diallylsulfide (DAS), diallyldisulfide (DADS), diallyltrisulfide (DATS), and Ajoene present in garlic, inhibits the progression of skin carcinoma when applied topically or ingested orally.
- Garlic and its selected components modulates PI3k/Akt, p53 and COX-2 pathway and regulate caspase3 and apoptosis to exert its function against skin cancer.
- Apart from skin cancer, garlic demonstrates potential to modulate prostate cancer, colon cancer, stomach cancer and leukemia.
- Based on the observation it is our recommendation that garlic should be included as a part of a healthy diet and lifestyle and not as an alternative.

18. Protective effect of garlic in skin cancer

I. Das[1]*, A. Acharya*[2] *and T. Saha*[3]
[1]*Department of Cancer Chemoprevention, Chittaranjan National Cancer Institute; Kolkata, India;* [2]*Genetarn Corporation, 5908 Holland Road, Rockville MD 20851, USA;* [3]*Department of Oncology, Lombardi Comprehensive Cancer Center, Georgetown University Medical Center, 3970 Reservoir Road, NW, Washington DC 20057, USA; ts283@georgetown.edu*

Abstract

The incidence of cancer is rising in almost all parts of the world because of changes in the environment, changes in life style and food habits as well as due to growing industrialization and modernization. UV exposures from the sun in the exposed areas of the body are the prime locations of developing skin cancer. The rate of skin cancer is rising in an alarming rate and more than 10 million cases of skin cancer patients register each year for treatments. Epidemiological evidences demonstrate that naturally occurring phytochemicals are highly capable of inhibiting or delay the progression of several human carcinogenesis. Garlic and its major organosulfur components have acquired a special position in the medicinal history due to its anti-viral, anti-bacterial and fungicidal properties. Cancer chemoprevention using garlic as a dietary phytochemical is an active area of research and here we discuss the potential role of garlic and its components to inhibit, delay or reverse skin carcinogenesis. Both oral and topical application of various forms of garlic showed protection against chemical carcinogen induced skin papillomagenesis in mice. The protection occurs best when garlic was used in the diet prior to induction of skin papillomas by the carcinogens. Garlic inhibits the deleterious action of the carcinogens by downregulating lipid peroxides and at the same time upregulates several antioxidants, antioxidation enzymes and phase II detoxification enzymes. Garlic also downregulates COX-2, which is a marker of carcinogenesis progression. Garlic exerts its chemopreventive activities by modulating p53 and PI3K/Akt signaling pathways as well as promotes ROS generation and apoptosis in the skin papilloma cells. Inclusion of garlic in regular diet will eliminate several physical abnormalities and will promote healthy lifestyle.

Keywords: skin papillomatogenesis, DMBA, GST, GPx, COX-2, antioxidants

Abbreviations

AGE	Aged garlic extract
AM	Allylmercaptan
AMS	Allyl methyl sulfide
BCC	Basal cell carcinoma
CAT	Catalase
COX-2	Cyclooxygenase-2
DAS	Diallylsulfide
DADS	Diallyldisulfide
DATS	Diallyltrisulfide
DMBA	7,12-dimethylbenz(a)anthracene
LPO	Lipid peroxides
MDA	Malondialdehyde
MSC	Melanoma skin cancer
NMSC	Non-melanoma skin cancer
PAH	Polycyclic aromatic hydrocarbon
ROS	Reactive oxygen species
SAC	S-allylcysteine
SAMC	S-allylmercaptocysteine
SCC	Squamous cell carcinoma
SOD	Superoxide dismutase
TPA	12-*O*-tetradecanoylphorbol-13-acetate
UV	Ultraviolet

18.1 Introduction

Skin carcinogenesis is a form of cancer that occurs on the tissues of the skin and is more prevalent on the body parts exposed directly to the sunlight. Skin cancer tends to be genetic (hereditary) and occurs frequently in certain ethnic groups, especially those with fair skin; individuals with weakened immune system are also at greater risk of developing skin cancer. The two most common types of skin cancers are NMSC and MSC. The incidence of NMSC has been increasing at an alarming rate over the past several decades all over the globe. In 2010, more than 10 million cases of skin cancer have registered in USA. While NMSC accounts for 95% of all skin cancers, melanoma is the most dangerous type of skin cancer and accounts for more than 75% of all deaths due to skin cancer. Since skin cancer generally develops in the epidermis, the outermost layer of skin, a tumor is usually clearly visible. This makes most skin cancers detectable in the early stages.

In recent years, significant interest has been generated to identify naturally occurring chemopreventive phytoproducts that are proficient to inhibit, delay, or reverse the process of carcinogenesis. Several epidemiological and *in vitro* cell culture studies suggest the protective role of phytochemicals in garlic, against the development of certain human cancers including skin cancer (Milner, 2001). Garlic is a spice, native to Asia and the Mediterranean region, used extensively in several cuisines for its strong flavor and taste. Garlic has antiviral, antibacterial and fungicidal properties and for years it has been used medicinally for prevention of common cold and coughs. Furthermore, it is also used to treat fungal infections such as thrush as well as for wounds and ulcers. Garlic helps reduce atherosclerotic buildup (plaque) within the arterial system, reducing possibilities of strokes, by regulating cholesterol levels. Garlic also helps regulate high blood sugar and blood pressure and is also believed to prevent certain types of cancer, including stomach and colon cancers (Fleischauer *et al.*, 2000; Sengupta *et al.*, 2004).

18.2 Garlic and its components

Garlic (*Allium sativum*) is a species in the onion family Alliaceae, belonging to the genus *Allium*. The word garlic originated from old English *garleac* that means 'spear leek'. The bulb is the most commonly used part of the garlic plant, which is divided into numerous fleshy sections called cloves. Garlic consists of 6-34 cloves surrounded by white or pinkish flimsy sheets. Garlic has two sub varieties, softneck and hardneck, with similar healing properties but differ in flavor, clove size and uses. Softneck garlic is the type, seen in the produce section of the grocery store. The entire bulb covered by multilayered parchment, which continues up to the neck. This type of garlic typically has several layers of cloves surrounding the central portion of the garlic bulb. The outermost layer's cloves are the stoutest; the cloves of the internal layers become smaller closer to the center of the bulb. Softneck garlics' are of several varieties namely Silverskin, Artichoke, Asiatic, Turban and Silverskin Creole (Figure 18.1A). Hardneck garlics do not have flexible stalk like the softneck garlic. Hardneck garlic forms scapes from the center and it contains several tiny cloves inside the parent bulb. There are three main types of hardneck garlic varieties namely Rocambole, Porcelain and Purple stripe (Figure 18.1B).

The cloves of garlic contain several water-soluble nutrients that include vitamins, enzymes, amino acids and natural sugars, as well as many oil soluble sulfur-containing compounds. The distinctive flavor and spiciness of garlic is due to the abundance of the organosulfur compounds, which are also likely accountable for garlic's several medicinal effects (Herman-Antosiewicz *et al.*, 2007). Garlic also contains several non-sulfur containing compounds including minerals, saponins, flavonoids, and maillard reaction products. A complete list of nutritional value of garlic per 100 gm is given in Table 18.1.

18.3 Skin cancer

Skin cancer is a disease where malignant cells are found in the outer layers of the skin. The skin has two main layers and several kinds of cells. The outermost layer of skin is epidermis. It contains three kinds of cells: flat, scaly cells on the surface called squamous cells; round cells called basal cells; and cells called melanocytes that give our skin its color. The second layer is dermis. Cut away view of a normal skin is shown in Figure 18.2. BCC is the most common form of skin cancer that

Figure 18.1. Two types of garlic. (A) Softneck (B) Hardneck garlic. Each type of garlic has several varieties as shown. Softneck garlic is the type that is seen in the produce section of the grocery store. The entire bulb is covered by multilayered parchment and continues up to the neck. Hardneck garlic forms scapes from the center and it contains several tiny cloves inside the parent bulb.

Table 18.1. Nutritional value of garlic per 100 g. Percentages are relative to US recommendations for adults.

Water	59 g	Thiamine (vitamin B1)	0.2 mg (15%)
Sugar	1.00 g	Riboflavin (vitamin B2)	0.11 mg (7%)
Calories	149 kcal	Niacin (vitamin B3)	0.7 mg (5%)
Protein	6.39 g	Pantothenic acid (vitamin B5)	0.596 mg (12%)
Lipids	0.5 g	Vitamin B6	1.235 mg (95%)
Carbohydrates	33.07 g	Vitamin C	31.2 mg (52%)
Fiber	2.1 g	Glutamic acid	0.805 g
Manganese	1.672 mg	Argenine	0.634 g
Potassium	401 mg (9%)	Aspertic acid	0.489 g
Sulfur	70 mg	Leucine	0.308 g
Calcium	181 mg (18%)	Lycine	0.273 g
Phosphorus	153 mg (22%)	selenium	4.2 μg
Magnesium	25 mg (7%)	Zinc	1.16 mg (12%)
Sodium	17 mg (1%)	Iron	1.7 mg (14%)
Beta-carotene	5 μg	Folate (vitamin B9)	3 μg (1%)

(USDA Nutrient database).

develops in the basal cells, which are the skin cells located in the lowermost layer of the epidermis. BCC accounts for more than 90% of all skin cancer in the U.S. They are present on sun-exposed areas of the skin, especially the face, ears, scalp, and upper trunk. These cancers almost never spread (metastasize) to other parts of the body. However, they can cause damage by growing and invading the surrounding tissue. SCC is accountable for about 16% of diagnosed skin cancers. This cancer begins in the squamous cells, found in the upper layer of the epidermis. While most commonly found on sun-exposed areas of the body, it can develop anywhere, including the inside of the mouth and the genitalia. Melanoma is the most dangerous type of skin cancer that occurs in melanocytes. Melanomas are the least frequent of the common skin cancers, but they frequently metastasize and are deadly once spread.

18.4 Mechanism of skin cancer development

Ultraviolet radiation from the sun is the main cause of skin carcinogenesis. Artificial source of UV radiation such as sunlamps, tanning booths and other carcinogenic exposures can also initiate skin carcinogenesis. Radiation exposure on the skin generates ROS mediated oxidative stress that irreversibly damage the regulatory cellular components, which results in altered cellular function. Accumulation of oxidatively damaged DNA or protein implicated in both acute and chronic cell injury leads to genetic instability and promotes the progression of skin cancer.

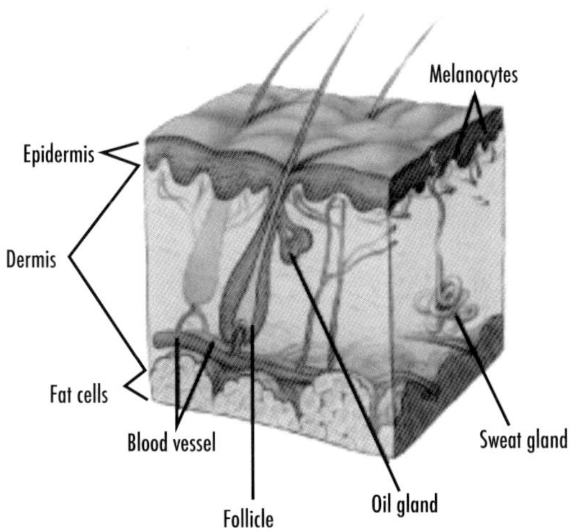

Figure 18.2. Cut away view of the skin showing epidermis, dermis, melanocytes, glands and blood vessel. Non melanoma skin cancer occurs in the epidermis region while melanoma occurs deep into the dermis (www.rationalmd.com).

18.4.1 Carcinogen mediated skin papilloma generation

Sir Percivol Pott (1775), the English surgeon was the first to relate the cause of scrotal skin cancer to a common history of occupational exposure to coal soot among his patients who were chimneysweepers when they were boys. The first animal model in skin cancer was successfully performed in 1918 (Yamagiwa and Ichikawa, 1918) using coal tar. Since then researchers have investigated several carcinogenic agents to understand the progression of skin cancer in mice. Although certain aspects of skin carcinogenesis differ between mouse and humans, mice do develop tumors in the same tissues and with similar histopathology as humans. In addition, the genetic events in human cancer are extremely similar to those observed in mice (Rice and O'Brien, 1980). It is well known that most human neoplasia results from abnormalities in the expression of critical proto-oncogenes, whose normal function is to control cell proliferation and differentiation. Activation of H-ras proto-oncogene was shown to be involved in tumor initiation in mouse skin by genotoxic carcinogens, PAHs (Di Giovanni, 1992). DMBA is one of the most potent carcinogen of the PAH family, which is metabolically activated for binding to cellular DNA through the generation of a genotoxic diol epoxide, or possibly through an intermediate. DMBA has been effectively used to generate skin papillomas in mice, which mimics NMSC in humans.

18.4.2 Multistep process of chemical carcinogenesis

Carcinogen mediated skin carcinogenesis follow three simple stages defined as 'Initiation', 'Promotion' and 'Progression'. The initiation stage is an irreversible step that promotes genetic mutations in the critical target genes resulting cellular malfunction. Initiation alone does not lead to the development of cancer unless initiation is followed by incubation with the cancer promoting agents such as croton oil, TPA etc. The promotion stage is mostly reversible and involves the clonal expansion of initiated cells. This genetic change apparently predisposes the cells to subsequent development of skin cancer. Following promotion, retention of the initial genetic abnormalities and simultaneously gaining additional genetic changes is the basis of the third step of experimental skin cancer progression in mice. The third stage of neoplastic development involves a loss of genotypic stability and development of heterogeneous cell population. The multi-stage model of mouse skin carcinogenesis has provided an important framework for understanding the nature of skin papilloma development in humans. A schematic diagram of multistage process of skin carcinogenesis is given in Figure 18.3.

18.5 Chemoprevention of skin carcinogenesis using dietary phytochemicals

Chemoprevention (Sporn *et al.*, 1976) is a rapidly growing area of oncology that focuses on the prevention of cancer using natural dietary phytochemicals as chemopreventive agents to interfere in the early stages of cancer growth. Since carcinogenesis is a multistage process, which usually takes many years in humans thus there are enough opportunities to intervene and prevent the development. It is estimated that more than two-thirds of cancer can be prevented through modification of lifestyle. Nearly one-third of these cancer occurrences can be attributed to diet

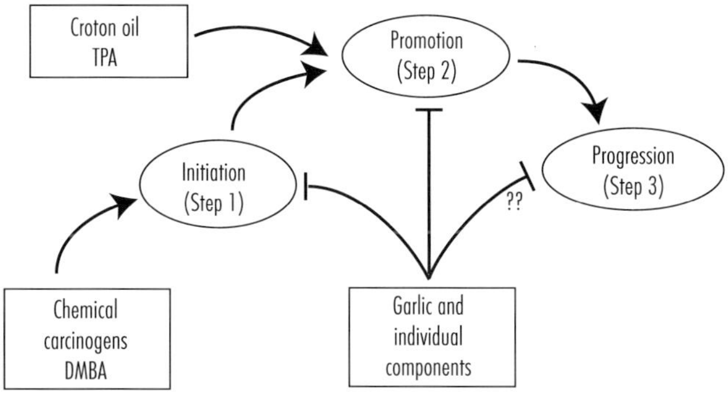

Figure 18.3. Schematic representation of multistep chemically induced carcinogenesis in the mice skin. DMBA act as a carcinogen to initiate skin papilloma growth on the applied skin. Croton oil or TPA helps in promotion of neoplastic skin cells. Garlic and its components are known to inhibit either the initiation or the promotional step of the skin cancer but its involvement in blocking the cancer progression is questionable.

alone. A number of studies support the hypothesis that a diet high in fat and low in carbohydrate increases the risk of cancer. Conversely, caloric restriction has shown to reduce the incidence of chemically induced tumor development in the skin of experimental animals. The risk of developing most types of cancers can also be reduced by avoiding tobacco, alcohols, pyrolysis products, nitrites, amines, and pesticides. Consumption of fruits, vegetables, and spices has been consistently shown to be associated with reduced risk of many cancers. A major chemoprevention strategy has been the "5 A Day for Better Health" program sponsored by the National Cancer Institute (NCI), encouraging the public to include five servings of fruits and vegetables in their regular diet.

18.5.1 Properties of chemopreventive agents

Many potential chemopreventive agents are under investigation all over the world. Chemopreventive compounds can be classified based on their mechanism of action or chemical structures. These compounds include antioxidant, anti-inflammatory, anti-proliferative and anti-hormonal agents with multiple mechanisms of action. Chemopreventive agents should be able to act in three ways: (1) prevention of genetic mutations that lead to genetic instability and cancer, (2) prevention of the processes that lead to excessive replication of damaged cells, and (3) prevention of occurrences of secondary tumors. An ideal chemopreventive agent must have minimal short-term and no long-term toxicity. Furthermore, it needs to be highly effective, easy to administer and inexpensive. Many chemopreventive agents are found in vegetables, fruits and spices, such as lycopene and quercetin in tomatoes, organosulfur compounds in onion and garlic, resveratrol and ellagic acid in grapes, β-carotene and vitamins in mango and crocin, crocetin in saffron, to name a few. It is estimated that there may be more than 100 different phytochemicals in just one serving of vegetables.

18.5.2 Garlic as a chemopreventive agent against skin carcinogenesis

Garlic is an exceptional plant with multiple beneficial effects such as anti-microbial, anti-arthritic, antithrombotic, hypoglycemic, hypolipidemic, and antitumor activity. Oral administration of garlic was shown to increase the lifespan of mice by 41.1% (Unnikrishnan and Kuttan, 1990). Several laboratory based animal studies and other epidemiological studies emphasis the preventive role of garlic and its constituents, against skin cancer when taken regularly in diet. Perchellet *et al.* back in 1990 started exploring the effect of garlic on skin using the mouse model. They demonstrated that in SENCAR mice, topical use of 2-5 mg of garlic oil prevents the progression of DMBA-initiated and TPA promoted skin papillomatosis. They have also shown that garlic oil was very effective when applied at least one hour before the promotion of carcinogenesis by TPA application. Moreover, a single 5 mg dose of garlic oil maximally inhibited DMBA-induced epidermal DNA synthesis by 86% when applied two hours before the carcinogen induction. Around the same time, Rao *et al.* (1990) also showed that topical application of garlic extracts, on DMBA induced skin papilloma, significantly reduced the incidences of tumors, in swiss albino mice.

Das and Saha (2009) reported the chemopreventive effects of aqueous extract of garlic on DMBA-induced skin carcinogenesis in female swiss albino mice. The goal was to investigate the effect of whole garlic in regular nutritional diet rather than the effect of its individual components. Mice were treated topically with DMBA to initiate the first stage of multistep chemical carcinogenesis and croton oil was used during the promotional step. Skin papillomas appeared on the site of the application from the fourth week of first DMBA application. Mice were divided in three groups. The first group of mice received oral administration of aqueous garlic suspension daily for 15 days before DMBA application (pre). The second group of mice received the garlic suspension daily concurrently with the DMBA treatments till the end of the experimental window (concomitant) and the third group of mice received the same garlic suspension daily for 15 days after the first DMBA treatment (post). Oral administration of an aqueous suspension of garlic was found to produce a remarkable reduction in the incidence of mouse skin papillomas especially in the concomitant group of mice. The number and the size of the papillomas were greatly reduced in all three groups of mice where garlic was administered orally but the effect was highly prominent in the group of mice where DMBA and garlic treatments were performed simultaneously. Histological analysis of the cross sections of all the treated skins (acetone and DMBA treated) showed similar results (Figure 18.4). DMBA treated skin-demonstrated hyperkeratosis, acanthosis with severe dysplasia and papillomatogenesis of the epidermis. Cross section from concomitant group of mice showed near normal histology with mild hyperkeratosis and moderate acanthosis of the epidermis. The other two groups of mice showed variable results although the effect of oral administration of garlic was evident in both groups of mice (Figure 18.4).

DMBA application produces free radicals on the skin that in turn promote lipid peroxidation. Free radicals and lipid peroxidation are known to cause initiation and promotion of carcinogenesis; MDA, a product of lipid peroxidation, was observed to be mutagenic and carcinogenic. The inhibition of lipid peroxides and the subsequent reduction of skin papillomas were associated

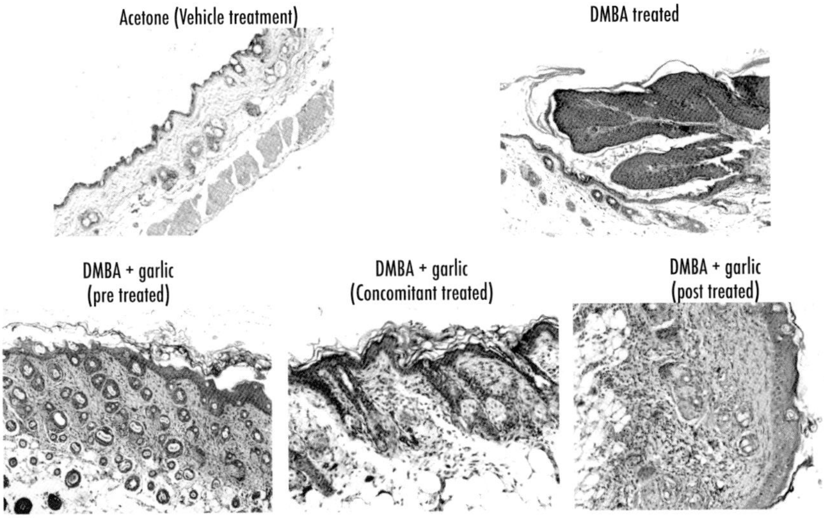

Figure 18.4. Histology of the papillomas that were treated with either DMBA or DMBA+ aqeous extract of garlic. Top left: representative H&E staining from acetone (vehicle) treated mice skin. Top right: representative stainig of a typical DMBA treated mice skin (skin of the papilloma). Bottom: H&E staining of the mice skins that were treated with DMBA as well as pretreated, concurrent treated and posttreated with aqeous extract of garlic. Concomitant treatment of garlic and DMBA showed the best protection agains the development of skin cancer in swiss albino mice.

with an induction of several phase II detoxification enzymes and other antioxidants. GST, a phase II enzyme, is thought to play a physiologic role in initiating the detoxification of many alkylating agents and environmental chemicals including mutagens and carcinogens. GPx, another phase II enzyme, also plays important role in cellular defense by eliminating H_2O_2 from the cells and is involved in maintenance of cellular membranes from free radical mediated oxidative damage. GST and GPx enzymes were upregulated in all garlic treated mice when compared to DMBA (carcinogen) only treated mice. Moreover, CAT and SOD are the antioxidation enzymes that are involved in the detoxification of free radicals in cells. Garlic treatments reduced the production of DMBA induced ROS generation in treated cells.

Additionally, aqueous extract of garlic induce downregulation of COX-2, which is considered as a marker of cancer progression, in DMBA + garlic treated mice, when compared to carcinogen treated mice only. Increased activity of GST, GPx, CAT, and SOD and decreased activity of COX-2 in skin tissues from the mice treated with garlic + DMBA indicates that garlic influences host cellular detoxification processes in order to exert its biological functions (Das and Saha, 2009). This study also indicates that garlic treatments could significantly reduce lipid peroxidation in mice exposed to DMBA and subsequently decrease the incidences of skin tumor formation. Thus, whole garlic shows enormous potential towards inhibition of skin carcinogenesis. The outcome of this preclinical research could be applied to human disease, although dose selection of garlic has to be modified according to the human body.

Melanoma, the aggressive form of skin cancer, has very poor prognosis. Despite ground breaking research the treatment of melanoma is still obscure. Most of the fatalities that occur in skin carcinogenesis are due to melanoma. Melanoma research has boosted due to the recent research outcome that aqueous extract of garlic induce cytotoxic effects on Sk-mel3 melanoma cell line (Hakimzadeh *et al.*, 2010). Although the authors conclude that whole garlic could be an excellent anti-tumor agent against melanoma treatments, more studies with melanoma mouse models as well as with other melanoma cell lines are required to fully understand the molecular mechanisms of action of garlic and its major active components. Garlic mainly inhibits the initiation and promotion stage of the multistep carcinogenesis development protocol and its effect on the progression step of the carcinogenesis model is not very well established (Figure 18.3).

18.5.3 Major individual components of garlic

The beneficial effects of garlic are attributed mostly to the organosulfur compounds present in garlic. Garlic contains at least 30 cancer preventive compounds including selenium. Major ingredients of garlic are given in Table 18.2. Alliin is one of the major components of whole fresh garlic and when chopped, alliin is converted to allicin (diallyl-thiosulfinate). Allicin is what gives the hot sensation in raw garlic. Transiently formed allicin, comprises 70-80% of the thiosulfinates and quickly decomposes to other compounds, such as DAS, DADS, DATS, dithine and ajoene. Allixin is another phytoalexin with a γ-pyrone skeleton structure and having antitumor, antimicrobial and antioxidative properties (Iciek *et al.*, 2009).

18.5.4 Diallyl sulfide

DAS is a major volatile, naturally occurring organosulfide substance believed to be the compound mostly responsible for garlic's anticancer properties. DAS has shown to protect against DMBA induced genomic strand break in mice skin. Mice pretreated with DAS before first application of DMBA induction showed superior protection than post treatment (Nigam and Shukla, 2007). Additionally, liposomized DAS application to mice also exhibit delayed onset of skin tumorigenesis initiated by DMBA and subsequently reduce the size and number of skin papillomas. DAS exert its biological effect by upregulating pro apoptotic protein, BAX1 and inhibiting anti apoptotic protein, Bcl2. The cycle dependent kinase inhibitor, p21/Waf1 and the p53 tumor suppressor protein was also modulated by DAS to demonstrate its beneficial role against skin cancer (Khan *et al.*, 2007). Moreover, DAS supplement significantly lowered DMBA induced expression of ras oncoprotein, PI3K/Akt and p38MAPK but did not alter the expression of JNK1 and ERK1/2 (Kalra *et al.*, 2006). DAS had also been shown to induce apoptosis as the possible mechanism of the anti-proliferative effect in solid tumors, which is evident by the appearance of a sub-G1 fraction in flow cytometry analysis (Arora and Shukla, 2002). So, DAS can act as a potent inhibitor of environmental toxicants induced genetic mutations and can emerge as a potential chemopreventive agent against skin tumor development.

Table 18.2. Major components of garlic. The molecular formulas of the major garlic components known for its beneficial action are given here.

Components	Chemical formula	Abbreviation
Alliin	$CH_2=CH-CH_2-S\ (O)-CH_2-CH\ (NH_2)-COOH$	
Allicin	$CH_2=CH-CH_2-S(O)-S-CH_2-CH=CH_2$	
Ajoene	$CH_2=CH-CH_2-S(O)-CH_2-CH=CH-S-S-CH_2-CH=CH_2$	
Diallylsulfide	$CH_2=CH-CH_2-S-CH_2-CH=CH_2$	DAS
Diallyldisulfide	$CH_2=CH-CH_2-S-S-CH_2-CH=CH_2$	DADS
Diallyltrisulfide	$CH_2=CH-CH_2-S-S-S-CH_2-CH=CH_2$	DATS
Allylmethylsulfide	$CH_2=CH-CH_2-S-CH_3$	AMS
Allylmethyldisulfide	$CH_2=CH-CH_2-S-S-CH_3$	AMD
Allylmethyltrisulfide	$CH_2=CH-CH_2-S-S-S-CH_3$	AMT
Allylmarcaptan	$CH_2=CH-CH_2-SH$	AM
S-allylcysteine	$CH_2=CH-CH_2-S-CH_2-CH\ (NH_2)-COOH$	SAC
S-allylmercaptocysteine	$CH_2=CH-CH_2-S-S-CH_2-CH(NH_2)-COOH$	SAMC
Allixin	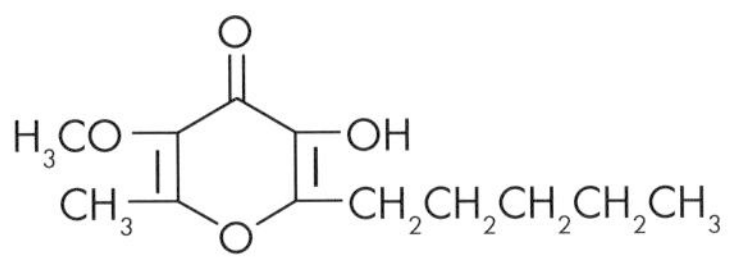	
1,3 Vinyldithiin	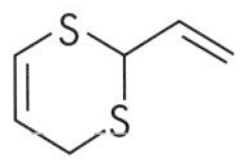	

18.5.5 Diallyl disulfide

Like DAS, topical application of DADS also showed considerable inhibition of the DMBA induced skin papilloma formation and significantly elevated the survivability of the DMBA treated SENCAR mice (Dwivedi *et al.*, 1992). DADS also showed effectiveness in preventing cell proliferation in SK-Mel-2 melanoma cell line. DADS treatment elevates intracellular free calcium in cells and decreases the activity of calcium dependent ATPase enzyme in a dose dependent manner. Alterations in calcium homeostasis are likely involved in the growth inhibition/cytotoxicity caused by DADS (Sundaram and Milner, 1996).

18.5.6 Diallyl trisulfide

DATS is reported to provide anticancer activity in several cancer types. DATS revealed superior growth inhibitory effect than DADS and DAS in A375 melanoma and BCC cells. DATS increases ROS production, maintains calcium homeostasis and decreases mitochondrial membrane potential to exert its beneficial action against skin carcinogenesis. Moreover, DATS arrest cancer cells in G2/M, induce p53 signaling pathway and promote apoptosis in response of oxidative damage of the cells. DATS also displayed selective target of growth inhibition between skin cancer cells and normal keratinocyte HaCaT cells (Wang *et al.*, 2010).

18.5.7 Ajoene

Ajoene was shown to induce apoptosis in human leukaemia cells. When ajoene was topically applied to the patients suffering from nodular and basal cell carcinoma, tumors in majority of the patients shrinked, demonstrating its anti-carcinogenic effects. Immunohistochemical analysis with the skin tumor samples and with the BCC tumor cell line, TE354T, showed induction of apoptosis in ajoene treated samples but there was no change in the expression of cell proliferation marker, Ki-67. Ajoene induced apoptosis in a dose and time dependent manner thus ajoene can also hold a high potential in reducing the size of BCC tumor, mainly by inducing the mitochondria-dependent route of apoptosis (Tilli *et al.*, 2003).

In general, the organosulfur compounds present in garlic have been shown to induce Phase II detoxification enzymes such as GST, NQO1 and UDP-glucuronosyl transferase in liver and colon tissue of rodents. Natural garlic with selenium fertilization demonstrates higher cancer preventive activities than garlic with normal fertilization. Diallyl selenide was at least 300-fold more effective than diallyl sulfide in protecting against DMBA-induced mammary adenocarcinomas in rats (El-Bayoumy *et al.*, 2006). In addition, benzyl selenocyanate inhibited the development of DMBA-induced mammary adenocarcinoma and azoxymethane-induced colon cancer in rats and benzo[a]pyrene-induced forestomach tumors in mice.

Garlic is often used as dietary supplements. There is considerable variation in the quantity of organosulfur compounds available in fresh and commercially available garlic products. Four major garlic supplements are available for use: (1) essential oil (2) dehydrated powder (3) oil macerate and (4) garlic extract. Essential oil is obtained by steam distillation of garlic that contains 0.2-5% cloves and variety of sulfides. Garlic powder is used as flavoring agent for condiments and processed foods. It is made by slicing or crushing the cloves of the garlic, dried and then pulverized into powder. The powder retains some ingredients of raw garlic although the constituents vary significantly. Oil macerate products are prepared of encapsulated mixtures of whole garlic cloves ground into vegetable oil. Two types of garlic extracts have been used according to the needs. Commonly whole or sliced garlic cloves are soaked for different times in an extracting solution (purified water and/or diluted alcohol) and following extraction, the extract is concentrated before being used. Another extract called AGE made via the procedure mentioned above, but aged for up to 20 months. During incubation, the initial unstable compounds in garlic are converted naturally into stable and water-soluble sulfur compounds such as SAC and SAMC, which are confirmed safe by toxicology. SAC is one of the biologically active transformation components of garlic that is present in every preparation.

18.6 Bioavailability of garlic components

All the beneficial action of garlic depends upon its bioavailability. Bioavailability is a term that describes the way chemicals are absorbed and circulated via blood stream. Intravenous administrations of medications are considered to have 100% bioavailability, as they are absorbed immediately in the body's circulatory system. However, when a medication is administered orally, its bioavailability decreases due to poor absorption and metabolism and varies between individuals. For dietary supplements, herbs and other nutrients the administration is mostly oral and bioavailability generally designates the percentage absorption of the starting material.

18.6.1 Alliin

Guo *et al.* (1990) found that distribution of alliin was prevalent in 10 mins in the stomach (7.2%), intestine (22.4%), and liver (2.5%) following oral administration of alliin (10 mg/mouse) in mice. In pharmacokinetic studies, 60-70% of ^{35}S-labeled alliin was absorbed in the blood stream of rats (Lachmann *et al.*, 1994). In both the cases no conversion of alliin to allicin and other organosulfur compounds was observed.

18.6.2 Allicin

Allicin could not be detected in liver passage of rat liver after allicin intake; rather presence of DADS and AM, metabolites of allicin was evident (Egen-Schwind *et al.*, 1992b). AMS is also considered a metabolite of allicin that is released into the breath (Rosen *et al.*, 2001) but its concentration in blood and its rate of formation has not been studied in detail.

18.6.3 Organosulfur volatiles

DAS, DADS, DATS and vinyldithiins are the major components of garlic oil and oil-macerate preparations. Vinyldithiins have been detected in the serum, kidney, and fat tissue 24 hours after application. No metabolites of vinyldithiins in the isolated perfused rat liver were identified (Egen-Schwind *et al.*, 1992a). Intraperitoneal injection of [^{35}S]-labeled DADS in rats reveals 70% of radioactivity in liver cytosol, of which 80% was metabolized to sulfate (Pushpendran *et al.*, 1980).

18.6.4 S-Allylcysteine

SAC is rapidly absorbed in the gastrointestinal tract following oral administration to rats, mice and dogs. It is distributed mainly in plasma, liver and kidney. Bioavailability of SAC was observed to be 103.0% in mice, 98.2% in rats, and 87.2% in dogs (Nagae *et al.*, 1994). Significant concentration of N-acetyl-S-allyl-L-cysteine is identified as a metabolite of SAC in the urine.

Taken together it can be hypothesized that whole garlic in diet or the selected major components of garlic could emerge as a potential anticancer compound against skin cancer progression. Figure 18.5 represents a schematic representation of molecular pathways involved en route to

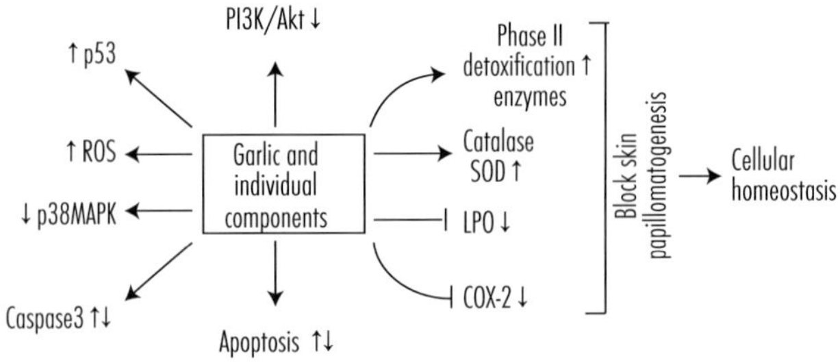

Figure 18.5. Schematic representation of the molecular mechanisms of action of garlic and its major constituents in mice. Garlic up regulates Phase II detoxification enzymes, antioxidants, ROS and p53 signaling pathway. It also modulates apoptosis and down regulate COX-2, which is a marker of carcinogenises progression, LPOs, PI3L/Akt and p38MAPK.
Up arrow: up regulation and down arrow: down regulation.

garlic's beneficial action against the development of skin cancer. Discovery of new tools to study the multiplicity of signaling pathways holds the optimism of understanding the pathways better in cancer chemoprevention by naturally occurring garlic and its components.

18.7 Garlic in other skin health, care and treatments

Garlic is rich in antioxidant and consumption of a small clove of garlic each day has shown to have enormous health benefits. Applied topically garlic can work as an anti-inflammatory to the outer surface of the skin. Following are the couple of common beauty complaints that can be treated with garlic and its extracts.

18.7.1 Acne

Acne is a skin disorder that occurs due to the excessive production of oily substances by the glands beneath the skin. The antibiotic, antifungal, antiseptic and germicidal properties of garlic acts against acne while the antioxidants help repair the ROS mediated skin damage. It also helps to fight facial spots, blemishes and scars.

18.7.2 Cold sores

Cold sores are painful small blisters or sores that appear on the lips, mouth, or nose caused by herpex simplex virus. The juice of garlic kills the virus, disinfects and heals the sore.

18.7.3 Psoriasis

Psoriasis is a common and chronic skin disease with patches of itchy red skin with thick scales like structure. A remarkable change in skin texture will be noticed in few days if garlic oil is directly applied to the psoriasis skin.

18.7.4 Hair loss

Garlic and its extracts have been used to prevent hair loss for ages. Use of garlic oil on the hair loss area for couple of weeks could stop hair loss and rejuvenate the hair follicles for new hair growth.

18.7.5 Athlete's foot

Athlete's foot (*tinea pedis*) is a rash on the skin of the foot, which is caused by fungal infection. Covering the affected area with raw garlic and gauze overnight followed by garlic oil on the affected area heals the disease.

Acknowledgements

TS thanks Lombardi Comprehensive Cancer Center, and acknowledges the support from the pilot grants from American Cancer Society (IRG #97-152-16-2) and Fisher Center for Familial Cancer Center, Georgetown University.

References

Arora, A. and Shukla, Y., 2002. Induction of apoptosis by diallyl sulfide in DMBA-induced mouse skin tumors. Nutrition and Cancer 44, 89-94.

Das, I. and Saha, T., 2009. Effect of garlic on lipid peroxidation and antioxidation enzymes in DMBA-induced skin carcinoma. Nutrition. 25, 459-471.

Di Giovanni, J., 1992. Multistage carcinogenesis in mouse skin. Pharmacology and Therapeutics 54, 68-128.

Dwivedi, C., Rohlfs, S., Jarvis, D. and Engineer, F.N., 1992. Chemoprevention of chemically induced skin tumor development by diallyl sulfide and diallyl disulfide. Pharmacology Research 9, 1668-1670.

Egen-Schwind, C., Eckard, R., Jekat, F.W. and Winterhoff, H., 1992a. Pharmacokinetics of vinyldithiins, transformation products of allicin. Planta Medica 58, 8-13.

Egen-Schwind, C., Eckard, R. and Kemper, F.H., 1992b. Metabolism of garlic constituents in the isolated perfused rat liver. Planta Medica 58, 301-305.

El-Bayoumy, K., Sinha, R., Pinto, J.T. and Rivlin, R.S., 2006. Cancer chemoprevention by garlic and garlic-containing sulfur and selenium compounds. Journal of Nutrition 136, 864S-869S.

Fleischauer, A.T., Poole, C. and Arab, L., 2000. Garlic consumption and cancer prevention: meta-analyses of colorectal and stomach cancers. American Journal of Clinical Nutrition 72, 1047-1052.

Guo, Z., Muller, D., Pentz, R., Kress, G. and Siegers, C.P., 1990. Bioavailability of sulphur containing ingredients of garlic in the rat. Planta Medica 56, 692-693.

Hakimzadeh, H., Ghazanfari, T., Rahmati, B. and Naderimanesh, H., 2010. Cytotoxic effect of garlic extract and its fractions on Sk-mel3 melanoma cell line. Immunopharmacology 32, 371-375.

Herman-Antosiewicz, A., Powolny, A.A. and Singh, S.V., 2007. Molecular targets of cancer chemoprevention by garlic-derived organosulfides. Acta Pharmacology Sinica 28, 1355-1364.

Iciek, M., Kwiecien, I. and Wlodek, L., 2009. Biological properties of garlic and garlic-derived organosulfur compounds. Environtal and Molecular Mutagenesis 50, 247-265.

Kalra, N., Arora, A. and Shukla, Y., 2006. Involvement of multiple signaling pathways in diallyl sulfide mediated apoptosis in mouse skin tumors. Asian Pacific Journal of Cancer Prevention 7, 556-562.

Khan, A., Shukla, Y., Kalra, N., Alam, M., Ahmad, M.G., Hakim, S.R. and Owais, M., 2007. Potential of diallyl sulfide bearing pH-sensitive liposomes in chemoprevention against DMBA-induced skin papilloma. Molecular Medicine 13, 443-451.

Lachmann, G., Lorenz, D., Radeck, W. and Steiper, M., 1994. The pharmacokinetics of the S35 labeled labeled garlic constituents alliin, allicin and vinyldithiine. Arzneimittelforschung 44, 734-743.

Milner, J.A., 2001. A historical perspective on garlic and cancer. Journal of Nutrition 131, 1027S-1031S.

Nagae, S., Ushijima, M., Hatono, S., Imai, J., Kasuga, S., Matsuura, H., Itakura, Y. and Higashi, Y., 1994. Pharmacokinetics of the garlic compound S-allyl cysteine. Planta Medica 60, 214-217.

Nigam, N. and Shukla, Y., 2007. Preventive effects of diallyl sulfide on 7,12-dimethylbenz[a]anthracene induced DNA alkylation damage in mouse skin. Molecular Nutrition and Food Research 51, 1324-1328.

Perchellet, J.P., Perchellet, E.M. and Belman, S., 1990. Inhibition of DMBA-induced mouse skin tumorigenesis by garlic oil and inhibition of two tumor-promotion stages by garlic and onion oils. Nutrition and Cancer 14, 183-193.

Pushpendran, C.K., Devasagayam, T.P.A., Chintalwar, G.J., Banerji, A. and Eapen, J., 1980. The metabolic fate of [-35S]-diallyl disulfide in mice. Experientia 36, 1000-1001.

Rao, A.R., Sadhana, A.S. and Goel, H.C., 1990. Inhibition of skin tumors in DMBA-induced complete carcinogenesis system in mice by garlic (*Allium sativum*). Indian Journal of Experimental Biology 28, 405-408.

Rice, M.C. and O'Brien, S.J., 1980. Genetic variance of laboratory outbred Swiss mice. Nature 283, 157-161.

Rosen, R.T., Hiserodt, R.D., Fukuda, E.K., Ruiz, R.J., Zhou, Z., Lech, J., Rosen, S.L. and Hartman, T.G., 2001. Determination of allicin, S allylcysteine and volatile metabolites of garlic in breath, plasma or simulated gastric fluids. Journal of Nutrition 131, 968S-971S.

Sengupta, A., Ghosh, S., Bhattacharjee, S. and Das, S., 2004. Indian food ingredients and cancer prevention - an experimental evaluation of anticarcinogenic effects of garlic in rat colon. Asian Pacific Journal of Cancer Prevention 5, 126-132.

Sporn, M.B., Dunlop, N.M., Newton, D.L. and Smith, J.M., 1976. Prevention of chemical carcinogenesis by vitamin A and its synthetic analogs (retinoids). Federation Proceedings 35, 1332-1338.

Sundaram, S.G. and Milner, J.A., 1996. Diallyl disulfide induces apoptosis of human colon tumor cells. Carcinogenesis. 17, 669-673.

Tilli, C.M., Stavast Kooy, A.J., Vuerstaek, J.D., Thissen, M.R., Krekels, G.A., Ramaekers, F.C. and Neumann, H.A., 2003. The garlic-derived organosulfur component ajoene decreases basal cell carcinoma tumor size by inducing apoptosis. Archives of Dermatological Research 295, 117-123. E-pub 2003 May 2020.

Unnikrishnan, M.C. and Kuttan, R., 1990. Tumour reducing and anticarcinogenic activity of selected spices. Cancer Letters 51, 85-89.

Wang, H.C., Yang, J.H., Hsieh, S.C. and Sheen, L.Y., 2010. Allyl sulfides inhibit cell growth of skin cancer cells through induction of DNA damage mediated G2/M arrest and apoptosis. Journal of Agricultural and Food Chemistry 58, 7096-7103.

Yamagiwa, K. and Ichikawa, K., 1918. Experimental study of the pathogenesis of carcinoma. Cancer Research 3, 1-29.

Key facts

- Skin barrier function resides in the *stratum corneum* and protects internal organs against environmental damage such as pathogens (bacteria, fungi, parasites, viruses), UV radiation and water loss.
- *Stratum corneum* is the uppermost thin layer of the skin.
- Atopic dermatitis is an inflammatory skin disease characterized by itching, redness, scaling, and excoriation.
- Molecular biology is the study of biology on a molecular basis and involves many other areas such as biochemistry and genetics. It represents the study of the interactions between the different types of DNA, RNA and protein biosynthesis including replication, transcription, translation and cell function.
- 16S rRNA gene is a phylogenetic marker.
- Defensins are a family of small cationic cysteine. They contribute to host defense against microbes and are active against pathogens, like bacteria, fungi and many viruses.
- β-defensins are antimicrobial peptides and represent the most widely distributed defensins, being secreted by leukocytes and epithelial cells of many kinds.
- The cathelicidin peptides belong to the family of antimicrobial peptides that are found in lysosomes, in macrophages and polymorphonuclear leukocytesin.
- Lactic acid bacteria are a group of Gram-positive acid-tolerant bacilli or cocci, which have the property of producing lactic acid as a result of carbohydrate fermentation. These bacteria are usually found in lactic products and are used in the food industry for several reasons.

Summary points

- The human skin provides a habitat for a variety of microorganisms, the skin microflora, which has a beneficial role.
- Skin microbiota plays a role in both maintenance of skin health and development of skin disease.
- Preservation of the resident microflora is a way to maintain healthy skin functions and because of its role in skin health and disease, the maintenance or restoration of healthy skin microbiota has become the aim of modern therapies.
- Pre- and probiotics have been suggested as important applications of such therapies and in the last few years have reached scientific popularity as safe and effective agents to regulate the body's micro-environment and the skin health. A constantly growing body of literature exists about their relevance in cosmetic and dermatology.
- Prebiotics rebalance the skin microflora.
- Probiotic concepts consist of applying inactivated microbial biomass of beneficial bacteria.
- Probiotics improve gut barrier function, restore healthier gut microecology, stimulate the host immune system, and antagonize the inflammatory alterations.
- In the case of diseases with some imbalance in microorganisms, such as impure skin/mild acne or dry skin/mild atopic dermatitis, pre- and probiotic concepts represent an effective alternative to strictly antibacterial products.

19. Pre- and probiotics for human skin

A. Marini and J. Krutmann
Institut für Umweltmedizinische Forschung (IUF), Leibniz Research Ctr for Envtl Medicine, Heinrich-Heine-University Düsseldorf, Auf'm Hennekamp 50, D-40225 Düsseldorf, Germany;alessandra.marini@uni-duesseldorf.de

Abstract

The human skin provides a habitat for a variety of microorganisms, the skin microflora. There is a complex network of interactions between the microbes and cells of the epidermis. The composition and function of the skin's microflora has become one of the most exciting and rapidly developing areas in cutaneous biology and the interest in strategies of a selective modulation of the skin microflora is constantly growing. Current research on the complex interplay between the microbiota, barrier function and innate immune system of the skin indicate that the skin's microbiota have a beneficial role, much like that of the gut microflora. Preservation of the microflora is a way to maintain healthy skin functions and because of its role in skin health and disease, the maintenance or restoration of healthy skin microbiota has become the aim of modern therapies. Pre- and probiotics have been suggested as important applications of such therapies and in the last few years have reached scientific popularity as safe and effective agents to regulate the body's micro-environment and the skin health. A constantly growing body of literature exists about their relevance in cosmetic and dermatology. Most of these examples deal with the orally administration of food grade, but recently their use in cosmetic applications direct to the skin has been proposed. Prebiotics rebalance the skin microflora while probiotic concepts consist of applying an inactivated microbial biomass. Probiotics improve gut barrier function, restore healthier gut microecology, stimulate the host immune system, and antagonize the inflammatory alterations. A regular consumption of fermented dairy products is likely to improve the natural skin barrier function and improved its cosmetic appearance. In the case of diseases with some imbalance in microorganisms, such as impure skin/mild acne or dry skin/mild atopic dermatitis, pre- and probiotic concepts represent an effective alternative to strictly antibacterial products.

Keywords: skin benefits, symbiotics, skin microflora, immunomodulation, skin microbiota

Abbreviations

FOS	Fructo-oligosaccharide
GOS	Galacto-oligosaccharide
hCAP	Human cathelicidin

19.1 Introduction

The human skin is an intricate habitat for a vast amount of bacteria, which reside within epidermis, dermis, and the skin-associated glands and follicles, know as skin microflora or microbiota. The total number of microbes on the skin surface is typically within the range of 10^4-10^6 cells/cm^2 (Ouwehand *et al.*, 2010). The skin plays an important role in protecting against dehydration and damage or insults from, external aggression. Composition, type, density, and complexity of the human microbiota differ depending on the anatomical location and on factors like temperature, local humidity, amount of sebum and sweat production, and the host's hormonal status and age (Simmering and Breves, 2011).

The composition and function of the skin's microflora has become one of the most exciting and rapidly developing areas in cutaneous biology (Cogen *et al.*, 2008). A major driving force for this development has been the discovery that epidermal keratinocytes have the potential to affect the cutaneous microflora by producing antimicrobial peptides (Gallo *et al.*, 2002).

Recent research on the close link between the microbiota, the barrier function and the innate immune system of the skin indicate that the skin's microbiota have a beneficial role, much like that of the gut microflora (Elias and Choi, 2005). Skin microbiota plays a role in both maintenance of skin health and development of skin disease. Recent years have seen a major change in how activities of the human gastrointestinal tract are perceived. Pre- and probiotics gained scientific and commercial popularity as safe and effective agents which regulate the body's micro-environment. After ingestion, they become transient constituents of the gut microbiota capable of exerting their systemical effects, thus giving a rationale for their use as functional foods (Figure 19.1).

Although known since a long time, only in this era of self-care and complementary medicine, pre- and probiotics have started to receive major attention, and several studies have been carried out on their effects, using different formulae and with numerous purposes of preventing or treating diseases. Prebiotics rebalance the skin microflora while probiotic concept consists of applying inactivated beneficial bacteria. The most frequently used dietary method of influencing the gut flora composition is that of probiotics, which are defined as living microorganisms found in fermented milk which upon ingestion in certain numbers exert direct health beneficial effects on the intestinal microbial balance (Guarner and Schaafsma, 1998). Probiotics have been widely used for gastrointestinal disorders, and a growing number of clinical studies suggest that probiotic strategies affect selected functions of the skin (Ouwehand *et al.*, 2003). Accordingly, modulation

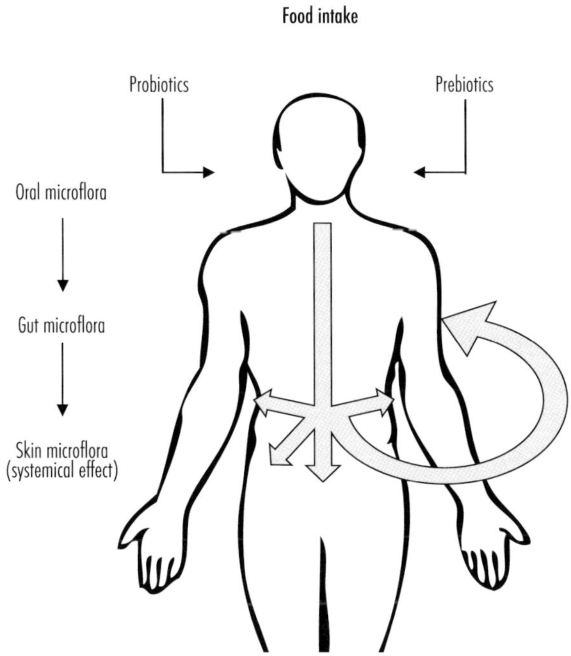

Figure 19.1. Oral ingestion of pre- and probiotics exerts beneficial effects on the skin. Modulation of the gut's microflora cause beneficial effects of any microbial community including the skin microflora. By boosting your immunity, maintaining a healthy gut's microflora and facilitating digestion, pre- and probiotics help restore a youthful vitality.

of the gut's microflora through probiotics appears to cause beneficial effects in healthy as well as diseased human skin (e.g. Guéniche *et al.*, 2006).

An additional approach is the prebiotic concept. This concept was introduced by Gibson and Roberfroid in 1995 who defined prebiotics as "non-digestible food ingredients that beneficially affect the host by selectively stimulating the growth and/or activity of one, or a limiting number of, bacteria in the colon" (Gibson and Roberfroid, 1995), that can improve the host health. This concept has originally been developed for the gut, but in principle can be applied to modulate the composition of any microbial community including the skin microflora to achieve beneficial effects (Krutmann, 2009). The concept of prebiotics was adapted also to cosmetic products (e.g. Bockmühl, 2006). Interesting, prebiotics, in contrast to antibiotics, may allow selective inhibition of detrimental and at the same time preservation and/or stimulation of beneficial bacteria, therefore became recently of a great interest in dermatology (Cogen *et al.*, 2008).

19.2 Skin microflora

The intensive research, the recent advances in microbiology and immunology and the use of new molecular techniques employing 16SrRNA-Gene strategies in the characterization of skin microbiota, revised and improved our understanding of the molecular mechanisms of microbial virulence and the specific events involved in the host-microbe interaction, increasing knowledge of the microflora composition and activities.

Current data suggest that the skin microbiota is considerably different from the historical classification based on traditional methods, and suggest that these organisms protect the host. The human skin microbiota is complex (Gao *et al.*, 2007, Grice *et al.*, 2008) and comprises 113 phylotypes that belong to six bacterial divisions (Krutmann, 2009). Vast majority (>90%) of the bacteria belong into three phyla: Actinobacteria, *Firmicutes*, and Proteobacteria (Ouwehand *et al.*, 2010). Among these, Proteobacteria dominate the skin microbiota. *Staphylococcus epidermidis* and *Propionibacterium acnes* consisted of less than 6% of the captured microbiota (Krutmann, 2009), in contrast to the commonly held notion representing those as the dominant aerobic bacteria resident in skin (Grice *et al.*, 2008). A small overview about the most important bacteria composing the skin microflora is presented in Table 19.1.

Microbial species are usually divided into three groups: transients, temporary residents and resident (Holland and Bojar, 2002). The term resident refers to viable, reproducing populations, whereas transient species are defined as contaminants with little or no capacity for sustained growth and reproduction, and are intermittently found in the cutaneous environment (Krutmann, 2009).

Table 19.1. Resident skin bacterial flora.

Microflora	
Actinobacteria	*A. johnsonii*
Corynebacteria	*C. minutissimum, C. tenuis, C. xerosis, C. jeikeium*
Enterococci	*E. faecalis*
Micrococci	*M. luteus, M. varians, M. lylae, M. kristinae, M. nishinomiyacnsis, M. roseus, M. sedentarius, M. agieis*
Propionibacteria	*P. acnes, P. granulosum, P. avidum*
Staphylococci	*S. epidermidis, S. aureus, S. warneri, S. hominis, S. haemolyticus, S. capitis, S. midis, S. saprophyticus, S. cohnii, S. xylosus, S. simulans, S. saccharolyticus*
Streptococci	*S. pyogenes, S. mitis, S. salivarius, S. mutans*

Some of the most important bacteria composing the skin microflora; among this, Proteobacteria dominate the skin microbiota.

Composition, density, and complexity of the human microbiota (microflora) differ depending on the region of the body and on factors like temperature, humidity, and sebum content (Simmering and Breves, 2011) and is determined by a variety of biochemical and physical factors, like the number and size of follicles and glands, gland function, the flow of secretions, the integrity of barrier function, skin pH and osmotic potential (Bojar *et al.*, 2002). A slightly acidic pH, favors *Propionibacterium* spp., whereas neutral and alkaline pH favors most other resident bacteria. Also, high hydration is associated with higher pH, which is again associated with high microbial population density, e.g. in the foot. Biochemical factors include chemical compounds such as soluble micronutrients derived from sebum (lipids and aminoacids) and sweat (vitamins, lactate and amino acids) as well as biochemical molecules which are produced as a consequence of the metabolic activity of microorganisms on the skin and which in turn function to influence colonization (Krutmann, 2009).

Of particular interest for cosmetic products, the different microbial species present on the human skin in the different anatomical regions has been analyzed in several studies. Primarily, the studies have been conducted to the types of microbes which colonize the skin, with little attention given to the functions. There is surprisingly little literature that has evaluated the influence of the skin microflora in skin health; therefore the role of the microflora is not yet fully understood. One of the most important functions of the skin is the protective barrier. Bacterial colonization belonging to the microflora inhibits the growth of aggressive pathogenic microorganisms (Krutmann, 2009). Accordingly, Propionibacteria have been shown to have adjuvant and antitumor activities and to contribute to a more efficient immunological response to general infections (Roszkowski *et al.*, 1990). Preservation of the resident microflora is a way to maintain healthy skin functions. The link between skin microorganisms and the skin immune system provides another level of complexity (Krutmann, 2009). Only when the host becomes compromised, the resident microflora may become dangerous and play a role in the development of skin disease. Examples associated with alterations of the skin microbiota include acne, psoriasis, atopic dermatitis and other inflammatory skin disorders. Pathogenic microbes associated with various skin diseases are numerous but an extensive review is not within the scope of the current chapter.

Skin microflora, skin barrier function and the skin immune system are closely linked to each other and appear to form a true symbiosis and highly regulated network that controls a variety of fundamental skin functions and is a key for well-being and health (Figure 19.2).

The immune system preserves self-integrity and fights against infection and it takes advantage of the capacity to mount adaptive and non-adaptive immune responses. Non-adaptive (=innate) immune responses are immediate and non-specific, whereas adaptive immune responses are secondary and selective. There is currently no evidence to suggest that adaptive immune responses of the skin influence the normal skin microflora, conversely there is some evidence that the skin microflora activates the adaptive immune system (Krutmann, 2009). Accordingly, microorganisms on the skin have been shown to be coated with immunoglobulins which are most likely derived from eccrine gland secretions (Metze *et al.* 1991). Also, numerous reports have described humoral and cell-mediated immune responses in the peripheral blood against

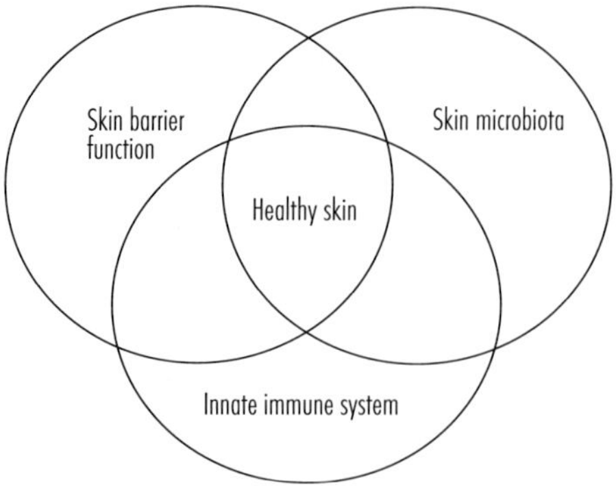

Figure 19.2. Skin microflora, skin barrier function and the skin immune system are closely linked to each other and appear to form a complex and highly regulated network that controls a variety of fundamental skin functions. Only when the host becomes compromised, the resident microflora may become dangerous and play a role for the development of skin disease.

skin microbials (e.g. Ashbee *et al.*, 1994, 1997). In marked contrast to adaptive immune responses, non-adaptive immune responses of the skin are clearly involved in the control of microbial colonization and have to be considered as an additional factor influencing the microbial equilibrium on the skin.

Because of its role in skin health and disease, the maintenance or restoration of healthy skin microbiota has become the aim of modern therapies. Pre- and probiotics, alone or together (synbiotics), have been suggested as important applications of such therapies and in the last few years have reached scientific popularity as safe and effective agents to regulate the skin health. The schematic illustration of the principal routes of prebiotic and probiotic skin microflora manipulation is presented in Figure 19.3.

The application of beneficial bacteria or inactivated biomass as probiotic or actives or nutrients as prebiotic will promote the growth of beneficial and inhibit harmful bacteria.

Especially many probiotic strains applications are known for their potential to modify the skin's immune responses. Accordingly, upon activation by micooorganisms, epidermal keratinocytes have the capacity to produce all four known β-defensins as well as cathelicidin hCAP-18 and by producing these antimicrobial peptides inhibit growth or kill microorganisms, i.e. bacteria, but also viruses (Krutmann, 2009). Donnarumma *et al.* (2007) have shown that an extract from avocado was able to influence the adherence of *Malassezia furfur* to keratinocytes and to induce the production of the human β-defensin 2. An even more complex reaction was induced by the probiotic strain *Streptococcus salivarius* K12 (Cosseau *et al.*, 2008). Elias (2007) and Simmering and Breves (2011) reviewed in more details the importance of these factors and their complex interplay.

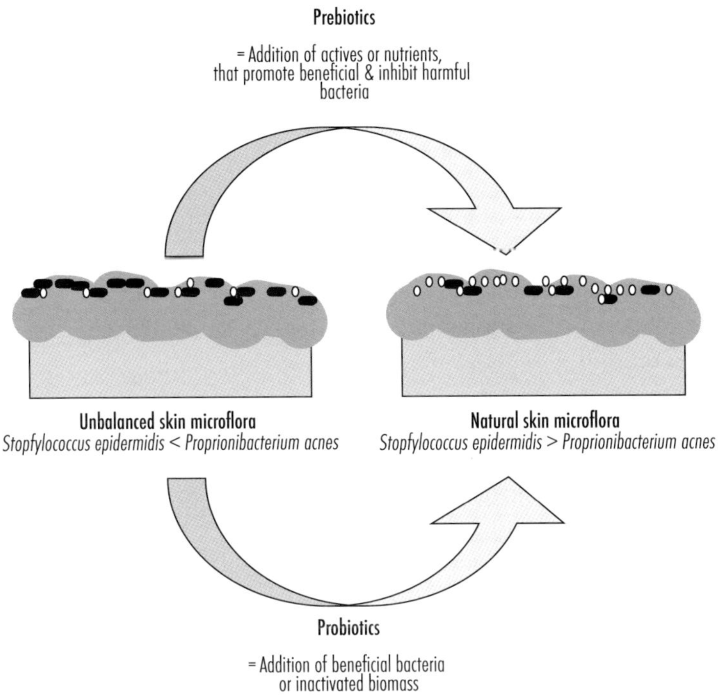

Figure 19.3. Schematic illustration of the principal routes of prebiotic and probiotic skin microflora manipulation. Topical application of beneficial bacteria or inactivated biomass as probiotic or actives or nutrients as prebiotic will promote the growth of beneficial and inhibit harmful bacteria. Modified from Krutmann (2009).

19.3 Prebiotics and skin

An approach to maintain the balance of the members of the protecting flora is using the prebiotic concept. Gibson and Roberfroid in 1995 introduced this concept and defined the prebiotics as "non-digestible food ingredients that beneficially affect the host by selectively stimulating the growth and/or activity of one, or a limiting number of, bacteria in the colon" (Gibson and Roberfroid, 1995). Most identified prebiotics (Figure 19.4) are carbohydrates (such as oligosaccharides), within these, there is a wide diversity of molecular structure. As prebiotics exploit non-viable food ingredients, their applicability in diets is wide ranging. Because of their selective metabolism, non-digestible oligosaccharides, and in particular the ingestion of FOS has been shown to stimulate bifidobacteria in the lower gut (Gibson *et al.*, 1995). GOS, animal-derived substance manufactured by extracting milk sugars from dairy products, are another class of prebiotics. It is increasingly common to distinguish other similar substances as "prebiotic" by using terms such as "possible" or "likely" prebiotics.

Traditional dietary sources of prebiotics, that contains them, include soybeans, inulin sources (such as Jerusalem artichoke and chicory root), raw oats, unrefined wheat, unrefined barley and

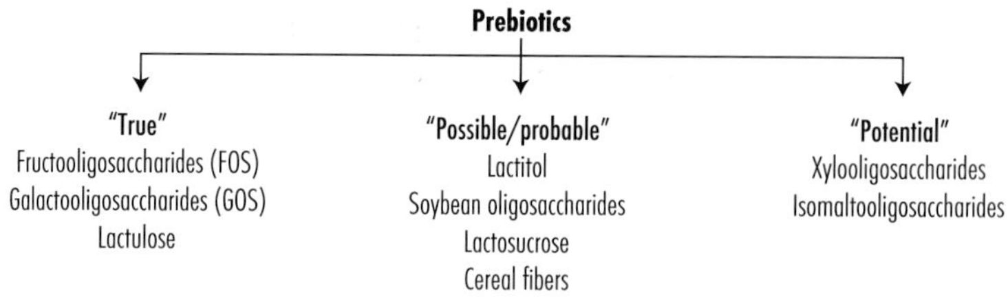

Figure 19.4. Most identified prebiotics are non-digestible oligosaccharides, and in particular fructose-containing oligosaccharides (FOS) and galacto-oligosaccharides (GOS), classified as "true" prebiotics. Other similar substances are classified as "possible" or "probable" prebiotics.

yacon. Some of the oligosaccharides that naturally occur in breast milk are believed to play an important role in the development of a healthy immune system in infants.

To rebalance the members of the microflora prebiotics are used also directly to the skin. There is an increasing amount of prebiotic strategies in cosmetic and dermatology that were developed in recent years (Simmering and Breves, 2011). A prominent example in which the bacterial equilibrium is disturbed is the skin of acne patients, overgrowth of *P. acnes* (Leyden *et al.*, 1998). Conventional cosmetic strategies make use of antibacterial agents which are effective in reducing the amount of *P. acnes*, but at the same time also affect other, beneficial bacteria such as *Staphylococcus epidermidis*, which protect human skin from infections and other environmental insults. It is therefore no surprise that attempts have been made beyond antibiotica to selectively manipulate the system in order to achieve beneficial effects for human skin. A prebiotic strategy would rebalance the composition of the skin's microflora by inhibiting the growth of *P. acnes* and at the same time preserving the growth of beneficial bacteria (Figure 19.3). Recent studies demonstrate the successful development of a prebiotic cosmetic approach to balance the composition of the cutaneous microflora (Bockmühl *et al.*, 2006). Twice daily application of a cosmetic product containing selected plant extracts from either Ginseng or Black currant or pine to human skin for a total of three weeks was effective in inhibiting the growth of *P. acnes*, whereas coagulase negative staphylococci were not affected (Bockmühl *et al.*, 2006). Janssen measured an improvement for papules, pustules, comedones, and sebum production using formulas containing 1% of the above-mentioned prebiotic actives (Janssen and Waldmann-Laue, 2008). Such a prebiotic cosmetic approach is clearly superior to antibacterial cosmetic products which unselectively reduce bacterial growth (Holland and Bojar, 2002). As a consequence the first prebiotic cosmetic product contained plant extracts like pine/blackcurrant and ginseng, has entered the market in 2005. Prebiotic strategies were found helpful also in the treatment of skin diseases with known deficiencies in barrier function and innate immunity such as atopic eczema. In patients with atopic dermatitis the antimicrobial barrier is compromised and colonization by *Staphylococcus aureus* is a common feature. Katsuyama *et al.* (2005a) used farnesol and xylitol to balance the skin microflora of patients with atopic dermatitis removing and preventing the adhesion of biofilm-producing species *S. aureus* on the skin. In a further study they demonstrated that the daily application of a skin-care cream including a combination of 0.02% farnesol and 5%

xylitol for 1 week decreased significantly the percentage of *S. aureus* (Katsuyama *et al.*, 2005b) but not the number of coagulase-negative staphylococci.

19.4 Probiotics and skin

Probiotics are defined as "live organisms which upon administration in adequate amounts confer a health benefit to the host" by improving the characteristics of the intestinal microflora (Krutmann, 2009). Probiotics can be consumed in various forms of fermented or nonfermented food products. Food vehicles include live yoghurts, fermented dairy drinks, freeze-dried supplements (capsules, pills, liquid suspensions, and sprays), cheese, fromage frais and fruit juices. Over the years, a wide range of bacteria has been proposed and used as probiotic, as indicated in Figure 19.5. However, only those classified as lactic acid bacteria have received major considerations in regard to food and nutrition.

Most probiotics currently used are microbes typical of healthy gastrointestinal microbiota aimed at promotion of gut health and improvement of immune system function. They include *Lactobacillus* (*L. casei, L. rhamnosus, L. Johnonii*) and *Bifidobacterium* species (*B. breve, B. longum,*

Probiotics

Lactobacillus	Bifdobacterium	Other lactic acid bacteria	Non lactic acid bacteria
L. acidophilus	*B. adolescentis*	**Pediococcus**	*Bacillus subtilis*
L. brevis	*B. animalis*		*Bacillus coaugulas*
L. casei	*B. bifidum*	**Enterococcus**	*Bacillus cereus*
L. crispatus	*B. breve*	*E. faecalis*	*Saccharomyces cerevisiae*
L. fermentum	*B. infantis*	*E. faecium*	*Saccharomyces boulardii*
L. gasseri	*B. lactis*		*Propionibacterium freudenreichii*
L. johnsonnii	*B. longum*	**Leuconostococcus**	*Clostridium butyricum*
L. paracasei		*L. mesenteroides*	*Escherichia coli*
L. plantarum			
L. reuteri		**Streptococcus**	
L. rhamnosus		*S. thermophilus*	
L. salivarius			
		Lactococcus	
		L. lactis	
		L. garvieae	
		L. plantarum	
		L. raffinolactis	
		L. piscium	

Figure 19.5. Major identified probiotics. These organisms are commonly used in the dairy industry in the manufacture of fermented dairy products like cheeses. They can be used in single strain starter cultures, or in mixed strain cultures with other lactic acid bacteria such as *Lactobacillus* and *Streptococcus* (*Lactococcus*).

B. bifidum), which belong to the lactic acid bacteria group, as well as *Enterococcus*, *Escherichia coli*, *Propionibacterium*, *Bacillus*, and some yeast. A constantly growing body of literature exists about the relevance of probiotics in cosmetics and dermatology. Most of these examples deal with the orally administration of food grade, mostly lactic acid bacteria, which are able to survive through the stomach and small intestine. Ouwehand *et al.* in 2003 demonstrated the antimicrobial activity of propionibacteria against the skin pathogens *M. furfur*, *Candida albicans*, and *S. aureus*. Inactivated *Lactobacillus acidophilus* was used as a treatment against mild to moderate vernal keratoconjunctivitis (Iovieno *et al.*, 2008).

Probiotic lactic acid bacteria have been linked to many effects including improving rates of recovery from gastroenteritis and diarrhoea of viral and bacterial origin (e.g. Krutmann, 2009; Saavedra *et al.*, 1994; Shornikova *et al.*, 1997). Specific strains have been shown to influence the composition and metabolic activity of the gut microflora and to inhibit the growth of a range of pathogens. They have also been suggested to modulate immunity in the gut but also systemically, and the latter property may be of relevance for human skin (e.g. Guéniche *et al.*, 2006; Link-Amster *et al.*, 1994). Accordingly, a significant improvement on the course of atopic dermatitis has been reported in infants given probiotic-supplemented elimination diets (e.g. Rosenfeldt *et al.*, 2003; Weston *et al.*, 2005). Preparation of different lactic acid bacteria, i.e. *L. paracasei*, *L. brevis*, or *L. fermentum*, were investigated as ingredients for skin-care products and were found to promote *S. epidermidis* and to block *S. aureus*, *E. coli*, or *Micrococcus luteus* (Lang *et al.*, 2006). Also, in mice, oral administration of *Lactobacillus casei* reduced contact hypersensitivity to a hapten only in the presence of CD4+ T cells, which control the size of the CD8+ effector pool (Chapat *et al.*, 2004). Further, cultured lactobacilli were proposed as actives for the reduction of axillary malodor in deodorants by *Corynebacterium jeikeium* (Lang *et al.*, 2007).

Also healthy skin may profit from the oral ingestion of probiotic bacteria. Nutritional supplementation of hairless mice with *Lactobacillus johnsonii* provided protection of the skin immune system against Ultraviolet B radiation-induced immunosuppressive effects (Guéniche *et al.*, 2006). Similar effects have recently been described in a human *in vivo* study and it has been proposed that oral consumption of probiotic bacteria may represent a novel approach to protect the skin immune system against ultraviolet radiation (Bouilly-Gauthier *et al.*, 2010). Another target for probiotics may be the skin barrier function. A recent double-blind, randomized clinical study has shown that a 24 weeks skin nutrition intervention with a fermented dairy product in female volunteers having dry and sensitive, but otherwise healthy skin significantly reduced transepidermal water loss and thus improved *stratum corneum* barrier function compared to a placebo product (Puch *et al.*, 2008). It should be noted, however, that in addition to the probiotic strains (*Lactobacillus casei*, *Lactobacillus bulgaris* and *Streptococcus thermophilus*) this dairy product contained a mixture of boretsch oil, green tea polyphenols and vitamin E and that skin barrier improvement may be at least in part due to these ingredients.

The use of probiotics in cosmetic applications has been proposed, although there are currently only a very few controlled scientific trials pursuing topical probiotic approach for the skin microflora (Ouwehand *et al.*, 2003). Probiotics may have a role e.g. as moisturizing agents. Studies results suggest that topical application of *Vitreoscilla filiformis* exerts beneficial effects in patients with seborrheic dermatitis and atopic eczema (Guéniche *et al.*, 2008).

Therefore, the topical application of probiotics for prophylaxis and therapy of cutaneous inflammatory immune reactions is very promising.

19.5 Conclusions

Thanks to the recent advances in microbiology and immunology, immense progress has been made in our understanding of the cutaneous microflora. The skin represents the first barrier between the body and its environment and requires appropriate nutrients for its integrity and functionalities. The skin reflects the general health status and age of the host. The health and functions of the skin are influenced by exogenous factors like lifestyle, climate conditions, the extent of UV exposure and food. The development of pre- and probiotic concepts in food has thrown new light on the positive effects of microorganisms on the skin health. A large number of studies have been performed that have demonstrated that an appropriate dietary consumption of prebiotic and probiotic food supplements can be useful for the prevention or treatment of dermatological and cosmetic problems or to maintain a healthy skin. Thereby the prebiotic and probiotic effect is now a well-established scientific fact. We briefly illustrated the beneficial effects of prebiotic and probiotic food supplements, not only for the intestinal tract but also for the skin. However, further researches are essential to confirm their benefits and their use in clinical practice.

References

Ashbee, H.R., Fruin, A., Holland, K.T., Cunliffe, W.J. and Ingham, E., 1994. Humoral immunity to *Malassezia furfur* serovars A, B and C in patients with pityriasis versicolor, seborrheic dermatitis and controls. Experimental Dermatology 3, 227-233.

Ashbee, H.R., Muir, S.R., Cunliffe, W.J. and Ingham, E., 1997. IgG subclasses specific to *Staphylococcus epidermidis* and *Propionibacterium acnes* in patients with acne vulgaris. British Journal of Dermatology 136, 730-733.

Bockmühl, D., Jasoy, C., Nieveler, S., Scholtyssek, R., Wadle, A. and Waldmann-Laue, M., 2006. Prebiotic cosmetics: an alternative to antibacterial products. IFSSC magazine 9, 1-5.

Bojar, R.A. and Holland, K.T., 2002. Review: the human cutaneous microflora and factors controlling colonisation. World Journal of Microbiology and Biotechnology 18, 889-903.

Bouilly-Gauthier, D., Jeannes, C., Maubert, Y., Duteil, Queille-Roussel, C., Piccardi, N., Montastier, C., Manissier, P., Piérard, G. and Ortonne, J.P., 2010. Clinical evidence of benefits of a dietary supplement containing probiotic and carotenoids on ultraviolet-induced skin damage. British Journal of Dermatology 163, 536-543.

Chapat, L., Chemin, K., Dubois, B., Bourdet-Sicard, R. and Kaiserlian, D., 2004. *Lactobacillus casei* reduces CD8+ T cell-mediated skin inflammation. European Journal of Immunology 34, 2520-2528.

Cogen, A.L., Nizet, V. and Gallo, R.L., 2008. Skin microbiota: a source of disease or defence? British Journal of Dermatology 158, 442-455.

Cosseau, C., Devine, D.A., Dullaghan, E., Gardy, J.L., Chikatamarla, A., Gellatly, S., Yu, L.L., Pistolic, J., Falsafi, R., Tagg, J. and Hancock, R.E.W., 2008. The commensal *Streptococcus salivarius* K12 downregulates the innate immune responses of human epithelial cells and promotes host-microbe homeostasis. Infection and Immunity 76, 4163-4175.

Donnarumma, G., Buommino, E., Baroni, A., Auricchio, L., Filippis, A.D., Cozza, V., Msika, P., Piccardi, N. and Tufano, M.A., 2007. Effects of AV119, a natural sugar from avocado, on *Malassezia furfur* invasiveness and on the expression of HBD-2 and cytokines in human keratinocytes. Experimental Dermatology 16, 912-919.

Elias, P.M. and Choi, E.H., 2005. Interactions among *stratum corneum* defensive functions. Experimental Dermatology 14, 719-726.

Elias, P.M., 2007. The skin barrier as an innate immune element. Seminars in Immunopathology 29, 3-14.

Gallo, R.L., Murakami, M., Ohtake, T. and Zaiou, M., 2002. Biology and clinical relevance of naturally occurring antimicrobial peptides. Journal of Allergy and Clinical Immunology 110, 823-831.

Gao, Z., Tseng, C., Pei, Z. and Blaser, M.J., 2007. Molecular analysis of human forearm superficial skin bacterial biota. Proceedings of the National Academy of Sciences USA 104, 2927-2932.

Gibson, G.R. and Roberfroid, M.B., 1995. Dietary modulation of the human colonic microbiota: introducing the concept of prebiotics. Journal of Nutrition 125, 1401-1412.

Gibson, G.R., Beatty, E.R., Wang, X. and Cummings, J.H., 1995. Selective stimulation of bifidobacteria in the human colon by oligofructose and inulin. Gastroenterology 108, 975-982.

Grice, E.A., Kong, H.H., Renaud, G., Young, A.C., Bouffard, G.G., Blakesley, R.W., Wolfsberg, T.G., Turner, M.L. and Segre, J.A., 2008. A diversity profile of the human skin microbiota. Genome Research 18, 1043-1050.

Guarner, F. and Schaafsma, G.J., 1998. Probiotics. International Journal of Food Microbiology 39, 237-238.

Guéniche, A., Benyacoub, J., Buetler, T.M., Smola, H. and Blum, S., 2006. Supplementation with oral probiotic bacteria maintains cutaneous immune homeostasis after UV exposure. European Journal of Dermatology 16, 511-517.

Guéniche, A., Cathelineau, A.C., Bastien, P., Esdaile, J., Martin, R., Queille Roussel, C. and Breton, L., 2008. *Vitreoscilla filiformis* biomass improves seborrheic dermatitis. Journal of the European Academy of Dermatology and Venereology 22, 1014-1015.

Holland, K.T. and Bojar, R.A., 2002. Cosmetics. What is their influence on the skin microflora? American Journal of Clinical Dermatology 3, 445-449.

Iovieno, A., Lambiase, A., Sacchetti, M., Stampachiacchiere, B., Micera, A. and Bonini, S., 2008. Preliminary evidence of the efficacy of probiotic eye-drop treatment in patients with vernal keratoconjunctivitis. Graefe's Archive for Clinical and Experimental Ophthalmology 246, 435-441.

Janssen, F. and Waldmann-Laue, M., 2008. Efficacy of prebiotic product combination against skin impurities. IFSCC Congress poster presentation at the IFSCC Congress.

Katsuyama, M., Ichikawa, H., Ogawa, S. and Ikezawa, Z., 2005a. A novel method to control the balance of skin microflora: Part 1. Attack on biofilm of *Staphylococcus aureus* without antibiotics. Journal of Dermatological Science 38, 197-205.

Katsuyama, M., Kobayashi, Y., Ichikawa, H., Mizuno, A., Miyachi, Y., Matsunaga, K. and Kawashima, M., 2005b. I: A novel method to control the balance of skin microflora: Part 2. A study to assess the effect of a cream containing farnesol and xylitol on atopic dry skin. Journal of Dermatological Science 38, 207-213.

Krutmann, J., 2009. Pre- and probiotics for human skin. Journal of Dermatological Science 54, 1-5.

Lang, C., Heilmann, A., Veen, M., Budde, E., Böttner, M., Reindl, A. and Knöll, R., 2006. Method and means for protecting the skin against pathogenic bacteria. US patent WO 2006/136420 A2.

Lang, C., Veen, M., Budde, E., Böttner, M., Reindl, A. and Knöll, R., 2007. Microorganisms inhibiting the formation of axillary malador. US patent WO 2007/031300.

Leyden, J.J., McGinley, K.J. and Vowels, B., 1998. *Propionibacterium acnes* colonization in acne and nonacne. Dermatology 196, 55-58.

Link-Amster, H., Rochat, F., Saudan, K.Y., Mignot, O. and Aeschlimann, J.M., 1994. Modulation of a specific humoral immune response and changes in intestinal flora mediated through fermented milk intake. FEMS Immunology and Medical Microbiology 10, 55-64.

Metze, D., Kersten, A., Jurecka, W. and Gebhart, W., 1991. Immunoglobulins coat microorganisms of skin surface: a comparative immunohistochemical and ultrastructural study of cutaneous and oral microbial symbionts. Journal of Investigative Dermatology 96, 439-445.

Ouwehand, A.C., Tühonen, K. and Lahtinen, S., 2010. The potential of probiotics and prebiotics for skin health. In: Farage, M.A., Miller, K.W. and Maibach, H.I. (eds.), Textbook of Aging Skin. Springer-Verlag, Berlin Heidelberg, Germany, pp. 799-810

Ouwehand, A.C., Batsman, A. and Salminen, S., 2003. Probiotics for the skin: a new area of potential application? Letters in Applied Microbiology 36, 327-331.

Puch, F., Samson-Villeger, S., Guyonnet, D., Blachon, J.L., Rawlings, A.V. and Lassel, T., 2008. The consumption of functional fermented milk containing borage oil, green tea and vitamin E enhances skin barrier function. Experimental Dermatology 17, 668-674.

Rosenfeldt, V., Benfeldt, E., Dam Nielsen, S., Fleichsler Michaelsen, K., Jeppesen, D.L., Valerius, N.H. and Paerregaard, A., 2003. Effect of probiotic *Lactobacillus* strains in children with atopic eczema. Journal of Allergy and Clinical Immunology 111, 389-395.

Roszkowski, W., Roszkowski, K., Ko, H.L., Beuth, J. and Jeljaszewicz, J., 1990. Immunomodulation by propionibacteria. Zentralblatt für Bakteriologie 274, 289-298.

Saavedra, J.M., Bauman, N.A., Oung, I., Perman, J.A. and Yolken, R.H., 1994. Feeding of *Bifidobacterium bifidum* and *Streptococus thermophilus* to infants in hospital for prevention of diarrhoea and sheding of rotavirus. Lancet 344, 1046-1049.

Shornikova, A.V., Casas, I., Mykkänen, H., Salo, E. and Vesikari, T., 1997. Bacteriotherapy with *Lactobacillus reuteri* in rotavirus gastroeneteritis. Pediatric Infectious Disease Journal 16, 1103-1107.

Simmering, R. and Breves, R., 2011. Prebiotic cosmetics. In: Krutmann, J. and Humbert P. (eds.), Nutrition for healthy skin. Strategies for clinical and cosmetic practice. Springer-Verlag, Berlin Heidelberg, Germany, pp. 137-147.

Weston, S., Halbert, A., Richmond, P. and Prescott, S.L., 2005. Effects of probiotics on atopic dermatitis: a randomised controlled trial. Archives of Disease in Childhood 90, 892-897.

Key facts

- Turmeric is a spice that has been used for centuries by the Middle Eastern and Asian cultures. The primary active constituent of turmeric is derived from the rhizome of the plant Curcumin *longa*.
- Curcumin uses range from being a condiment in the famous curry sauce, a coloring agent, as well as to cure many diseases and conditions in traditional medicine.
- Extensive research within the last century has demonstrated that curcumin has an unprecedented number of molecular targets giving it a unique power to control many molecular pathways that could lead to disease processes
- Curcumin is capable of affecting many of anti-inflammatory pathways. These include: cyclooxygenase-2 (COX-2), lipooxygenase and glutathione-S-tranferase.
- Curcumin can also regulate anti-oxidative and anti-carcinogenic pathways such as: transcription factors, growth factors, growth factor receptors and their associated signaling pathways, epidermal growth factor and fibroblast growth factor, and nuclear factors such as nuclear kappa beta (NF-κB).
- Curcumin has an evolving role in the treatment and prevention of multiple dermatologic conditions that include: skin cancer, wound healing and keloids, acne, psoriasis, and photoaging to name a few.
- Researchers around the world are paying attention to this ancient spice and are trying to explore all of its potential molecular targets to outline therapy and intervention with its application to human health.

Summary points

- Curcumin, the active constituent of turmeric has been shown to have a wide range of therapeutic actions.
- Multiple molecular pathways mediate curcumin's effect: inhibition of nuclear factor-kappa beta (NF-κB) and activator protein-1 (AP-1); inhibition of the inflammatory cascade; promotion of plasminogen activation among others.
- It protects against tumor formation by inhibiting the transcription of oncogenes mediated by the nuclar factor kappa beta (NF-κB) and the activator protein-1 (AP-1) pathways.
- It reduces inflammation by inhibiting cyclooxygenase-2 (COX-2) and lipooxygenases.
- Curcumin has an active role in accelerating the process of wound healing, it does that by inhibiting the formation of matriz metalloproteinases (MMP) and enhancing the activity of urokinase plasminogen activator (uPA) in the healing process.
- Curcuimin, by targeting the tumor growth factor-beta receptor represents an unusual targeted therapy for keloids as well as a chemopreventive agent
- The well known anti-inflammatory effects of curcumin make it a promising agent in the treatment of acne.
- Curcimun's properties have caught the scientific attention of researchers for the development of a low toxicity treatment option and chemopreventive agents.

20. Curcumin (turmeric) and its evolving role in skin health

T. Gonzalez[1] *and A. Sethi*[2]
[1]*University of Puerto Rico, School of Medicine,161 C, Cesar Gonzalez #35, San Juan, PR 00918, USA;* [2]*University of Chicago, Section of Dermatology, 5841 S. Maryland MC 5067, Chicago, IL 60637, USA; asethi@medicine.bsd.uchicago.ed*

Abstract

Turmeric is a spice that has been used for centuries by the Middle East and Asian cultures. The primary active constituent of turmeric is derived from the rhizome of the plant *Curcuma longa* L from the genus *Zingiberaceae* (Ginger family). This tropical plant is widely cultivated in the South Asia region, especially India, where it has been used for a wide variety of diseases and conditions. Its uses range from being a condiment in the famous curry sauce, a coloring agent, as well as to cure many diseases and conditions in traditional medicine. Extensive research within the last century with this tropical root have demonstrated that its medical powers are linked to curcumin, the main and most active constituent of the root. Curcumin effects are mediated through the regulation of various transcription factors, growth factors, inflammatory cytokines, and other enzymes. It has an unprecedented number of molecular targets giving it a unique power to control many molecular pathways that could lead to diseases. Its effects on these pathways are particularly well documented in the scientific literature. These targets include nuclear factor kappa beta (NF-κB) and its associated protein kinases, AP-1, lipooxigenases, plasminogen activator, cyclooxygenase-2, tumor growth factor beta (TGF-β) and many others. It is by blocking or promoting these molecular targets that curcumin is being studied as a potential anti-carcinogenic, anti-proliferative, anti-inflammatory and chemo preventive agent. In dermatology, curcumin has been utilized in diseases such as: skin cancers, psoriasis, and acne, wound healing and keloids. By affecting different but related pathways, curcumin has shown great potential not only for the treatment of skin diseases but also for their prevention. The main objective for this chapter is to describe the different molecular pathways targeted by curcumin and explain, by reviewing the scientific literature, how these are potential remedies for skin diseases.

Keywords: tumeric, curry, cancer, psoriasis, acne, wound healing, keloids, AP-1, NF-κB, ROS, antioxidant

Abbreviations

AP-1	Activator protein 1
ATP	Adenosine triphosphate
cAMP	Cyclic adenosine monophosphate
COX-2	Cyclooxygenase-2
EGF	Epidermal growth factor
EGFR	Endothelial growth factor
JNK	Jun N-terminal kinase
KF	Keloid fibroblasts
MAPK	Mitogen-activated protein kinase
MMP	Matrix metalloproteinase
NF-κB	Nuclear factor kappa beta
PAI	Plasminogen activator inhibitor
PhK	Phophorylase kinase enzyme
RKC	Receptors protein kinases
ROS	Reactive oxygen species
TGF-β	Tumor growth factor beta
TIMP	Tissue inhibitors of metalloproteinase
TNF-α	Tumor necrosis factor-alpha
TNFR	Tumor necrosis factor receptor
tPA	Tissue plasminogen activator
TPA	12-*O*-tetradecanoylphorbol-13-acetate
uPA	Urokinase plasminogen activator
UV	Ultraviolet

20.1 Background

The spice turmeric has been used for centuries by the Middle Eastern and Asian cultures. The primary active constituent of turmeric is derived from the rhizome of the plant *Curcumin longa* from the genus *Zingiberaceae* (Ginger family). Vogel and Pelletier first described it in 1815 but its uses have been documented in records dating back to the 600 BC (Aggarwal *et al.*, 2006a). This tropical plant is widely cultivated in the South Asian region, especially India, where it has multiple uses. Its uses range from being a condiment in the famous curry sauce, a coloring agent, as well as to cure many diseases and conditions in traditional medicine. In Indian and Chinese medicine, turmeric was used, as an anti-inflammatory agent, to treat gas, indigestion, toothaches, chest pains and dysmenorrhea. It was also used for wound healing and and skin lightening.

Turmeric is an orange-yellow crystalline hydrophobic powder composed of three curcuminoids: the polyphenol curcumin (diferuloymethane) which is the main and most active constituent, demethoxy-curcumin, and bisdemethoxycurcumin, as well as volatile oils, sugars, proteins and resins. Polyphenols, such as curcumin, are a large class of chemicals, which are found in plants and have gathered attention to the medical community given their chemo preventive properties (Singltary, 2010). As a phenol curcumin targets multiple molecular signaling pathways, which have been studied to be potential treatments for well-known human diseases. In the current literature, curcumin has been studied and proved to be an effective potent anti-inflammatory and antioxidant agent. These properties contribute to its great potential to prevent conditions such as different types of cancer, skin conditions, heart diseases, neurological diseases, arthritis, diabetes and renal diseases to mention a few (Nochols *et al.*, 2010; Sziliszka *et al.*, 2010).

Although a promising agent for the treatment of many diseases, its lipophilic nature gives curcumin a limited systemic bioavailability. On the other hand, the rapid degradation of curcumin in the gastrointestinal tract gives this plant a relatively low potential for toxicity. Human trials have demonstrated that intakes as high as 8 grams a day are well tolerated and intakes of 12 grams a day have not demonstrated adverse consequences. Oral intakes in amounts greater than these are unlikely to be applicable to the common food recipes or usual dietary uses. This describes the primary obstacle for the therapeutic use of curcumin. In most skin diseases, the route of curcumin administration is topical making this barrier less significant in terms of delivering the active agent to the targeted disease. Currently, there is active research ongoing on curcumin pharmacokinetics and its systemic bioavailability.

It has been described that when absorbed, curcumin can affect more than one hundred different molecular targets. The major medically relevant studied biological molecular pathways that are targeted by curcumin are the anti-oxidative and anti-inflammatory ones. These include: (1) interaction in the expression and activity of a variety of enzymes such as COX-2, lipooxygenase and glutahthione-S-tranferase; (2) modulation of transcription factors, growth factors, growth factor receptors and their associated signaling pathways; (3) specific activity for epidermal growth factor and fibroblast growth factor 2, as well as for nuclear factors such as NFkB; (4) alteration of cell proliferation by influencing cyclin protein actions; and (5) affects cytokine expression such as IL-6 and 8, tumor necrosis factor. Its anti-inflammatory effects are believed to be secondary to its effects on cytokine and its modulation of transcription factors; its beneficial effect as an anticancer agent is found in its influences on oncogenes and tumor suppressor gene signaling. It is known that curcumin affects the transformation, proliferation and invasion of cancer cells. This great variety of curcumin's targets gives this old spice its numerous medical uses and the potential for the development of innovative therapies (Aggarwal *et al.*, 2007b).

In terms of skin health, it has been shown that curcumin has an evolving role in the treatment of multiple skin diseases that includes: skin cancer, wound healing, keloids, acne, alopecia, psoriasis and photoaging. As with other diseases, curcumin's primary role on dermatologic conditions are its anti-inflammatory and antioxidant characteristics described above. It is not a single pathway but a combination of mechanisms that makes curcumin a promising agent for the development of dermatologic treatments. In this chapter, we have described the current documented benefits of curcumin for certain skin conditions and the potential of this ancient spice as a developing therapeutic agent.

20.2 Curcumin: its evolving role in skin diseases

20.2.1 The role of curcumin in skin cancers

According to the world health organization (WHO), the incidence of both non-melanoma and melanoma skin cancers has been increasing over the past few decades. Currently, between two and three million non-melanoma skin cancers and 132,000 melanoma skin cancers occur globally each year. The leading identified risk factor for the development of melanoma and non-melanoma skin cancers is repeated exposure of the skin to ultraviolet light. Prolonged and repeated exposure to the sun, which is the principal source of UV light, leads also to photoaging. The mechanism for this damage is known to be at the molecular level and has an important role in tumorigenesis (Figure 20.1). More specifically modulation on the expression of "UV response genes", which include the NF-κB and AP-1. These molecular pathways are the two principal mechanisms that lead to the development of photoaging and skin cancer (Cooper and Bowden, 2007).

NF-κB is a transcription factor that controls the expression of genes involved in the regulation of apoptosis and the cell cycle for example cyclin D1, proto-oncogenes such as c-myc and Bcl-2 (Figure 20.1). It also helps regulating pro-inflammatory factors such as COX-2 and cytokines (Figure 20.2). Incorrect regulation (increased activity) of NF-κB can result in autoimmune diseases and cancer. A variety of stimuli that activate the NF-κB include cytokines, viral protein and most importantly for the development of skin cancer, UV radiation. A dysregulated NF-κB is constitutively active and therefore turns on the expression of genes that keep the cells proliferating giving place to the development of tumors, including those of the skin. The cells proliferation is then perpetuated under situations that would otherwise induce apoptosis. For this reason, NF-κB is an attractive transcription factor when studying skin cancers.

On the other hand, another "UV response gene". AP-1 is another transcription factor composed of the products of the *jun* and *fos* oncogene families that also regulates gene expression in response to cytokines, growth factors, stress, including that of prolonged exposure to UV light, and infections. The activation of AP-1 requires the phosphorylation of c-jun through activation of JNK. When activated, AP-1 is involved in the differentiation, proliferation and anti apoptotic mechanisms. It has been demonstrated that constitutively elevated levels of AP-1 are present in skin cancers (Hopper *et al.*, 2009). AP-1 pathway is similar to that of NF-κB (Figure 20.1). The application of natural compounds such as curcumin which can interact and block the above-described mechanism of action has the potential to be a chemopreventive agent for photoaging and carcinogenesis.

Figure 20.1 illustrates the proposed action mechanisms for the inhibition of carcinogenesis. Extracellular growth factors, cytokines, stress, UV light, tumor promoter agents such as tumor promoter 12-O-tetradeca-noylphorbol-13-acetate and many other offending agents bind to membrane RKCs such as EGFR, TNFR, JAK2, MAPK. The activation of these receptors leads to a chained activation of NF-κB causing activation of the c-myc, iNOS and other cellular proliferation genes. It also promotes the JNK and c-jun protein, which forms a heterodimer with c-fos thus

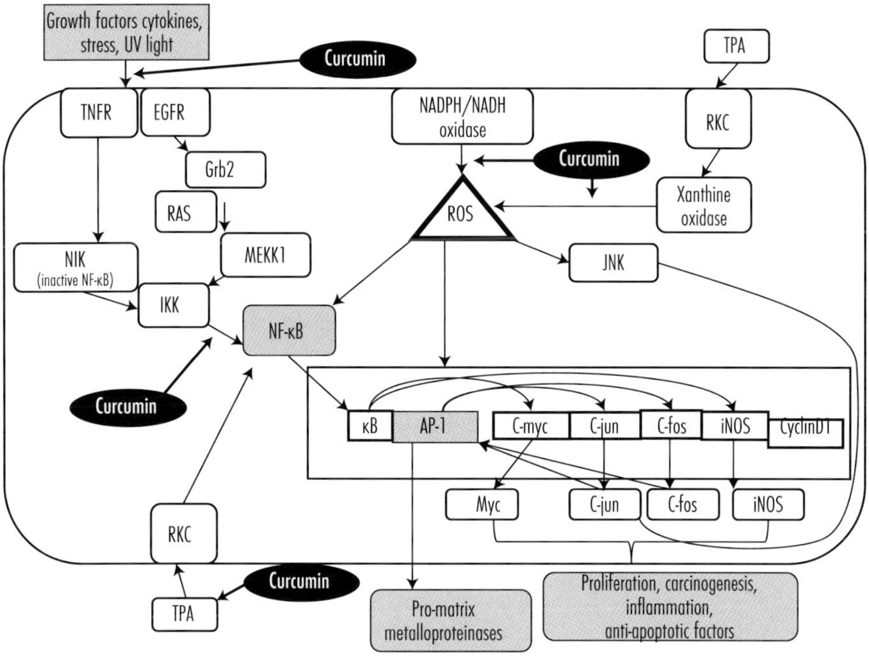

Figure 20.1. Proposed mechanisms for the development of skin cancer and curcumin interaction in different steps of this pathway. Curcumin has been shown to block many of these pathways as indicated by the double lined black arrows.

TPA: 12-O-tetradeca-noylphorbol-13-acetate, EGFR: endothelial growth factor, TNFR: tumor necrosis factor receptor, NF-κB: nuclear factor kappa beta, JNK: c-jun N-terminal kinase, AP-1: activated protein-1, ROS: reactive oxygen species, IKK: IκB kinase, RKC: receptors protein kinases.

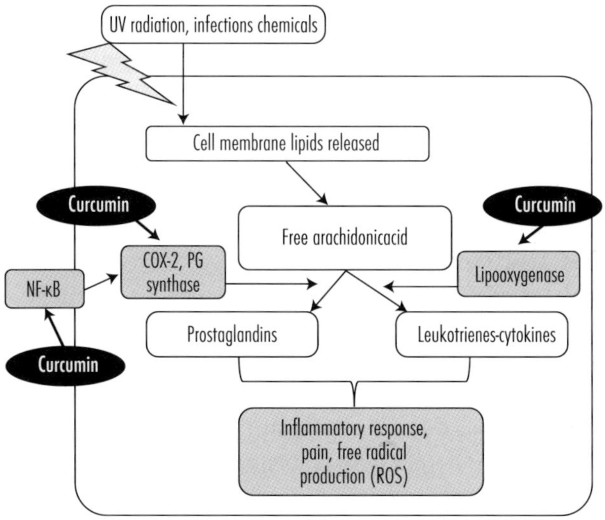

Figure 20.2. Role of curcumin on inhibiting various steps of the inflammatory process.

enhancing the activity of AP-1. ROS can also activate NF-κB and other nuclear transcription factors. ROS are considered endogenous mitogenic agents. The activation of all the illustrated receptors help in the processes of cell proliferation, carcinogenesis, inflammation and apoptosis. Curcumin has been shown to block many of these pathways.

The effect of curcumin in the above mentioned gene expression pathways has been studied in cells and animal models (Aggrawal and Kaur, 2010; Heng, 2010). In a study done by Marin *et al.* with melanoma cells specifically, curcumin induced melanoma cells to undergo and decrease NF-κB activity, which by all means, blocking the mechanism explained above, decreased the tumorogenicity properties of these cells (Marin *et al.*, 2007). Another study on keratinocytes demonstrated that curcumin reduced the AP-1 factor transcriptional activity and activated apoptosis and suppressed keratinocyte proliferation, all of which have an active role in the development of malignant tumors such as squamous cell carcinoma of the skin (Balasubramanian and Eckert, 2007). Park *et al.* also demonstrated the role of curcumin in skin cancer. Their research led to the discovery that curcumin was not only an inhibitor of the UV induced AP-1 DNA binding activity, but that it was also capable to inhibit mRNA and protein expression of COX-2 and the activation of p38 MAPK and JNK, pathways which are also responsible for cancer development and metastasis. All of these discovered mechanisms contribute to the current curcumin benefits on skin cancer pathogenesis and photoaging (Park and Lee, 2010). By blocking these pathways (in any way) curcumin prevents the generation of proto-oncogenes and therefore interferes with the mechanisms that would otherwise lead to tumor growth and metastasis.

Skin cancer prevention public health campaigns include messages such as the avoidance of sun exposure, wearing protective clothing and application of sunscreen. People are exposed to UV light on a daily basis and current scientific efforts are targeting to find an effective agent for the prevention and treatment of skin cancer. Curcumin is a promising chemopreventive agent. It has been reported that the topical use of curcumin extracts inhibits tumor initiation and progression by TPA (Figure 20.1) in mouse skin (Huang *et al.*, 1997). More research in human subjects will be needed to determine the most appropriate route of delivery and specific effective curcumin doses. Meanwhile, in our efforts to prevent skin cancer, curcumin represents a potentially inexpensive option for our patients in the future.

This image represents the role of curcumin as an anti-inflammatory agent. Human cells are exposed to diverse mechanisms of injury including exposure to UV radiation, chemicals and infections. These external factors lead to the release of cell membrane lipids, which become free arachidonic acids. The enzymes COX-2, PG synthase and lipooxygenase mediate the activation of the inflammatory cascade by producing prostaglandins, leukotrienes and cytokines. This process will form an inflammatory response including pain and free radical production. Free radicals or reactive oxygen species are known to have an essential role in the development of skin cancers and other systemic diseases. Curcumin blocks (double lined dark arrows as shown in Figures 20.1 and 20.2) the activity of COX-2 and lipooxygenase, which by consequence inhibits the inflammatory response. Curcumin also blocks NF-κB COX-2 induction.

20.2.2 Curcumin: its evolving role in psoriasis

Psoriasis is a common chronic hyperproliferative skin disorder with its most classic presentation characterized by erythematous plaques with a silvery scale more prominent on the elbows, knees and scalp. There is some evidence pointing to genetics as the main predisposing factor for the development of psoriasis but it is known that environmental and behavioral factors can affect the course of this disease. The triggers for psoriasis include: wounds, sunburns, allergic reactions and infections, all representing injurious stimuli for the skin (Tomfohrde *et al.*, 1994). The healing process leads to the generation of T lymphocytes, which further leads to the production of cytokines, scar tissue formation, cell migration and epidermal proliferation. This process requires a source of energy that is mediated by a phophorylase kinase enzyme (ATP phosphorylase b phosphotransferase). After the injurious stimulus is removed and the wound is healed, there is a switch-off mechanism for the phosphorylase kinase pathway that culminates the inflammatory response and consequently terminates the process of cell migration and proliferation. It has been demonstrated that in psoriatic patients there is a defect in the switching off mechanism due to a defective type II cAMP kinase, one of the mediators in the inflammatory response process. Once the PhK is activated there is a chained reaction that activates inflammatory cells: T lymphocytes and macrophages as well as epidermal growth factor. These cells then produce cytokines, including TNF-α, which are then responsible for the continuous inflammatory response (Heng *et al.*, 1994). In psoriasis, the defective cAMP is unable to stop the healing process leading to a continuous source of energy supply, which promotes excessive epidermal proliferation and the development of psoriatic plaques (Figure 20.3).

There is evidence of elevated activity of phosphorylase kinase in psoriatic lesions that correlates with disease activity. This production of proinflammatory mediators and interaction with the adaptive immune system is responsible for perpetuating the development of skin lesions. Reddy and Aggarwal (1994) demonstrated in 1994 that curcumin is a non-competitive and selective inhibitor of phophorylase kinase (Reddy *et al.*, 1994). By inhibiting the phosphorylase kinase curcumin has been shown to improve and in some cases resolve psoriatic lesions. In a study performed by Heng *et al.* psoriatic lesions treated with topical curcumin gels resulted in the resolution of psoriasis as demonstrated by clinical, histopathological and immunohistologic criteria (Heng *et al.*, 2000). The role of oral curcumin for the treatment of psoriatic lesions is less clear and more placebo-controlled trials are needed before recommending oral curcumin for psoriasis. Curcumin is also capable of reducing the inflammatory response by blocking TNF-α and reducing the intracellular production of ROS.

Currently there are multiple available topical and systemic treatments for psoriasis. The type of treatment is chosen on the basis of disease severity, relevant comorbidities, efficacy, and also the evaluation of distinct patient response. Although systemic immunomodulatory, biological and immunosupressive agents have increased the treatment options for moderate to severe psoriasis, it is also known that prolonged use of these agents may be detrimental to human health in terms of their multiple side effects and interactions with other drugs and diseases (Menter *et al.*, 2009). The decision of starting these new agents remains complex at times since comorbidities; toxicity,

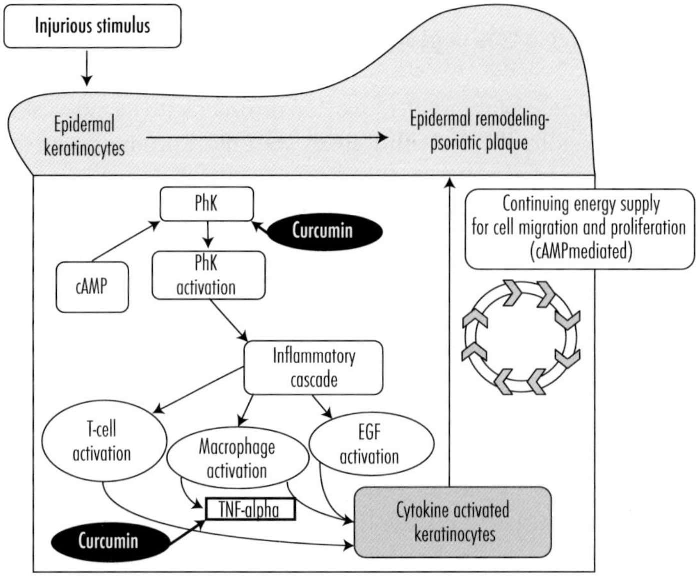

Figure 20.3. Curcumin relationship with PhK inhibition and the development of psoriatic plaques. cAMP: cyclic adenosine monophosphate, PhK: phophorylase kinase enzyme, TNF: tumor necrosis factor, EGF: epidermal growth factor.

accessibility and efficacy of therapy, along with patient preference represent a challenge. For example, infliximab and etanercept are two known effective biological treatments for psoriasis. The mechanism of action of these two drugs, is mediated ultimately through TNF-α blockage. By blocking the TNF-α there is a normalization of keratinocyte proliferation and decreased inflammatory response. Curcumin, which can block the production of TNF-α in two different steps and also provides this latter effect: earlier by blocking PhK and later by blocking TNF-α itself. These mechanisms of curcumin should be further studied in depth since this spice shows promising results in the treatment of psoriasis. Meanwhile, considering that multitargeted therapy is better than monotargeted therapy for most diseases, curcumin may be considered as a potential treatment option for psoriasis that could be used in combination with other agents.

This image illustrated the proposed mechanism for the development of psoriatic lesions and curcumin's role on preventing them. External injurious stimuli to the skin (wounds, sunburns, allergic reactions and infections) are belief to be the trigger for developing psoriasis. The healing process for any of these stimuli leads to the activation of inflammatory cells including: T lymphocytes, macrophages and epidermal growth factor. To activate these inflammatory response the cells utilizes a PhK, which is activated by cAMP. In psoriasis there is a defect in the swithch off of cAMP leading to a continuous activation of the inflammatory process. The activation of these cells produces release of cytokines, including TNF-α, which are known to be the main cause for the development of psoriatic lesions. Curcumin is a non-competitive and selective inhibitor of phophorilase kinase and is also capable of reducing the inflammatory response by blocking

TNF-α (produced by macrophages). The benefits of curcumin in psoriasis are by blocking these two important processes (double-lined dark arrows in Figure 20.3) in the inflammatory response.

20.2.3 Curcumin: its evolving role in acute wound healing

The skin is the largest organ in our body and as with any other organ it deteriorates with aging. Other factors that contribute to skin deterioration include chronic metabolic diseases such as diabetes, chronic exposure to the sun, and extended use of corticosteroids. The responsible mechanism for skin damage in all of these situations is a decrease in collagen synthesis with an increase in the collagen degrading MMP. Collectively the MMP's are capable of degrading all kinds of extracellular matrix proteins. When there is an imbalance in the collagen synthesis to MMP ratio the skin is prone to bruising and wounds tend to heal more slowly. Due to the these mentioned characteristics, when wounded, chronically damaged or atrophic skin has the potential to develop into chronic ulcers that consequently represent a challenge for treatment in current medical practice (Lund *et al.*, 1999).

Another important factor in the phases of wound healing is degradation of the initial fibrin matrix that forms immediately after the wound is formed. This fibrin matrix is replaced with granulation tissue, which is essential for the process of proper healing. The fibrin matrix is degraded by plasmin that is a product of plasminogen cleaved by the cell's uPA and tPA. uPA is responsible for fibrinolysis in tissue and tPA has the same role in the circulation. The proteolytic degradation of the provisional fibrin matrix and subsequent substitution by fibroblasts is essential for proper wound repair. The uPA system's roles in wound healing have been described in the literature and it is known to affect cell migration, adhesion, and proliferation. Furthermore, as revealed by several studies, inhibition of the uPA system leads to complete arrest of wound healing (Madhyastha *et al.*, 2010). Other factors that may contribute to the failure of some wounds to heal include elevated levels of reactive oxygen species.

Curcumin represents a treatment option as well as a chemopreventive measure for impairments in wound healing. In a study done by Madhyastha *et al.* (2010) curcumin was shown to: (1) have a cytotoxic and antioxidant profile by decreasing the amount of reactive oxygen species; (2) up regulate the expression of uPA and enhance the uPA dependent-fibrinolytic activity necessary to accelerate the healing process; and (3) promote cell migration towards the wounded region in a uPA dependent manner (Madhyastha *et al.* 2010). All these identified mechanisms promote the process of fibrinolysis and cell migration, which as explained previously, are essential in the process of wound healing. In another study performed by Bhagavathula *et al.* the effectiveness of curcumin on wound healing was demonstrated by the topical application of curcumin and ginger extracts to the skin of corticosteroid-impaired hairless rat skin. As demonstrated by histological studies, curcumin treated rats were able to increase the amounts of collagen type 1 (major connective tissue in the dermal matrix) and decrease MMP-9 (Bhagavathula *et al.*, 2009). The pathway of inhibition of MMP is believed to be mediated by the AP1 pathway, previously illustrated in Figure 20.1. Although, it is known that activation of plasmin leads to the cleavage of pro-MMP with a subsequent activation of these enzymes; curcumin exerts its effect at the genotypic expression of MMP. Therefore, although curcumin enhances uPA and one would expect to have increased levels of MMP, the building blocks (pro-MMP) have been blocked by curcumin at the nuclear level (Figure 20.1). Curcumin effects on wound healing are also demonstrated by its known potential effect on reducing ROS (Panchatcharam *et al.*, 2006). Furthermore, topical curcumin, acting through the mentioned pathways, appears to be one of the only chemicals known to protect skin after exposure to radiation or during radiotherapy (Kerr, 2002).

Figure 20.4 illustrates the mechanisms of wound healing and curcumin's effects on the different molecular pathways. Early, after the skin is wounded an acute inflammatory response is generated including the formation of a fibrin matrix. Degradation of the initial fibrin matrix is essential for proper injury repair and it is undertaken by uPA. This enzyme is enhanced by curcumin, which promotes a rapid and effective response. uPA also cleaves TIMP for inhibition of MMPs. Curcumin inhibits the formation of MMP through the AP-1 pathway, preventing MMP from degrading the ECM. By affecting these molecular pathways, curcumin enhances and accelerates the healing process.

20.2.4 Curcumin: its evolving role in keloids

Keloids represent an exaggerated scar formation response to the healing process that may be disfiguring and a major cause of discomfort. It is estimated that 15 to 20 percent of African-Americans, Hispanics, and Asians develop keloids with a suggested genetic predisposition. In black and Hispanic populations, the incidence of keloid formation is as high as 16%, with higher

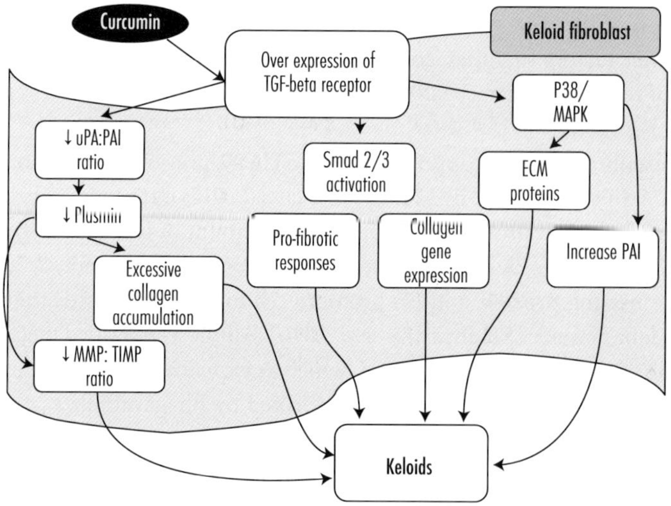

Figure 20.4. Curcumin role on promoting wound healing.
Double black arrow: inhibition; single black arrow: promotion. uPA: urokinase plasminogen activator, MMP: matrix metalloproteinase, TIMP: tissue inhibitors of metalloproteinase, PAI: plasminogen activator inhibitor, TGF: tumor growth factor, MAPK: mitogen-activated protein kinase, ECM: extracellular matrix.

frequencies during puberty and pregnancy (Chike-Obi *et al.*, 2009). It has been shown that keloid fibroblasts exhibit an altered phenotype of growth factor that stimulates collagen, fibronectin, elastin, and proteoglycan accumulation. These phenotypic alterations are responsible for keloid formation since it has been shown that these defects not only increase the amount of collagen deposition in keloids but also have a direct effect on fibrin degradation inhibition. The principal mechanism of keloid formation is illustrated in Figure 20.3. The overexpression of growth factors, such as TGF-β appears to play a major role in the formation of these lesions.

The over expression of TGF-β leads to a variety of proposed mechanisms that are implicated in keloid formation. Tuan *et al.* (2003) proposed decreased levels of uPA, induced by over expression of TGF-β as a possible mechanism for keloid formation. Decrease levels of uPA leads to impairment in wound repair and excessive collagen accumulation. By decreasing the uPA there is a consequent decrease in plasmin levels, which decreases fibrinolysis leading to accumulation of fibrin and other proteins in the extracellular matrix. A decrease in uPA also decreases the MMP, which plays an important role in the degradation of the extracellular matrix and has been found to be one of the major factors in keloid formation. High levels of plasminogen activator inhibitors have been found in keloids and it was attributed as the main factor for keloid development in keloid fibroblasts (Tuan *et al.*, 2003). Other proposed mechanisms for keloid formation have been documented to be secondary to TGF-β over-expression. It is known that there is a continuous activation of Smad2/3 and the p38/MAPK pathway that leads to collagen gene expression and decreased fibrinolysis, respectively (Daian *et al.*, 2010; He *et al.*, 2010). By either of these pathways TGF-β is known to be an important factor in the development of keloids. In conclusion, in normal fibroblasts, TGF-β is one of the main contributors for proper wound healing and its activity diminishes upon the completion of wound repair. In keloids, TGF-β is overproduced and poorly regulated.

Curcumin can directly affect expression of TGF-β in KF. Curcumin, in a dose dependent manner, is capable of blocking this growth factor (Hsu *et al.*, 2010). The results have led to the conclusion that the excessive production of extracellular matrix seen in keloid formation could be blocked and/or rapidly decreased by curcumin. More research on KF is needed to determine which of these proposed mechanisms is the most essential in the pathogenesis of keloid formation. Curcumin, by targeting the TGF-β receptor, the first step in the development of lesions, if applied early could represent an unusual targeted therapy for keloids as well as a chemopreventive agent, independent of the described pathways. Nonetheless, further controlled trials are needed to determine its exact effective dose and forms of delivery.

Figure 20.5 illustrates the role of curcumin in the development of improper wound healing and the formation of keloids at the fibroblast level. It has been shown that in patients who have a predisposition to keloid formation or have factors that may develop into poor wound healing there is an overexpression of TGF-β receptor. This over expression leads to the activation of three proposed pathways that are responsible for the development of lesions. These are: (1) A decrease in the uPA: PAI ratio that by decreasing the active uPA leads to a decrease in plasmin. Plasmin is essential for tissue remodeling and when decreased can lead to excessive fibroplasia and impaired wound healing. Decreased plasmin levels can also lead to the decrease in MMP, which has been shown to be one of the major factors in keloid formation. (2) Activation of the Smad2/3 pathway leading to the transcription of collagen genes and a pro-fibrotic response, and (3) activation of the p38/MAPK pathway. This pathway is similar to the uPA in which increased levels of PA-I result in decreasing fibrinolysis and contribute to excessive collagen accumulation. Curcumin inhibits TGF-β (doubled-lined dark arrow in Figure 20.5) affecting all the proposed mechanisms and inhibiting, early in the process, keloid formation.

20.2.5 Curcumin: its evolving role in acne

Acne vulgaris is a common inflammatory disorder that has been linked to multiple factors such as increased sebum production, inflammation of hair follicles and sebaceous glands due to the colonization of bacteria such as *Propionibacterium acnes*. ROS, especially superoxide anions (O^{2-}) are linked to this chronic inflammation and these are rapidly produced by keratinocytes upon stimulation of *P. acnes*. ROS are chemically reactive molecular structures that are generated during the course of normal human life but in times of stress ROS levels can increase dramatically leading to significant damage to cell structure. Retinoic acid derivatives are among the most effective therapies for acne since they prevent the production of ROS suggesting the relevance of this pathway in acne (Grange *et al.*, 2009). Once ROS are produced these molecules can induce activation of NF-κB or AP-1 leading to further inflammation by the recruitment of cytokines and

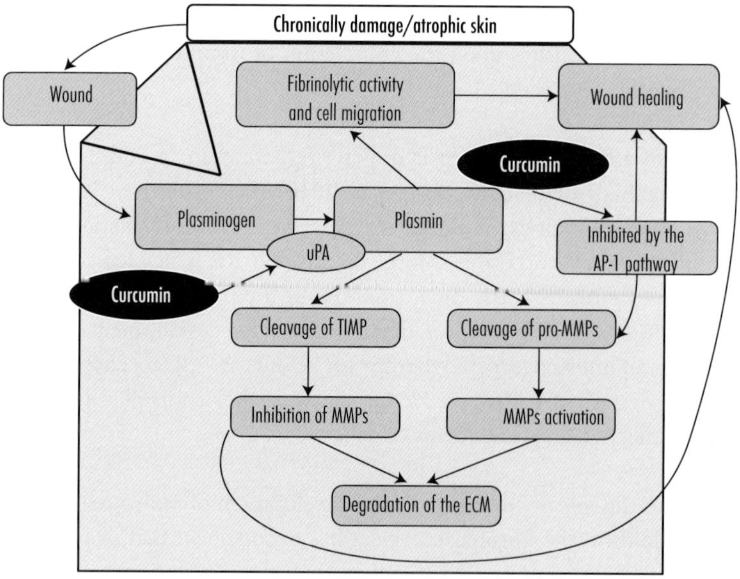

Figure 20.5. The TGF-beta receptor and its proposed mechanism for the development of keloids in keloid fibroblasts.
uPA: urokinase plasminogen activator, MMP: matrix metalloproteinase, TIMP: tissue inhibitors of metalloproteinase, ECM: extracellular matrix, AP-1: activator protein 1.

the modulation of various signaling pathways such as activation of MMP. The MMPs, especially an increase level of MMP-9 with an associated reduction in the TIMPs have been also linked to the development of chronic inflammatory acne (Papakonstantinou *et al.*, 2005). These processes are similar to the ones described above and represent important pathways in the development of skin diseases.

As mentioned previously in this chapter curcumin has a direct effect on the inhibition of ROS and the NF-κB and AP-1 pathways (Figures 20.1 and 20.2). It is expected then that curcumin would have a potential benefit in the treatment of acne vulgaris. This was demonstrated in a study performed in India in which inhibition of *P. acnes* mediators of inflammation was obtained by applying different herbal extracts to cell cultures, including that of *Curcumin longa* (Jain and Bsal, 2003). Although the relationship of diet and the development of acne is not absolutely clear it is expected that curcumin's great potential as an anti-inflammatory agent might be beneficial in acne treatment.

20.2.6 Other potential benefits of curcumin on skin health

The discovery of the different molecular mechanisms of curcumin have given further insight into its potential uses in a variety of disease processes. It has been used with success as an antibacterial, antiparasitic and anti-viral agent as for example for the Human Immunodeficiency Virus (Barthelemy, *et al.*, 1998). This could implicate the development of new and inexpensive microbidical agents to treat skin infections such as human papillomavirus, bacterial skin infections such as impetigo, cellulitis and many others. An increase in bacterial resistance to topical and oral antibiotics has been noticed in the past decade accounting for their decreased efficacy. In acne therapy, for example, the addition of alternative therapies such as curcumin as an antibacterial agent with less potential for toxicity may be of great interest for the scientific community. Curcumin has been also investigated for the potential treatment of cutaneous T cell lymphoma (Zhang *et al.*, 2010). This effect is also explained by the NF-κB as mentioned previously, which is a mechanistic rationale for curcumin usage in a selection of cancers. Curcumin has also been investigated for the treatment of atopic dermatitis, anti-aging processes as well as for cutaneous anaphylactoid reactions and allergies (Choi *et al.*, 2010). Cosmetics containing the powdered rhizome are available worldwide, particularly in India. Samoans sprinkle the powdered rhizome on newborns to mend the belly button after severing the umbilical cord. Indian cultures have used it for many years for the prevention of diaper rash, skin ulcers and bacterial skin infections. Ayurvedic medicine has utilized curcumin for many years in the treatment of acne and as a sun protective agent.

Researchers around the world are paying much attention to this ancient spice and are trying to explore all of its potential molecular targets and applications to human health. More research is needed to further determine the specific benefits, routes of administration and effectiveness of curcumin in the dermatological world.

References

Aggarwal, B.B., Bhatt, I.D., Ichikawa, H., Ahn, K.S., Sethi, G., Sandur, S.K., Natarajan, C., Seeram, S. and Shishodia, S., 2006. Curcumin biological and medicinal properties. In: Ravindran, P.N., Nirmal babu, K. and Sivaraman, K. (eds.), Turmeric: The genus *Curcuma*; Medicinal and aromatic plant: industrial profiles. CRC Press, New York, NY, USA, pp. 297-368.

Aggarwal, B.B., Sundaram, C., Malani, N. and Ichikawa, H., 2007. Curcumin the Indian solid gold. Advances in experimental medicine and biology 595, 1-75.

Aggrawal, R. and Kaur, I.P., 2010. Inhibitory effect of encapsulated curcumin on ultraviolet- induced photoaging in mice. Rejuvenation Research 2010; 13(4), 397-410.

Balasubramanian, S. and Eckert, R.L., 2007. Keratinocyte proliferation, differentiation and apoptosis: differential mechanisms of regulation by curcumin, EGCG and apigenin. Toxicology and Applied Pharmacology 224, 214-219.

Barthelemy, S., Vergens, L., Moyneir, M., Guyot, D., Labidalle, S. and Bahraoui, E., 1998. Curcumin and curcumin derivatives inhibit Tat-mediated transactivation of Type 1 human immunodeficiency virus long terminal repeat. Research in Virology 149, 43-52.

Bhagavathula, N., Warner, R.L., DaSilva, M., McClintock, S.D., Barron, A., Aslam, M., Johnson, K.J. and Varani, J., 2009. A combination of curcumin and ginger extract improves abrasion wound healing in corticosteroid-impaired hairless rat skin. Wound Repair and Regeneration 17, 360-336.

Chike-Obi, C.J., Cole, P.D. and Brissett, A.E., 2009. Keloids: pathogenesis, clinical features, and management. Seminars in Plastic Surgery 23, 178-184.

Choi, Y.H., Yan, G.H., Chai, O.H. and Song, C.H., 2010. Inhibitory effects of curcumin on passive cutaneous anaphylactoid response and compound 48/80- induced mast cell activation. Anatomy and Cell Biology 43, 36-43.

Cooper, S.J. and Bowden, G.T., 2007. Ultraviolet B regulation of transcription factor families: roles of nuclear factor-kappa B (NF-κB) and activator protein-1 (AP-1) in UVB induced skin carcinogenesis. Currrent Cancer Drug Targets 4, 325-334.

Daian, T., Ohtsuru, A., Rogounovitch, T., Ishihara, H., Hirano, A., Akiyama-Uchida, Y., Saenko, V., Fujii, T. and Yamashita, S., 2010. Insulin-like growth factor-I enhances transforming growth factor-b-induced extracellular matrix protein production through the P38/activating transcription factor-2 signaling pathway in keloid fibroblasts. The Journal of Investigative Dermatology 120, 956-962.

Grange, P.A, Chéreau, C., Raingeaud, J., Nicco, C., Weill, B., Dupin, N. and Batteux, F., 2009. Production of superoxide anions by keratinocytes initiates P. acnes-induced inflammation of the skin. PLoS Pathogens 5: e1000527.

He, S., Liu, X., Yang, Y., Huang, W., Xu, S., Zhang, X. and Roberts, M.S., 2010. Mechanisms of transforming growth factor beta (1)/Smad signalling mediated by mitogen-activated protein kinases in keloid fibroblasts. Brithish Journal of Dermatology 162, 538-546.

Heng, M.C, 2010. Curcumin targeted signaling pathways: basis for anti-photoaging and anti-carcinogenic therapy. International Journal of Dermartology 49, 608-622.

Heng, M.C., Song, M.K., Harker, J. and Heng, M.K., 2000 Drug-induced suppression of phosphorylase kinase activity correlates with resolution of psoriasis as assessed by clinical, histological and immunohistochemical parameters. The British Journal of Dermatology 143, 937-949.

Heng, M.C., Song, M.K. and Heng, M.K., 1994. Elevated phosphorylase kinase activity in psoriatic epidermis: correlation with increased phosphorylation and psoriatic activity. The British Journal of Dermatology 130, 298-306.

Hopper, B., Przybyszewski, J., Chen, H.W., Hammer, K. and Birt, D.F., 2009. Effect of ultraviolet B radiation on activator protein-1 constituent proteins and modulation by dietary energy restriction in SKH-1 mouse skin. Molecular Carcinogenesis 48, 843-852.

Hsu, Y.C., Chen, M.J., Yu, Y.M., Ko, S.Y. and Chang, C.C., 2010. Suppression of TGF-β1/SMAD pathway and extracellular matrix production in primary keloid fibroblasts by curcuminoids: its potential therapeutic use in the chemoprevention of keloid. Archived of Dermatological Research 302, 717-724.

Huang, M.T., Ma, W., Yen, P., Xie, J.G., Han, J., Frenkel, K., Grunberger, D. and Conney, A.H., 1997. Inhibitory effects of topical application of low doses of curcumin on 12-O-tetradecanoylphorbol-13- acetate-induced tumor promotion and oxidized DNA bases in mouse epidermis. Carcinogenesis 18, 83-88.

Jain, A. and Bsal, E., 2003. Inhibition of *Propionibacterium acnes*-induced mediators of inflammation by Indian herbs. Phytomedicine: International Journal of Phytotherapy and Phytopharmacology 10, 34-38.

Kerr, C., 2002. Curry ingredient protects skin against radiation. Lancet Oncology 3, 713.

Marin, Y.E., Wallas, B., Wanga, S., Namkoonga, J., Martino, J., Suh, J., Lee, H.J., Rabson, A., Yang, C., Chen, S. and Ryu, J.H., 2007 Curcumin downregulates the constitutive activity of NF-κB and induces apoptosis in novel mouse melanoma cells. Melanoma Research 17, 274-283.

Park, K. and Lee, J.H., 2010. Photosensitizer effect of curcumin on UVB-irradiated HaCaT cells through activation of caspase pathways. Oncology reports 17, 537-540.

Madhyastha, R., Madhyastha, H., Nakajima, Y., Omura, S. and Maruyama, S., 2010. Curcumin facilitates fibrinolysis and cellular migration during wound healing by modulating urokinase plasminogen activator expression. Pathophysiology of Haemostasis and Thrombosis 37, 59-66.

Menter, A., Chair, Korman, N.J., Elmets, C.A., Feldman, S.R., Gelfand, J.M., Gordon, K.B., Gottlieb, A., Koo, J.Y., Lebwohl, M., Lim, H.W., Van Voorhees, A., Beutner, K.R. and Bhushan, R., 2009. Guideline of care for the management of psoriasis and psoriatic arthritis. Journal of the American Academy of Dermatology 62, 114-135.

Morris, L., Wuepper, K.D., Stastny, P., Menter, A. and Bowcock, A., 1994. Gene for familial psoriasis susceptibility mapped to the distal end of human chromosome 17q. Science 264, 1141-1145.

Nochols, J.A. and Katiyar, S.K., 2010. Skin photoprotection by natural polyphenols: Anti-inflammatory, anti-oxidant and DNA repair mechanisms. Archieves of Dermatological Research 302, 71-83.

Panchatcharam, M., Miriyala, S., Gayathri, V.S. and Suguna, L., 2006 Curcumin improves wound healing by modulating collagen and decreasing reactive oxygen species. Molecuar and Cellular Biochemistry 290, 87-96.

Papakonstantinou, E., Aletras, A.J., Glass, E., Tsogas, P., Dionyssopoulos, A., Adjaye, J., Fimmel, S., Gouvousis, P., Herwig, R., Lehrach, H., Zouboulis, C.C. and Karakiulakis, G., 2005. Matrix metalloproteinases of epithelial origin in facial sebumof patients with acne and their regulation by isotretinoin. The Journal of Investigative Dermatology 125, 673-684.

Reddy, S. and Aggarwal, B., 1994. Curcumin is a non-selective inhibitor of phorphorilase kinase. Federation of European Biochemical Societies Journal 341, 19-22.

Singltary, K., 2010. Turmeric: An overview of potential health benefits. Nutrition Today 45, 216-225.

Szliszka, E. and Krol, W., 2010. The role of dietary polyphenols in tumor necrosis factor-related apoptosis inducing ligand (TRAIL)-induced apoptosis for cancer chemoprevention. European Journal of cancer Prevention 20, 63-69.

Tomfohrde, J., Silverman, A., Barnes, R., Fernandez-Vina, M.A., Young, M., Lory, D., Lund L.R., Romer, J., Bugge T.H., Nielsen, B.S., Frandsen, T.L., Degen, J.L., Stephens, R.W. and Dano, K., 1999. Functional overlap between two classes of matrix-degrading proteases in wound healing. European Molecular Biology Organization Journal 18, 4645- 4656.

Tuan, T.L., Wu, H., Huang, E.Y., Chong, S.S., Laug, W., Messadi, D., Kelly, P. and Le, A., 2003. Increased plasminogen activator inhibitor-1 in keloid fibroblasts may account for their elevated collagen accumulation in fibrin gel cultures. The American Journal of Pathology 162, 1579-1589.

Zhang, C., Li, B., Zhang, X., Hazarika, P., Aggarwal, B.B. and Duvic, M., 2010. Curcumin selectively induces apoptosis in cutaneous T-cell lymphoma cell lines and patients' PBMCs: potential role for STAT-3 and NF-kappaB signaling. The Journal of Investigative Dermatology 130, 2110-2119.

Key facts

- Superoxide dismutase (SOD) are antioxidant enzymes that catalyze cellular O_2^- to H_2O_2 and play a central role in antioxidative systems.
- *Sod1*$^{-/-}$ mice exhibit common phenotypes of premature aging, skin atrophy and loss of elasticity. Administration of L-ascorbyl 2-phosphate 6-palmitate trisodium salt (APPS), one of the VC derivatives, rescues skin atrophy of *Sod1*$^{-/-}$ mice.
- *Sod1*$^{-/-}$ dermal fibroblasts show decreased cell viability associated with increased O_2^- production. Treatment of APPS for *Sod1*$^{-/-}$ fibroblasts suppresses intracellular O_2^- production and improves the cell viability.

Summary points

- Reactive oxygen species (ROS), including O_2^- and H_2O_2, are generated as secondary products of cellular metabolism and induced oxidative stress. The intrinsic aging of skin and other organs suffer from increased oxidative stress by accumulated ROS. Intracellular ROS are eliminated by multiple antioxidant systems.
- SOD1 is antioxidant enzyme localized in the cytoplasm. *Sod1* deficiency led to skin atrophy associated with collagen malformation. Furthermore, this dermatrophia has progressed in the aged *Sod1*$^{-/-}$ mice, suggesting that skin atrophy might advance with increasing age.
- *Sod1*$^{-/-}$ dermal fibroblasts show low cell viability and increased O_2^- accumulation in 20% O_2 condition but not 1% O_2, indicating that decreased cell viability of *Sod1*$^{-/-}$ cells is caused by increased oxidative stress.
- Skin atrophy caused by *Sod1* deficiency is rescued by APPS administration. Furthermore, APPS decreases intracellular O_2^- production and improves cell viability of *Sod1*$^{-/-}$ fibroblasts.
- Redox regulation might be one of the powerful anti-aging strategies. Antioxidative materials, including APPS, also could be beneficial for delaying age-related changes.

21. Protective effects of vitamin C derivatives on skin atrophy caused by *Sod1* deficiency

S. Shibuya[1,2], *K. Kinoshita*[1,2] *and T. Shimizu*[1,3]
[1]*Molecular Gerontology, Tokyo Metropolitan Institute of Gerontology, 35-2 Sakae-cho, Itabasi-ku, Tokyo 173-0015, Japan;* [2]*JAC Co., Ltd., 1-2-7 Higashiyama, Meguro-ku, Tokyo, 153-0043, Japan;* [3]*Department of Aging Control Medicine, Chiba University Graduate School of Medicine, 1-8-1 Inohana, Chuo-ku, Chiba 260-8670, Japan; shimizut@chiba-u.jp*

Abstract

Skin aging is defined by two skin phenotypes: photoaging-induced hypertrophy and intrinsic aging-associated atrophy. Accumulating evidence suggests that reactive oxygen species (ROS) play a pivotal role in intrinsic aging of skin. Recently, we reported that genetic inactivation of *Sod1* (CuZn-SOD), which a major superoxide dismutase (SOD) in the cytoplasm, resulted in significant skin atrophy accompanied by the degeneration of collagen and elastic fibers. The skin phenotypes in these mice resemble skin atrophy during physiological aging in humans. We investigated the effects of vitamin C derivatives on skin atrophy in *Sod1*-deficient ($Sod1^{-/-}$) mice. We used three agents; vitamin C (VC), L-ascorbyl 2-phosphate trisodium salt (APS), which is conjugated to a phosphate group to improve stability, and L-ascorbyl 2-phosphate 6-palmitate trisodium salt (APPS), which is conjugated to a long hydrophobic chain to improve liposolubility. VC and its derivatives were transdermaly administered to $Sod1^{-/-}$ mice once per day for 4 weeks. APPS significantly rescued the skin thinning of the $Sod1^{-/-}$ mice, while VC and APS failed to produce any significant change. In *in vitro* experiments using primary dermal fibroblasts, *Sod1* deficiency decreased cell viability due to increased apoptosis and superoxide (O_2^-) production. To estimate the antioxidant activity of APPS *in vitro*, intracellular O_2^- generation was detected by dihydroethidium in $Sod1^{-/-}$ dermal fibroblasts treated with APPS. APPS strikingly repressed intracellular O_2^- generation in $Sod1^{-/-}$ fibroblasts. Furthermore, the viability of $Sod1^{-/-}$ fibroblasts was significantly improved by treatment with APPS. These results suggest that a novel VC derivative, APPS suppressed excess O_2^- production in skin cells, thus leading to prevention of skin atrophy in mice.

Keywords: L-ascorbyl 2-phosphate 6-palmitate trisodium salt, skin atrophy, superoxide dismutase, reactive oxygen species

Abbreviations

ANX V-FITC	FITC-conjugated annexin V
APPS	L-ascorbyl 2-phosphate 6-palmitate trisodium salt
APS	L-ascorbyl 2-phosphate trisodium salt
DHE	Dihydroethidium
PI	Propidium iodide
ROS	Reactive oxygen species
SOD	Superoxide dismutase
SOD1	Copper/zinc superoxide dismutase
UV	Ultraviolet
VC	Vitamin C

21.1 Introduction

Skin aging leads to two main symptoms: photoaging induced by environmental stress such as sunlight leads to skin hypertrophy, and the intrinsic aging induced by chronological or intrinsic factors leads to skin atrophy (Glogau, 1997). There have been many studies of the former using UV radiation (such as UV-A or UV-B) (Ichihashi *et al.*, 2003). The principle cause of photoaging is not only UV-induced direct DNA damage, but also UV-induced ROS production (de Gruijl *et al.*, 2001; Scharffetter-Kochanek *et al.*, 1997). Studies on the intrinsic aging process have been reported using representative senescence models such as Klotho mice (Kuro-o *et al.*, 1997), p53 mutant mice (Tyner *et al.*, 2002), and Ku86 knockout mice (Vogel *et al.*, 1999). These model mice exhibited common phenotypes of premature aging, skin atrophy and loss of elasticity.

ROS, including O_2^- and H_2O_2, are generated as secondary products of cellular metabolism and induced oxidative stress. The intrinsic aging of the skin and other organs included a progressive accumulation of oxidatively modified proteins, DNA, and lipids (Finkel and Holbrook, 2000), suggesting that ROS are strongly associated with skin aging. To protect them from oxidative stress, cells possess multiple antioxidative systems. SOD is thought to play a central role in antioxidative systems because of its ability to catalyze cellular O_2^- to H_2O_2. H_2O_2 is further degraded to O_2 and H_2O by catalase, glutathione peroxidase, and peroxiredoxin. There are three isozymes of SOD: the copper/zinc superoxide dismutase (SOD1) is localized in the cytoplasm; the manganese superoxide dismutase (SOD2) is distributed in the mitochondrial matrix; and the extracellular superoxide dismutase (SOD3) is secreted in extracellular plasma. Our previous studies indicated that *Sod1*-deficient ($Sod1^{-/-}$) mice showed various aging phenotypes such as age-related macular degeneration (Imamura *et al.*, 2006), fatty deposits in the liver (Uchiyama *et al.*, 2006), and skin atrophy (Murakami *et al.*, 2009). Moreover, other groups also reported that $Sod1^{-/-}$ mice exhibited hepatic carcinoma (Elchuri *et al.*, 2005), hemolytic anemia (Iuchi *et al.*, 2007), and muscle atrophy (Muller *et al.*, 2006). These observations suggest that the $Sod1^{-/-}$ mouse is a suitable model for studying human aging.

Vitamin C is a soluble vitamin that has strong antioxidant activity, suggesting that VC may delay skin aging. However, since VC has low stability and liposolubility, it is difficult to absorb into the skin. For this reason, VC has not been successfully used for skin care materials. To increase the stability and liposolubility of VC, various VC derivatives were developed for dermatologic application. APPS, one of the VC derivatives, is conjugated to a phosphate group and a long hydrophobic chain to enhance its stability and liposolubility. It has been reported that APPS showed the ability to prevent tumor invasion, as well as metastasis, and to resist auto-oxidation through intracellular ascorbic enrichment (Liu *et al.*, 1999, 2000). APPS also eliminated ROS production as detected by electron spin resonance spectrometry (Liu *et al.*, 2003). Perifollicular pigmentation was found to be the first target of APPS in clinical studies (Inui and Itami, 2007).

21.2 Skin atrophy in *Sod1*-deficient mice

To investigate the pathological consequences of *Sod1* deficiency on murine skin, we compared the thickness of the back skin in $Sod1^{-/-}$ and $Sod1^{+/+}$ mice. In hematoxylin and eosin staining, the thicknesses of the epidermis and dermis of the backs were significantly decreased, by 37.2%, in the $Sod1^{-/-}$ mice compared to the $Sod1^{+/+}$ mice (Figure 21.1B). The skin weight was also diminished by 22.2% in the $Sod1^{-/-}$ mice compared to the $Sod1^{+/+}$ mice (Figure 21.1C). In addition, to examine the skin collagen and elastin contents in $Sod1^{-/-}$ mice, we measured the hydroxyproline content in skin by amino acid analysis. The hydroxyproline content was decreased by 17.3% in $Sod1^{-/-}$ mice compared to the $Sod1^{+/+}$ mice (Figure 21.1D). These results indicate that the skin thinning of $Sod1^{-/-}$ mice was associated with a decreased hydroxyproline content.

It should be noted that this skin atrophy has progressed in the aged $Sod1^{-/-}$ mice (86 weeks of age), especially in the fat layer (Figure 21.2A). However, atrophy of the back skin at two days old was not showed in the $Sod1^{-/-}$ mice (data not shown). These results suggest that skin atrophy might advance with increasing age and not be related to developmental anomalies. Furthermore, the ear skin of the $Sod1^{-/-}$ mice also became atrophied at 16 weeks of age as occurred in the back skin, and the upper lipid layer of ears in the $Sod1^{+/+}$ mice disappeared in the $Sod1^{-/-}$ mice (Figure 21.2B).

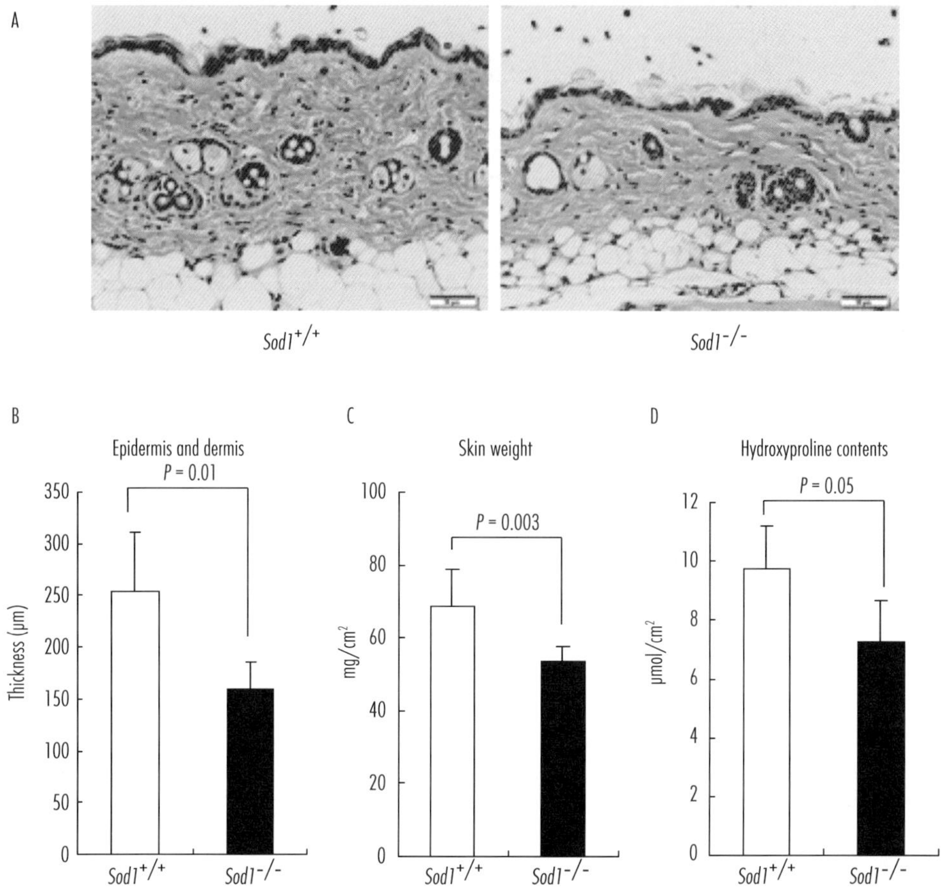

Figure 21.1. (A) The hematoxylin and eosin staining of the back skin of *Sod1*$^{+/+}$ and *Sod1*$^{-/-}$ mice (20 weeks of age). The scale bar represents 50 µm. (B) The thickness of the back skin in *Sod1*$^{+/+}$ and *Sod1*$^{-/-}$ mice was measured using the Leica Qwin V3 image software program (Leica, Germany) (19-21 weeks of age, n=5). (C) The back skin weight of *Sod1*$^{+/+}$ and *Sod1*$^{-/-}$ mice (19-21 weeks of age, n=6-8). (D) Amino acid analysis of the hydroxyproline in the back skin of *Sod1*$^{+/+}$ and *Sod1*$^{-/-}$ mice (25 weeks of age, n=5-9).

21.3 A novel vitamin C derivative improves the skin atrophy of *Sod1*-deficient mice

In order to investigate the effects of VC treatment on the skin symptoms in *Sod1*$^{-/-}$ mice, we administered 1% (w/v) VC in drinking water to *Sod1*$^{-/-}$ and *Sod1*$^{+/+}$ mice from 4 weeks of age for 12 weeks. The skin thicknesses of the *Sod1*$^{-/-}$ mice treated with VC was not significantly different from that of untreated mice (data not shown).

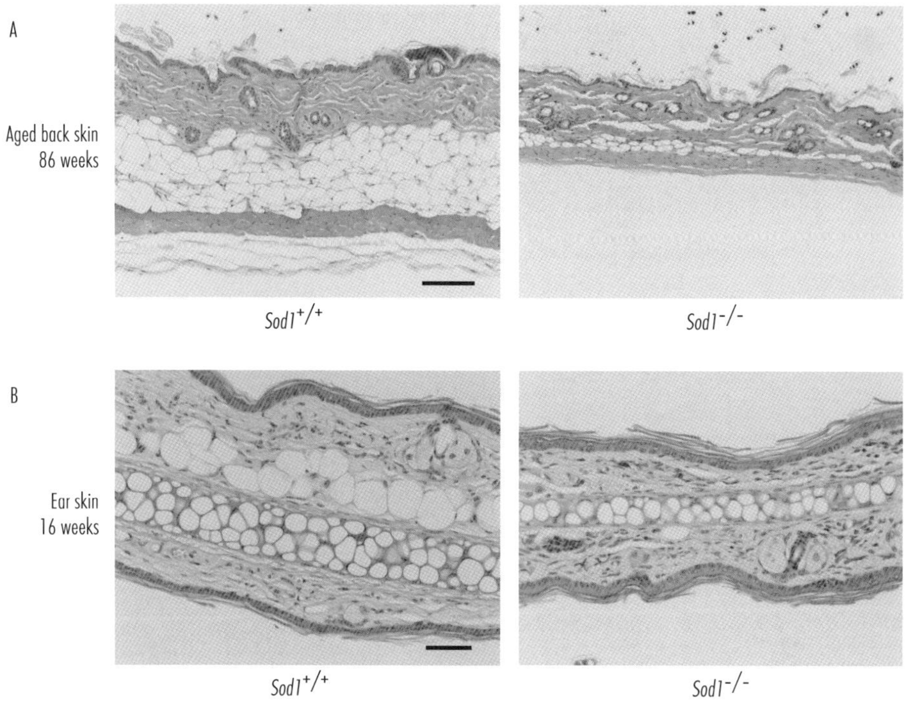

Figure 21.2. Representative micrographs of hematoxylin and eosin staining (A) back skin of *Sod1*$^{+/+}$ and *Sod1*$^{-/-}$ mice from 86 weeks of age, and (B) ear skin of *Sod1*$^{+/+}$ and *Sod1*$^{-/-}$ mice from 16 weeks of age. The scale bar represents 100 μm in (A) and 50 μm in (B).

Because it was hypothesized that the reason for the lack of efficacy could have been insufficient drug delivery to the skin, we next examined the effects of transdermal administration of VC and its derivatives, APS and APPS. APS is conjugated to a phosphate group to improve its stability, and APPS is conjugated to a long hydrophobic chain that increases its membrane permeability (Figure 21.3A). The VC derivative solutions were transdermally administered to the skin of the *Sod1*$^{-/-}$ and *Sod1*$^{+/+}$ mice daily for 4 weeks beginning at 17 weeks of age. As shown in Figure 21.3B and 21.3C, the skin of the *Sod1*$^{-/-}$ mice that had been treated with the 1% APPS solution was significantly thicker (by 22.8%) compared to that of *Sod1*$^{-/-}$ treated with the vehicle, although no significant differences in skin thickness were observed in *Sod1*$^{-/-}$ mice treated with 1% APS or 1% VC. These findings demonstrated that transdermal application of APPS reversed the skin atrophy of *Sod1*$^{-/-}$ mice.

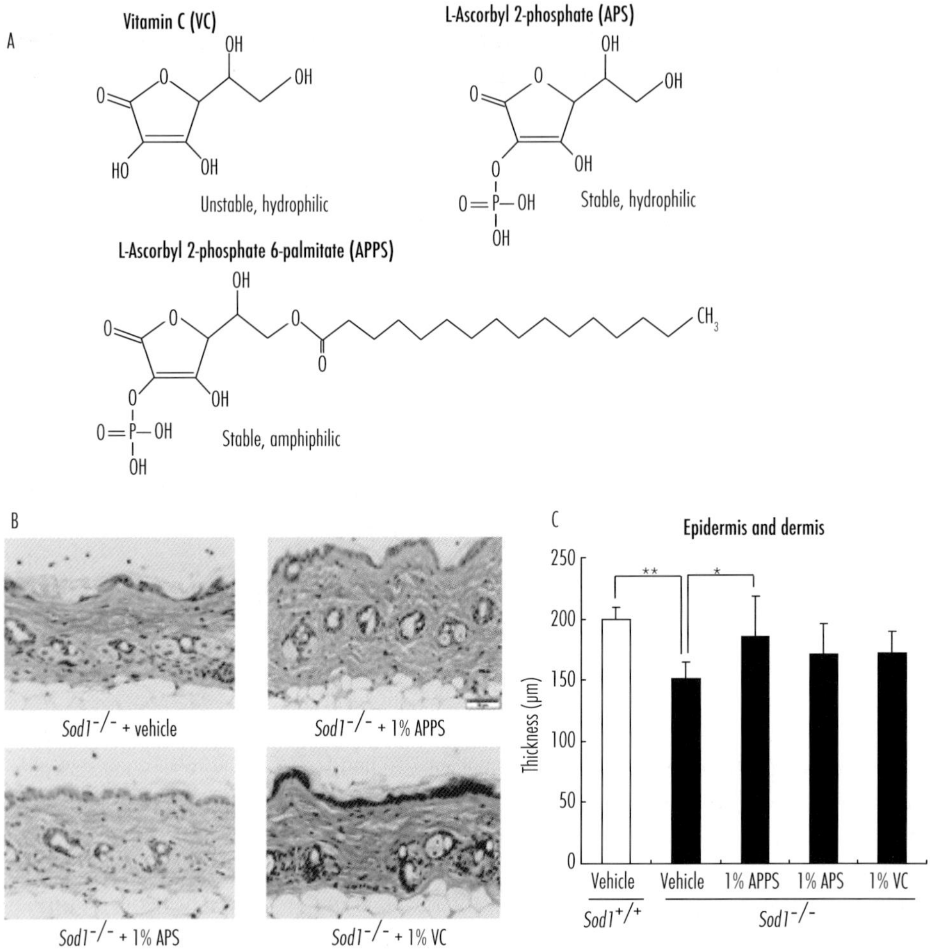

Figure 21.3. (A) The structure and characteristics of VC, APS and APPS. VC has low stability and liposolubility. APS is additionally conjugated with a phosphate group to enhance stability. APPS is additionally conjugated with a phosphate group and a long hydrophobic chain to enhance stability and liposolubility. (B, C) The *Sod1*$^{-/-}$ and *Sod1*$^{+/+}$ mice were treated transdermally with the VC derivatives at 17 weeks of age for 4 weeks. 200 µl of VC, APS and APPS solution dissolved in MilliQ water at a concentration of 1% (w/v) were painted daily onto the back skin of the mice after cutting their hair on first day of every week. (B) The hematoxylin and eosin staining of the back skin of Sod1-/- mice treated with 1% VC, 1% APS, 1% APPS. MilliQ water was used as a vehicle control. The scale bar represents 50 µm. (C) The thickness of the epidermis and dermis of the back skin of *Sod1*$^{+/+}$ and *Sod1*$^{-/-}$ mice treated with 1% VC, 1% APS, or 1% APPS (n=5-6). MilliQ water was used as a vehicle control. VC was obtained from Sigma (St. Louis, MO, USA). APS and APPS were provided from Showa Denko K.K. (Tokyo, Japan). *P<0.05, **P<0.01.

21.4 *Sod1*-deficient fibroblasts showed decreased cell viability associated with increased superoxide production

To investigate the cellular phenotypes of dermal fibroblasts lacking the *Sod1* gene, we compared the viability of *Sod1*$^{-/-}$ and *Sod1*$^{+/+}$ fibroblasts *in vitro*. Primary dermal fibroblasts were incubated in 1% or 20% O_2 and analyzed for cell numbers as well as cell viability (Figure 21.4A and 21.4B). The number of *Sod1*$^{+/+}$ fibroblasts cultured in 20% O_2 were increased by Day 5, and a large number of cells was still maintained on day 8. In contrast, the numbers of *Sod1*$^{-/-}$ fibroblasts cultured in 20% O_2 were not increased, and most of the cells were dead by day 4.

When *Sod1*$^{+/+}$ and *Sod1*$^{-/-}$ fibroblasts were incubated in 1% O_2, both types of fibroblasts had proliferated by day 3, but the populations gradually declined by day 8 (Figure 21.4A). When *Sod1*$^{+/+}$ fibroblasts were incubated in 20% O_2, the cell viability was not decreased by Day 8 (Figure 21.4B). In contrast, the viability of *Sod1*$^{-/-}$ fibroblasts incubated in 20% O_2 was severely decreased by day 3. When *Sod1*$^{+/+}$ and *Sod1*$^{-/-}$ fibroblasts were incubated in 1% O_2, the viability of both types of fibroblasts was increased on Day 3. However, both types of fibroblasts had almost completely died and the cell viabilities were decreased severely by day 8 (Figure 21.4B).

Next, *Sod1*$^{+/+}$ and *Sod1*$^{-/-}$ fibroblasts were incubated in 20% O_2 for 24 hr, and the number of apoptotic cells was measured by double supravital staining with recombinant ANX V-FITC and PI using flow cytometry. After a 24 hr incubation under 20% O_2, the apoptotic *Sod1*$^{-/-}$ fibroblasts were significantly increased (56.0%) compared with *Sod1*$^{+/+}$ fibroblasts (11.8%, Figure 21.4C). Furthermore, to evaluate cellular O_2^- generation, *Sod1*$^{-/-}$ fibroblasts were incubated with a DHE fluorescent probe for 30 min in a 1% O_2 condition. On day 0, we could not observe any differences between *Sod1*$^{+/+}$ and *Sod1*$^{-/-}$ fibroblasts (Figure 21.5A). In contrast, *Sod1*$^{-/-}$ fibroblasts had generated a significant amount of O_2^- in a time-dependant manner after 20% O_2 exposure compared to the *Sod1*$^{+/+}$ fibroblasts (Figure 21.5A). These results suggested that intracellular O_2^- generation decreased cell viability and increased apoptotic cell death in *Sod1*$^{-/-}$ dermal fibroblasts.

21.5 APPS suppresses superoxide production and rescues the viability of *Sod1*-deficient dermal fibroblasts

To assess the antioxidant activity of APPS on dermal fibroblasts of *Sod1*$^{-/-}$ mice, the *Sod1*$^{-/-}$ fibroblasts were pre-treated with APPS (10-100 μM) for 24 hr in 1% O_2, followed by incubation for 16 hr under 20% O_2 to induce oxidative stress. After incubation, these fibroblasts were incubated with the DHE fluorescent probe for 20 min under 20% O_2 conditions. O_2^- was significantly increased in the *Sod1*$^{-/-}$ fibroblasts compared to the *Sod1*$^{+/+}$ cells (Figure 21.5B and 21.5C). In addition, the population of *Sod1*$^{-/-}$ fibroblasts treated with 10 μM APPS was significantly decreased (by 70.1%) due to O_2^- generation (Figure 21.5B and 21.5C). Pre-treatment with 100 μM APPS suppressed O_2^- generation in *Sod1*$^{-/-}$ fibroblasts by 90%, resulting in production at the same level as in the *Sod1*$^{+/+}$ fibroblasts (Figure 21.5B and 21.5C).

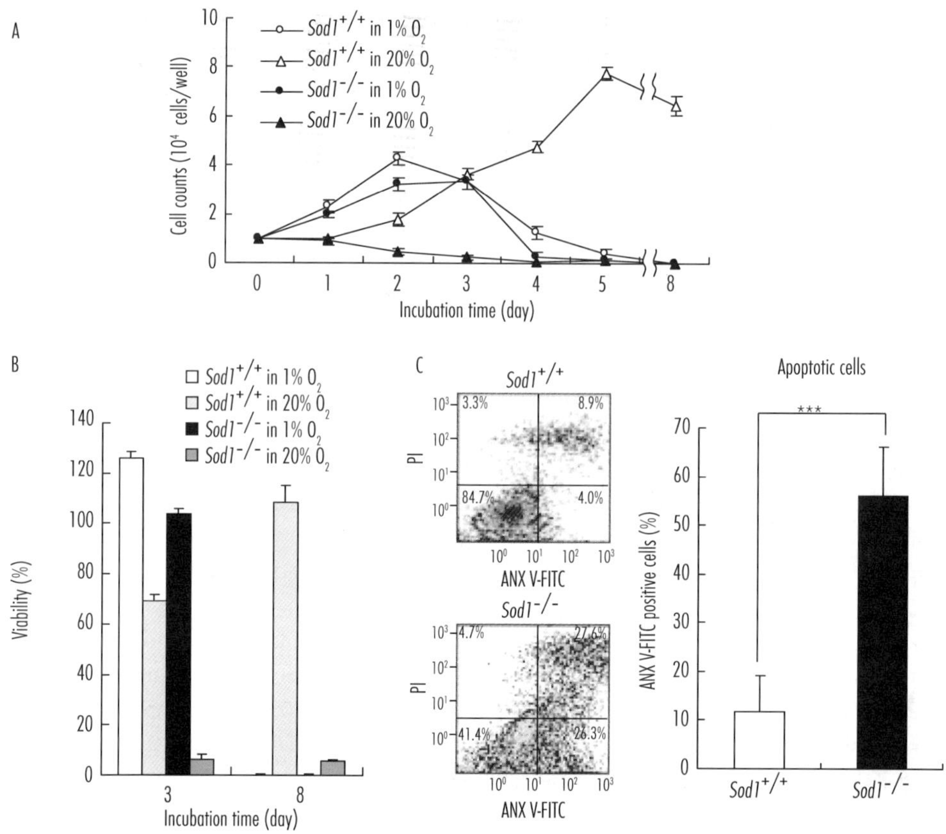

Figure 21.4. Primary dermal fibroblasts were isolated from the skin of pups (4-6 days after birth) and grown in a-MEM (Invitrogen, Carlsbad, CA, USA) supplemented with 20% FBS (Invitrogen), penicillin (Sigma) (100 U/ml), and streptomycin (Sigma) (100 μg/ml). *Sod1*$^{-/-}$ dermal fibroblasts were grown at 37 °C under a humidified atmosphere of 95% air and 5% CO_2 with the same medium as *Sod1*$^{+/+}$ cells. The cells were cultivated at a density of 1 x 10^4 cells/well on 96 well culture plates. (A) The *Sod1*$^{+/+}$ and *Sod1*$^{-/-}$ dermal fibroblasts were incubated under 1% or 20% O_2 and the cell numbers were counted for the indicated times. (B) *Sod1*$^{+/+}$ and *Sod1*$^{-/-}$ dermal fibroblasts were incubated under 1% or 20% O_2 and the cell viability was measured at the indicated times using a WST-8 kit (Dojindo, Kumamoto, Japan). (C) Apoptotic *Sod1*$^{+/+}$ and *Sod1*$^{-/-}$ dermal fibroblasts were detected using an Annexin V (ANX V)-FITC apoptosis detection kit I (BD Pharmingen, USA). *Sod1*$^{+/+}$ and *Sod1*$^{-/-}$ dermal fibroblasts were incubated at a density of 5 x 10^4 cells/well on 24 well culture plated under 20% O_2. Then, the cells were incubated with ANX V-FITC and propidium iodide (PI) in 100 μl annexin-binding buffer at room temperature for 15 min in the dark. After incubation, cells were analyzed by flow cymotemtry (Beckman Coulter, USA), and data are shown as two-color dot plots with ANX V-FITC (X axis) versus PI (Y axis). ***P<0.001. The scale bars represents 100 μm.

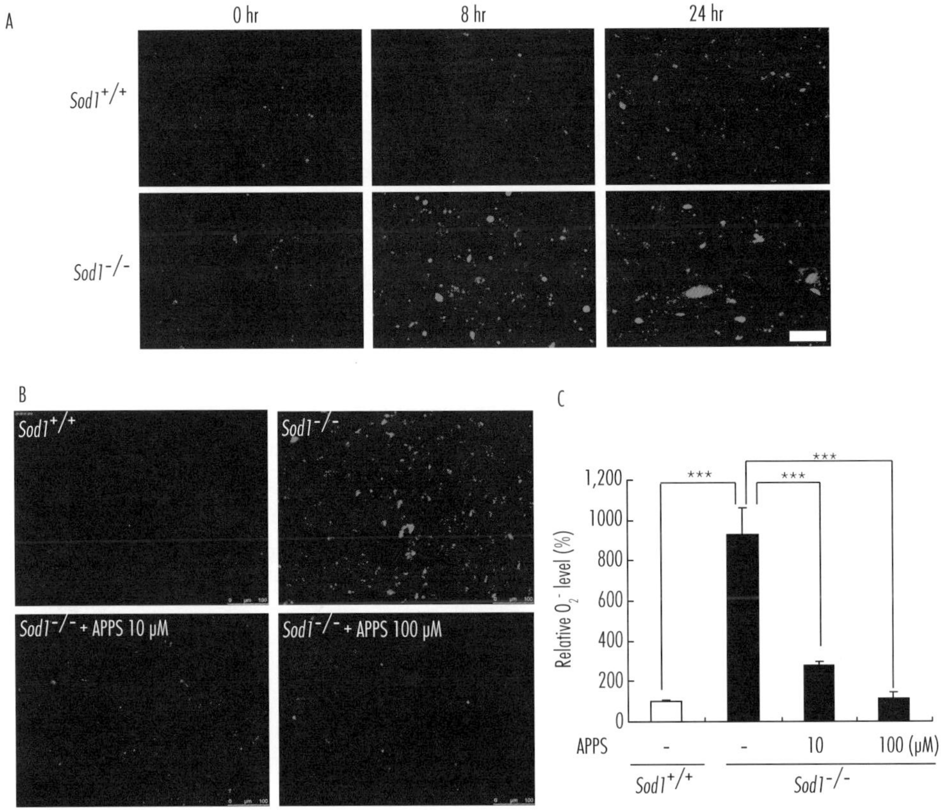

Figure 21.5. (A) *Sod1*$^{+/+}$ and *Sod1*$^{-/-}$ dermal fibroblasts were incubated under 20% O_2 to induce oxidative tress. After incubation, the fibroblasts were incubated with the dihydroethidium (DHE) fluorescent probe (Invitrogen) (10 μM) for 20 min under 20% O_2 conditions. Intracellular O_2^- generation was detected using fluorescence microscopy (Leica). Fluorescent images of intracellular O_2^- were captured through a band pass filter at 662-737 nm. The time indicates the incubation period (hr) after cells were transferred from 1% to 20% O_2 condition. (B, C). *Sod1*$^{-/-}$ dermal fibroblasts were incubated with APPS (10-100 μM) or vehicle for 24 hr in 1% O_2, followed by incubation for 16 hr under 20% O_2 to induce oxidative stress. After this incubation, the fibroblasts were incubated with a DHE fluorescent probe (10 μM) and Hoechst 33342 (10 μM) (Calbiochem, Germany) for an additional 20 min under 20% O_2. Intracellular O_2^- generation and nuclei were detected using fluorescent microscopy (B). Fluorescent image of nuclei were captured through a band pass filter at 450-490 nm. The DHE fluorescent area and Hoechst 33342 fluorescence as an indicator of the cell number were measured using a Leica QWin V3 image software program, and intracellular O_2^- generation was calculated as the DHE fluorescent area per cell (C). The scale bar represents 100 μM. ****P*<0.001.

Finally, we investigated whether APPS rescues the viability of *Sod1*$^{-/-}$ dermal fibroblasts. After APPS pre-treatment, fibroblasts were incubated in 20% O_2 and the cell numbers and cell viability were examined at the indicated times (Figure 21.6A and 21.6B). The number of *Sod1*$^{-/-}$ fibroblasts was not dramatically altered after 48 hr. In contrast, APPS treatment led to a marked increase in the number and the viability of *Sod1*$^{-/-}$ cells (Figure 21.6A and 21.6B). These findings suggested that APPS preserved the viability of *Sod1*$^{-/-}$ dermal fibroblasts via suppression of O_2^- production.

21.6 VC derivatives in other areas of skin health, care and treatments

A novel VC derivative, APPS prevented skin atrophy in *Sod1*$^{-/-}$ mice and improved the viability of *Sod1*$^{-/-}$ dermal fibroblasts by suppression of intracellular O_2^- production, suggesting that APPS is a beneficial reagent for cosmetic and medical applications. To overcome the low stability and liposolubility of VC, various other VC derivatives have been developed and examined for application for skin diseases. Shibayama *et al.* (2008a,b) reported that disodium isostearyl 2-O-L-ascorbyl phosphate (VCP-IS-2Na) suppressed the loss of type I collagen induced by UV-A

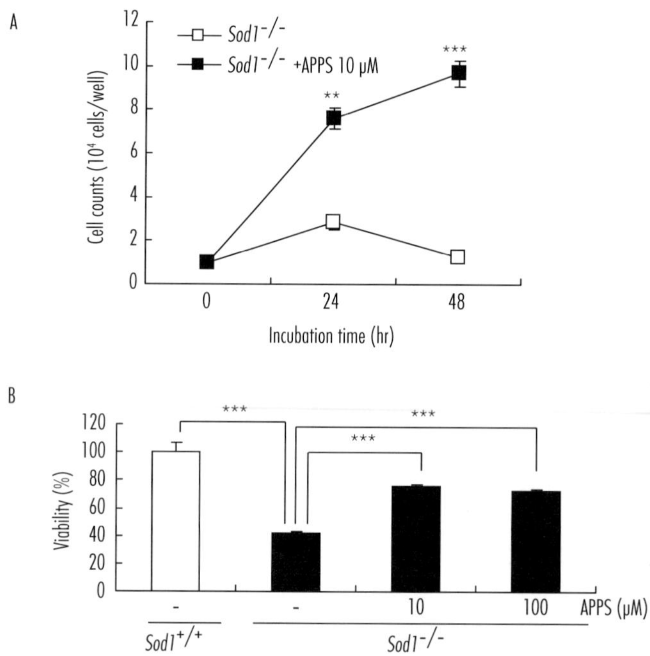

Figure 21.6. (A) *Sod1*$^{-/-}$ dermal fibroblasts were incubated with 10 μM APPS or vehicle for 24 hr in 1% O_2 followed by incubation in 20% O_2 to induce oxidative stress. The number of cells was counted at the indicated times. ***P*<0.01, ****P*<0.001. (B) *Sod1*$^{+/+}$ and *Sod1*$^{-/-}$ dermal fibroblasts were incubated with APPS (10-100 μM) or vehicle for 24 hr in 1% O_2 followed by incubation for 16 hr under 20% O_2, and cell viability was assessed at the indicated times using a WST-8 kit. ****P*<0.001.

irradiation via increased permeability and conversion to VC *in vitro*. Matsuda *et al.* (2008) reported that VCP-IS-2Na inhibited melanogenesis in human melanoma cells and epidermal melanocytes. In addition, Xiao *et al.* (2009) reported that 2,3,5,6-O-tetra-2'-hexyldecanoyl L-ascorbic acid in a liquid form showed cytoprotective effects against UV-A ray-induced injuries in human skin cells. Saitoh *et al.* (2009) reported that a fucoidan-vitamin C complex suppressed tumor invasion through the basement membrane via decreased oxidative stress and metalloproteinase activity in human fibrosarcoma cells. It has been reported that APS attenuated age-dependent telomere shortening in human skin keratinocytes via reduction of intracellular oxidative stress (Yokoo *et al.*, 2004). Furthermore, APS prevented apoptotic cell death and DNA strand cleavage in mouse epidermal keratinocytes (Sugimoto *et al.*, 2006). *In vitro* experiments showed that APS suppressed intracellular O_2^- generation and cell death (data not shown). In contrast, APS failed to induce any significant improvement in skin thickness of *Sod1*$^{-/-}$ mice, suggesting that the permeability of VC derivatives is an essential factor for transdermal treatment.

It is known that VC is also required for hydroxylation of proline residues in procollagen to stabilize the collagen triple helical structure (Peterkofsky, 1991). Therefore, VC deficiency induces the loss of skin elasticity due to collagen degeneration, thus resulting in scurvy symptoms (Peterkofsky, 1991). Interestingly, Massip *et al.* reported that VC supplementation rescued the mice from a shorter mean life span and reversed several age-related abnormalities in adipose tissues and liver endothelial defenestration, genomic integrity, and inflammatory status in a mouse model for Werner syndrome (Massip *et al.*, 2009). Furthermore, Esteban *et al.* reported that VC treatment also enhanced the induced pluripotent stem cell generation from both mouse and human somatic cells, at least partly by alleviating cell senescence (Esteban *et al.*, 2009). VC derivatives, including APPS, are also likely to behave more efficiently in terms of the pharmacological action of VC, due to their higher stability and permeability.

Sod1$^{-/-}$ mice showed various aging-like phenotypes such as age-related macular degeneration (Imamura *et al.*, 2006), fatty liver (Uchiyama *et al.*, 2006), hepatic carcinoma (Elchuri *et al.*, 2005), hemolytic anemia (Iuchi *et al.*, 2007), and muscle atrophy (Muller *et al.*, 2006). Recently, we found that *Sod1* deficiency also caused bone loss associated with low turnover in mice. Since one of the main components of skin and bone is collagen, some common molecular mechanism may regulate collagen metabolism in both skin and bone. In fact, we also found that VC treatment improved the bone phenotypes of *Sod1*$^{-/-}$ mice. In this context, our results suggested that intracellular oxidative stress induced by *Sod1* deficiency caused collagen malfunction leading to skin and bone atrophy in mice. Further analysis will be needed to clarify the molecular mechanism(s) responsible for the collagen abnormality due to *Sod1* deficiency in skin and bone.

The presence of hair often prevents precise analysis of skin. To prevent bias that might result from the thickness of hair during skin analysis of mice, hairless mice (Hos: HR-1, (Inomata *et al.*, 2003; Naganumaa *et al.*, 2001) are useful for experiments of transdermal administration. Recently, our group generated hairless-*Sod1*$^{-/-}$ (HR-1$^{hr/hr}$, *Sod1*$^{-/-}$) mice. The HR-1$^{hr/hr}$, *Sod1*$^{-/-}$ mice showed no hair growth, and maintained the same phenotypes as *Sod1*$^{-/-}$ mice (Shibuya *et al.*, unpublished results). Thus, the HR-1$^{hr/hr}$, *Sod1*$^{-/-}$ mouse will be a suitable mouse model for studying age-related skin disorders, and may be useful for further clarifying the effects of damage due to other factors, such as UV exposure, in *Sod1*-deficient mice.

Acknowledgements

This research was supported by Showa Denko K.K. We thank Dr. Yoh-ichi Koyama, Mr. Keisuke Tanaka, and Mr. Takayuki Ogura from Nippi Inc. for amino acid analysis, and Dr. Eiichi Kitakuni from Showa Denko K.K. for providing the APPS and APS. We also thank Dr. Kazuma Murakami, Mr. Jun Inagaki, Mr. Yusuke Ozawa, Mr. Kenji Watanabe, Ms. Miyuki Ito, Mr. Toshihiko Toda, and Mr. Yoshihiro Noda from the Tokyo Metropolitan Institute of Gerontology for providing technical assistance.

References

De Gruijl, F.R., Van Kranen, H.J. and Mullenders, L.H., 2001. UV-induced DNA damage, repair, mutations and oncogenic pathways in skin cancer. Journal of Photochemistry and Photobiology B 63, 19-27.

Elchuri, S., Oberley, T.D., Qi, W., Eisenstein, R.S., Jackson Roberts, L., Van Remmen, H., Epstein, C.J. and Huang, T.T., 2005. CuZnSOD deficiency leads to persistent and widespread oxidative damage and hepatocarcinogenesis later in life. Oncogene 24, 367-380.

Esteban, M.A., Wang, T., Qin, B., Yang, J., Qin, D., Cai, J., Li, W., Weng, Z., Chen, J., Ni, S., Chen, K., Li, Y., Liu, X., Xu, J., Zhang, S., Li, F., He, W., Labuda, K., Song, Y., Peterbauer, A., Wolbank, S., Redl, H., Zhong, M., Cai, D., Zeng, L. and Pei, D., 2009 Vitamin C enhances the generation of mouse and human induced pluripotent stem cells. Cell Stem Cell 6, 71-79.

Finkel, T. and Holbrook, N.J., 2000. Oxidants, oxidative stress and the biology of ageing. Nature 408, 239-247.

Glogau, R.G., 1997. Physiologic and structural changes associated with aging skin. Clinics in Dermatology 15, 555-559.

Ichihashi, M., Ueda, M., Budiyanto, A., Bito, T., Oka, M., Fukunaga, M., Tsuru, K. and Horikawa, T., 2003. UV-induced skin damage. Toxicology 189, 21-39.

Imamura, Y., Noda, S., Hashizume, K., Shinoda, K., Yamaguchi, M., Uchiyama, S., Shimizu, T., Mizushima, Y., Shirasawa, T. and Tsubota, K., 2006. Drusen, choroidal neovascularization, and retinal pigment epithelium dysfunction in SOD1-deficient mice: a model of age-related macular degeneration. Proceeding of the National Academy of Sciences of the United States of America 103, 11282-11287.

Inomata, S., Matsunaga, Y., Amano, S., Takada, K., Kobayashi, K., Tsunenaga, M., Nishiyama, T., Kohno, Y. and Fukuda, M., 2003. Possible involvement of gelatinases in basement membrane damage and wrinkle formation in chronically ultraviolet B-exposed hairless mouse. Journal of Investigative Dermatology 120, 128-134.

Inui, S. and Itami, S., 2007. Perifollicular pigmentation is the first target for topical vitamin C derivative ascorbyl 2-phosphate 6-palmitate (APPS): randomized, single-blinded, placebo-controlled study. Journal of Dermatology 34, 221-223.

Iuchi, Y., Okada, F., Onuma, K., Onoda, T., Asao, H., Kobayashi, M. and Fujii, J., 2007. Elevated oxidative stress in erythrocytes due to a SOD1 deficiency causes anaemia and triggers autoantibody production. Biochemical Journal 402, 219-227.

Kuro-o, M., Matsumura, Y., Aizawa, H., Kawaguchi, H., Suga, T., Utsugi, T., Ohyama, Y., Kurabayashi, M., Kaname, T., Kume, E., Iwasaki, H., Iida, A., Shiraki-Iida, T., Nishikawa, S., Nagai, R. and Nabeshima, Y.I., 1997. Mutation of the mouse klotho gene leads to a syndrome resembling ageing. Nature 390, 45-51.

Liu, J.W., Nagao, N., Kageyama, K. and Miwa, N., 1999. Antimetastatic and anti-invasive ability of phospho-ascorbyl palmitate through intracellular ascorbate enrichment and the resultant antioxidant action. Oncology Research 11, 479-487.

Liu, J.W., Nagao, N., Kageyama, K. and Miwa, N., 2000. Anti-metastatic effect of an autooxidation-resistant and lipophilic ascorbic acid derivative through inhibition of tumor invasion. Anticancer Research 20, 113-118.

Liu, J.W., Kayasuga, A., Nagao, N., Masatsuji-Kato, E., Tuzuki, T. and Miwa, N., 2003. Repressions of actin assembly and RhoA localization are involved in inhibition of tumor cell motility by lipophilic ascorbyl phosphate. Internation Journal of Oncology 23, 1561-1567.

Massip, L., Garand, C., Paquet, E.R., Cogger, V.C., O'Reilly, J.N., Tworek, L., Hatherell, A., Taylor, C.G., Thorin, E., Zahradka, P., Le Couteur, D.G. and Lebel, M., 2009. Vitamin C restores healthy aging in a mouse model for Werner syndrome. FASEB Journal 24, 158-172.

Matsuda, S., Shibayama, H., Hisama, M., Ohtsuki, M. and Iwaki, M., 2008. Inhibitory effects of a novel ascorbic derivative, disodium isostearyl 2-O-L-ascorbyl phosphate on melanogenesis. Chemical and Pharmaceutical Bulletin (Tokyo) 56, 292-297.

Muller, F.L., Song, W., Liu, Y., Chaudhuri, A., Pieke-Dahl, S., Strong, R., Huang, T.T., Epstein, C.J., Roberts, L.J., 2nd, Csete, M., Faulkner, J.A. and Van Remmen, H., 2006. Absence of CuZn superoxide dismutase leads to elevated oxidative stress and acceleration of age-dependent skeletal muscle atrophy. Free Radical Biology and Medicine 40, 1993-2004.

Murakami, K., Inagaki, J., Saito, M., Ikeda, Y., Tsuda, C., Noda, Y., Kawakami, S., Shirasawa, T. and Shimizu, T., 2009. Skin atrophy in cytoplasmic SOD-deficient mice and its complete recovery using a vitamin C derivative. Biochemical and Biophysical Research Communications 382, 457-461.

Naganumaa, M., Yagi, E. and Fukuda, M., 2001. Delayed induction of pigmented spots on UVB-irradiated hairless mice. Journal of Dermatological Science 25, 29-35.

Peterkofsky, B., 1991. Ascorbate requirement for hydroxylation and secretion of procollagen: relationship to inhibition of collagen synthesis in scurvy. American Journal of Clinical Nutrition 54, 1135S-1140S.

Saitoh, Y., Nagai, Y. and Miwa, N., 2009. Fucoidan-Vitamin C complex suppresses tumor invasion through the basement membrane, with scarce injuries to normal or tumor cells, via decreases in oxidative stress and matrix metalloproteinases. International Journal of Oncology 35, 1183-1189.

Scharffetter-Kochanek, K., Wlaschek, M., Brenneisen, P., Schauen, M., Blaudschun, R. and Wenk, J., 1997. UV-induced reactive oxygen species in photocarcinogenesis and photoaging. Journal of Biological Chemistry 378, 1247-1257.

Shibayama, H., Hisama, M., Matsuda, S. and Ohtsuki, M. 2008a. Permeation and metabolism of a novel ascorbic acid derivative, disodium isostearyl 2-O-L-ascorbyl phosphate, in human living skin equivalent models. Skin Pharmacology and Physiology 21, 235-243.

Shibayama, H., Hisama, M., Matsuda, S., Kawase, A., Ohtsuki, M., Hanada, K. and Iwaki, M., 2008b. Effect of a novel ascorbic derivative, disodium isostearyl 2-O-L-ascorbyl phosphate, on normal human dermal fibroblasts against reactive oxygen species. Bioscience, Biotechnology and Biochemistry 72, 1015-1022.

Sugimoto, M., Okugawa, Y. and Miwa, N., 2006. Preventive effects of phosphorylated ascorbate on ultraviolet-B induced apoptotic cell death and DNA strand cleavage through enrichment of intracellular vitamin C in skin epidermal keratinocytes. Free Radical Research 40, 213-221.

Tyner, S.D., Venkatachalam, S., Choi, J., Jones, S., Ghebranious, N., Igelmann, H., Lu, X., Soron, G., Cooper, B., Brayton, C., Hee, Park. S,, Thompson, T., Karsenty, G., Bradley, A. and Donehower, L.A., 2002. p53 mutant mice that display early ageing-associated phenotypes. Nature 415, 45-53.

Uchiyama, S., Shimizu, T. and Shirasawa, T., 2006. CuZn-SOD deficiency causes ApoB degradation and induces hepatic lipid accumulation by impaired lipoprotein secretion in mice. Journal of Biological Chemistry 281, 31713-31719.

Vogel, H., Lim, D.S., Karsenty, G., Finegold, M. and Hasty, P., 1999. Deletion of Ku86 causes early onset of senescence in mice. Proceedings of the National Academy of Science United States of America 96, 10770-10775.

Xiao, L., Kaneyasu, K., Saitoh, Y., Terashima, Y., Kowata, Y. and Miwa, N., 2009. Cytoprotective effects of the lipoidic-liquiform pro-vitamin C tetra-isopalmitoyl-ascorbate (VC-IP) against ultraviolet-A ray-induced injuries in human skin cells together with collagen retention, MMP inhibition and p53 gene repression. Journal of Cellular Biochemistry 106, 589-598.

Yokoo, S., Furumoto, K., Hiyama, E. and Miwa, N., 2004. Slow-down of age-dependent telomere shortening is executed in human skin keratinocytes by hormesis-like-effects of trace hydrogen peroxide or by anti-oxidative effects of pro-vitamin C in common concurrently with reduction of intracellular oxidative stress. Journal of Cellular Biochemistry 93, 588-597.

Skin cancer, nutrition and diet

Key facts

- Dietary ω-6 fatty acids (FA) exacerbate ultraviolet radiation (UVR) carcinogenic expression, with respect to both shortened tumor latent period and increased tumor multiplicity.
- Dietary ω-3 FA inhibits UVR carcinogenic expression.
- ω-6 FA exerts its principal effect on the post-initiation, or promotion stage of carcinogenesis.
- The influence of ω-3 FA on carcinogenic expression is exerted across the carcinogenic continuum, i.e. initiation and promotion.
- Pro-inflammatory and immunosuppressive prostaglandin E_2 (PGE_2) levels are increased linearly as ω-6 FA concentrations increase.
- Pro-inflammatory and immunosuppressive PGE_2 levels are dramatically reduced by dietary ω-3 FA intake.
- Dietary ω-6 FA suppresses the immunologic responses involved in tumor transplant rejection and the immunologic pathway(s) manifested in delayed hypersensitivity (DTH).
- Dietary ω-3 FA inhibits the UVR suppression of contact hypersensitivity (CHS) and DTH responses.
- ω-3 FA supplementation significantly increases the erythema threshold to UVR.
- Decreased sensitivity to UVB is accompanied by a significant increase in the threshold to UVA-provocation in photosensitivity disorders.
- ω-3 FA modulate a number of cytokines and prostaglandins (PG) that mediate inflammatory and immune responses.
- ω-3 FA inhibits certain genotoxic markers of UVR-induced DNA damage, e.g. UVR-induced cutaneous p53 expression.

Summary points

- A major mode of action of dietary fat on UVR carcinogenic expression is via modulation of immune pathways, which appear to be closely related to differential influence of ω-6 and ω-3 FA on inflammatory and immune-active products of the eicosanoid cascade. The effects of these products occur at a time when the host animal is being immune-compromised by UVR.
- There is strong evidence that support a role for ω-3 FA in the prevention of NMSC. Animal studies have provided direct proof that ω-3 FA inhibits UVR carcinogenic expression. A probable mechanism for this effect rests in ω-3 FA competitive effects on the cyclooxygenase pathway that results in reduced levels of PGE_2. Clinical studies have found that ω-3 FA supplementation increases the erythema threshold (reduces the inflammatory response) to UVB. UVB treatment of human keratinocytes grown in the presence of ω-3 FA resulted in lower levels of PGE_2 than cells grown in the presence of ω-6 FA. The levels of other cytokines, expected to have pronounced stimulatory effects on cellular immune function, were also modulated by ω-3 FA.
- Because of the promising evidence from animal studies; the influence of ω-3 FA on cyclooxygenase and lipoxygenase products; and upon cancer biomarkers in humans, it is important that the potential of ω-3 FA as a preventive agent to NMSC be fully explored.

22. Omega-3 fatty acids and non-melanoma skin cancer

H.S. Black
Department of Dermatology, Baylor College of Medicine, 811 Briar Park Dr., Houston, Texas 77042 USA; hblack@bcm.tmc.edu

Abstract

Considerable experimental evidence has accrued from animal studies that implicate dietary fats as important modulators of ultraviolet (UVR)-carcinogenic expression. These fats are composed of essential fatty acids (EFA) and are grouped into two families of polyunsaturated fatty acids (PUFA), omega (ω)-6 and omega (ω)-3 fatty acids (FA). They are essential in that they are precursors of a group of highly active hormones, eicosanoids, that are necessary for normal growth and maintenance of health. Relatively minor differences in molecular structure of these FA not only cause them to act differently in the body, but also to produce diverse effects in modulation of carcinogenic expression. Dietary ω-6 FA has been shown to *exacerbate* UVR carcinogenic expression (tumor latent period and tumor multiplicity) in a manner that is directly related to level of FA intake. Dietary ω-3 FA *inhibits* UVR carcinogenic expression. Pro-inflammatory and immunosuppressive eicosanoids, e.g. prostaglandin E_2 (PGE_2) levels increase in a linear fashion with the concentration of ω-6 FA intake. These levels are dramatically reduced with ω-3 FA intake. ω-6 FA intake *suppresses* the immunologic pathways manifested in delayed hypersensitivity (DTH) and the responses involved in tumor transplant rejection. ω-3 FA *sustains* the DTH response in UVR-irradiated animals. Clinical studies have found that supplementary ω-3 FA significantly raise the UVR-mediated erythema threshold and reduce the level of pro-inflammatory and immunosuppressive PGE_2 in UVR-irradiated human skin. Taken *in toto*, these studies suggest that ω-3 supplementation could be beneficial in reducing the occurrence of non-melanoma skin cancer, particularly in those individuals who are at highest risk.

Keywords: essential fatty acids, polyunsaturated fatty acids, skin carcinogenesis, eicosanoids, cancer prevention

Abbreviations

AA	Arachidonic acid
ALA	α-linolenic acid
BCC	Basal cell carcinoma
CHS	Contact hypersensitivity
COX	Cyclooxygenase
DHA	Docosahexaenoic acid
DTH	Delayed type hypersensitivity
EFA	Essential fatty acid
EPA	Eicosapentaenoic acid
FA	Fatty acid
LA	Linoleic acid
MED	Minimal erythema dose
NMSC	Non-melanoma skin cancer
NSAID	Non-steroidal anti-inflammatory drugs
ODC	Ornithine decarboxylase
PUFA	Polyunsaturated fatty acid
PGE_2	Prostaglandin E_2
P/S	Polyunsaturated to saturated fatty acid ratio
PUVA	8-methoxypsoralen + UVA radiation
SCC	Squamous cell carcinoma
SPF	Sun protection factor
TNF-α	Tumor necrosis factor-alpha
UVR	Ultraviolet radiation

22.1 Introduction

Essential fatty acids are, as the designation implies, essential for normal growth and sustenance of good health. LA, an ω-6 FA, and ALA, an ω-3 FA, cannot be synthesized by humans and are, thus, considered essential and must be supplied in the diet. These EFA, LA and ALA, are the precursors of the ω-6 and ω-3 FA series, respectively. FA are often abbreviated by their chemical designation, e.g. LA as 18:2n-6 where 18 indicates the length of the carbon chain; the 2, the number of *cis* double bonds; and the n-6 indicates that the first of these double bonds begins at the sixth carbon atom from the methyl end of the carbon chain. ALA is abbreviated as 18:3n-3, the n-3 indicates that the first double bond is at the third carbon from the methyl end of the chain. The longer chain FA, i.e. AA (20:4n-6), EPA (20:5n-3) and DHA (22:6n-3), can be synthesized in humans from their respective presursor EFA through a series of desaturation (addition of a double bond) and elongation (addition of two carbon atoms) enzymatic reactions (Figure 22.1). Western diets may contain 15-20 times more LA than ALA and, thus, competition for these shared enzymes results in greater levels of AA than EPA and DHA. Consequently, under certain conditions supplementation with EPA and DHA may be essential for maintenance of good health. Indeed,

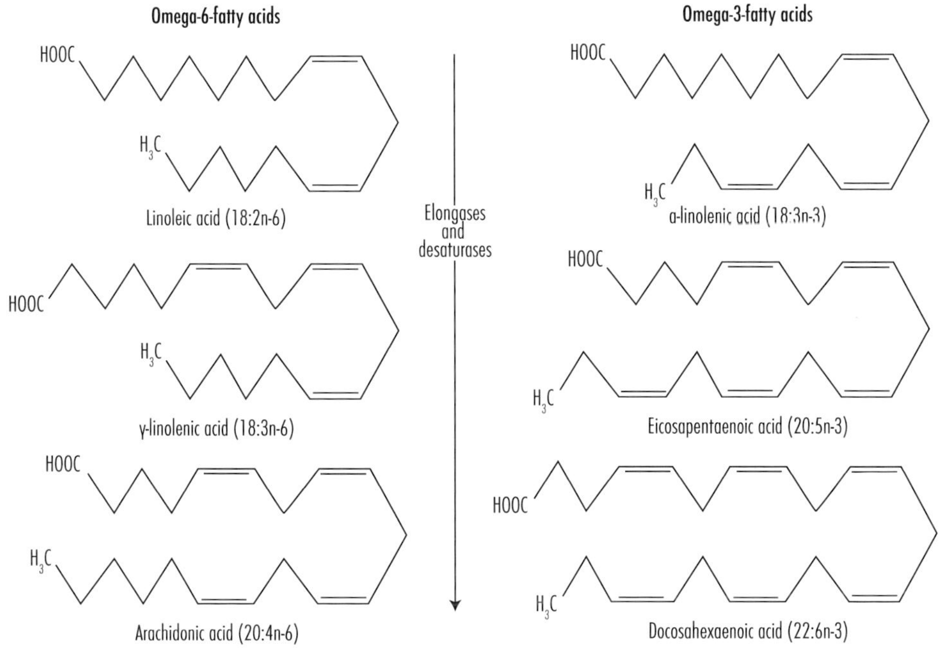

Figure 22.1. Structural diagram of the essential fatty acids, linoleic acid and α-linolenic acid, precursors to the long chain ω-6 and ω-3 fatty acids.

the high level of interest in the potential health benefits of ω-3 FA has resulted from the seminal reports that these FA were associated with low incidence of ischemic heart disease, as well as other general inflammatory symptoms, among Greenlandic Eskimos (Bang and Dyerberg, 1972; Dyerberg and Bang, 1978). Subsequently, randomized control and direct intervention trials have provided strong evidence that increased ω-3 FA intake is associated with significant reductions in cardiovascular disease risk, myocardial infarction, and sudden cardiac death (Breslow, 2006). It has been suggested that an intake of 400-500 mg/day of EPA plus DHA would significantly reduce the risk of death from cardiovascular heart disease in healthy adults (Harris *et al.*, 2008).

Not only do the two series of EFA compete for elongase and desaturase enzymes, but they cannot be inter-converted in humans. Thus, each series differentially influences the flux of metabolites through the COX and lipoxygenase pathways (Figure 22.2). These oxidation products differ in hormonal potency, with the ω-6 derived products being more active than their ω-3 counterparts (Lees, 1990). Several of these metabolites are known to influence tumor biology. PGE_2, derived from ω-6 FA oxidation via the COX pathway, may act as a tumor promoter and increased levels of PGE_2 synthetase, and hence PGE_2, have been associated with aggressive tumor growth patterns in both BCC and SCC in humans (Vanderveen *et al.*, 1986). ω-3 FA compete with ω-6 FA for binding sites on COX and inhibits the production of PGE_2, resulting in higher levels of the less potent PGE_3. ω-3 FA may also shunt potential PG precursors through the lipoxygenase pathway, resulting in products that inhibit tumor growth and in some products that are involved in

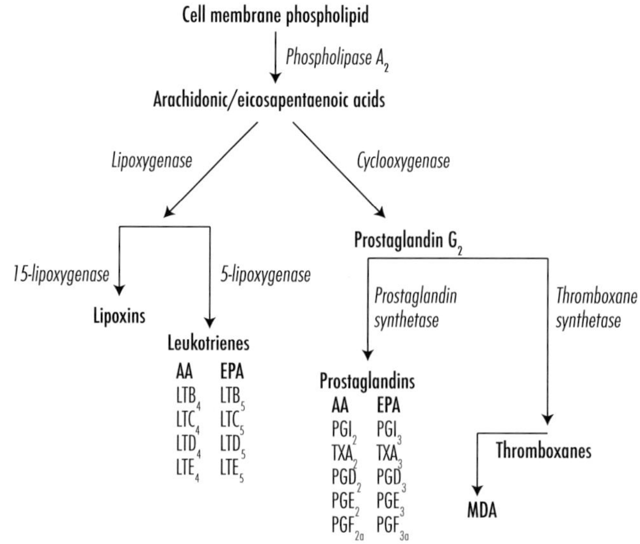

Figure 22.2. Simplified diagram of eicosanoid metabolism. AA is metabolized via lipoxygenase and cyclooxygenase pathways to produce lipoxins, leukotrienes, prostaglandins, and thromboxanes. EPA, an ω-3 FA, competes as a competitive inhibitor to the cyclooxygenase enzyme complex with AA, an ω-6 FA, and produces different oxidation products as shown for the leukotrienes and prostaglandins. The 3-series prostaglandins resulting from EPA oxidation are usually much less active than the 2-series derived from AA. Malondialdehyde (MDA) is a product of prostaglandin and thromboxane metabolism and is commonly used as a measure of lipid peroxidation. Reproduced from Black and Rhodes (2006).

immunosurveillance (Malmsten, 1984; Werner *et al.*, 1985). These pathways, with their bioactive intermediates, are prime targets for potential cancer chemopreventive strategies. Epidemiological studies of NSAID have failed to provide conclusive evidence that these agents reduce the risk for BCC and SCC (Grau *et al.*, 2006). However, experimental animal studies, employing levels of Celecoxib (an inhibitor of the inducible COX-2 isozyme) equivalent to those that are prescribed in humans, showed a significant reduction in UVR-induced skin cancer (Orengo *et al.*, 2002). Recently, Elmets *et al.* (2010), in a randomized, double-blind, placebo-controlled trial, found significantly fewer BCC and SCC in the Celecoxib arm of the study. As ω-3 FA compete for COX active sites and alter the type and level of bioactive intermediates, there is a strong rationale for the cancer chemopreventive potential of these EFA (Black and Rhodes, 2006).

22.2 Evidence for involvement of dietary PUFA in UVR-induced skin cancer

22.2.1 Animal studies

The first indication that dietary fat could potentiate UVR-carcinogenesis came in 1939 (Baumann and Rusch). It was nearly 45 years later when this clue was followed-up in a series of studies that demonstrated that an approximate linear relationship existed between PUFA (dietary oils that contained roughly 50% ω-6 FA) and UVR-carcinogenic expression (Black *et al.*, 1983, 1985). Tumor latency (time, in weeks, for 50% of the animals to develop tumors) *decreased* and tumor multiplicity (number of tumors per animal) *increased* in an approximate linear fashion as the level of ω-6 FA in the diet increased. In the initial study there was some indication that PUFA was required for UVR-carcinogenic expression. When PUFA, employed in that study, was partially hydrogenated, there was a marked inhibition of carcinogenesis (Black *et al.*, 1983). Indeed, Reeve *et al.* (1996), in a dramatic demonstration, found that feeding a totally hydrogenated PUFA completely abolished UVR-carcinogenic expression whereas those animals fed the normal PUFA exhibited 100% tumor incidence. Furthermore, when the diet of animals fed the hydrogenated fat was reconstituted with a normal mixed fat, large numbers of skin tumors rapidly appeared. These data suggested that UVR initiation had not been prevented by lack of PUFA, but that an EFA deficiency held the tumors in abeyance, probably at the promotion stage. Confirmation that ω-6 FA exerted their influence principally at the post-initiation stage of carcinogenesis came from crossover feeding studies (Black *et al.*, 1992). Animals were placed on defined iso-caloric rations containing low (0.75%. w/w) and high (12%. w/w) levels of corn oil (roughly 50% ω-6 LA). Upon completion of a regimen of UVR, and before tumors appeared, some diets were crossed to the contravening diet, i.e. high to low fat and low to high fat. The resulting tumor incidence curves and tumor multiplicity analysis provided confirmation that diets containing high levels of ω-6 FA enhanced UVR-carcinogenic expression and that enhancement occurred during the post-initiation, or promotion, stage of carcinogenesis. Importantly, crossing from a high fat to a low fat diet after a cancer inducing dose of UVR had been administered, negated the exacerbating influence of high fat diets. The latter finding provided a rationale upon which a low-fat dietary intervention could act to ameliorate cancer expression.

Contrary to the promoting effects of ω-6 FA, animals fed a diet containing menhaden oil (rich in ω-3 FA) as lipid source, exhibited a marked *inhibition* of UVR-carcinogenic expression (Table 22.1) (Orengo *et al.*, 1989). Also, unlike ω-6 FA, crossover studies suggested that ω-3 FA exert their principal anti-carcinogenic effects during the initiation phase of carcinogenesis. Alternatively, ω-3 FA failed to produce its ameliorating effect on tumor multiplicity when crossing from ω-6 to ω-3 FA, suggesting that perhaps promotion events had progressed beyond the point where ω-3 FA could intervene (Black *et al.*, 1992). Nevertheless, animals receiving ω-3 FA rich menhaden oil throughout the study exhibited significant inhibition of carcinogenesis (an increased tumor latent period and reduced tumor multiplicity) when compared to animals receiving an equivalent level of corn oil rich in ω-6 FA.

An earlier study with an ω-6 FA source had implicated lipid radical mediated reactions as participants in at least a part of the carcinogenic response to UVR (Black *et al.*, 1985). The level of UVR-induced cutaneous lipid peroxidation, and carcinogenic expression, was found to be a direct function of ω-6 FA intake. However, EPA, an ω-3 FA, has a higher proportion of unsaturated linkages than does LA and, thus, represents an even more vulnerable target for free radical attack. There were no significant differences in forward scattering transmission of UVR through epidermis of ω-3 FA fed animals. It was concluded, from these observations, that the

Table 22.1. Comparison of the influence of ω-6 and ω-3 FA on UVR induced carcinogenic, inflammatory, and immunological responses. Compilation of data from references (Black *et al.*, 1992; Orengo *et al.*, 1989; Fischer and Black, 1991).

Dietary lipid	Tumor latency[1]	Tumor multiplicity[2]	Edema[3]	ODC[4]	Inflammatory response[5]	Hapten response[6]
0.75% ω-6 FA	22.30	1.23				
2.0% ω-6 FA	21.2	1.75				
4.0% ω-6 FA	20.5	2.10	0.263	2.72		
12% ω-6 FA	19.20	2.5		2.86	40.8	4.6
4.0% ω-3 FA	23.31	0.31	0.269	2.80		
12% ω-3 FA	26.14	0.23	0.109	0.78	29.7	21.1

[1] Median tumor time (wk).
[2] Number of tumors per animal.
[3] Measured as double skin fold thickness at 0 and 24 h post UVR. Reported as change in thickness (mm).
[4] Ornithine decarboxylase activity 24 h post UVR (nmol $^{14}CO_2$/mg protein/h).
[5] Inflammatory response to dimethyl sulfoxide (DMSO). Increase in footpad thickness in mm x 10^{-2}.
[6] Net hapten response to dinitrochlorobenzene (DNCB). Total delayed-type hypersensitivity (DTH) minus inflammatory response to DMSO vehicle. Footpad thickness, mm x 10^{-2}.
(Reproduced from Black and Rhodes, 2006).

differences in carcinogenic or acute responses to UVR could not be attributed to UVR dose diminution or reduction in lipid peroxidation.

As previously noted, earlier studies had suggested that carcinogenesis might be modulated immunologically and that this influence might occur at the promotion stage (Malmsten, 1984). Indeed, the systemic alteration induced by UVR that suppresses an animal's ability to reject highly antigenic UVR-induced tumor transplants occurs during the chemical-promotion stage of carcinogenesis (Strickland *et al.*, 1985). It had already been shown that feeding mice an EFA (ω-6 FA) deficient diet inhibited the appearance of UVR-induced tumors and it was then suggested that this inhibition might be due to the lack of eicosanoid precursors that might, in turn, prevent UVR induction of the immune suppressed state. This could account for the protection observed from UVR-initiated tumor outgrowth (Reeve *et al.*, 1996). It had been shown that suppressor T-cell function was PGE_2 dependent and that UVR-induced suppression of CHS was abrogated by treatment with an inhibitor of PG synthesis (Chung *et al.*, 1986). Subsequent studies showed that plasma PGE_2 levels were directly related to the level of ω-6 FA intake, with the highest level of PGE_2 occurring with the highest level of ω-6 FA intake that, in turn, induced the greatest exacerbation of UVR carcinogenic expression (Fischer and Black, 1991). ω-3 FA reduced the plasma PGE_2 level below that exhibited by the lowest level of ω-6 FA intake. More importantly, ω-3 FA provided striking protection against UVR-induced immunosuppression (Fischer and

Black, 1991; Moison and Beijersbergen van Henegouwen, 2001). DTH and CHS are regulated by T-cell function and share common pathways with immunological tumor rejection. DTH in UVR-irradiated animals is dramatically suppressed in animals fed high levels of ω-6 FA when compared to those receiving low levels of ω-6 FA or those receiving ω-3 FA (Black *et al.*, 1995; Fischer and Black, 1991). The ability of an animal to reject a transplanted tumor was related to the level of ω-6 FA intake. Indeed, the tumor rejection time for animals fed high levels of ω-6 FA was three times longer than animals fed low levels of ω-6 FA and occurred at a time when high ω-6 FA had been shown to exacerbate primary tumor expression (Black *et al.*, 1995).

In summary, these studies suggest that one potential mechanism of ω-3 FA inhibition of carcinogenic expression is via immune modulation. ω-3 FA, as a substrate analogue, is an effective competitor of ω-6 FA and inhibits synthesis of 2-series PG through the cyclooxygenase pathway. Some immune pathways, e.g. CHS, DTH, and transplant rejection, are known to be PGE_2 dependent. By lowering the PGE_2 levels, ω-3 FA effectively protects against UVR-induced immune suppression.

22.2.2 Clinical and human cell culture studies

Unlike the experimental animal studies that have shown a strong influence of dietary fat on UVR-carcinogenesis, epidemiological studies have failed to provide consistent evidence for a link between level or type of fat intake and NMSC occurrence in humans.

However, a low-fat dietary intervention trial clearly demonstrated that level of fat intake (as percent of calories consumed as fat) exerted significant influence on actinic keratosis and NMSC occurrence in skin cancer patients (Black *et al.*, 1994; Jaax *et al.*, 1997). The dietary parameters involved only a reduction in the calories consumed as fat, while maintaining total calorie intake and body weight. Efforts were made to maintain the P/S ratio and there were no significant increases in ω-3 FA intake. Thus, the influence of fat in this study was primarily that resulting from lowering fat intake, primarily ω-6 FA. The parameters of the experimental design of this tightly controlled study have been discussed in relation to NMSC prevention and provides a framework for an intervention trial with ω-3 FA (Black, 2005; Black and Rhodes, 2006).

While lower intake of ω-6 FA reduces the risk of NMSC occurrence in skin cancer patients, a population-based case control study showed a consistent tendency toward a lower risk of SCC with higher intakes of ω-3 FA (Hakim *et al.*, 2000). Importantly, a tendency toward decreased risk of SCC with increased intake of diets with high ω-3 to ω-6 FA ratios was also found. Whereas this study was suggestive that ω-3 FA could influence NMSC risk, a number of human studies have provided a physiologic rationale to support such an hypothesis. Animal studies had shown that ω-3 FA strongly inhibited inflammatory responses, e.g. erythema and edema. Encouraged by these results, a short term supplementation study of mixed ω-3 FA was conducted in humans (Orengo *et al.*, 1992). The patients received either 4 g/day of mixed ω-3 FA (2.8 g EPA + 1.2 g DHA) or a gelatin placebo. After four weeks there was a small, but significant, increase in the MED (equivalent to a SPF of 1.15). Triglyceride levels had decreased by 40 mg/dl.

A second study was undertaken to examine the effect of ω-3 FA supplementation upon UVB-induced erythema and lipid peroxidation (Rhodes *et al.*, 1994). This study employed 3 g/day of mixed ω-3 FA (1.8 g EPA + 1.2 g DHA) administered over a 3-6 month period. The MED to UVB rose progressively with increasing time of ω-3 FA supplementation and had more than *doubled* at 6 months. The increase in MED was accompanied by an increase in epidermal ω-3 FA composition and increased susceptibility to lipid peroxidation. The MED returned to baseline two and a half months after cessation of ω-3 FA supplementation.

A number of cytokines and PG have been shown to be modulated by ω-3 FA. When human keratinocytes were cultured in the presence of ω-3 FA, TNF-α and IL-1α secretion was induced and PGE_2 and IL-6 levels reduced (Pupe *et al.*, 2002). Subsequent studies revealed that ω-3 FA inhibited IL-8, a chemokine pivital to UVB-induced skin inflammation and which exhibits pro-carcinogenic activity (Storey *et al.*, 2005).

A double-blind, randomised study of 28 patients supplemented with 4 g/day of 95% ethyl esters of EPA or oleic acid for three months found no evidence that the MED response evoked by the ω-3 FA was mediated by the pro-inflammatory cytokines IL-8, TNF-α, IL-6 or IL-1β. There was, however, a marked and significant reduction in PGE_2, a pro-inflammatory and immune suppressive mediator (Shahbakhti *et al.*, 2004).

Whereas ω-3 FA protect against the clinical sunburn response, there was no evidence of an effect of EPA on direct UVR-induced DNA damage, e.g. cyclobutane thymine dimer formation or upon basal oxidative damage to DNA. However, there was protection against the UVR-induction of p53. The latter is considered to be a biomarker of DNA damage and which acts as a tumor suppressor gene (Rhodes *et al.*, 2003). Further, in *ex vivo* UVR-treated peripheral blood lymphocytes, ω-3 FA protected against single strand breaks. The influence of dietary ω-3 FA on early genotoxic markers could predict a long term reduction in NMSC in humans. Indeed, based upon age-adjusted cancer incidence/UVR exposure plots, an ω-3 FA enhanced SPF of the magnitude reported could significantly reduce NMSC incidence by as much as 30%.

22.2.3 Future directions

Thus far, observational studies have failed to provide clear evidence that dietary ω-3 FA reduce the risk for NMSC. Nevertheless, the direct evidence from animal studies and the promising evidence from cell culture and clinical studies make it important that the potential of ω-3 FA as a preventive agent to NMSC be explored. The most direct way to address this issue is through intervention trials in populations with high, and known, risk for NMSC. It has been proposed that a study design be adopted similar to that in which a reduction in the percentage of calories consumed as fat was shown to reduce NMSC occurrence in NMSC patients (Black and Rhodes, 2006). Some caveats should be considered when designing such as study. First, it should be noted that in animal studies, in which ω-6 FA exacerbated or ω-3 FA inhibited UVR carcinogenic expression, these were the sole sources of dietary fat in the animals' diet. This will not be the case for a human study. The relative ω-6 FA/ω-3 FA ratios will determine response. Thus, any intervention trial must also carefully monitor the diets of study participants to assure that any potential beneficial effect of ω-3 FA supplementation is not being diminished by increased ω-6 FA intake. Dietary assessment and protocol compliance should be indexed to ω-6 FA/ω-3 FA that can be monitored to an easily determined parameter such as red blood cell membrane ω-6 FA/ω-3 FA ratios.

As NMSC is known to have a relatively long latent period, NMSC occurrence in any short-term study period is assumed to have arisen from previously transformed (initiated) cells and outgrowth of tumors is related to post-initiation, or promotion, events, e.g. immune surveillance. Although the animal studies suggest that, unlike ω-6 FA that exert their principal effect during the promotion stage of carcinogenesis, ω-3 FA exert their effect across the entire carcinogenic continuum. Considering that tumor outgrowth is related to the action of immunomodulatory eicosanoids, an intervention in skin cancer patients (known to have a high risk for subsequent NMSC development) with a two-year follow-up should provide a definitive answer of whether ω-3 FA exerts beneficial effects on NMSC occurrence. In this regard, it is important that an adequate level of supplementation be employed. ω-3 FA, being EFA, have a high safety profile and supplementation with 4-5 g/day of mixed ω-3 FA should be adequate.

22.3 Applications to other areas of skin health, care, and treatments

ω-3 FA influence acute UVR-induced responses in animals, e.g. ODC-induction and edema (Table 22.1). The relationship of these responses to carcinogenesis remains speculative as low-levels of ω-3 FA intake significantly inhibited carcinogenesis while producing no observable effect upon these acute responses. Only at higher levels of ω-3 FA intake were significant effects on these acute responses observed. It was suggested that inflammation and ODC-induction might simply be manifestations of accompanying, but not requisite, ω-3 FA dose-dependent responses (Black and Rhodes, 2006).

Evidence that inflammation is not necessarily a requisite event for UVR-carcinogenesis comes from studies of ω-3 FA influence on PUVA-induced tumorigenesis (Yen *et al.*, 1994). Dietary ω-3 FA resulted in a dramatic decrease in inflammatory response and a rapid repair of PUVA-toxicity. Yet, there was no inhibition of PUVA-tumorigenesis. Whereas, there was no significant effect of ω-3 FA on tumor latent period, at week 40 of the experiment there was a significant *increase* in tumor multiplicity when compared to animals receiving an equivalent level of ω-6 FA. Cutaneous photoxic reactions to various photosensitizers are mediated by the release of inflammatory agents from dermal mast cells and the generation of AA (ω-6 FA) metabolites (Black, 1989).

PUVA tumorigenesis does not exhibit time-dose reciprocity, i.e. cumulative dose, *per se*, is not an accurate index of tumorigenic risk. Because of the dramatic reduction of PUVA toxicity by ω-3 FA, it was suggested that ω-3 FA supplementation might be used in conjunction with PUVA therapy. More aggressive PUVA might therefore be tolerated by ameliorating PUVA toxicity.

Taking advantage of the non-reciprocative nature of PUVA tumorigenesis, this adjunct therapy might reduce the risk of skin cancer resulting from PUVA therapy alone.

As previously reported, ω-3 FA supplementation that led to decreased sensitivity to UVB was accompanied by a significantly increased threshold to UVA-provocation of a photosensitivity disorder, polymorphic light eruption. Another photosensitivity condition, hydroa vacciniforme was partially ameliorated with mixed ω-3 FA supplementation (Rhodes and White, 1998). Thus, it is possible that ω-3 FA supplementation may have a beneficial effect on a range of photosensitivity disorders.

References

Bang, H. and Dyerberg, J., 1972. Plasma lipids and lipoprotein pattern in Greenlandic west-coast Eskimos. Acta Medical Scandinavia 192, 85-94.

Baumann, C. and Rusch, H., 1939. Effect of diet on tumors induced by ultraviolet light. American Journal of Cancer 35, 213-221.

Black, H., 1989. Role of reactive oxygen species in inflammatory process. In: Lowe, N.J. and Hensby C.N. (eds.), Nonsteroidal anti-inflammatory drugs. Karger, Basel, Germany, pp. 1-20.

Black, H., 2005. Can diet prevent nonmelanoma skin cancer progression? Expert Review of Anticancer Therapy 5, 801-808.

Black, H., Herd, J., Goldberg, L., Wolf, J., Thornby, J., Rosen, T., Bruce, S., Tschen, J., Foreyt, J., Scott, L., Jaax, S. and Andrews, K., 1994. Effect of a low-fat diet on the incidence of actinic keratosis. New England Journal of Medicine 330, 1272-1275.

Black, H., Lenger, W., Gerguis, J. and Thornby, J., 1985. Relation of antioxidants and level of dietary lipid to epidermal lipid peroxidation and ultraviolet carcinogenesis. Cancer Research 45, 6254-6259.

Black, H., Lenger, W., Phelps, A. and Thornby, J., 1983. Influence of dietary lipid upon ultraviolet light-carcinogenesis. Nutrition and Cancer 5, 59-68.

Black, H. and Rhodes, L., 2006.The potential of omega-3 fatty acids in the prevention of non-melanoma skin cancer. Cancer Detection and Prevention 30, 224-232.

Black, H., Thornby, J., Gerguis, J. and Lenger, W., 1992. Influence of dietary omega-6, -3 fatty acid sources on the initiation and promotion stages of photocarcinogenesis. Photochemistry and Photobiology 56, 195-199.

Breslow, J., 2006. N-3 fatty acids and cardiovascular disease. American Journal of Clinical Nutrition 83, 1477S-1482S.

Chung, H., Burnham, D., Robertson, B., Roberts, L. and Daynes, R., 1986. Involvement of prostaglandins in the immune alterations caused by the exposure of mice to ultraviolet radiation. Journal of Immunology 137, 2478-2884.

Dyerberg, J. and Bang, H., 1978. Eicosapentaenoic acid and prevention of thrombosis and atherosclerosis? Lancet 15, 117-119.

Elmets, C., Viner, J., Pentland, A., Cantrell, W., Lin, H., Bailey, H., Kang, S., Linden, K., Hefferman, M., Duvic, M., Richmond, E., Elewski, B., Umar, A., Bell, W. and Gordon G., 2010. Chemoprevention of nonmelanoma skin cancer with Celecoxib: A randomized, double-blind, placebo-controlled trial. Journal of the National Cancer Institute 102, 1-10.

Fischer, M. and Black, H., 1991. Modification of membrane composition, eicosanoid metabolism, and immunoresponsiveness by dietary omega-3 and omega-6 fatty acid sources, modulators of ultraviolet-carcinogenesis. Photochemistry and Photobiology 54, 381-387.

Grau, M., Baron, J., Langholz, B., Karagas, M., Greenberg, R., Stukel, T. and Mandel, J., 2006. Effect of NSAIDs on the recurrence of nonmelanoma skin cancer. International Journal of Cancer 119, 682-686.

Hakim, I., Harris, R. and Ritenbaugh, C., 2000. Fat intake and risk of squamous cell carcinoma of the skin. Nutrition and Cancer 36, 155-162.

Harris, W, Kris-Etherton, P. and Harris, K., 2008. Intakes of long-chain omega-3 fatty acid associated with reduced risk for death from coronary heart disease in healthy adults. Current Atherosclerosis Reports 10, 503-509.

Jaax, S., Scott, L., Wolf, J., Thornby, J. and Black, H., 1997. General guidelines for a low-fat diet effective in the management and prevention of nonmelanoma skin cancer. Nutrition and Cancer 27, 150-156.

Lees, R., 1990. Impact of dietary fat on human health. In: Lees, R. and Karel, M. (eds.), Omega-3 fatty acids in health and disease. Marcel Dekker, Inc., New York, NY, USA, pp. 1-38.

Lowe, N.J. and Hensby, C.N. (eds.), 1989. Nonsteroidal anti-inflammatory drugs. Karger, Basel, Germany, pp. 1-20.

Malmsten, C., 1984. Leukotrienes: mediators of inflammation and immediate hypersensitivity reactions. Critical Reviews in Immunology 4, 307-334.

Moison, R. and Beijersbergen Van Henegouwen, G., 2001. Dietary eicosapentaenoic acid prevents systemic immunosuppression in mice induced by UVB radiation. Radiation Research 156, 36-44.

Orengo, I., Black, H., Kettler, A. and Wolf Jr., J., 1989. Influence of dietary menhaden oil upon photocarcinogenesis and various cutaneous responses to ultraviolet radiation. Photochemistry and Photobiology 49, 71-77.

Orengo, I., Black, H. and Wolf Jr., J., 1992. Influence of fish oil supplementation on the minimal erythema dose in humans. Archives of Dermatology Research 284, 219-221.

Orengo, I., Gerguis, J., Phillips, R., Guevara, A., Lewis, A. and Black, H., 2002. Celecoxib, a cyclooxygenase-2 inhibitor as a potential chemopreventive to UV induced skin cancer: A study in the hairless mouse model. Archives of Dermatology 138, 751-755.

Pupe, A., Moison, R., De Haes, P., Beijersbergen van Henegouwen, G., Rhodes, L., Degreef H. and Garmyn, M., 2002. Eicosapentaenoic acid, a n-3 polyunsaturated fatty acid differentially modulates TNF-α, IL-1α, IL-6 and PGE_2 expression in UVB-irradiated human keratinocytes. Journal of Investigative Dermatology 118, 692-698.

Reeve, V., Bosnic, M. and Boehm-Wilcox, C., 1996. Dependence of photocarcinogenesis and photoimmunosuppression in the hairless mouse on dietary polyunsaturated fat. Cancer Letters 108, 271-279.

Rhodes, L., O'Farrell, S., Jackson, M. and Friedmann, P., 1994. Dietary fish-oil supplementation in humans reduces UVB-erythemal sensitivity but increases epidermal lipid peroxidation. Journal of Investigative Dermatology 103, 151-154.

Rhodes, L., Shahbakhti, H., Azurdia, R., Moison, R., Steenwinkel, M-JST., Homburg, M., Dean, M., McArdle, F., Beijersbergen van Henegouwen, G., Epe, B. and Vink, A., 2003. Effect of eicosapentaenoic acid, an omega-3 polyunsaturated fatty acid, on UVR-related cancer risk in humans. An assessment of early genotoxic markers. Carcinogenesis 24, 919-925.

Rhodes, L. and White, S., 1998. Dietary fish oil as a photoprotective agent in hydroa vacciniforme. British Journal of Dermatology 138, 173-178.

Shahbakhti, H., Watson, R., Azurdia, R., Ferreira, C., Garmyn, M. and Rhodes, L., 2004. Influence of eicosapentaenoic acid, an omega-3 fatty acid, on ultraviolet-B generation of prostaglandin-E2 and proinflammatory cytokines interleukin-1 beta, tumor necrosis factor-alpha, interleukin-6 and interleukin-8 in human skin *in vivo*. Photochemistry and Photobiology 80, 231-235.

Storey, A., McArdle, F., Friedmann, P., Jackson, M. and Rhodes, L., 2005. Eicosapentaenoic acid and docosahexaenoic acid reduce UVB- and TNF-alpha-induced IL-8 secretion in keratinocytes and UVB-induced IL-8 in fibroblasts. Journal of Investigative Dermatology 124, 248-255.

Strickland, P., Creasia, D. and Kripke, M., 1985. Enhancement of two-stage skin carcinogenesis by exposure of distant skin to UV-radiation. Journal of the National Cancer Institute 74, 1129-1134.

Vanderveen, E., Grekin, R., Swanson, N. and Kragballe, K.,1986. Arachidonic acid metabolites in cutaneous carcinomas. Evidence suggesting that elevated levels of prostaglandins in basal cell carcinomas are associated with an aggressive growth pattern. Archives of Dermatology 122, 407-412.

Werner, E., Walenga, R., Dubowy, R., Boone, S. and Stuart, M., 1985. Inhibition of human malignant neuroblastoma cell DNA synthesis by lipoxygenase metabolites of arachidonic acid. Cancer Research 45, 561-563.

Yen, A., Black, H. and Tschen, J., 1994. Effect of dietary omega-3 and omega-6 fatty acid sources on PUVA-induced cutaneous toxicity and tumorigenesis in the hairless mouse. Archives of Dermatology Research 286, 331-336.

Key facts

- Skin cancer is the most common cancer in the Caucasian population in the Western world.
- Individual risk for skin cancer is due to a combination of risk factors. Well known risk factors include ultraviolet exposure and skin type.
- Nutrition factors in the pathogenesis of skin cancer are less well characterized.
- Overall evidence suggests that a low fat diet reduces the occurrence of nonmelanoma skin cancer (NMSC) and that moderate intake of wine and oily fish reduce the acquisition of actinic keratoses (premalignant skin lesions).
- Adequate cellular folate depends on the competence of the folate metabolic pathway.
- Alterations in the folate pathway are associated with cutaneous squamous cell carcinoma in the form of genetic variation in folate metabolism and also functional alterations in methylation status of squamous cell carcinoma (SCC).

Summary points

- Folate is a nutrient that occurs naturally in food as a water-soluble B vitamin. It occurs in its natural form in leafy greens such as spinach, broccoli, peas and fermented dairy products and synthetically in supplements and fortified foods as folic acid.
- Squamous cell carcinoma is the predominant type of skin cancer in renal transplant patients and these patients have a 65-250 fold increased risk of development of SCC compared to the non transplanted community.
- Folate is an essential nutrient crucial for deoxyribonucleic acid (DNA) methylation, synthesis and repair.
- Folate is necessary for the regeneration of methionine, the precursor of *S*-adenosyl-l-methionine (SAM), which is the methyl donor in the methylation of DNA.
- Aberrant DNA methylation occurs in cancer cells and has an important role in tumorigenesis.
- Aberrant DNA methylation occurs in two ways: global DNA hypomethylation and gene specific hypermethylation.
- The methylenetetrahydrofolate reductase (MTHFR) enzyme is important in the methylation pathway and alterations in this enzyme cause aberrant methylation which is important in oncogenesis.
- A genetic variant in the methylation and folate metabolic pathway (the MTHFR C677T polymorphism) was found to confer risk for the development of skin cancer in renal transplant patients.
- The MTHFR variant is also associated with aberrant methylation in tumors from patients with the MTHFR polymorphism compared to patients without the MTHFR polymorphism.
- There is evidence to show that folate metabolism and methylation are linked to the pathogenesis of skin cancer and this may indicate a role for intervention in the form of demethylating agents as therapeutic options or folate supplementation as prophylaxis.

23. Skin cancer and folate metabolism

M. Laing
Dermatology Specialist Registrar, Beaumont Hospital, P.O. Box 1297, Beaumont Road, Dublin 9, Ireland; mrylaing@yahoo.co.uk

Abstract

Skin cancer is the most common malignancy in the Caucasian population in the Western world. The incidence of the three major types of skin cancer – basal cell carcinoma, squamous cell carcinoma and malignant melanoma – continues to increase. Skin cancers are broadly divided into melanoma and nonmelanoma skin cancers. Melanoma accounts for 4% of all skin cancers however it is responsible for 80% of deaths from skin cancer. Non melanoma skin cancers are locally destructive and the expense of their removal is a significant economic burden to the healthcare system. Individual risk for skin cancer is due to a combination of risk factors. Well known risk factors include ultraviolet exposure and skin type. In transplant patients immunosuppression plays a major role and squamous cell carcinoma is an important cause of mortality in this group. In the following chapter the relevance of folate metabolism and skin cancer is discussed including our recent findings on the folate metabolic pathway and squamous cell carcinoma in renal transplant patients.

Keywords: skin cancer, folate, genetic risk, squamous cell carcinoma, methylation

Abbreviations

8-MOP	8-methoxypsoralen
AK	Actinic keratosis
BCC	Basal cell carcinoma
CT	Heterozygotes
DNA	Deoxyribonucleic acid
EBV	Epstein-Barr virus
ECP	Extracorporeal photopheresis
GvHD	Graft-versus-host disease
HPV	Human papilloma virus
MGMT	Methyl guanine DNA methyltransferase
MTHFR	Methylenetetrahydrofolate reductase
NMSC	Non-melanoma skin cancer
NTD	Neural tube defects
RTR	Renal transplant recipients
SAM	*S*-adenosyl-l-methionine
SCC	Squamous cell carcinoma
SNP	Single nucleotide polymorphism
TMECG	Trimethoxybenzoyl-(-)-epicatechin
UVA	Ultraviolet A
UVB	Ultraviolet B
UVR	Ultraviolet radiation

23.1 Introduction

UVR is the principle cause of skin cancer in Caucasians. The majority of SCCs (90%) occur at sun exposed sites. UVR in itself is a complete carcinogen which has the ability to initiate, promote and allow progression of skin cancers. Tumor initiation can occur by cumulative mutations in DNA induced by UVR. UV mutations in the p53 tumor suppressor gene convert it into an oncogene, which promotes mutated cell proliferation. UVR, predominantly UVB, results in the formation of cyclobutane pyrimidine dimers (thymine dimers) which are mutagenic DNA photoproducts. Skin cancer has become one of the most common causes of mortality in RTRs. These patients have an increased risk of all types of skin cancer including NMSC, melanoma(Laing *et al.*, 2006), Kaposi's sarcoma and Merkel cell carcinoma. NMSC, predominantly SCC and BCC account for 90% of all skin cancer in transplant patients (Bouwes Bavinck *et al.*, 1996). Squamous cell carcinoma is the predominant type of skin cancer in RTRs and these patients have a 65-250 fold increased risk of development of SCC compared to the nontransplanted community (Moloney *et al.*, 2006). Fifty per cent of patients with a first SCC develop a second SCC within 3-5 years. SCC in organ transplant patients has a more aggressive clinical course, a greater tendency to metastasize and a higher mortality compared to that of the general population (Martinez *et al.*, 2003). Established risk factors for skin cancer in these patients include skin type, sun exposure and immmunosupppressive therapy (Figure 23.1). Such risk factors are well described in the literature. They do not adequately explain however why some transplant patients develop NMSC which develop early, may be very numerous or behave aggressively, whilst others with equivalent recognized risk factors do not. Individual risk in transplant patients is a result of the interaction of a combination of genetic factors, epigenetic factors, immunosuppression and environmental exposure to carcinogens and co-carcinogens such as UV irradiation and HPV. Nutrition factors in the pathogenesis of skin cancer are less well characterized. Overall evidence suggests that a low fat diet reduces the occurrence of NMSC (Black *et al.*, 1994) and that moderate intake of wine and oily fish reduce the acquisition of actinic keratoses (premalignant skin lesions) (Hughes *et al.*, 2009). Therefore, cutaneous carcinogenesis is multifaceted with many intertwined risk factors.

23.2 Folate and UV

Folate is known to be sensitive to UVR (Off *et al.*, 2005). Folate can be readily degraded by natural sunlight or UVR (Branda and Eaton, 1978). Photolysis of folate has been demonstrated by a significant decline in folate levels when serum (*in vitro*) (Cohn, 2002) of light-skinned subjects (*in vivo*) were exposed to ultraviolet radiation. PUVA therapy refers to the use of a photosensitising agent (i.e. an agent which reacts with light to produce biological effects) or psoralen taken orally or applied topically before exposure to UVA light (320-400 nm). Branda and Eaton demonstrated that post-pUVA exposure, serum levels of folate were significantly lower than the serum folate levels of 64 light-skinned persons who had not had any PUVA therapy (Branda and Eaton, 1978). This study suggests that increased UVA exposure leads to reduced folate levels. Folate depletion is an UVA effect. ECP is a form of phototherapy where plasma is directly exposed to UVA. ECP is a well-established treatment modality for different disorders including Sézary syndrome (a form of cutaneous T-cell lymphoma), systemic sclerosis, and GvHD. During ECP a fraction of leucocytes and plasma is subjected extracorporeally to 8-MOP plus UVA and given back to the patient. A recent study has shown that ECP leads to photodegradation of folic acid

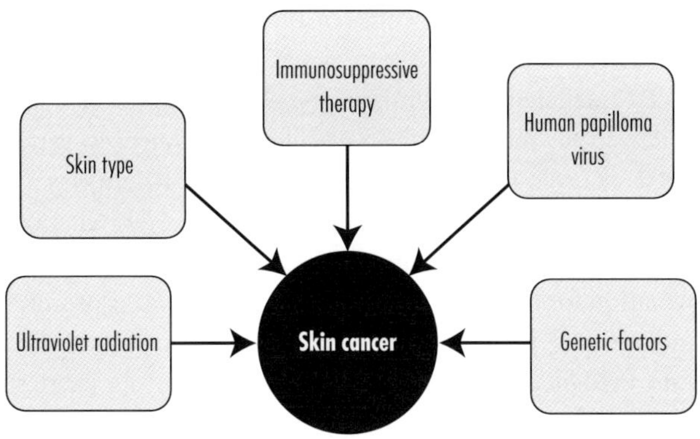

Figure 23.1 Pathogenesis of skin cancer in organ transplant recipients.

(Der-Petrossian *et al.*, 2007). The important association between folate deficiency and NTDs is widely appreciated. A causal relationship between *in vivo* photolysis in humans and NTDs, first suggested by Jablonski and colleagues (Jablonski, 1992, 1999) has recently been supported by the report of NTDs in three unrelated infants whose mothers had spent time in tanning salons (mainly UVA devices) during the first weeks of their pregnancies (Lapunzina, 1996). In addition, experimental UVR exposure leading to NTDs in some amphibian populations has been shown by Jablonski (Jablonski, 1998). A recent *in vitro* study has found strong indications that folate is photodegraded when exposed to UVR. The photodegradation is divided into three phases and results in the formation of photoproducts including p-aminobenzoyl-l-glutamic acid, 6-formylpterin and pterin-6-carboxylic acid. More recent evidence suggests that high doses of broadband UVB phototherapy may slightly decrease blood folates in psoriasis patients. Another study showed that low-dose nUVB treatment showed no significant decrease in the folate level in psoriasis patients. Human studies on folate photodegradation are difficult to carry out because many parameters have to be taken into account, including the diet and food fortification. The cumulative effect of UV-induced DNA damage coupled with UV-induced folate degradation might be responsible for aberrant methylation of DNA and its association with skin cancer. This is a new hypothesis for a further carcinogenic effect for NMSC in Caucasian skin.

23.3 Folate and cancer

Folate is essential for nucleotide synthesis, DNA replication, and as a methyl group donor. Low-folate status has been associated with increased risks of several cancer types including cancer of the colon, lung, oesophagus, cervix, breast and pancreas (Kim, 2004), suggesting a chemopreventive role of folate. It is now well recognized that taking folic acid supplements before pregnancy and in the first 12 weeks can reduce the risk of neural tube defects such as spina bifida. In recent years the US, Canada, Chile, the Czech Republic and Australia have all introduced mandatory fortification of flour with folic acid. Despite the fact that the amount of folic acid added into the food supply does not provide enough to fully protect unborn children, recent reports from Canada and the USA indicate fortification of cereal grains has been very effective, with reduction in neural tube defects of between 27% and 50%. There is concern that supraphysiological doses of folic acid may enhance the development of undiagnosed pre-malignant and malignant lesions and that it may also mask vitamin B12 deficiency anaemia in the elderly population. Results from the first randomized trial of folic acid for the prevention of colorectal cancer in genetically predisposed patients (Cole *et al.*, 2007) showed that at the second follow-up, there was a 67% increased risk of advanced lesions with a high malignant potential (RR: 1.67; 95% CI: 1.00, 2.80), along with a >2-fold increased risk of having ≥3 adenomas (RR: 2.32; 95% CI: 1.23, 4.35) in the folic acid group. A possible explanation for this result is that folic acid may have promoted the progression of already existing, undiagnosed preneoplastic lesions (e.g. aberrant crypt foci or microscopic adenomas). There is no evidence to date to demonstrate that folic acid supplementation is chemopreventive in skin cancer development.

23.4 Folate, methylation and MTHFR

Folate is a water soluble essential B vitamin. It occurs in its natural form in leafy greens such as spinach, broccoli, peas and fermented dairy products. It exists in synthetic form as folic acid which is a nutritional supplement and used in food fortification. Folate deficiency is well known to cause neural tube defects and macrocytic anaemia. Adequate cellular folate depends on the competence of the folate metabolic pathway. The rate limiting enzyme on which folate levels depend is MTHFR. MTHFR catalyses the irreversible conversion of 5, 10 methylenetetrahydrofolate (5, 10 methylene THF), the central metabolite for folate metabolism, to 5-methyltetrahydrofolate (5-methyl THF), the main circulating form of folate, which serves both as a cofactor and a substrate for regeneration of methionine (Figure 23.2). Folate metabolism is central to DNA methylation and MTHFR is a rate limiting enzyme for this process. Alterations in this enzyme may result in hyper or hypomethylation. Aberrant DNA methyaltion is a common event in carcinogenesis. Aberrant DNA methylation includes global DNA hypomethyaltion and gene specific hypermethylation. Hypermethylation of promoters of tumor suppressor genes – causes

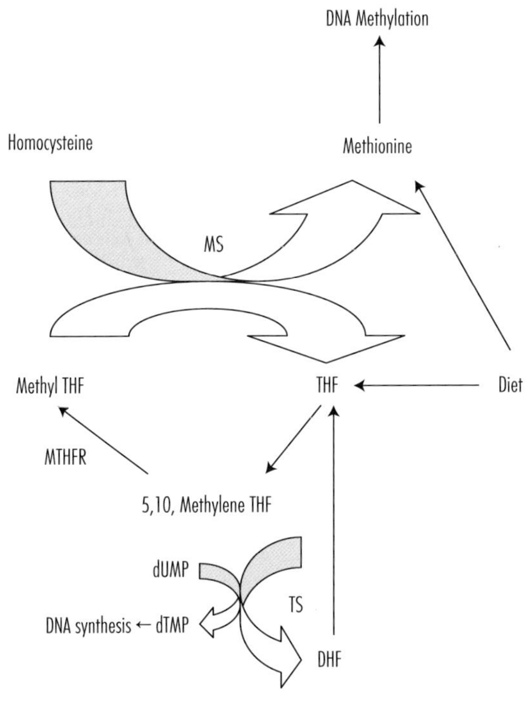

Figure 23.2. Methylenetetrahydrofolate reductase (MTHFR) and methylation pathway. Folate metabolism is central to DNA methylation and MTHFR is a rate limiting enzyme for this process. Alterations in this enzyme may result in hyper or hypomethylation. DHF: dihydrofolate, THF: tetrahydrofolate; MS: methionine synthase, TS: thymidylate synthase, dUMP: deoxyuridine monophosphate (deoxyuridylate), dTMP: deoxythymidine monophosphate (deoxythymidylate).

silencing of these genes. Methylation has been shown to be important in cutaneous squamous cell carcinoma. Promotors of tumor suppressor genes p14 ARF and p16 – encoded by CDKN2A gene – are hypermethylated in SCC (Table 23.1). The E-cadherin gene, which functions in cell adhesion is most frequently hypermethylated in SCC. Therefore methylation is an essential function of the folic acid metabolic pathway, important in switching on and off genes.

23.5 MTHFR C677 T and squamous cell carcinoma in renal transplant patients

The gene for MTHFR is located on the shot arm of chromosome 1. We studied variation in this gene in the form of genetic polymorphisms which are the most common form of genetic variation accounting for 90% of human genetic variation. Single nucleotide polymorphisms result in substitution of a single base for another in the DNA sequence. Some of these are functional polymorphisms leading to alteration of proteins and metabolic pathways. One polymorphism of the MTHFR gene was identified at nucleotides position 677 on chromosome 1 with a C-T substitution, which leads to amino acid change of alanine to valine and significantly reduced enzyme activity (Frosst *et al.*, 1995). Reduced enzyme activity results in insufficient synthesis of 5-methyl THF, the active form of tissue folate and hypomethylation of genomic DNA (Shellnut *et al.*, 2004; Stern *et al.*, 2000). MTHFR C677T (also known as rs1801133) is a functional polymorphism in the MTHFR gene. Individuals with the 677TT genotype (variant homozygotes) have no more than 30% of normal enzyme activity, and heterozygotes have 65% of normal enzyme activity (Kono and Chen, 2005). This polymorphism is associated with major pathologies ranging from neural tube defects to cardiovascular diseases. MTHFR C677T polymorphism is also implicated in colorectal cancer, pancreatic cancer, endometrial cancer, gastric cancer and leukaemia. It is associated with hypermethylation of CDKN2A gene in colorectal carcinoma. It has been shown to be associated with increased homocysteine and lower folate levels in renal transplant patients. Whether this functional polymorphism was relevant to the development of skin cancer in transplant patients was previously unknown.

Table 23.1. Risk factors for squamous cell carcinoma (SCC) in renal transplant recipients (RTR).

Multivariate analysis				
Exposure (factor)	**P-value**	**Odds ratio**	**Lower 95% CI**	**Upper 95% CI**
Male gender	0.9700	0.99	0.55	1.80
Age >50 years	<0.0001	7.47	3.94	14.16
Time since transplant ≥8 years	<0.0001	3.31	1.84	5.97
Skin type	0.0262	1.47	1.05	2.06
Sun exposure score >8	0.0069	2.45	1.28	4.70
MTHFR_677T carrier	0.0019	2.54	1.41	4.56

We therefore examined this polymorphism to determine risk for squamous cell carcinoma in a group of renal transplant patients. We genotyped patients for the MTHFR polymorphism amd examined risk of squamous cell carcinoma. We found an association between the MTFHR polymorphisnm and squamous cell carcinoma which was of the order of magnitude of some of the known risk factors, including sun exposure with an odds ratio of 2.34 (Table 23.1). In addition to an association with SCC we also found a relationship between the MTHFR polymorphism and timing of onset, multiplicity and time to subsequent tumors in renal transplant recipients (Laing *et al.*, 2007). We reported an association between this genetic variant in the folate pathway and squamous cell carcinoma. In the same year our finding was replicated in the non transplant Boston population with low folate (Han *et al.*, 2007). Therefore two studies showed an association with a genetic variant in the folate pathway and skin cancer. Our next study determined the functional link between this variation and skin cancer in relation to DNA methylation.

23.6 MTHFR linked to methylation in squamous cell carcinoma

DNA methylation is a naturally occurring event in both prokaryotic and eukaryotic organisms. In prokaryotes DNA methylation provides a way to protect host DNA from digestion by restriction endonucleases that are designed to eliminate foreign DNA, and in higher eukaryotes DNA methylation functions in the regulation of gene expression. Aberrant DNA methylation is a widespread phenomenon in cancer and may be one of the earliest changes to occur during oncogenesis. Our aim in this study was firstly to determine the methylation profile of SCC compared to adjacent non-neoplastic skin. Secondly we aimed to elucidate whether the MTHFR polymorphism, which we previously found to be associated with SCC, impacts upon the methylation patterns in SCC (Table 23.2). To do this we examined the methylation status in skin tumors and non neoplastic skin biopsies from patients with and without the MTHFR polymorphism. The MTHFR polymorphism has been linked with hypermethylation of methyl

Table 23.2. Genes that exhibit hypermethylation in cutaneous squamous cell carcinoma.

Gene	Function
Cyclin dependant kinase inhibitor 2 (p14)	negative regulation of cell cycle
Cyclin dependant kinase inhibitor 2 (p16)	negative regulation of cell cycle
E-Cadherin	cell adhesion
T-Cadherin	cell adhesion
Insulin like growth factor binding protein 3	insulin like growth factor receptor signaling pathway
Death associated protein kinase	pro-apoptotic kinase
Methyl guanine DNA methyltransferase	DNA repair
RB1	negative regulation of cell cycle
RASSF1	negative regulation of mitosis

guanine DNA MGMT, a gene essential for DNA repair, in squamous cell carcinoma of the oesophagus. The tumor suppressor gene p16 has been shown to be hypermethylated in SCC. We therefore also examined these genes with respect to methylation in our group.

Our first novel finding, was that SCC is hypomethylated in renal transplant patients compared with adjacent non neoplastic skin (Laing *et al.*, 2010). Global hypomethylation has been shown to be associated with melanoma, head and neck SCC , neuroendocrine tumors, colorectal cancer and tumors of liver, kidney and prostate. Global hypomethylation is also associated with tumors of the lung, breast, oesophagus and stomach. The global methylation status of cutaneous SCC however had not been hitherto studied. The main oncogenic consequence of global hypomethylation appears to be genomic instability. It has been shown that a high degree of chromosomal instability at 13q14.2-13q14.3 exists in cutaneous SCCs of renal transplant recipients. Global hypomethylation is also thought to increase expression of proto-oncogenes and transposable elements, both of which are oncogenic. In mice it has been shown that genome wide DNA hypomethylation is sufficient to induce T-cell lymphomas. Hypomethylation of latent viral sequences is thought to allow their re-expression and consequent tumor progression. For example, cervical cancer latency has been linked to hypermethylation of the HPV genome, and activation of the HPV 16 genome has been linked to progressive hypomethylation that results in cervical dysplasia. EBV latency follows a similar pattern in EBV associated lymphoma. Perhaps hypomethylation in SCC of RTRs allows a similar activation of oncogenic subtypes of HPV which are thought to be co-carcinogenic in the aetiology of skin cancer in transplant patients. The presence of aberrant methylation in SCC may give potential for intervention with folic acid.

The second novel finding in this study was that patients with the MTHFR polymorphism had increased levels of methylation in both SCC and non neoplastic skin compared with those without the polymorphism. It has been shown that DNA methylation promotes UV radiation-induced DNA damage. Perhaps therefore in normal/UV exposed skin increased methylation predisposes to skin cancer. The presence of a methyl-group in a CpG dinucleotide is known to increase the rate at which C-T mutations are induced by ultraviolet radiation by shifting the absorption spectrum for cytosine. Cytosine methylation induces a red shift of the nucleoside absorption spectrum which renders it more vulnerable to solar irradiation. Cyclobutane pyrimidine dimers, which are mutagenic DNA photoproducts are induced by UV irradiation. Cyclobutane pyrimidine dimers form preferentially at dipyrimidine sequences with methylated cytosines. Methylation of CpG sequences can therefore create preferential targets for ultraviolet radiation-induced DNA damage. This enhancement has been demonstrated for CpG dinucleotides in the coding region of the p53 tumor suppressor gene during the development of skin cancer where a disproportionately higher number of mutations in p53 are found at methylated cytosines. Patients with MTHFR polymorphism have increased levels of methylation in tumor prone skin which may create preferential targets for UV radiation damage, delete tumor suppressor genes and enable the tumor to develop more readily. The tumors developing in these individuals also carry more methylated CpG sites and so this could be one mechanism whereby tumors in some renal transplant patients grow more rapidly. Of interest levels of methylation for MTHFR heterozygotes and homozygotes did not differ significantly which implies a lack of dose effect of the reduced MTHFR enzyme level on methylation.

23.7 Age and methylation

Epigenetic signalling including DNA methylation is essential for normal development and becomes altered during aging and by cancer. It has been shown that epigenetic changes such as global hypomethylation and CpG island hypermethylation, are progressively accumulated during aging and directly contribute to cell transformation. Both colon and gastric cancer exhibit global hypomethylation directly correlated with age. On analysis of over 150 random CpG loci by methylation-sensitive amplified fragment length polymorphism (MS-AFLP) in human colon and gastric carcinomas Suzuki *et al.* (2006) showed that DNA hypomethylation alterations increased with cancer patient age. Our study is the first study to demonstrate this finding in cutaneous SCC. It is proposed that the gradual and age-dependent increase in global DNA hypomethylation may increase the probability of occurrence of genomic alterations accompanying tumor development and progression. Age dependent progressive hypomethylation is thought to occur by accumulation of errors of DNA methylation replication by constitutive methyl transferases. Once genomic hypomethylation occurs, the risk of errors of chromosome segregation increases, leading to accumulation of genomic damage and eventual cancer development. Perhaps aberrant methylation in combination with other age induced risk mechanisms such as reduced DNA repair and reduced immunosurveillance is important in skin cancer development in advancing age. The relevance of methylation to these known risk factors has not yet been established. It is of interest that the MTHFR polymorphism was of most risk to the younger RTR with SCC thus other mechanisms may be relevant in older patients with increased hypomethylation. The patients with the MTHFR polymorphism were slightly younger than patients without the MTHFR polymorphism (TT/CT-61 years, CC-64 years) although this was not statistically significant and after age correction methylation differences were still significant. In summary our analysis revealed that SCC is hypomethylated compared with adjacent non neoplastic skin, and moreover, that skin samples from patients with the MTHFR polymorphism had higher levels of methylation in tumors and non neoplastic skin compared with those without the MTHFR polymorphism (Laing, 2010). The results from this study extend our previous findings that MTHFR polymorphism confers risk for skin cancer in a transplant population. We have shown a functional link between the MTHFR polymorphism and the development of squamous cell carcinoma. The exact mechanism of aberrant methylation in the development of skin cancer is unknown, however it may relate to creation of preferential targets for UV irradiation.

23.8 Folate and methylation in other skin diseases

23.8.1 Psoriasis

Prosiasis is a chronic inflammatory skin disease characterized by epidermal hyperproliferation. A mild folate deficiency in patients with psoriasis has been suspected for many years and was recently confirmed in two studies (Kural *et al.*, 2003; Malerba *et al.*, 2006). The mechanism of deficiency in patients with psoriasis is thought to be due to reduced absorbtion in the gastrointestinal tract or increased vitamin utilization in the skin due to rapid turnover rate of epidermal cells. Psoriasis patients on methtrexate, the folate antagonist, are routinely supplemented with folic acid. Psoriasis patients have an increased standardized mortality ratio for coronary heart disease (Mallbris *et al.*, 2004). There is also a higher prevalence of obesity and smoking habit among psoriatic patients. Increased plasma homocysteine concentration is thought to play a role in the development of atherothrombotic events in patients with psoriasis. Hyperhomocysteinemia is linked to folate deficiency. Folic acid is therefore a therapeutic option in patients with psoriasis who have concomitant hyperhomocysteinemia, low folate and additional risk factors for cardiovascular disease.

23.8.2 Oral atrophy

The oral mucosa undergoes atrophic changes in relation to folate deficiency. Among a large number of symptoms and clinical findings that were traditionally linked to B12 or folate deficiency, a recent study has shown that only changes in tongue mucosa and angular stomatitis were significantly associated with elevated homocysteine and low folate levels (Bjorkegren and Svardsudd, 2003). Changes in oral mucosa are found in early deficiency of folate and B12 indicating that these may be the very early markers of metabolic defects.

23.8.3 Melanoma

At present there is no evidence that folate is important in the pathogenesis or prevention of malignant melanoma. There is cellular evidence however that TMECG, which downregulates dihydrofolate reductase (DHFR) shows high antiproliferative activity against melanoma cells (Sánchez-del-Campo *et al.*, 2010). Recent work has shown that aberrant DNA methylation, lead to alterations in gene expression and are important in the development of malignant melanoma (Conway *et al.*, 2011). Demethylating agents including 5 azacytidine or 5 aza 2 deoxycytidine (decitabine) are used in the treatment of metastatic melanoma. A recent report has shown that a DNA-methylation signature discriminates melanomas from nevi indicating that DNA methylation may be an additional tool for enhancing melanoma diagnosis (Schinke *et al.*, 2010).

23.9 Conclusion

Well established risk factors for skin cancer include skin type, sun exposure and immunosuppression. These factors however do not adequately explain why some patients develop skin cancers early, which may be very numerous or behave aggressively, whilst others with equivalent recognized risk factors do not. In this chapter I have explored genetic and epigenetic mechanisms of the folate pathway in skin carcinogenesis in renal transplant patients. In summary our findings suggest intricate regulation of MTHFR in the pathogenesis of skin cancer and may indicate a role for intervention in the form of demethylating agents as therapeutic options or folate supplementation as prophylaxis. The subject of folate supplementation in the prevention of cancer is complex and we should take a lesson from recent randomized controlled trials in colon cancer. It is likely that folate supplementation for the prevention of skin cancer may be appropriate in patients with the MTHFR polymorphism however this requires confirmation in a well designed randomized controlled clinical trial setting.

References

Bjorkegren, K. and Svardsudd, K., 2003. Reported symptoms and clinical findings in relation to serum cobalamin, folate, methylmalonic acid and total homocysteine among elderly Swedes: a population-based study. Journal of Internal Medicine 254, 343-352.

Black, H.S., Herd, J.A., Goldberg, L.H., Wolf, J.E. Jr, Thornby, J.I., Rosen, T., Bruce, S., Tschen, J.A., Foreyt, J.P., Scott, L.W., Jaax, S. and Andrews, K., 1994. Effect of a low-fat diet on the incidence of actinic keratosis. New England Journal of Medicine 330, 1272-1275.

Bouwes Bavinck, J.N., Hardie, D.R., Green, A., Cutmore, S., MacNaught, A., O'Sullivan, B., Siskind, V., Van Der Woude, F.J. and Hardie, I.R., 1996. The risk of skin cancer in renal transplant recipients in Queensland, Australia. A follow-up study. Transplantation. 61, 715-721.

Branda, R.F. and Eaton, J.W., 1978. Skin color and nutrient photolysis: an evolutionary hypothesis. Science 201, 625-626.

Cohn, B.A., 2002. Sunlight, skin color, and folic acid. Journal of American Academy of Dermatology 46, 317-318.

Cole, B.F., Baron, J.A., Sandler, R.S., Haile, R.W., Ahnen, D.J., Bresalier, R.S., McKeown-Eyssen, G., Summers, R.W., Rothstein, R.I., Burke, C.A., Snover, D.C., Church, T.R., Allen, J.I., Robertson, D.J., Beck, G.J., Bond, J.H., Byers, T., Mandel, J.S., Mott, L.A., Pearson, L.H., Barry, E.L., Rees, J.R., Marcon, N., Saibil, F., Ueland, P.M., Greenberg, E.R. and Polyp Prevention Study Group, 2007. Folic acid for the prevention of colorectal adenomas: a randomized clinical trial. Journal of the American Medical Association 297, 2351-2359.

Conway, K., Edmiston, S.N., Khondker, Z.S., Groben, P.A., Zhou, X., Chu, H., Kuan, P.F., Hao, H., Carson, C., Berwick, M., Olilla, D.W. and Thomas, N.E., 2011. NEDNA-methylation profiling distinguishes malignant melanomas from benign nevi. Pigment Cell Melanoma Research 24, 352-360.

Der-Petrossian, M., Födinger, M., Knobler, R., Hönigsmann, H. and Trautinger, F., 2007. Photodegradation of folic acid during extracorporeal photopheresis. The British Journal of Dermatology 156, 117-121.

Frosst, P., Blom, H.J., Milos, R., Goyette, P., Sheppard, C.A., Matthews, R.G., Boers, G.J., den Heijer, M., Kluijtmans, L.A., Van den Heuvel, L.P. and Rozen, R., 1995. Candidate genetic risk factor for vascular disease: a common mutation in methylenetetrahydrofolate reductase. Nature Genetics 10, 111-113.

Han, J., Colditz, G.A. and Hunter, D.J., 2007. Polymorphisms in the MTHFR and VDR genes and skin cancer risk. Carcinogenesis. 28, 390-397.

Hughes, M.C., Williams, G.M., Fourtanier, A. and Green, A.C., 2009. Food intake, dietary patterns, and actinic keratoses of the skin: a longitudinal study. The American Journal of Clinical Nutrition 89, 1246-1255.

Jablonski, N.G., 1992. Sun, skin colour and spina bifida: an exploration of the relationship between ultraviolet light and neural tube defects. Proceedings of the Australasian Society for Human Biology 5, 455-462.

Jablonski, N.G., 1998. UV light induced neural tube defects in amphibian larvae and their implications for the evolution of melanized pigmentation and declines in amphibian populations. Journal of Herpetology 32, 455-457.

Jablonski, N.G., 1999. A possible link between neural tuble defects and ultraviolet light exposure. Medical Hypotheses 52, 581-582.

Kim, Y.I., 2004. Folate and DNA methylation: A mechanistic link between Folate Deficiency and Colorectal Cancer. Cancer Epidemiology and Biomarkers and Prevention 13, 511-519.

Kono, S. and Chen, K., 2005 Genetic polymorphisms of methylenetetrahydrofolate reductase and colorectal cancer and adenoma. Cancer Science 96, 535.

Laing, M.E., Cummins, R., O'Grady, A., O'Kelly, P., Kay, E.W. and Murphy, G.M., 2010. Aberrant DNA methylation associated with MTHFR C677T genetic polymorphism in cutaneous squamous cell carcinoma in renal transplant patients. British Journal of Dermatology 163, 345-352.

Laing, M.E., Dicker, P., Moloney, F.J., Ho, W.L., Murphy, G.M., Conlon, P., Whitehead, A.S. and Shields, D.C., 2007. Association of methylenetetrahydrofolate reductase polymorphism and the risk of squamous cell carcinoma in renal transplant patients. Transplantation 84, 113-136.

Laing, M.E., Moloney, F.J., Kay, E.W., Conlon, P. and Murphy, G.M., 2006. Malignant melanoma in renal transplant recipients. British Journal of Dermatology 155, 857.

Lapunzina, P., 1996. Ultraviolet light-related neural tube defects? American Journal of Medical Genetics 67, 106.

Malerba, M., Gisondi, P., Radaeli, A., Sala, R., Calzavara Pinton, P. and Girolomoni, G., 2006. Plasma homocysteine and folate levels in patients with chronic plaque psoriasis. British Journal of Dermatology 155, 1165-1169.

Mallbris, L., Akre, O., Granath, F., Yin, L., Lindelöf, B., Ekbom, A. and Ståhle-Bäckdahl, M., 2004. Increased risk for cardiovascular mortality in psoriasis inpatients but not in outpatients. European Journal of Epidemiology19, 225-230.

Martinez, J.C., Otley, C.C., Stasko, T., Euvrard, S., Brown, C., Schanbacher, C.F., Weaver A.L. and Transplant-Skin Cancer Collaborative, 2003. Defining the clinical course of metastatic skin cancer in organ transplant recipients: A multicentre collaborative study. Archives of Dermatology 139, 301-306.

Moloney, F.J., Comber, H., O'Lorcain, P., O'Kelly, P., Conlon, P.J. and Murphy, G.M., 2006. A population-based study of skin cancer incidence and prevalence in renal transplant recipients. British Journal of Dermatology 154, 498-504.

Off, M.K., Steindal, A.E., Porojnicu, A.C., Juzeniene, A., Vorobey, A., Johnsson, A. and Moan, J., 2005 Ultraviolet photodegradation of folic acid. Journal of Photochemistry and Photobiology B: Biology 80, 47-55.

Sánchez-del-Campo, L., Chazarra, S., Montenegro, M.F., Cabezas-Herrera, J. and Rodríguez-López, J.N., 2010. Mechanism of dihydrofolate reductase downregulation in melanoma by 3-O-(3,4,5-trimethoxybenzoyl)-(-)-epicatechin. Journal of Cellular Biochemistry 110, 1399-1409.

Schinke, C., Mo, Y., Yu, Y., Amiri, K., Sosman, J., Greally, J. and Verma, A., 2010. Aberrant DNA methylation in malignant melanoma. Melanoma Research 20, 253-265.

Shelnutt, K.P., Kauwell, G.P., Gregory, J.F., 3rd, Maneval, D.R., Quinlivan, E.P., Theriaque D.W., Henderson, G.N. and Bailey L.B., 2004. Methylenetetrahydrofolate reductase 677C→T polymorphism affects DNA methylation in response to controlled folate intake in young women. Journal of Nutritional Biochemistry 15, 554-560.

Stern, L.L., Mason, J.B., Selhub, J. and Choi, S.W., 2000. Genomic DNA hypomethylation, a characteristic of most cancers, is present in peripheral leukocytes of individuals who are homozygous for the C677T polymorphism in the methylenetetrahydrofolate reductase gene. Cancer Epidemiology, Biomarkers and Prevention 9, 849-853.

Suzuki, K., Suzuki, I., Leodolter, A., Alonso, S., Horiuchi, S., Yamashita, K. and Perucho, M., 2006. Global DNA demethylation in gastrointestinal cancer is age dependant and precedes genomic damage. Cancer Cell 3, 199-207.

Vanizor Kural, B., Orem, A., Cimşit, G., Uydu, H.A., Yandi, Y.E. and Alver, A., 2003. Plasma homocysteine and its relationship with atherothrombotic markers in psoriatic patients. Clinica Chimica Acta 332, 23-30.

Key facts

- The sun emits ultraviolet radiation in the UVA, UVB and UVC wavelengths, but UVC radiation is filtered by the ozone layer.
- Vitamin D was discovered in 1920 and has since been more correctly termed a hormone rather than a vitamin since it is produced in the body through the action of UVB.
- Biologically active 1,25$(OH)_2D_3$ can be formed in skin cells following sun exposure.
- Vitamin D compounds protect against many of the harmful effects of sun exposure, such as DNA damage, that ultimately lead to skin cancer.
- The 1,25$(OH)_2D_3$ produced in skin cells following sun exposure may protect against subsequent sun damage.
- Low calcemic analogs of vitamin D may be suitable for use as skin cancer preventative agents.

Summary points

- Exposure to ultraviolet radiation causes many harmful effects in skin, including sunburn (erythema), apoptosis, DNA damage and immunosuppression. Both DNA damage and immunosuppression are key factors in the initiation of skin cancer.
- While ultraviolet radiation can be harmful, it is required for production of vitamin D3 in the skin, necessary for optimal bone and muscle function. The active metabolite 1,25$(OH)_2D_3$ can also be produced in skin cells. Continued exposure to ultraviolet radiation results in the formation of several overirradiation products, including lumisterol.
- 1,25$(OH)_2D_3$ and structurally related compounds, such as 1,25-dihydroxylumisterol$_3$ (JN) can inhibit DNA damage in human skin cells, in mouse models and in human subjects.
- Ultraviolet radiation increases levels of p53 and nitric oxide products in cells. The increase in p53 following ultraviolet radiation may facilitate DNA repair in cells yet to undergo replication, or may facilitate apoptosis of cells damaged beyond repair. Nitric oxide products have been implicated in ultraviolet radiation-induced DNA damage and can inhibit repair. Vitamin D compounds can further increase p53 levels after ultraviolet radiation and can reduce nitric oxide products in skin cells.
- There are two main signal transduction pathways for vitamin D: a genomic pathway and a rapid, non-genomic pathway. It is this latter pathway through which the photoprotective effects of vitamin D compounds appear to be mediated.
- Production of vitamin D and metabolites takes some hours after sun exposure. Thus, build up of vitamin D compounds in skin, like epidermal thickening and pigmentation, is an adaptation that protects from the next dose of ultraviolet radiation, not the immediate exposure.
- There must be a balance between some sun exposure to produce vitamin D and the need for sun avoidance or protection to reduce risks of skin cancer. The required amount of varies with latitude, altitude, time of day, season of the year, cloud cover, skin pigmentation, clothing and sunscreens amongst other factors.

24. Vitamin D and skin cancer

K.M. Dixon[1], V.B. Sequeira[1], A.J. Camp[2] and R.S. Mason[1]
[1]Discipline of Physiology, Bosch Institute, School of Medical Sciences, University of Sydney, NSW 2006, Australia; [2]Discipline of Biomedical Science, Bosch Institute, School of Medical Sciences, University of Sydney, NSW 2006, Australia; katie.dixon@sydney.edu.au

Abstract

Exposure to ultraviolet radiation from sunlight causes many harmful effects in skin, including sunburn, apoptosis, DNA damage and immunosuppression. Both DNA damage and immunosuppression are key factors in the initiation of skin cancer. While exposure to sunlight is harmful, it is necessary for production of vitamin D3 and the biologically active metabolite $1{,}25(OH)_2D_3$ in the skin. Both $1{,}25(OH)_2D_3$ and analogs, such as 1,25-dihydroxylumisterol$_3$ can inhibit ultraviolet radiation-induced DNA damage, apoptosis and immunosuppression in human skin cells and in mouse models. Levels of p53 and nitric oxide products also increase in skin cells following ultraviolet radiation. Increases in p53 may facilitate DNA repair in cells yet to undergo replication, or may facilitate apoptosis of irreparably damaged cells. Nitric oxide products have been implicated in ultraviolet radiation-induced DNA damage and can inhibit DNA repair. Interestingly, vitamin D compounds can further increase p53 levels and can reduce nitric oxide products in skin cells. There are two main signal transduction pathways for vitamin D: a genomic pathway and a rapid, non-genomic pathway. It is this latter pathway through which the photoprotective effects of vitamin D compounds appear to be mediated. ultraviolet radiation-induced production of vitamin D and metabolites requires several hours. Thus, build up of vitamin D compounds in skin is likely to protect from the next dose of ultraviolet radiation rather than the immediate exposure. Clearly, there must be a balance between some sun exposure to produce vitamin D and the need for sun avoidance or protection to reduce risks of skin cancer.

Keywords: 1,25-dihydroxyvitamin D3, photoprotection, DNA damage, immunosuppression, photocarcinogenesis, ultraviolet radiation.

Abbreviations

$1,25(OH)_2D_3$	1,25-dihydroxyvitamin D3
AK	Actinic keratosis
CPD	Cyclobutane pyrimidine dimer
JM	1α,25-dihydroxy-7-dehydrocholesterol
JN	1,25-dihydroxylumisterol$_3$
MARRS	Membrane-associated rapid response steroid binding
MED	Minimal erythemal dose
NER	Nucleotide excision repair
PDIA3	Protein disulfide isomerase family A, member 3
SBC	Sunburn cell
SCC	Squamous cell carcinoma
UVR	Ultraviolet radiation
VDR	Vitamin D receptor

24.1 Introduction

While sunlight exposure is a major cause of skin cancer, it is also the major source of vitamin D. Exposure of 7-dehydrocholesterol in skin to UVR in the UVB range (290-320 nm) results in formation of pre-vitamin D3 (Holick *et al.*, 1980), which undergoes a rapid rearrangement of its three double bonds to form vitamin D3 (cholecalciferol) (Tian *et al.*, 1994).

The relatively inactive vitamin D3 is transported in the bloodstream bound partly to the vitamin D binding protein and converted to its active form in two sequential hydroxylation reactions, the first in the liver and the second in the kidneys, resulting in the formation of the active metabolite $1,25(OH)_2D_3$ (calcitriol; Figure 24.1A, B) (Malloy *et al.*, 1999). There is evidence that the enzymes required for both hydroxylation steps are present in epidermal keratinocytes, enabling production of active $1,25(OH)_2D_3$ in the 2-5 nM range (Bikle *et al.*, 1986; Lehmann *et al.*, 2001). Dendritic cells are also capable of metabolizing vitamin D3 to the active form (Sigmundsdottir *et al.*, 2007). Prolonged exposure to UVR cannot result in excessive production of vitamin D3 since both previtamin D3 and vitamin D3 can absorb solar UVR during exposure and are converted to several "overirradiation" photoproducts including lumisterol and tachysterol (Figure 24.1C, D) (Holick, 2004).

The UV radiation from sunlight is also the cause of many pathological responses in the skin including sunburn (erythema), oedema, melanin pigmentation, actinic keratosis, photoaging, cataracts, DNA damage, immunosuppression and skin cancer (Figure 24.2) (Kochevar *et al.*, 1999; McGregor and Hawk, 1999; Pathak *et al.*, 1999).

A number of groups have reported protection against UV-induced skin cell death by vitamin D compounds in cultured human skin cells (Mason *et al.*, 2002; Wong *et al.*, 2004; Youn, 1997).

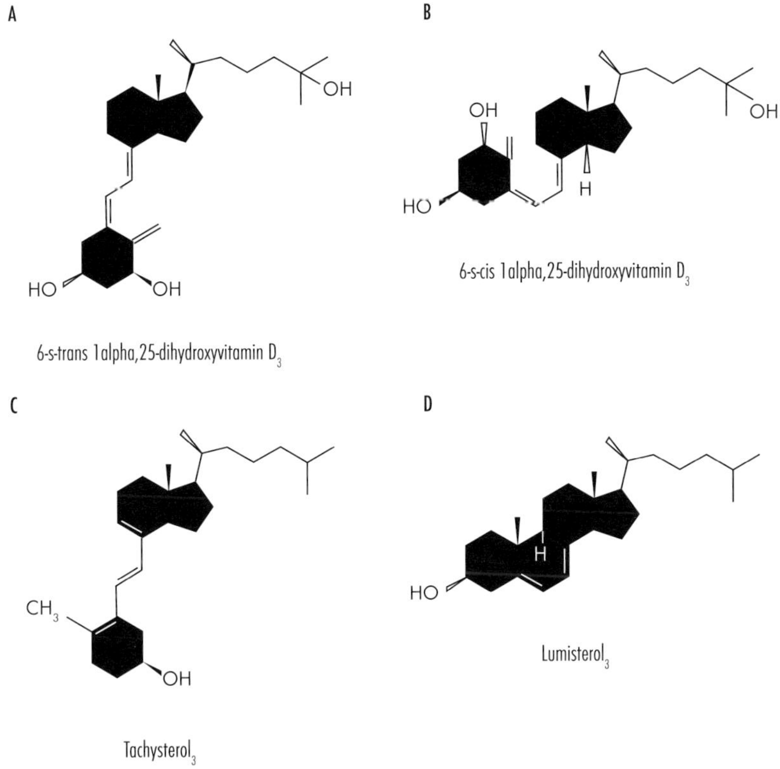

Figure 24.1. Structure of vitamin D compounds. Vitamin D and its metabolites are flexible, resulting in rapid conformational changes. The active metabolite 1,25-dihydroxyvitamin D3 can form a number of different ligand shapes, from the 6-s-*trans* extended steroid conformation (A) to the 6-s-*cis* steroid-like conformation (B). Overirradiation products, including tachysterol (C) and lumisterol (D) are similarly flexible compounds.

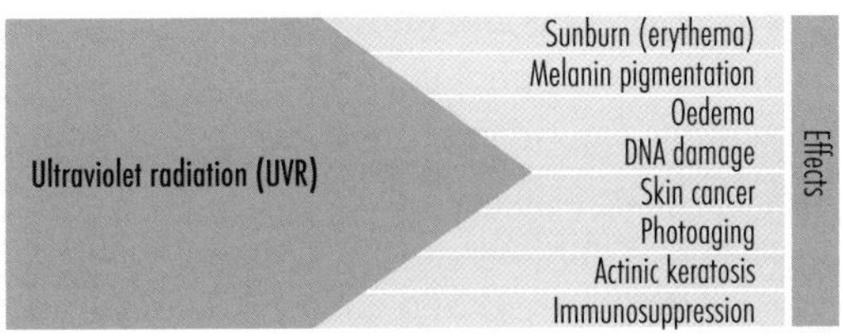

Figure 24.2. Pathological consequences of ultraviolet radiation exposure.

Moreover, several groups have reported protection against a variety of events leading to skin cancer *in vivo* in murine (Dixon *et al.*, 2005, 2007; Hanada *et al.*, 1995) and human (Damian *et al.*, 2010; Youn, 1997) models.

24.2 Vitamin D compounds attenuate UVB-induced skin cell death

The active compound 1,25$(OH)_2D_3$ has been shown to inhibit UVR-induced cell loss when added to cultured human keratinocytes, melanocytes and dermal fibroblasts prior to and/or immediately after UVR (Mason *et al.*, 2002; Wong *et al.*, 2004). Treatment with 1,25$(OH)_2D_3$ has also been shown to protect against doxorubicin-induced cell death (Mason *et al.*, 2002; Ravid *et al.*, 2002). Doxorubicin is known to induce apoptosis via oxidative stress and therefore the protective properties of 1,25$(OH)_2D_3$ cannot be simply due to UVR absorption, but may work to enhance cellular defenses against oxidative damage, possibly through an increase in metallothionein (Lee and Youn, 1998).

Studies showing protection against UVR-induced cell loss by 1,25$(OH)_2D_3$ *in vitro* are supported by evidence of reductions in sunburn cells (apoptotic keratinocytes) *in vivo*. Topical application or intraperitoneal injection of 1,25$(OH)_2D_3$ prior to administration of UVR (Hanada *et al.*, 1995), or topical application immediately after UVR only (Gupta *et al.*, 2007) reduced apoptotic sunburn cells in mouse skin. Furthermore, topical 1,25$(OH)_2D_3$ reduced sunburn cells in skin biopsies from UV-irradiated human subjects when applied before and immediately after UVR (Damian *et al.*, 2010). The presence of SBCs following UVR indicates the DNA has been damaged beyond repair, and it has been suggested that the function of SBCs is to eliminate DNA-damaged cells before they undergo replication (Claerhout *et al.*, 2006).

24.3 Vitamin D compounds reduce UVR-induced DNA photolesions

Protection against skin cell loss with 1,25$(OH)_2D_3$ is paralleled by reductions in UVR-induced DNA photolesions (De Haes *et al.*, 2005; Wong *et al.*, 2004). This implies that the skin cell protection by 1,25$(OH)_2D_3$ does not come at the expense of increased DNA damage in remaining cells.

Exposure of the skin to UVR causes damage to DNA, resulting in several types of DNA photolesions (Brash, 1988). The CPD, (Figure 24.3), is the most frequently occurring promutagenic UVR-induced DNA lesion. The absorption of photons from UVB radiation by DNA opens the 5-6 double bond of pyrimidines. In the case where two adjacent pyrimidines undergo this opening, a stable ring structure (CPD) can be formed (McGregor, 1999). CPDs can be formed between the 5-6 bonds of any two adjacent pyrimidines, forming predominantly thymine-thymine dimers (T<>T), with thymine-cytosine (C<>T or T<>C) and cytosine-cytosine (C<>C) occurring to a lesser extent (Cooke *et al.*, 2003).

There are several reports of reductions in UVR-induced CPD in human skin cells treated with 1,25$(OH)_2D_3$ (De Haes *et al.*, 2005; Dixon *et al.*, 2005, 2007; Gupta *et al.*, 2007; Wong *et al.*, 2004). Furthermore, a reduction in skin CPD after UVR has been noted in mice (Dixon *et al.*, 2005) and in human subjects (Damian *et al.*, 2010) treated topically with 1,25$(OH)_2D_3$. Interestingly, low calcemic vitamin D analogs have also shown protection against UVR-induced CPD (De Haes *et al.*, 2005; Dixon *et al.*, 2005; Wong *et al.*, 2004).

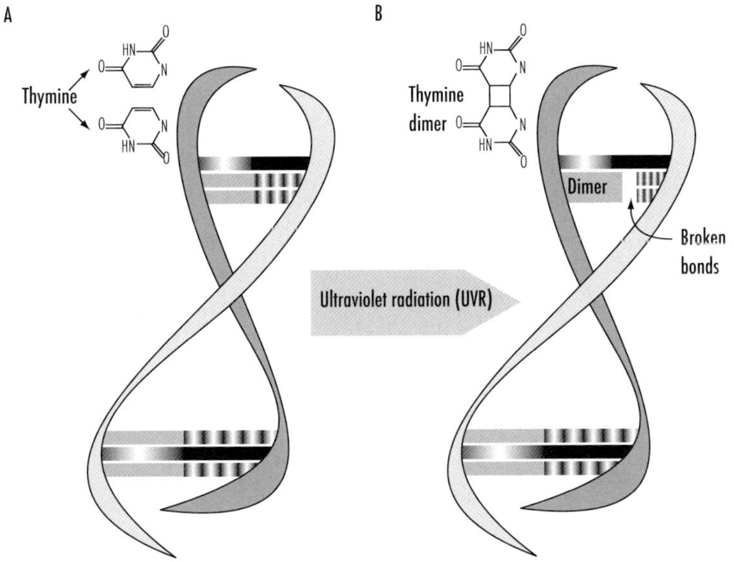

Figure 24.3. The cyclobutane pyrimidine dimer (CPD) is the most frequently occurring DNA photolesion. They form when the absorption of photons from UVB radiation by DNA opens the 5-6 double bond of adjacent pyrimidines, forming a stable ring structure. CPDs can be formed between the 5-6 bonds of any two adjacent pyrimidines, forming predominantly thymine-thymine dimers (T<>T), with thymine-cytosine (C<>T or T<>C) and cytosine-cytosine (C<>C) occurring to a lesser extent.

UVR-induced CPD are primarily repaired by NER, a complex multi-step process involving roughly 30 proteins. The significance of NER in the prevention of skin cancer is highlighted in patients with xeroderma pigmentosum, an inherited disorder in which there is a defect in some aspect of NER. Incidence of skin cancers in these patients is greater than one thousand times that of normal individuals (Matsumura and Ananthaswamy, 2002b; McGregor, 1999). In a recent study by Mason *et al.* (2010) $1,25(OH)_2D_3$ was shown to have no effect on a key gene involved in NER. XPG (xeroderma pigmentosum, complementation group G) mRNA was not altered in untreated or $1,25(OH)_2D_3$-treated cells up to five hours after UVR. It should be noted, however, that other NER genes were not investigated in this study.

Not all the protective effects of $1,25(OH)_2D_3$ may be due to enhanced repair. A significant reduction in CPD was noted 30 min after UVR in the presence of $1,25(OH)_2D_3$ (Gupta *et al.*, 2007). This time course is inconsistent with the reported half-life of repair of thymine dimers of around 7-12 h (Katiyar *et al.*, 2000; Mitchell *et al.*, 1990). Notably, there is also evidence of a reduction in UVR-induced CPD over a short time course in Skh:hr1 mice treated with silibinin (one hour) (Dhanalakshmi *et al.*, 2004). These results support the hypothesis of inhibition of CPD production, though the mechanism remains to be determined.

24.4 p53

The p53 tumor suppressor gene has been described as the guardian of the genome (Lane, 1992) and is frequently mutated in UVR-induced skin cancer amongst other cancers (Brash *et al.*, 1996; Nataraj *et al.*, 1995). A number of groups have detected p53 mutations in BCCs, SCCs and premalignant AK lesions, considered precursors for SCCs (Matsumura and Ananthaswamy, 2002a; Ziegler *et al.*, 1994).

The p53 gene codes for a 53 kDa protein that plays key roles in cell cycle regulation and facilitation of DNA repair. It is normally present in cells at very low levels due to its short half-life of less than 30 mins (Berg *et al.*, 1996; Hall *et al.*, 1993), and is bound to its negative regulator, MDM2. Exposure to UVR results in up-regulation of p53 transcription followed by the rapid accumulation and activation of p53 protein in the skin, mediated through phosphorylations and acetylations of the protein, peaking between 8-24 h following UVR exposure (Hall *et al.*, 1993; Saito *et al.*, 2003). Following accumulation and activation of p53, it is translocated to the nucleus where it acts on downstream genes including the cyclin-dependent kinase inhibitor p21, to mediate G1 arrest. This disruption would presumably provide enough time for DNA repair before the cell replicates (Matsumura and Ananthaswamy, 2004; Ouhtit *et al.*, 2000). p53 also functions to regulate the expression of genes involved in apoptosis, including pro-apoptotic genes such as Bax and Fas/Apo-1 or anti-apoptotic genes such as Bcl-2 (Mullauer *et al.*, 2001). This would facilitate the elimination of cells that are damaged beyond repair.

Treatment of UV-irradiated human keratinocytes (Gupta *et al.*, 2007) or melanocytes (Dixon *et al.*, 2005) has been shown to further increase p53 expression, beyond the increase induced by UVR, which may facilitate repair.

24.5 Vitamin D and nitric oxide

UVR stimulates production of reactive nitrogen species, including NO, and reactive oxygen species, including superoxide radicals, in skin cells (Bruch-Gerharz *et al.*, 1998; Chang *et al.*, 2003; Deliconstantinos *et al.*, 1996). NO and its products have been implicated in UVR-induced DNA damage and prevention of nucleotide excision repair mechanisms in skin (Bau *et al.*, 2001; Jaiswal *et al.*, 2000). A preliminary report has described a reduction in nitrite, a product of NO, after $1{,}25(OH)_2D_3$ treatment in keratinocytes exposed to UVR, however this was measured by the relatively insensitive Griess reaction. The reduction in nitrite by $1{,}25(OH)_2D_3$ was similar to that seen following treatment of keratinocytes with aminoguanidine, a potent NO inhibitor (Gupta *et al.*, 2007). It is proposed that the decrease in NO products by $1{,}25(OH)_2D_3$ may be involved in the mechanism of reduction of thymine dimers.

24.6 Vitamin D compounds and UVB-induced immunosuppression

UVR exposure facilitates skin carcinogenesis by suppressing cell-mediated immune reactions that normally inhibit skin tumor development (Kripke and Fisher, 1976; Nghiem *et al.*, 2002). The many cells involved in the skin immune system include keratinocytes, Langerhans cells, T cells and mast cells. UVR can alter the secretion of cytokines from keratinocytes (Nishigori *et al.*, 1996) and mast cells (Grimbaldeston *et al.*, 2007). UVR significantly reduces Langerhans cell numbers (Toews *et al.*, 1980) and compromises the antigen-presenting capabilities of remaining cells (Simon *et al.*, 1991; Tang and Udey, 1991). Moreover, UVR causes both local and systemic immunosuppression, inhibiting both contact hypersensitivity and delayed-type hypersensitivity reactions to antigens (reviewed in Dixon *et al.*, 2010).

The role of 1,25$(OH)_2D_3$ in the skin immune system remains controversial, and may depend on the concentration or immune models used (Dixon *et al.*, 2010). High concentrations of vitamin D have been reported to be immunosuppressive, while vitamin D deficiency also causes immune suppression (Yang *et al.*, 1993).

Topical application of 1,25$(OH)_2D_3$ led to increased suppressive capacity of CD4+CD25+ regulatory T cells in the skin-draining lymph nodes of BALB/c mice (Gorman *et al.*, 2007). A similar finding was reported in C57 BL/6 mice treated topically with calcipotriene (Ghoreishi *et al.*, 2009). In human subjects, the vitamin D analog calcipotriene suppressed contact hypersensitivity responses to dinitrochlorobenzene to levels induced by solar simulated UVR (Hanneman *et al.*, 2006), and a more recent study showed that 1,25$(OH)_2D_3$ suppressed the recall delayed type hypsersensitivity reaction when total doses of 1 μg or higher were applied topically to human skin (Damian *et al.*, 2010). In other models, topical 1,25$(OH)_2D_3$ did not cause immunosuppression by itself, but protected against UVR-induced systemic immunosuppression in immune competent, hairless mice, tested for contact hypersensitivity (Dixon *et al.*, 2005, 2007; Mason *et al.*, 2010) . In more recent studies, topical treatment with 1,25D protected against UVR-induced immunosuppression in other mouse species with an intact *hr* gene (W Tongkao-on, V.E. Reeve and R.S. Mason, unpublished observations). Low calcemic vitamin D analogs also inhibited UV-induced immunosuppression in these studies (Dixon *et al.*, 2005, 2007; Mason *et al.*, 2010). 1,25$(OH)_2D_3$ was also shown to reduce the inflammatory cytokine, interleukin-6, in human keratinocytes (De Haes *et al.*, 2003) and *in vivo* in mouse skin (Mason *et al.*, 2010).

24.7 Skin carcinogenesis and vitamin D

UVR-induced skin carcinogenesis is initiated by DNA damage resulting from exposure to UVR and subsequent genetic mutations if the DNA is not repaired. Normally, DNA damage results in an increase in p53 phosphorylation and translocation to the nucleus to facilitate DNA repair or apoptosis, however excessive UVR can disrupt this process, leading to p53 mutations, clonal expansion, and subsequent formation of premalignant actinic keratoses or SCC. In fact a number of studies have highlighted the significance of p53 mutations in the initiation of skin carcinogenesis (Brash *et al.*, 1996; Ziegler *et al.*, 1994) (Figure 24.4).

Of particular significance is the finding that UVR-induced sunburn cells are reduced in the skin of p53 knockout (–/–) mice (Ziegler *et al.*, 1994). This suggests that cells containing inactivated p53 due to mutations, have a reduced capacity to remove potentially mutagenic cells. The elimination of such cells prior to replication is an important step in preventing skin carcinogenesis. This is further supported by a study in which $p53^{-/-}$ and $p53^{+/-}$ mice showed accelerated UVR-induced skin tumor development compared with wild-type mice. Heterozygous mice showed greater susceptibility to skin cancer induction compared with wild-type mice, and homozygous mice were even more susceptible (Jiang *et al.*, 1999).

UVR-induced immunosuppression has been long known to play a role in skin carcinogenesis. In a murine study by Kripke and Fisher (1976) chronic UVR-induced skin tumors were translplanted into both UVR-irradiated and non-irradiated immune-competent mice. The tumors were rejected from the non-irradiated mice, but grew progressively in UV-irradiated mice. Based on this, it is not surprising that renal transplant patients and others undergoing immunosuppressive therapy are at an increased risk of developing skin cancer (Moloney *et al.*, 2006).

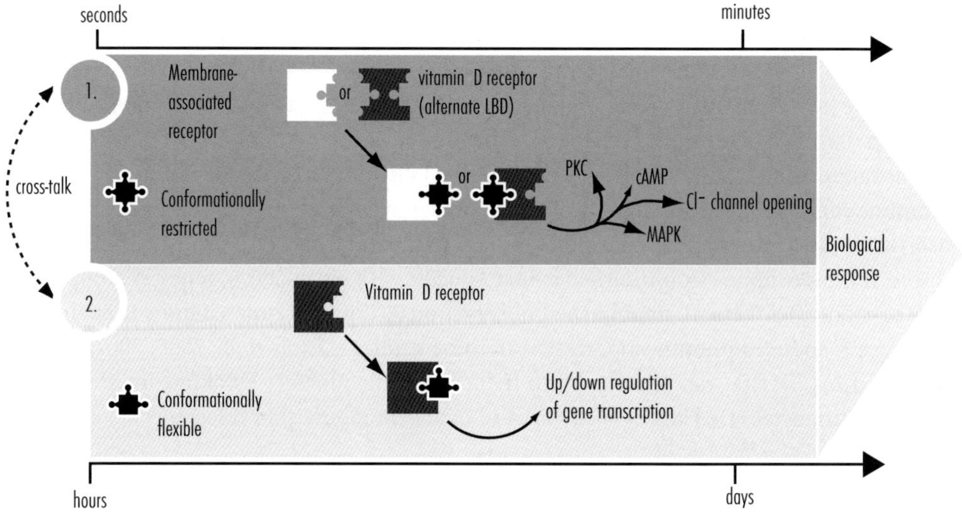

Figure 24.4. Initiation and progression of human squamous cell carcinoma. Initiation commences with UVR-induced DNA damage. Excessive UVR exposure can lead to p53 mutations, allowing damaged cells to resist apoptosis, and thereby facilitating initiation of skin carcinogenesis. Such cells can undergo clonal expansion, leading to formation of premalignant actinic keratoses (AKs) and squamous cell carcinomas.

24.8 Mechanisms of action of vitamin D compounds

There is substantial evidence for two signal transduction pathways for 1,25(OH)$_2$D$_3$; a classic/genomic pathway mediated by a nuclear VDR and a non-genomic pathway mediated by a membrane-associated receptor, which is not well defined (Figure 24.5) (Nemere *et al.*, 1994; 2004; Norman, 2001; Zanello and Norman, 2004b).

The genomic pathway has been studied for several years and is well characterized. Activation of this pathway occurs when 1,25(OH)$_2$D$_3$ binds deep into the genomic ligand pocket of the nuclear VDR. Once this complex heterodimerises with the retinoid X receptor, it attaches to the vitamin D response element on the promoter region of target genes (Malloy and Feldman, 1999). Transcription is modulated resulting in a delayed biological effect.

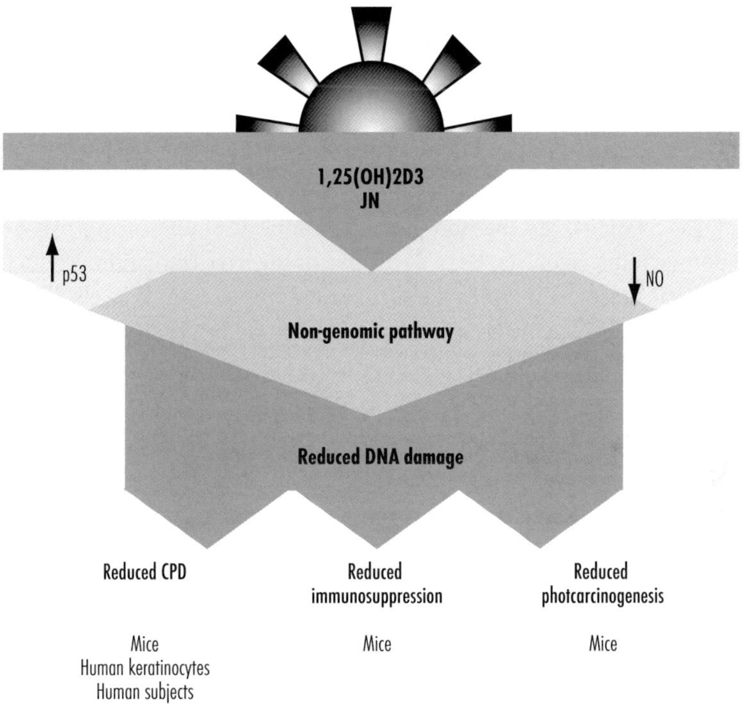

Figure 24.5 Vitamin D mechanism of action. The nature of the receptor involved in vitamin D non-genomic responses remains unclear. The presence of a separate membrane-associated receptor or an alternate ligand binding domain within the classical vitamin D receptor located at the membrane have been proposed. The molecular shape preferred for non-genomic responses is the conformationally restricted 6-s-*cis* conformation. The active compound, 1,25(OH)$_2$D$_3$, mediates genomic biological responses by binding to the VDR. Genomic processes usually require a longer time frame than non-genomic responses, though the timeline is not exact.

The non-genomic pathway is much more rapid, with responses occurring within seconds to minutes. Up-regulation of a number of molecular cascades has been shown, including the activation of protein kinase C, phosphatidylinositol 3-kinase and opening of calcium-activated chloride channels (Norman *et al.*, 2004). Current studies are investigating the membrane-associated receptor(s) that mediate the non-genomic pathways for 1,25$(OH)_2D_3$. There is evidence for the involvement of the VDR as well as the 66 kDa membrane-associated MARRS protein, also known as ERp57 and PDIA3 (Nemere *et al.*, 2010). Neutralising the MARRS protein with Ab099, which targets the N-terminus of MARRS, prevents non-genomic actions of 1,25$(OH)_2D_3$, such as calcium currents and protein kinase C activity in epithelial cells (Nemere *et al.*, 2004). There are recent reports suggesting the involvement of the membrane-located nuclear VDR in the non-genomic response of 1,25$(OH)_2D_3$. An alternate ligand binding domain on the nuclear VDR has been proposed to interact with 1,25$(OH)_2D_3$ and the 6-s-*cis*-locked analog, 1α,25 dihydroxylumisterol$_3$ (Mizwicki *et al.*, 2004, 2010). This alternate pocket enables the VDR to form a complex with a large number of 1,25$(OH)_2D_3$ conformations, enabling the VDR to mediate both genomic and non-genomic responses. Support for this hypothesis was observed when 1,25$(OH)_2D_3$-induced chloride currents, which were present in osteoblasts from wildtype and heterozygous mice, were completely abolished in osteoblasts which lacked a functional nuclear VDR (Zanello and Norman, 2004a).

There is evidence from a number of studies that protection against UVR-induced skin damage by 1,25$(OH)_2D_3$ is mediated via the non-genomic pathway. Studies using two 6-s-*cis* locked low calcemic non-genomic agonists, JN, a derivative of the overirradiation product, lumisterol, and JM showed that these compounds conferred photoprotection at levels similar to that of 1,25$(OH)_2D_3$. The significance of the non-genomic pathway in photoprotection by vitamin D compounds is further supported by the reversal of these effects in the presence of a non-genomic antagonist, but not with a genomic antagonist (Dixon *et al.*, 2005; Wong *et al.*, 2004). Topical application of JN has also shown promising results in an *in vivo* murine model, inhibiting SBCs, CPD and immunosuppression following exposure to UVR (Dixon *et al.*, 2005).

24.9 Potential for use of vitamin D compounds to reduce skin cancer

The use of low calcemic vitamin D compounds is a novel approach to skin cancer prevention. The agent 1α-hydroxymethyl-16-ene-24,24-difluoro-25-hydroxy-26,27-bis-homovitaminD3 (QW-1624F2-2 or QW), which is 80-100 times less calciuric than 1,25$(OH)_2D_3$, is a potential candidate for further investigation given that it has been fast-tracked by the United States Food and Drug Administration for approval for clinical use. This compound is transcriptionally active and has demonstrated anti-proliferative and pro-differentiating ability (Posner *et al.*, 2004). Significantly, it has been shown to inhibit formation of chemically-induced skin tumors, and delay onset (latency) to tumor outgrowth (Kensler *et al.*, 2000). Other low calcemic vitamin D compounds may also prove to be promising ingredients in sunscreen lotions or after-sun preparations to reduce sun damage to DNA.

24.10 Applications to other areas of skin health, care and treatments

In evolutionary terms, the capacity to produce vitamin D in skin by UVR preceded the move out of the sea to a low calcium environment, where calcium absorption became a major issue. Since derivatives of overirradiation products such as lumisterol contribute to reduced DNA damage after UVR, it is possible that the vitamin D system in skin was primarily involved as one of the key adaptive mechanisms against UV damage. It is worth noting that even vitamin D production takes some hours after UV, and metabolite production even more time. Thus, build up of vitamin D compounds in skin, like epidermal thickening and pigmentation, is an adaptation that protects from the next dose of UVR, not the immediate exposure.

24.11 Guidelines for sun exposure

A key issue in advising optimal sun exposure for maintenance of vitamin D levels is that there must be a balance between some sun exposure to produce vitamin D and the need for sun avoidance or protection to reduce risks of skin cancer. While sun exposure causes DNA damage and immunosuppression, leading to skin cancers, there is some evidence that small amounts of sun damage are relatively better repaired (De Winter *et al.*, 2001). Moreover, pre-vitamin D and vitamin D are broken down with continued UVR, suggesting that short but frequent sun exposure might be appropriate for vitamin D production with less likelihood of sun damage.

The amount of UVR required for vitamin D production varies with latitude, altitude, time of day, season of the year, cloud cover as well as skin pigmentation and other factors such as clothing and sunscreens (Springbett *et al.*, 2010; Webb and Engelsen, 2006). These variations, together with the fact that the relationship between sun exposure and vitamin D synthesis has not been fully elucidated, further complicates advice pertaining to safe sun exposure. A biological endpoint of UV irradiation that can be assessed with reasonable convenience is MED, which refers to the UVR exposure required to just cause faint redness of skin. It is possible, but not optimal, to use MED as a guide to a vitamin D dose, but the limitation is that the UV action spectra for erythema and for vitamin D production are similar only at low wavelengths, below 315 nm (Webb and Engelsen, 2006). For example, UVA, which makes up the bulk of UV in sunlight at the earth's surface, does not produce vitamin D but it can cause erythema.

Complex tables, based on atmospheric measurements, of the exposure times required to produce 1/3 of an MED in people with white skin at different times of day in different seasons and at various latitudes have been produced (Diamond *et al.*, 2005; Webb and Engelsen, 2006). In practice, though, this is difficult advice to convey to health practitioners or the public. Sufficient vitamin D should be maintained by, in summer, going out for a walk with sleeves rolled up around mid-morning or mid-afternoon for around 6-8 minutes (longer may be needed at high latitudes), most days and in winter, by trying to get out at lunchtime most days for about 20-50 minutes or so (depending on latitude). Walk briskly, so that you can roll your sleeves up a bit. While this is not practical for everyone, the idea of getting outside to get a bit of exercise, perhaps clear the head and maybe improve people's mood is a goal worth considering.

24.12 Conclusion

It is likely that the vitamin D compounds produced in skin cells following ultraviolet radiation contribute to the photoprotective effects observed in a number of experimental systems, including DNA damage, immunosuppression and skin carcinogenesis (Figure 24.6). Since low calcemic analogs, such as JN and QW, are capable of similar levels of photoprotection, this raises the possibility of the incorporation of such compounds into after sun lotions, and highlights a role for vitamin D compounds in prevention of skin cancer.

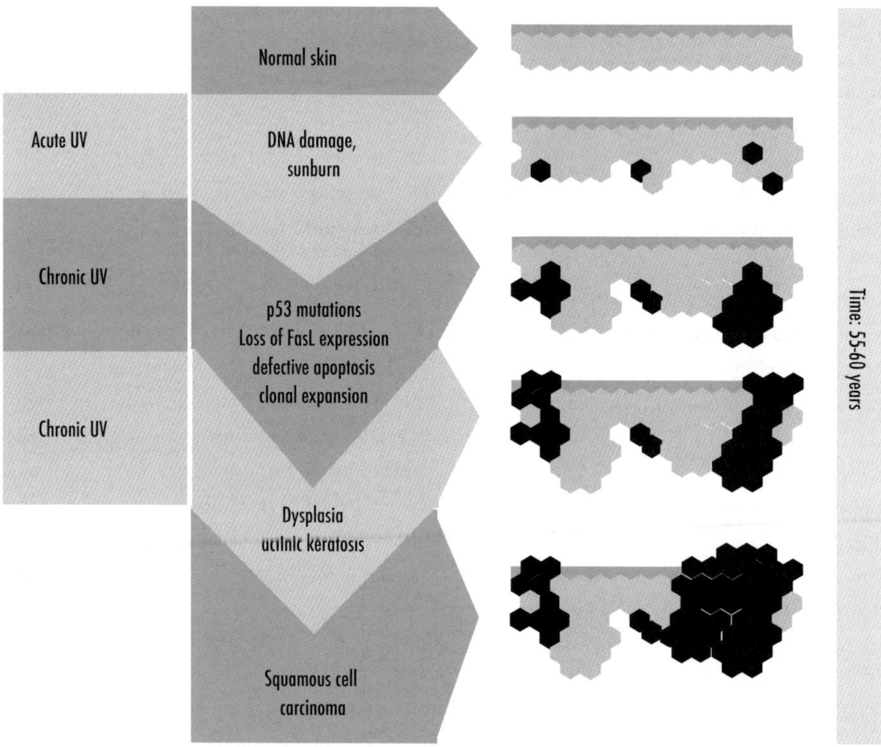

Figure 24.6 Photoprotection by vitamin D compounds. Ultraviolet radiation produces $1,25(OH)_2D_3$ and other compounds, some similar in structure to the analog JN. Vitamin D compounds produced in skin may contribute to the photoprotective effects of reduced DNA damage in the form of CPD, reduced immunosuppression and ultimately inhibition of photocarcinogenesis. These effects are thought to be mediated through the non-genomic pathway for vitamin D compounds and involve further increases in p53 and reductions in nitric oxide products.

References

Bau, D.T., Gurr, J.R. and Jan, K.Y., 2001. Nitric oxide is involved in arsenite inhibition of pyrimidine dimer excision. Carcinogenesis 22, 709-716.

Berg, R.J., Van Kranen, H.J., Rebel, H.G., De Vries, A., Van Vloten, W.A., Van Kreijl, C.F., van der Leun, J.C. and De Gruijl, F.R., 1996. Early p53 alterations in mouse skin carcinogenesis by UVB radiation: immunohistochemical detection of mutant p53 protein in clusters of preneoplastic epidermal cells. Proceedings of the National Academy of Sciences of the United States of America 93, 274-278.

Bikle, D.D., Nemanic, M.K., Whitney, J.O. and Elias, P.W., 1986. Neonatal human foreskin keratinocytes produce 1,25-dihydroxyvitamin D3. Biochemistry 25, 1545-1548.

Brash, D.E., 1988. UV mutagenic photoproducts in Escherichia coli and human cells: a molecular genetics perspective on human skin cancer. Photochemistry and Photobiology 48, 59-66.

Brash, D.E., Ziegler, A., Jonason, A.S., Simon, J.A., Kunala, S. and Leffell, D.J., 1996. Sunlight and sunburn in human skin cancer: p53, apoptosis, and tumor promotion. Journal of Investigative Dermatology Symposium Proceedings 1, 136-142.

Bruch-Gerharz, D., Ruzicka, T. and Kolb-Bachofen, V., 1998. Nitric oxide in human skin: current status and future prospects. Journal of Investigative Dermatology 110, 1-7.

Chang, H.R., Tsao, D.A., Wang, S.R. and Yu, H.S., 2003. Expression of nitric oxide synthases in keratinocytes after UVB irradiation. Archives of Dermatology Research 295, 293-296.

Claerhout, S., Van Laethem, A., Agostinis, P., Garmyn, M., Claerhout, S., Van Laethem, A., Agostinis, P. and Garmyn, M., 2006. Pathways involved in sunburn cell formation: deregulation in skin cancer. Photochemical & Photobiological Sciences 5, 199-207.

Cooke, M.S., Podmore, I.D., Mistry, N., Evans, M.D., Herbert, K.E., Griffiths, H.R. and Lunec, J., 2003. Immunochemical detection of UV-induced DNA damage and repair. Journal of Immunological Methods 280, 125-133.

Damian, D., Kim, Y.J., Dixon, K., Halliday, G.M., Javeri, A. and Mason, R.S., 2010. Topical calcitriol protects from UV-induced genetic damage but suppresses immunity in humans. Experimental Dermatology 19, e23-e30.

De Haes, P., Garmyn, M., Degreef, H., Vantieghem, K., Bouillon, R. and Segaert, S., 2003. 1,25-Dihydroxyvitamin D3 inhibits ultraviolet B-induced apoptosis, Jun kinase activation, and interleukin-6 production in primary human keratinocytes. Journal of Cellular Biochemistry 89, 663-673.

De Haes, P., Garmyn, M., Verstuyf, A., De Clercq, P., Vandewalle, M., Degreef, H., Vantieghem, K., Bouillon, R. and Segaert, S., 2005. 1,25-Dihydroxyvitamin D3 and analogues protect primary human keratinocytes against UVB-induced DNA damage. Journal of Photochemistry & Photobiology. B Biology 78, 141-148.

De Winter, S., Vink, A.A., Roza, L. and Pavel, S., 2001. Solar-simulated skin adaptation and its effect on subsequent UV-induced epidermal DNA damage. Journal of Investigative Dermatology 117, 678-682.

Deliconstantinos, G., Villiotou, V. and Stavrides, J.C., 1996. Alterations of nitric oxide synthase and xanthine oxidase activities of human keratinocytes by ultraviolet B radiation. Potential role for peroxynitrite in skin inflammation. Biochemical Pharmacology 51, 1727-1738.

Dhanalakshmi, S., Mallikarjuna, G.U., Singh, R.P. and Agarwal, R., 2004. Silibinin prevents ultraviolet radiation-caused skin damages in SKH-1 hairless mice via a decrease in thymine dimer positive cells and an up-regulation of p53-p21/Cip1 in epidermis. Carcinogenesis 25, 1459-1465.

Diamond, T.H., Eisman, J.A.E., Mason, R.S., Nowson, C.A., Pasco, J.A., Sambrook, P.N. and Wark, J.D., 2005. Vitamin D and adult bone health in Australia and New Zealand: a position statement. Medical Journal of Australia 182, 281-285.

Dixon, K.M., Deo, S.S., Norman, A.W., Bishop, J.E., Halliday, G.M., Reeve, V.E. and Mason, R.S., 2007. *In vivo* relevance for photoprotection by the vitamin D rapid response pathway. Journal of Steroid Biochemistry and Molecular Biology 103, 451-456.

Dixon, K.M., Deo, S.S., Wong, G., Slater, M., Norman, A.W., Bishop, J.E., Posner, G.H., Ishizuka, S., Halliday, G.M., Reeve, V.E. and Mason, R.S., 2005. Skin cancer prevention: a possible role of 1,25dihydroxyvitamin D3 and its analogs. Journal of Steroid Biochemistry and Molecular Biology 97, 137-143.

Dixon, K.M., Sequeira, V.B., Camp, A.J. and Mason, R.S., 2010. Vitamin D-fence. Photochemical & Photobiological Sciences 9, 564-570.

Ghoreishi, M., Bach, P., Obst, J., Komba, M., Fleet, J.C., Dutz, J.P., Ghoreishi, M., Bach, P., Obst, J., Komba, M., Fleet, J.C. and Dutz, J.P., 2009. Expansion of antigen-specific regulatory T cells with the topical vitamin d analog calcipotriol. Journal of Immunology 182, 6071-6078.

Gorman, S., Kuritzky, L.A., Judge, M.A., Dixon, K.M., McGlade, J.P., Mason, R.S., Finlay-Jones, J.J. and Hart, P.H., 2007. Topically applied 1,25-dihydroxyvitamin D3 enhances the suppressive activity of CD4+CD25+ cells in the draining lymph nodes. Journal of Immunology 179, 6273-6283.

Grimbaldeston, M.A., Nakae, S., Kalesnikoff, J., Tsai, M. and Galli, S.J., 2007. Mast cell-derived interleukin 10 limits skin pathology in contact dermatitis and chronic irradiation with ultraviolet B. Nature Immunology 8, 1095-1104.

Gupta, R., Dixon, K.M., Deo, S.S., Holliday, C.J., Slater, M., Halliday, G.M., Reeve, V.E. and Mason, R.S., 2007. Photoprotection by 1,25dihydroxyvitamin D is associated with an increase in p53 and a decrease in nitric oxide products. Journal of Investigative Dermatology 127, 707-715.

Hall, P.A., McKee, P.H., Menage, H.D., Dover, R. and Lane, D.P., 1993. High levels of p53 protein in UV-irradiated normal human skin. Oncogene. 8, 203-207.

Hanada, K., Sawamura, D., Nakano, H. and Hashimoto, I., 1995. Possible role of 1,25-dihydroxyvitamin D3-induced metallothionein in photoprotection against UVB injury in mouse skin and cultured rat keratinocytes. Journal of Dermatological Science 9, 203-208.

Hanneman, K.K., Scull, H.M., Cooper, K.D. and Baron, E.D., 2006. Effect of topical vitamin D analogue on in vivo contact sensitization. Archives of Dermatology 142, 1332-1334.

Holick, M.F., 2004. Sunlight and vitamin D for bone health and prevention of autoimmune diseases, cancers, and cardiovascular disease. American Journal of Clinical Nutrition 80, 1678S-1688S.

Holick, M.F., MacLaughlin, J.A., Clark, M.B., Holick, S.A., Potts Jr., J.T., Anderson, R.R., Blank, I.H., Parrish, J.A. and Elias, P., 1980. Photosynthesis of previtamin D3 in human skin and the physiologic consequences. Science 210, 203-205.

Jaiswal, M., LaRusso, N.F., Burgart, L.J. and Gores, G.J., 2000. Inflammatory cytokines induce DNA damage and inhibit DNA repair in cholangiocarcinoma cells by a nitric oxide-dependent mechanism. Cancer Research 60, 184-190.

Jiang, W., Ananthaswamy, H.N., Muller, H.K. and Kripke, M.L., 1999. p53 protects against skin cancer induction by UV-B radiation. Oncogene 18, 4247-4253.

Katiyar, S.K., Matsui, M.S. and Mukhtar, H., 2000. Kinetics of UV light-induced cyclobutane pyrimidine dimers in human skin in vivo: an immunohistochemical analysis of both epidermis and dermis. Photochemistry & Photobiology 72, 788-793.

Kensler, T.W., Dolan, P.M., Gange, S.J., Lee, J.K., Wang, Q. and Posner, G.H., 2000. Conceptually new deltanoids (vitamin D analogs) inhibit multistage skin tumorigenesis. Carcinogenesis. 21, 1341-1345.

Kochevar, I.E., Pathak, M.A. and Parrish, J.A., 1999. Photophysics, Photochemistry and Photobiology. In: I.M. Freedberg, Eisen, A., Wolff, K., Austen, K.F., Goldsmith, L., Katz, S. and Fitzpatrick, T.B. (eds.). Fitzpatrick's Dermatology in General Medicine. McGraw-Hill, New York, NY, USA, pp. 220-230.

Kripke, M.L. and Fisher, M.S., 1976. Immunologic parameters of ultraviolet carcinogenesis. Journal of the National Cancer Institute 57, 211-215.

Lane, D.P., 1992. Cancer. p53, guardian of the genome. Nature 358, 15-16.

Lee, J. and Youn, J.I., 1998. The photoprotective effect of 1,25-dihydroxyvitamin D3 on ultraviolet light B-induced damage in keratinocyte and its mechanism of action. Journal of Dermatological Science 18, 11-18.

Lehmann, B., Genehr, T., Knuschke, P., Pietzsch, J. and Meurer, M., 2001. UVB-induced conversion of 7-dehydrocholesterol to 1alpha,25-dihydroxyvitamin D3 in an *in vitro* human skin equivalent model. Journal of Investigative Dermatology 117, 1179-1185.

Malloy, P.J. and Feldman, D., 1999. Vitamin D resistance. The American Journal of Medicine 106, 355-370.

Malloy, P.J., Pike, J.W. and Feldman, D., 1999. The vitamin D receptor and the syndrome of hereditary 1,25-dihydroxyvitamin D-resistant rickets. Endocrine Reviews 20, 156-188.

Mason, R.S., C.J. Holliday and Gupta, R., 2002. 1,25 Dihydroxyvitamin D and photoprotection in skin cells. In: Tsambaos, D. and Merk, H. (ed.), Modern Trends in Skin Pharmacology. Parissianos Medical Publications S.A. Athens, Athens, Greece, pp. 59-66.

Mason, R.S., Sequeira, V.B., Dixon, K.M., Gordon-Thomson, C., Pobre, K., Dilley, A., Mizwicki, M.T., Norman, A.W., Feldman, D., Halliday, G.M. and Reeve, V.E., 2010. Photoprotection by 1,25-dihydroxyvitamin D and analogs: Further studies on mechanisms and implications for UV-damage. Journal of Steroid Biochemistry and Molecular Biology 121, 164-168.

Matsumura, Y. and Ananthaswamy, H.N., 2002a. Molecular mechanisms of photocarcinogenesis. Frontiers in Bioscience 7, d765-783.

Matsumura, Y. and Ananthaswamy, H.N., 2002b. Short-term and long-term cellular and molecular events following UV irradiation of skin: implications for molecular medicine. Expert Reviews in Molecular Medicine 4, 1-22.

Matsumura, Y. and Ananthaswamy, H.N., 2004. Toxic effects of ultraviolet radiation on the skin. Toxicology and Applied Pharmacology 195, 298-308.

McGregor, J.M. and Hawk, J.L.M., 1999. Acute effects of ultraviolet radiation on the skin. In: Freedberg, I.M., Eisen, A., Wolff, K., Austen, K.F., Goldsmith, L., Katz, S. and Fitzpatrick, T.B. (eds.). Fitzpatrick's Dermatology in General Medicine. McGraw- Hill, New York, NY, USA, pp. 1555-1561.

McGregor, W.G., 1999. DNA repair, DNA replication, and UV mutagenesis. Journal of Investigative Dermatology. Symposium Proceedings 4, 1-5.

Mitchell, D.L., Cleaver, J.E. and Epstein, J.H., 1990. Repair of pyrimidine (6-4) pyrimidone photoproducts in mouse skin. Journal of Investigative Dermatology 95, 55-59.

Mizwicki, M.T., Keidel, D., Bula, C.M., Bishop, J.E., Zanello, L.P., Wurtz, J.-M., Moras, D. and Norman, A.W., 2004. Identification of an alternative ligand-binding pocket in the nuclear vitamin D receptor and its functional importance in 1alpha,25(OH)2-vitamin D3 signaling. Proceedings of National Acadamy of Sciences of the United States of America 101, 12876-12881.

Mizwicki, M.T., Menegaz, D., Yaghmaei, S., Henry, H.L. and Norman, A.W., 2010. A molecular description of ligand binding to the two overlapping binding pockets of the nuclear vitamin D receptor (VDR): structure-function implications. Journal of Steroid Biochemistry and Molecular Biology 121, 98-105.

Moloney, F.J., Comber, H., O'Lorcain, P., O'Kelly, P., Conlon, P.J. and Murphy, G.M., 2006. A population-based study of skin cancer incidence and prevalence in renal transplant recipients. British Journal of Dermatology 154, 498-504.

Mullauer, L., Gruber, P., Sebinger, D., Buch, J., Wohlfart, S., Chott, A., Mullauer, L., Gruber, P., Sebinger, D., Buch, J., Wohlfart, S. and Chott, A., 2001. Mutations in apoptosis genes: a pathogenetic factor for human disease. Mutation Research 488, 211-231.

Nataraj, A.J., Trent, J.C., 2nd and Ananthaswamy, H.N., 1995. p53 gene mutations and photocarcinogenesis. Photochemistry and Photobiology 62, 218-230.

Nemere, I., Dormanen, M.C., Hammond, M.W., Okamura, W.H. and Norman, A.W., 1994. Identification of a specific binding protein for 1 alpha,25-dihydroxyvitamin D3 in basal-lateral membranes of chick intestinal epithelium and relationship to transcaltachia. Journal of Biological Chemistry 269, 23750-23756.

Nemere, I., Farach-Carson, M.C., Rohe, B., Sterling, T.M., Norman, A.W., Boyan, B.D. and Safford, S.E., 2004. Ribozyme knockdown functionally links a 1,25(OH)2D3 membrane binding protein (1,25D3-MARRS) and phosphate uptake in intestinal cells. Proceedings of National Academy of Science of the United States of America 101, 7392-7397.

Nemere, I., Garbi, N., Hammerling, G.J. and Khanal, R.C., 2010. Intestinal cell calcium uptake and the targeted knockout of the 1,25D3-MARRS (membrane-associated, rapid response steroid-binding) receptor/PDIA3/Erp57. Journal of Biological Chemistry 285, 31859-31866.

Nghiem, D.X., Kazimi, N., Mitchell, D.L., Vink, A.A., Ananthaswamy, H.N., Kripke, M.L. and Ullrich, S.E., 2002. Mechanisms underlying the suppression of established immune responses by ultraviolet radiation. Journal of Investigative Dermatology 119, 600-608.

Nishigori, C., Yarosh, D.B., Ullrich, S.E., Vink, A.A., Bucana, C.D., Roza, L. and Kripke, M.L., 1996. Evidence that DNA damage triggers interleukin 10 cytokine production in UV-irradiated murine keratinocytes. Proceedings of the National Academy of Sciences of the United States of America 93, 10354-10359.

Norman, A.W., Henry, H.L., Bishop, J.E., Song, X., Bula, C. and Okamura, W.H., 2001. Different shapes of the steroid hormone 1α,25(OH)$_2$-vitamin D$_3$ act as agonists for two different receptors in the vitamin D endocrine system to mediate genomic and rapid responses. Steroids 66, 147-158.

Norman, A.W., Mizwicki, M.T. and Norman, D.P.G., 2004. Steroid-hormone rapid actions, membrane receptors and a conformational ensemble model. Nature Reviews Drug Discovery 3, 27-41.

Ouhtit, A., Muller, H.K., Davis, D.W., Ullrich, S.E., McConkey, D. and Ananthaswamy, H.N., 2000. Temporal events in skin injury and the early adaptive responses in ultraviolet-irradiated mouse skin. American Journal of Pathology 156, 201-207.

Pathak, M.A., Ngheim, P. and Fitzpatrick, T.B., 1999. Acute and chronic effects of the sun. In: Freedberg, I.M., Eisen, A., Wolff, K., Austen, K.F., Goldsmith, L., Katz, S. and Fitzpatrick, T.B. (eds.). Fitzpatrick's Dermatology in General Medicine. McGraw-Hill, New York, NY, USA, pp. 1598-1607.

Posner, G.H., Jeon, H.B., Sarjeant, A., Riccio, E.S., Doppalapudi, R.S., Kapetanovic, I.M., Saha, U., Dolan, P. and Kensler, T.W., 2004. Low-calcemic, efficacious, 1a,25-dihydroxyvitamin D$_3$ analog QW-1624F$_2$-2: calcemic dose-response determination, preclinical genotoxicity testing, and revision of A-ring stereochemistry. Steroids 69, 757-762.

Ravid, A., Rubinstein, E., Gamady, A., Rotem, C., Liberman, U.A. and Koren, R., 2002. Vitamin D inhibits the activation of stress-activated protein kinases by physiological and environmental stresses in keratinocytes. Journal of Endocrinology 173, 525-532.

Saito, S., Yamaguchi, H., Higashimoto, Y., Chao, C., Xu, Y., Fornace Jr., A.J., Appella, E. and Anderson, C.W., 2003. Phosphorylation site interdependence of human p53 post-translational modifications in response to stress. Journal of Biological Chemistry 278, 37536-37544.

Sigmundsdottir, H., Pan, J., Debes, G.F., Alt, C., Habtezion, A., Soler, D., Butcher, E.C., Sigmundsdottir, H., Pan, J., Debes, G.F., Alt, C., Habtezion, A., Soler, D. and Butcher, E.C., 2007. DCs metabolize sunlight-induced vitamin D3 to 'program' T cell attraction to the epidermal chemokine CCL27. Nature Immunology 8, 285-293.

Simon, J.C., Cruz Jr., P.D., Tigelaar, R.E., Sontheimer, R.D. and Bergstresser, P.R., 1991. Adhesion molecules CD11a, CD18, and ICAM-1 on human epidermal Langerhans cells serve a functional role in the activation of alloreactive T cells. Journal of Investigative Dermatology 96, 148-151.

Springbett, P., Buglass, S. and Young, A.R., 2010. Photoprotection and vitamin D status. Journal of Photochemistry and Photobiology. B, Biology 101, 160-168.

Tang, A. and Udey, M.C., 1991. Inhibition of epidermal Langerhans cell function by low dose ultraviolet B radiation. Ultraviolet B radiation selectively modulates ICAM-1 (CD54) expression by murine Langerhans cells. Journal of Immunology 146, 3347-3355.

Tian, X.Q., Chen, T.C., Lu, Z., Shao, Q. and Holick, M.F., 1994. Characterization of the translocation process of vitamin D3 from the skin into the circulation. Endocrinology 135, 655-661.

Toews, G.B., Bergstresser, P.R. and Streilein, J.W., 1980. Epidermal Langerhans cell density determines whether contact hypersensitivity or unresponsiveness follows skin painting with DNFB. Journal of Immunology 124, 445-453.

Webb, A.R. and Engelsen, O., 2006. Calculated ultraviolet exposure levels for a healthy vitamin D status. Photochemistry and Photobiology 82, 1697-1703.

Wong, G., Gupta, R., Dixon, K.M., Deo, S.S., Choong, S.M., Halliday, G.M., Bishop, J.E., Ishizuka, S., Norman, A.W., Posner, G.H. and Mason, R.S., 2004. 1,25-Dihydroxyvitamin D and three low-calcemic analogs decrease UV-induced DNA damage via the rapid response pathway. Journal of Steroid Biochemistry and Molecular Biology 89-90, 567-570.

Yang, S., Smith, C., Prahl, J.M., Luo, X. and DeLuca, H.F., 1993. Vitamin D deficiency suppresses cell-mediated immunity *in vivo*. Archives of Biochemistry & Biophysics 303, 98-106.

Youn, J.I., 1997. Photoprotective effect of calcipotriol upon skin photoreaction to UVA and UVB. Photodermatology, Photoimmunology and Photomedicine 13, 109-114.

Zanello, L.P. and Norman, A.W., 2004a. Electrical responses to 1alpha,25(OH)2-Vitamin D3 and their physiological significance in osteoblasts. Steroids 69, 561-565.

Zanello, L.P. and Norman, A.W., 2004b. Rapid modulation of osteoblast ion channel responses by 1alpha,25(OH)2-vitamin D3 requires the presence of a functional vitamin D nuclear receptor. Proceedings of the National Acadamy of Sciences of the United States of America 101, 1589-1594.

Ziegler, A., Jonason, A.S., Leffell, D.J., Simon, J.A., Sharma, H.W., Kimmelman, J., Remington, L., Jacks, T. and Brash, D.E., 1994. Sunburn and p53 in the onset of skin cancer. Nature 372, 773-776.

Specific skin conditions in relation to diet and nutrition

Key facts

- Androgen excess, peroxisome proliferator-activated receptors and inflammation are the main pathogenetic mechanisms of acne.
- Nutrition seems to play an important role in skin biology and pathology, affecting the onset and clinical manifestation of several dermatologic disorders, including acne.
- The typical Western diet consists of numerous dairy sources and foods with high glycemic indices.
- A study performed by Adebamowo *et al.* (2005) demonstrated the association between dairy products and acne.
- A study provided by Smith *et al.* (2007) showed the link between high glycemic load intake of carbohydrates and acne.
- High glycemic carbohydrates and milk appear to raise serum insulin levels, free IGF-1 and insulin resistance, thus contributing to the pathogenesis of acne.
- IGF-1 seems to be the most important acneigenic factor contained in diet.
- At the genomic level, the effects of insulin and IGF-1 are mediated by the nuclear concentration of the transcription factor FoxO1.
- At the promoter level, SREBP-1c expression is suppressed by nuclear FoxO1, which is an important co-repressor of the retinoid X receptor and liver X receptor.
- Dermatologists should be able to include dietary restriction in acne therapy management.

Summary points

- Skin reflects individual age, health and beauty.
- Nutritional customs affect several skin diseases including psoriasis, atopic dermatitis and acne.
- Epidemiological studies with milk and dairy products support the association of milk consumption with acne onset and clinical course.
- High glycemic load diets are also considered to be involved in acne pathogenesis because of the consequent hyperglycemia and hyperinsulinemia.
- Dermatologists should include restrictive dietary management in acne therapy in their daily clinical practice.

25. Acne and nutrition

A.I. Liakou[1], C.I. Liakou[2] and C.C. Zouboulis[1]
[1]Departments of Dermatology, Venereology, Allergology and Immunology, Dessau Medical Center, Auenweg 38, 06847 Dessau, Germany; [2]Departments of Internal Medicine, University of Kentucky, 740 S. Limestone K529 Kentucky Clinic, Lexington, KY 40536, USA; christos.zouboulis@klinikum-dessau.de

Abstract

Acne vulgaris, the most common dermatological disease, has currently been associated with androgen excess, peroxisome proliferator-activated receptors and inflammation. There are several factors contributing to these mechanisms, including genetics, lifestyle and nutrition. Historically, nutrition has been considered a critical factor in skin biology, health and beauty. Nutrition habits have been shown to affect the onset and clinical course of various skin diseases, including atopic dermatitis, hair loss and acne. The relationship between acne and diet was first reported in the 19th century. The typical Western diet of numerous dairy sources and foods with high glycemic indices has been associated with acneigenic activity. Diets composed of foods with a high glycemic index produce hyperglycemia, reactive hyperinsulinemia and increased formation of free insulin-like growth factors that contribute to acne pathogenesis. In addition, milk and further dairy products seem to consist of more than 60 molecules able to cause a glycemic increase and an insulinemic response. The current challenge is to prove the correlation between diet and severity of acne as well as to provide a restrictive dietary management for assistance of every acne- and non-acne patient.

Keywords: acne, diet, milk, high glycemic load, insulin, IGF-1, FoxO1

Abbreviations

FoxO1	Forkhead BoxO1
IGF	Insulin-like growth factor
IGFBP	Insulin-like growth factor binding protein
PPAR	Peroxisome proliferator-activating receptors
SREBP	Sterol regulatory element-binding protein

25.1 Introduction

The association between acne and nutrition was first reported in 1887, and the dairy restriction was part of standard acne therapy until the 1960s. The second half of the 20th century has tended to minimize the relationship between diet and acne, and this relationship remains controversial today (Figure 25.1 and Figure 25.2). The first studies of Fulton *et al.* in 1969 and Anderson in 1971 failed to show any associations between diet and acne, resulting in the rejection of the role of diet in acne by the scientific dermatological community. The first randomized placebo-controlled study was provided by Smith *et al.* (2007). This study demonstrated a link between high intake of carbohydrates and acne. High glycemic carbohydrates as well as milk appeared to raise serum insulin levels, serum insulin-like growth factor-1 and insulin resistance, contributing to the pathogenesis of acne (Figure 25.3).

Prior to 2005, case-control and clinical studies failed to provide adequate results. But in 2005, Adebamowo *et al.* (2005) showed a solid association between dairy and acne, and the challenging task of elucidating the link remains on-going. A well-designed meta-analysis by Spencer *et*

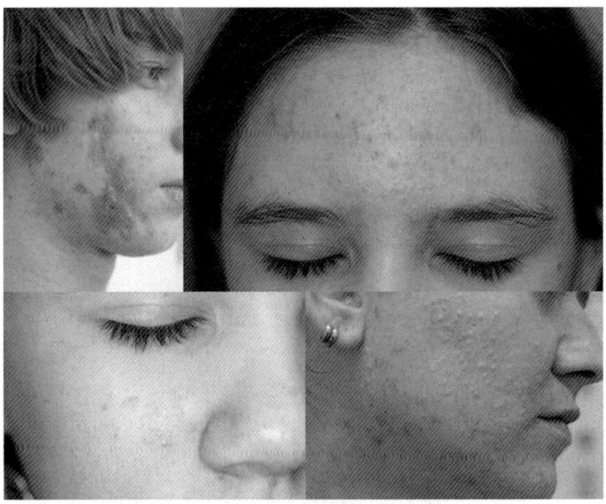

Figure 25.1. The different "faces" of acne. Acne is mostly localized on face and includes many different types, such as acne comedonica, papulopustulosa and conglobata.

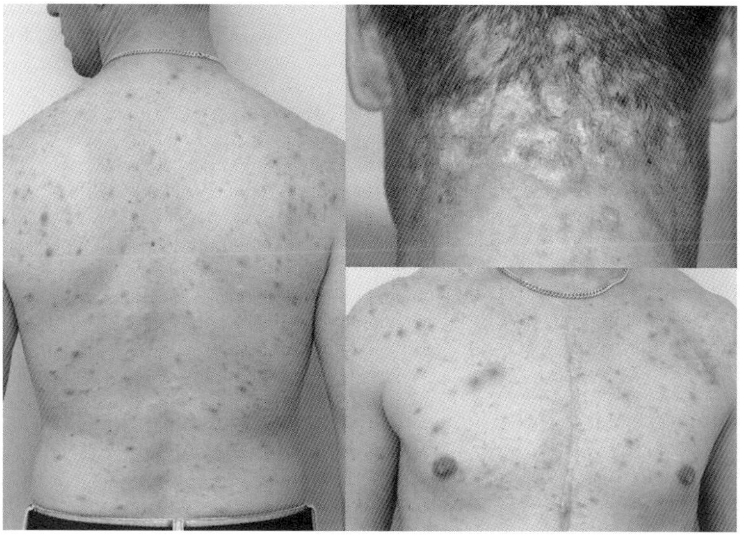

Figure 25.2. Acne at different body locations.

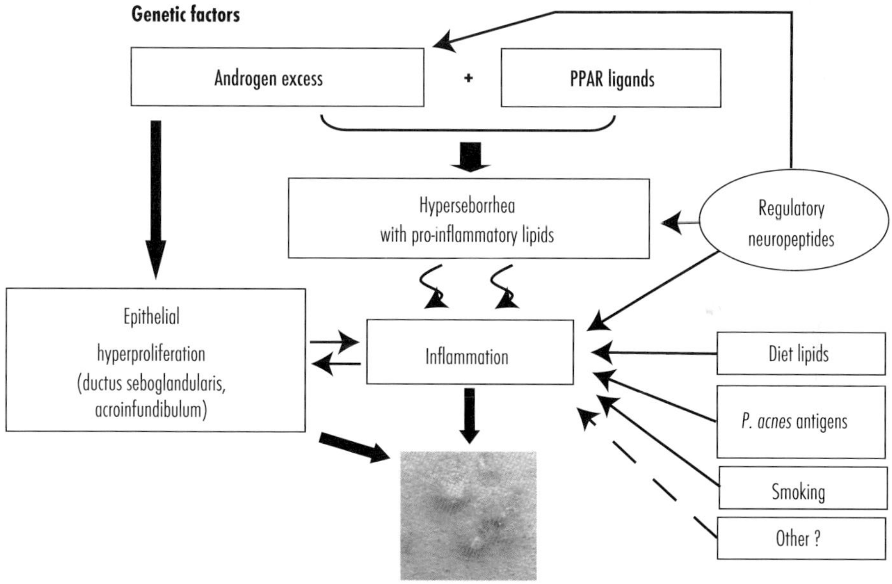

Figure 25.3. Modern aspects of acne pathogenesis. Androgens, peroxisome proliferator-activating receptor (PPAR) ligands, regulatory neuropeptides with hormonal and non-hormonal activity and environmental factors induce hyperseborrhoea, epithelial hyperproliferation in the ductus seboglandularis and the acroinfundibulum and expression of proinflammatory chemokines/cytokines with comedones and inflammatory acne lesions.

al. in 2009 provided substantial evidence for an association between acne and high glycemic carbohydrates as well as dairy products. Bowe *et al.* (2010) recently provided compelling evidence for the association of acne and high glycemic load diets as well as weak evidence for an association between dairy ingestion and acne (Table 25.1).

It is interesting to notice that Eskimos, Chinese and residents of Okinawa Island have shown an elevated acne incidence after changing their dairy habits (Nast *et al.*, 2010). The typical Western diet consists of numerous dairy sources and foods with high glycemic indices. The fact that an introduction to the Western diet contributed to the elevation of acne supports the theory of the diet being associated with the pathogenesis of acne. The Western diet includes a lower amount of ω3-fattic acids and antioxidant vitamins as well as a larger amount of proinflammatory ω6-fattic acids. This diet along with high glycemic carbohydrates and milk, are considered the main aggravating factors in acne pathogenesis. Of these three factors, milk recently provoked an ongoing debate over its level of contribution to acne development.

25.2 Acne and milk

Milk and other dairy products contain more than 60 molecules including prolactin, somatostatin, gonadotropin-releasing hormone, luteinizing hormone, thyroid-stimulating and thyreotropin-releasing hormones, insulin, epidermal growth factor, nerve growth factor, IGF-1 and 2, transforming growth factors, vitamin D, transferrin, lactoferrin and prostaglandins (Koldovsky, 1995). This makes it difficult to distinguish which of these factors could have an acneigenic effect, especially when this fact is combined with the broad range of dairy products (Table 25.2, Figure 25.4).

The most important factor of the ones mentioned above is the insulin-like growth factor. The IGFs are proteins with high sequence similarity to insulin. IGFs are part of a complex system that cells use to communicate with their physiologic environment. Cow milk contains IGF-1 and -2, even after pasteurization and homogenization, and bovine and human IGF-1 share exactly the same amino acid sequence (Melnik and Schmitz, 2009). High milk consumption increases

Table 25.1. Clinical studies that showed an association between acne and nutrition.

Fulton *et al.*, 1969
Anderson *et al.*, 1971
Adebamowo *et al.*, 2005
Smith *et al.*, 2007
Spencer *et al.*, 2009
Bowe *et al.*, 2010

Table 25.2. Numerous molecules are contained in milk and dairy products. This fact induces difficulties regarding which of them is critical for acne pathogenesis..

Prolactin
Somatostatin
Gonadotropin-releasing hormone
Luteinizing hormone
Thyroid-stimulating hormone
Thyreotropin-releasing hormone
Epidermal growth factor
Insulin-like growth factor 1
Insulin-like growth factor 2
Insulin
Vitamin D
Transferrin
Lactoferrin
Prostaglandins

Figure 25.4. The broad range of dairy products makes it difficult to distinguish which of the containing molecules most impair acne onset and clinical course.

IGF-1 levels 10%-20% in adults and 20%-30% in children (Hoppe *et al.*, 2004 a, b) and milk and dairy products raise IGF-1 levels more than dietary proteins such as meat (Hoppe *et al.*, 2005).

Milk also contains carbohydrates, including lactose, and therefore its consumption produces a glycemic response and an insulinemic response. The insulinemic response to ingested milk is actually three to six times what would be predicted from the carbohydrate load in the milk serving (Ostman *et al.*, 2001). This happens for skimmed and full-fat milk, but not for cheese

(Holt *et al.*, 1997; Hoyt *et al.*, 2005). The reasons are not yet understood, but they may relate to the insulinotropic effects of some of the other multiple hormones that are present in milk (Koldovsky, 1995).

A glass of milk added to a low glycemic index meal can boost the insulin response up to 300% of the level produced by a high glycemic index meal and cow milk-formula does this even better than human breast milk (Liljeberg and Bjorck, 2001; Lucas *et al.*, 1980). Different studies suggest that insulin rises in response to the whey component (20% of milk protein), whereas casein is responsible for the IGF-1 increase (Hoppe *et al.*, 2006). Because whey and casein are both involved in stimulating androgen production, there is little point in further differentiating them in dietary restriction, since both should be avoided.

25.3 Acne and high glycemic load

High glycemic load diets may result in increased androgen activity and IGF-1, thereby promoting the development of acne. That was the conclusion of the randomized controlled trial of Smith *et al.* (2007) that demonstrated significant improvement of acne severity in 23 Australian males aged 15-25 who adhered to a low glycemic load diet. The low glycemic load diet resulted in significant reductions in weight, body mass index, free androgen index, increased IGF-binding protein (IGFBP)-1 serum levels with reduced bioavailability of free IGF-1 and improved insulin sensitivity (Smith *et al.*, 2006). Serum IGFBP-1 and IGFBP-3 increased from baseline in the low glycemic load group, whereas sex hormone-binding globulin decreased from baseline in the high glycemic load group (Smith *et al.*, 2007).

At the genomic level, the effects of insulin and IGF-1 are mediated by the nuclear concentration of the transcription factor FoxO1 that functions as a metabolic sensor (Essaghir *et al.*, 2009; Van der Heide *et al.*, 2004). Increased insulin/IGF-1-signaling is known to reduce the nuclear content of FoxO1, and it has been suggested FoxO1 plays an important role in the regulation of acne (Melnik, 2010). At the promoter level, SREBP-1c expression is suppressed by nuclear FoxO1, which is an important co-repressor of the retinoid X receptor and liver X receptor. Insulin/IGF-1-mediated regulation of nuclear FoxO1 activity observed in various cell types perfectly matches with the diet-induced changes and degree of fatty acid desaturation of sebaceous lipids in response to either high or low glycemic load diets (Melnik and Schmitz, 2009; Smith *et al.*, 2008).

25.4 Conclusion

The skin is the largest organ in the body. As such, it provides the first impression of individual biologic condition, age and beauty. It has long been believed that nutrition impacts overall skin well-being and health in such a manner that predisposes the onset or recurrence of various dermatologic disorders and is related to their pathogenesis and clinical manifestations. The typical Western diet, consisting of numerous dairy sources and foods with high glycemic indices appears to effect serum insulin and IGF-1 levels, thereby promoting the androgens that are at the basis of acne pathogenesis and development. Understanding the relationship between acne and nutrition remains challenging, but the first data oblige dermatologists to discuss restrictive dietary management with acne patients and include it in their daily clinical practice.

References

Adebamowo, C., Spiegelman, D., Danby W., Frazier, L., Willett, W. and Holmes, M., 2005. High school dietary dairy intake and teenage acne. Journal of the American Academy of Dermatology 52, 207-214.

Anderson, P.C., 1971. Foods as the cause of acne. American Family Physician 3, 102-103.

Bowe, W.P., Joshi, S.S. and Shalita, A.R., 2010. Diet and acne. Journal of the American Academy of Dermatology 63, 124-141.

Essaghir, A., Dif, N., Marbehant, C.Y., Coffer, P.J. and Demoulin J., 2009. The transcription of FOXO genes is stimulated by FOXO3 and repressed by growth factors. Journal of Biological Chemistry 284, 10334-10342.

Fulton, J., Plewig, G. and Kligman A., 1969. Effect of chocolate on acne vulgaris. The Journal of the American Medical Association 210, 2071-2074.

Holt, S.H., Miller, J.C. and Petocz, P., 1997. An insulin index of foods: the insulin demand generated by 1000-kJ portions of common foods. American Journal of Clinical Nutrition 66, 1264-1276.

Hoppe, C., Mølgaard, C., Juul, A. and Michaelsen, K.F., 2004. High intakes of skimmed milk, but not meat, increase serum IGF-I and IGFBP-3 in eight-year-old boys. European Journal of Clinical Nutrition 58, 1211-1216.

Hoppe, C., Udam, T.R., Lauritzen, L., Mølgaard, C., Juul, A. and Michaelsen, K.F., 2004. Animal protein intake, serum insulin-like growth factor I, and growth in healthy 2.5-y-old Danish children. American Journal of Clinical Nutrition 80, 447-452.

Hoppe, C., Mølgaard, C., Vaag, A., Barkholt, V. and Michaelsen, K.F., 2005. High intakes of milk, but not meat, increase s-insulin and insulin resistance in 8-year-old boys. European Journal of Clinical Nutrition 59, 393-398.

Hoppe, C., Mølgaard, C. and Michaelsen, K.F., 2006. Cow's milk and linear growth in industrialized and developing countries. Annual Review of Nutrition 26, 131-173.

Hoyt, G., Hickey, M.S. and Cordain, L., 2005. Dissociation of the glycaemic and insulinaemic responses to whole and skimmed milk. British Journal of Nutrition 93, 175-177.

Koldovsky, O., 1995. Hormones in milk. Vitamins and Hormones 50, 77-149.

Liljeberg, E.H. and Bjorck, I., 2001. Milk as a supplement to mixed meals may elevate postprandial insulinaemia. European Journal of Clinical Nutrition 55, 994-999.

Lucas, A., Blackburn, A.M., Aynsley-Green, A., Sarson, D.L., Adrian, T.E. and Bloom, S.R., 1980. Breast vs. bottle: endocrine responses are different with formula feeding. The Lancet 1, 1267-1269.

Melnik, B.C. and Schmitz, G., 2009. Role of insulin, insulin-like growth factor-1, hyperglycemic food and milk consumption in the pathogenesis of acne vulgaris. Experimental Dermatology 18, 833-841.

Melnik, B.C., 2010. FoxO1 – the key for the pathogenesis and therapy of acne? Journal der Deutschen Dermatologischen Gesellschaft 8, 105-113.

Nast, A., Bayerl, C., Borelli, C., Degitz, K., Dirschka, T., Erdmann, R., Fluhr, J., Gieler, U., Hartwig, R., Meigel, E., Möller, S., Ochsendorf, F., Podda, M., Rabe, T., Rzany, B., Sammain, A., Schink, S., Zouboulis C.C. and Gollnick, H., 2010. S2k-Leitlinie zur Therapie der Akne. Journal der Deutschen Dermatologischen Gesellschaft 2, 1-59.

Ostman, E.M., Liljeberg H. and Björck, I., 2001. Inconsistency between glycemic and insulinemic responses to regular and fermented milk products. American Journal of Clinical Nutrition 74, 96-100.

Smith, R.N., Mann, N.J., Braue, A. and Mäkeläinen, H., 2007. The effect of a high-protein, low glycemic-load diet versus a conventional, high glycemic-load diet on biochemical parameters associated with acne vulgaris: a randomized, investigator-masked, controlled trial. Journal of the American Academy of Dermatology 57, 247-256.

Smith, R.N., Mann, N.J., Braue, A., Mäkeläinen, H. and Varigos G.A., 2007. A low-glycemic-load diet improves symptoms in acne vulgaris patients: a randomized controlled trial. American Journal of Clinical Nutrition 86, 107-115.

Smith, R.N., Braue, A., Varigos, G.A. and Mann, J.N., 2008. The effect of a low glycemic load diet on acne vulgaris and the fatty acid composition of skin surface triglyceride. Journal of Dermatological Science 50, 41-52.

Smith, T.M., Cong, Z., Gilliland, K.L., Clawson, G.A. and Thiboutot, D.M., 2006. Insulin-like growth factor-1 induces lipid production in human SEB-1 sebocytes via sterol response element binding protein-1. The Journal of Investigative Dermatology 126, 1226-1232.

Spencer, E.H., Ferdowsian, H.R. and Barnard, N.D., 2009. Diet and acne: a review of the evidence. International Journal of Dermatology 48, 339-347.

Van der Heide, L.P., Hoekman, M.F. and Smidt, M.P., 2004. The ins and outs of FoxO shuttling: mechanisms of FoxO translocation and transcriptional regulation. Biochemical Journal 380, 297-309.

Key facts

- Food allergies are an essential cause for atopic dermatitis (AD) and should be excluded in AD patients.
- Non-IgE-mediated food allergy (NFA) is also an critical cause for atopic dermatitis.
- In 2000, IFN-γ was tried to induce immune tolerance for allergy to house dust mites.
- In 2003, oral immunotherpy for food allergy was successfully tried for the first time in NFA by Noh and Lee. Specific oral tolerance induction (SOTI) and tolerance induction for food allergy (TIFA) was termed in this trial.
- Since 2004, several invesitgators have tried SOTI but with limited success.
- In 2009, SOTI for IFA was completed using IFN-γ as an adjuvant.
- In 2010, a differential diagnosis and a relevant treatment of IFA and NFA were analysed and established.

Summary points

- Over 50% of atopic dermatitis patients have food allergies to more than one food, with an average of 2.8 items of food.
- Food allergies should be assessed in atopic dermatitis to facilitate effective disease control. Through basic laboratory tests, the dynamic status and prediction of food allergies should be suspected in atopic dermatitis patients.
- Diagnosis is the first step for the management of food allergies in atopic dermatitis. IgE-mediated food allergies and non-IgE-mediated food allergies should be discriminated during the diagnosis and treatment.
- IgE-based laboratory data should be applied only to IgE-mediated food allergies.
- Non-IgE-mediated food allergies should be considered in atopic dermatitis, and their diagnosis should be made by properly.
- The presence of food allergies should be suspected with high blood eosinophil counts.
- Elimination diets are the easy and simple way to suspect the presence of food allergies. An initial elimination diet and laboratory test may confirm the causative role of food allergies in atopic dermatitis.
- Food challenge tests should be conducted for the exact diagnosis of food allergies.
- Oral food challenges should be conducted properly for IgE-mediated food allergy and non-IgE-mediated food allergy with relevant protocols. Oral food challenges for non-IgE-mediated food allergy generally precede those for IgE-mediated food allergy.
- IFN-γ improves atopic dermatitis with the proper control of allergy provocation. IFN-γ does not affect oral food challenge results because it does not prevent allergy provocation. Moreover, IFN-γ has tolerogenic effects for allergens.
- Once a food allergy diagnosis is made, the therapeutic and management plan should be articulated.
- Tolerance induction is now available and effective. Moreover, tolerance induction for non-IgE-mediated food allergy is a simple, easy and safe. If it is indicated, tolerance induction should be applied; it is simple and safe for non-IgE-mediated food allergy and can be actively considered as a treatment in non-IgE-mediated food allergy.

26. Food allergy and atopic dermatitis

G. Noh[1], J.H. Lee[1,2] and S.S. Lee[3]
[1]Subdivision of Allergy and Clinical Immunology, Department of Paediatrics, Chungnam National University Hospital, 33 Munwha-ro, Jung-gu, Daejoen 301-721, Korea;[2]Department of Paediatrics, Chungnam National University, College of Medicine, 640 Daesa-dong, Jung-gu, Daejeon 301-721, Korea; [3]Department of Food and Nutrition, College of Human Ecology, Hanyang University, 17 Haengdang-dong, Seongdong-gu, Seoul 133-791, Korea; admyth@naver.com

Abstract

Food allergies are an important cause of atopic dermatitis. While the association between eczema and food allergies is well established, effective guidelines for dietary modifications in children with eczema remain unavailable. Food allergies can be IgE or non-IgE mediated and food allergies should be approached for both IgE-mediated food allergy and non-IgE-mediated food allergy. But tests based on IgE levels cannot predict the likelihood of all eczematous reactions, as some reactions occur in the absence of detectable food-specific IgE levels. Oral food challenges are the gold standard for food allergy testing and can allow for recognition of both early and late cutaneous reactions, including eczematous reactions. An elimination diet should precede the oral food challenge, along with appropriate control of other allergens such as dust mites, aggravating factors and infections. Reduced serum eosinophil levels following an elimination diet protocol strongly suggest food allergy as a cause of atopic dermatitis. After allergenic foods resulting in atopic dermatitis are determined through oral food challenge using a relevant protocol for IgE-mediated food allergy and non-IgE-mediated food allergy, allergenic foods can be managed by avoidance or by tolerance induction of the allergenic foods. Oral immunotherapy for food allergies is remarkably advanced, and specific oral tolerance induction is actively recommended for food allergies when indicated. Specific oral tolerance induction using IFN-γ has been particularly well established. A definitive distinction between IgE-mediated food allergy and non-IgE-mediated food allergy is absolutely necessary because the diagnostic and therapeutic procedures differ significantly. This chapter highlights the conceptual immunologic background and the practical effective control of food allergies in atopic dermatitis, including the diagnosis and management of these conditions.

Keywords: food allergy, atopic dermatitis, IgE-mediated food allergy, non-IgE-mediated food allergy, tolerance, tolerance induction, IFN-γ

Abbreviations

AD	Atopic dermatitis
APT	Atopic patch testing
DBPCFC	Double-blinded placebo-controlled food challenge
ECP	Eosinophil cationic protein
FA	Food allergy
IFA	IgE-mediated food allergy
NFA	Non-IgE-mediated food allergy
OFC	Oral food challenge
sIgE	Allergen-specific IgE
SPT	Skin prick test
TIFA	Tolerance induction for food allergy
TIgE	Serum total IgE
SOTI	Specific oral tolerance induction

26.1 Introduction

Food allergy is an important cause of atopic dermatitis (Suh, 2010). Food allergies can be IgE or non-IgE mediated. A lack of diagnostic modalities, such as laboratory tests for NFA (Jyonouchi, 2008), contributes to situations in which many clinicians and scientists are confused about role of food allergy in atopic dermatitis. Diagnostic IgE values determined for specific foods may be helpful in determining the foods that are highly likely to result in immediate-type clinical manifestations, such as urticaria, erythema, or pruritus, which can also lead to excessive scratching and exacerbation of preexisting AD. However, the clinical symptoms and signs of atopic dermatitis are late eczematous lesions that are compatible with those seen in non-IgE-mediated allergy. Tests based on IgE levels cannot predict the likelihood of all eczematous reactions, as some occur in the absence of detectable food-specific IgE levels.

OFCs are the gold standard for food allergy testing and allow for the recognition of both early and late cutaneous reactions, including eczematous reactions. Recently, oral tolerance induction protocols for food allergies were shown to be successful, and they have subsequently become an essential part of the treatment for atopic dermatitis. OFC is absolutely necessary for the management of atopic dermatitis as a basic step. Fortunately, effective management protocols for the diagnosis and oral immunotherapy of food allergies in atopic dermatitis have recently become available.

Avoidance has been suggested as the main principle for the treatment of both food allergies and atopic dermatitis. With the advent of successful tolerance induction for food allergies using IFN-γ, the causative therapy for food allergy has been suggested in atopic dermatitis. The clinical and laboratory characteristics of IFA and NFA in atopic dermatitis have recently been well-characterized, and the relevant treatment for tolerance induction for food allergy in atopic dermatitis is markedly advanced.

Based on large-scale studies, a number of guidelines have been suggested for food allergies in atopic dermatitis. However, these guidelines are primarily focused on IFA, despite observations that NFA also plays an important role in atopic dermatitis. Therefore, guidelines focused exclusively on IFA not only have critical limitations in their application to food allergy in atopic dermatitis, but are also clinically ineffective in controlling the symptoms of these conditions. The principles of appropriate diagnosis and management of food allergy in atopic dermatitis based on immunologic concepts for both IFA and NFA are reviewed here, and integrated guidelines for dietary modifications to control atopic dermatitis in both of these diseases are suggested in this chapter.

26.2 Atopic dermatitis

26.2.1 Definition and general description of atopic dermatitis

Atopic dermatitis is a chronic relapsing inflammatory skin disease commonly associated with atopy that affects 15% to 30% of children (Bieber, 2008) AD is frequently the initial manifestation of atopy and often affects very young children early within the 1st year of life. Many children who develop AD go on to develop other allergic diseases such as food allergy, asthma, or allergic rhinitis (the "atopic march") (Hahn and Bacharier, 2005). The clinical features of AD are well described and include, pruritus, a relapsing eczematous rash typically found over flexor surfaces, and a personal history of atopy. Minor features include elevated serum IgE levels, cutaneous skin infections (*Staphylococcus*, *Streptococcus*, or fungal infections); Dennie-Morgan lines, keratoconus, allergic shiners, white dermatographism, ichthyosis, non-specific hand/food dermatitis, and cheilitis (Hanifin and Rajka, 1980).

26.2.2 Immunopathogenesis

The role of genetic factors in AD has been clearly demonstrated by twin studies (Schultz Larsen and Holm, 1985). The importance of genetic factors in AD is further underlined by the finding that a positive parental history is the strongest risk factor for AD; the incidence rate is doubled if AD is present in one parent and tripled if both parents are affected. One of the major hallmarks of AD is xerosis, which affects lesional and non-lesional skin areas and is characterized by increased transepidermal water loss. Recently, physical injury due to various causes (most commonly dry skin (xerosis) and pruritus) changes the immunologic skin status into allergic tendency (Th2) immunologically. In this situation, a patient may acquire an allergy to an exposed allergen if the skin is injured due to itching or dryness. For these reasons, physicians, patients and parents should keep in mind to care skin barrier functions as well as xerosis and pruritis (Figure 26.1).

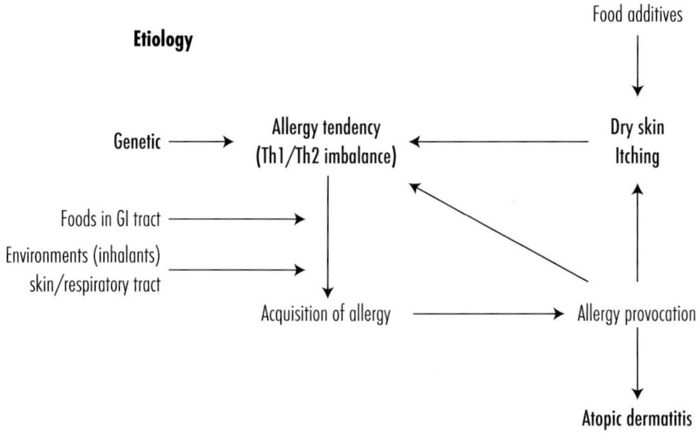

Figure 26.1. Structure of atopic dermatitis. Due to their genetic background or acquired conditions, patients display allergies to foods or environmental allergens. Allergy provocation by re-exposure to allergens aggravates the Th1/Th2 imbalance, xerosis and pruritis. Late eczematous reactions of non-IgE-mediated type allergies are characteristic of atopic dermatitis.

A predominant systemic Th2 imbalance with increased IgE levels and eosinophilia is widely accepted in the pathogenesis of atopic diseases (Ricci *et al.*, 2010). The production of Th2-mediated cytokines, notably IL-4, IL-5, and IL-13, can be detected in lesional and non-lesional skin during the acute phase of disease. IL-4 and IL-13 are implicated in the initial phase of tissue inflammation and may mediate an isotype switching to IgE synthesis. IL-5 increases the survival of eosinophils and results in a systemic eosinophilia, with an increase in ECP correlating to AD disease severity.

26.2.3 Two kinds of food allergy

There are two kinds of food allergies (IFA and NFA), although these two food allergies are also known to participate simultaneously in the pathogenic roles of AD (mixed type). IFA is mediated through IgE via antibody-mediated allergic responses, while NFA is a cellular-mediated event that is dependent on T cells (Figure 26.2). Although the cytokine profile expressed by these T cells in AD is characterized by IL-4, IL-5, and IL-13, a Th2 milieu similar to that seen in IgE-mediated reactions has also been noted (Hamid *et al.*, 1994).

Immediate-type, IgE-mediated cutaneous reactions such as urticaria and erythema have been overestimated because these reactions are more visible and readily attributable to food exposure due to their rapid development, but they do not represent flares of AD (Suh, 2010). By contrast, eczematous flares that occur as delayed-type hypersensitivity reactions are generally non-IgE mediated. These reactions may be overlooked because they develop 2 or more hours after food challenges, and this delay may render a correlation between the food exposure and the reaction more difficult.

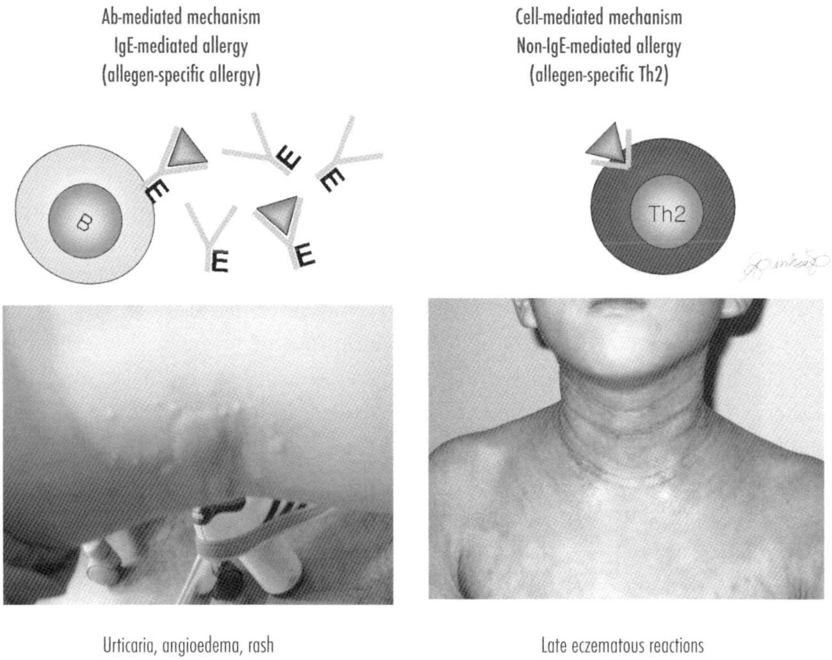

Figure 26.2. Two types of food allergy: IFA and NFA. (1) IFA occurs early and can be easily correlated to food exposure. The skin reactions commonly reported include urticaria, erythema, and pruritus; on occasion "papulation" was reported. These reactions are mistakenly overinterpreted as flares of AD. (2) NFA, a true food-allergy-related flare of AD, occurs when food exposure results in an eczematous reaction. This usually occurs late after a food challenge, but can infrequently occur early.

26.2.4 Tolerance as the third component of food allergy and its implications in AD

Food is a foreign body that is necessary for nutrition. Inevitably, food antigens present a continuous challenge throughout life. Humans have adapted to food via mechanisms of immune tolerance. Food allergies should be approached on this conceptual background.

26.3 Clinical features of AD

The clinical phenotype of AD varies with age and may differ during the course of the disease (Ricci *et al.*, 2010). The eczematous lesions may present with acute (oozing, crusted, eroded vesicles or papules on erythematous plaques), subacute (thick and excoriated plaques), or chronic (lichenified, slightly pigmented, excoriated plaques) forms. Furthermore, xerosis and a lowered threshold for itching are usual hallmarks of AD. Pruritus attacks can occur throughout the day and worsen during the night, causing insomnia, exhaustion, and substantial impairment in quality of life.

26.4 Food allergy in AD

Food allergy is strongly associated with AD (Suh, 2010). Food allergy is much more common in children with AD, with an association that ranges from 20% to 80%, but is generally accepted to be around 30%. Not all patients with AD have elevated IgE levels as many as 40% to 80% and with high food-specific IgE levels. Sensitization to food occurs early, peaking at approximately 6 to 9 months of age, and does not generally increase during later childhood. The delayed onset of symptoms in AD (6 hours to as long as 48 hours after challenge) makes observation difficult, although up to 25% of cases will present within 1-2 hours of ingestion (Breuer *et al.*, 2004). Most studies focus mostly on IFA. Although clinical improvements are well described following appropriate elimination diets and OFC for IFA (Lee *et al.*, 2001a), the major role of non-IgE-mediated allergy has come to light during the last decade. Episodes of AD as the late eczematous reactions of NFA are well documented during OFC (Lee *et al.*, 2001b).

As confirmed through several studies, cow's milk, hen's eggs, wheat, soy, and peanuts are responsible for ~75% of food-associated AD (Burks *et al.*, 1988; Burks *et al.*, 1998) in IFA, while for NFA, meats, as well as cow's milk, hen's eggs, soy and wheat are the major causes of AD. Therefore, in determining the appropriate approach for food allergies in AD, the statistics for IFA and NFA should be considered together.

26.5 Food additives in AD as aggravation factors

In a study of AD conducted in Korea, food additives were shown to provoke xerosis (dry skin) and resultant pruritus consistent with that seen in AD (Anonymous, 2008). Food additives are mainly pseudoallergens rather than allergens and do not directly provoke allergy frequently (Vieluf *et al.*, 1999). Among suspected patients with AD, over 30% showed pseudoallergies to food additives in Korean official reports (Anonymous, 2008). Moreover, scratching resulting from dry skin and itching is well known to lead to sensitization to environmental allergens such as dust mites, fungi, pollen, dander etc. These signs and symptoms caused by food additives interfere in the interpretation of OFC and therapeutic progress.

Despite these observations, there is no consensus on whether food additives provoke allergy or are aggravating factors in AD. However, there have been several reports that food additives play a role in AD, at least as aggravating factors. In a recent Korean official report, intake of excessive food additives or oral food challenges aggravated the clinical severity of AD as shown by increased levels of serum ECP (Anonymous, 2008; Worm *et al.*, 2000). Thus, elevated ECP has been used as a marker of pseudoallergic reactions due to food additives, especially when blood eosinophil fractions are normal. Based on these data, it seems that food additives may aggravate the clinical severity of AD, and it is important to restrict food additives during the management of food allergies to avoid confusion of OFC results before the safety of the additives is confirmed.

26.6 IFN-γ in AD and its use in food allergy

IFN-γ has been discarded in the treatment of AD for some time because of Th1 dominance in the chronic lesions of AD; to date, there has been no clarification regarding whether IFN-γ in the lesions is beneficial or harmful. However, IFN-γ is beneficial in AD when it is used properly and with adequate control of allergy provocation (Lee *et al.*, 2001a). In this report, it is demonstrated that IFN-γ therapy improves AD clinically and does not interfere with OFC because IFN-γ does not inhibit allergy provocation.

In most reports, the recurrence of AD is a controversial point regarding the effects of IFN-γ because IFN-γ dose not inhibit allergy provocation. However, this is actually an advantage in the diagnosis and treatment of food allergy in AD because IFN-γ treatment does not interfere with the interpretation of OFC results and with the therapeutic effects of oral tolerance induction. Thus, IFN-γ can be used simultaneously in OFC or oral immunotherapy.

26.7 Diagnosis of food allergy in AD

Food allergy should be approached with the goal of discriminating between IgE- and non-IgE-mediated allergies. Diagnostic protocols for IFA have been proposed by various researchers, institutes and academic societies. However, following successful immunotherapy for both IFA and NFA, not only is a differential diagnosis between IFA and NFA absolutely necessary, but to determine the initial therapeutic dose and incremental doses as well as to follow-up the natural course of food allergy. The precise initial dosage for allergy provocation and its clinical severity must be evaluated precisely especially in IFA.

In conclusion, in the diagnosis of food allergy: (1) the distinction should be made between IFA and NFA and (2) the treatment dosage must be calibrated and the clinical severity assessed for the purposes of therapy and follow-up.

26.7.1 Allergy history

The first suspicion of food allergies usually arises from the parents. Unless major immediate anaphylactic reactions have occurred, history has proven to be an unreliable way to diagnose food allergies (Suh, 2010). Studies have reported a poor correlation of food allergy with medical history, with only a 25%-48% sensitivity and a 72%-97% specificity for IFA. NFA is more difficult for patients and parents to recognize because of delayed reactions, and a history of NFA is inconsistent with the final diagnosis of NFA (Noh *et al.*, 1999).

26.7.2 Skin prick test and specific IgEs for IFA

For IFA, initial testing in an outpatient setting is determined on the basis of the presence of food-specific IgEs. The skin prick test has a high negative predictive value (~95%) and is most informative when it is negative. However, the positive predictive value ranges between 30% and 50% (Suh, 2010). Therefore, the skin prick test is useful for excluding immediate food hypersensitivity, but a positive result may only suggest hypersensitivity. Laboratory testing for food-specific IgEs also has a high negative predictive value, estimated to be 75%, but the positive predictive value can be low, ranging from 20% to 60%. Recently, the diagnostic levels of food-specific IgEs have been determined, and specific IgEs above these levels may offer a positive predictive value of ~95% for food allergy. Although skin prick tests and serum IgE measurements can confirm sensitization, neither on its own can prove clinical allergy to a specific food with reliability or consistency. Moreover, IgE-based laboratory tests will only predict IFA. IgE-based tests are not correlated significantly with NFA in AD (Noh *et al.*, 1999).

26.7.3 Allergy patch test

Eczematous allergic reactions to food have similarities to allergic contact dermatitis in that both are T-lymphocyte mediated; they are associated with food-specific T-lymphocytes (Suh, 2010). These shared features have led to APT as a way to investigate food-induced eczema. APT is carried out similarly to patch tests that are performed in dermatology clinics, with the application of small amounts of food allergens to a clear area on a patient's back. The application sites are checked for contact urticaria at 20 minutes and again at 48 and 72 hours. Smaller reports have supported the use of APT in combination with IgE testing to increase the positive predictive value for diagnosing food allergy, thereby bypassing the need for OFCs.

26.7.4 NFA

The decision points of skin prick tests and specific IgEs for food allergies are based on immediate-type reactions and are not meant to predict risk for late eczematous reactions to foods (Suh, 2010). Clinical statistics for NFA are absolutely necessary in AD. The prevalence and food items for the cause of NFA exhibit different clinical characteristics in AD. It is noticeable that meats (beef, chicken, pork and its related meats) also are important causes for NFA in AD (Noh *et al.*, 1999). Allergy patch tests are also useful for the diagnosis of NFA in AD.

26.7.5 Multiple food allergies in AD

In many cases, multiple foods are the cause of AD, although food allergies are not the only causes of AD. Over 50% of AD patients also exhibited NFA to a mean of 2.8 food items (Noh *et al.*, 1999). Moreover, IFA co-exists simultaneously in AD in addition to NFA. Interestingly, patients with isolated IFA did not have AD in most cases. From these results, it seems clear that NFA is the major type of food allergy in AD, although IFA is the aggravating cause of AD.

26.7.6 Oral food challenge

The "gold standard" test to confirm or disprove food allergy is the OFC, particularly double-blinded, placebo-controlled OFCs (Suh, 2010). OFCs are safe in cases of NFA, while OFCs are time-consuming with the potential for severe reactions in IFA, and they should be performed by experienced health-care professionals who have access to emergency equipment. Despite the mentioned caveats, OFCs are especially useful because observation after food exposure for 24 hours or more can allow for assessment of both early and late reactions to the food, thereby picking up both IgE and non-IgE-mediated processes.

26.7.7 Elimination diets as a first step for the diagnosis of food allergies

Over 50% of AD patients have NFA. Without the control of allergy provocation, AD cannot be controlled effectively. As a first step toward the diagnosis of food allergy, dietary restrictions should be initiated. If a patient shows improvement on an elimination diet, a food allergy to the restricted foods may be strongly suspected. An optimal elimination diet can be achieved by providing a list of the food to be restricted, which in turn can be determined by clinical statistics.

In the case of breast-fed infants, maternal elimination diets are recommended. Once a maternal elimination diet is implemented, improvements are observed in the infant (Rance, 2008). If eczema improves during the elimination diet, the clinical relevance of the eliminated foods should be confirmed by standardized, physician-supervised oral food challenges.

26.8 Laboratory approaches to food allergy in AD

26.8.1 Blood eosinophils

Blood eosinophils are increased within a week by allergic provocation of NFA in AD (Noh *et al.*, 2010). Considering the half-life of eosinophils, recent provocation of non-IgE-mediated allergy within ~1 week is suspected in the setting of increased blood eosinophil levels. Increases in blood eosinophil counts due to NFA provocation by OFC have been well demonstrated.

26.8.2 Total IgE

Total IgE is increased by allergy provocation. IL-4 from immunocytes, including Th2 cells, induces isotype switching to IgE. From the overall data, the strength, frequency, and duration of AD are suspected from a high serum total IgE. Thus, high serum total IgE does not indicate a current allergy provocation or AD disease activity.

26.8.3 IgE-based tests: allergen-specific IgE and skin prick test for food allergens

IgE-based tests imply the status of IgE sensitization described above. IgE sensitization does not necessarily indicate IFA or NFA.

26.8.4 ECP

Eosinophil cationic protein is produced from activated eosinophils. Therefore, eosinophil activity can be extrapolated from blood ECP levels. ECP is uniquely elevated in the patients who ingest artificial chemical food additives. Food-additive hypersensitivity should be suspected when patients show increased blood ECP levels with normal blood eosinophil counts.

26.9 Oral food challenge procedures in AD

26.9.1 Principle of oral food challenge in AD (Goldmann's triad): intake, elimination diet and challenge test

The diagnosis of food allergy follows the standard principle of Goldmann's triad (Figure 26.3) (Lee *et al.*, 2001a). When patients show symptoms and signs after intake of the suspected causative foods, an elimination diet is initiated. If the patient shows improvement with this diet, the eliminated food is strongly suspected as the cause for AD; if the patient shows the clinical manifestation again with subsequent food challenges, the challenged food is confirmed as the cause of AD.

26.9.2 Proper elimination of foods as an absolute condition for OFC

An elimination diet is the first effective way to confirm a food allergy in AD. However, the proper elimination diet is necessary. Unless such a diet is performed, it is likely that elimination diet may be ineffective. A proper elimination diet is essential and should not be omitted.

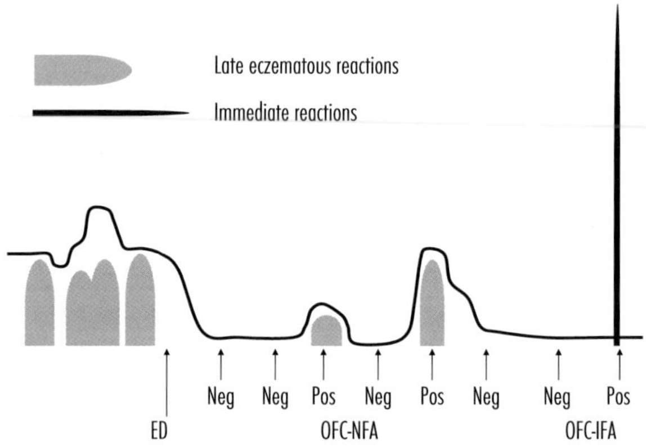

Figure 26.3. OFC in AD. An elimination diet should precede OFC. After the clinical improvement of AD by the elimination diet, OFC is initially conducted for NFA, followed by OFC for IFA.
ED: elimination diet; Pos: positive reaction; Neg: negative reactions; OFC: oral food challenge; IFA: IgE-mediated food allergy; NFA: non-IgE-mediated food allergy.

26.9.3 Indications for an OFC

Indications for an OFC in AD are listed as below (Nowak-Wegrzyn *et al.*, 2009):

- Identify foods causing AD for the initial diagnosis of a food allergy and for monitoring resolution of the food allergy.
- Discriminate between IFA and NFA.
- Calibrate the provoking dose and determine the clinical severity for therapeutic purposes in IFA in order to monitor progress in IFA treatment.
- Expand the diet in persons who have multiple dietary restrictions because of subjective complaints such as headaches or hyperactive behavior.
- Assess the status of tolerance to cross-reactive foods.
- Assess the effect of food processing on food tolerability, e.g. fruits and vegetables that may be tolerated in cooked form in the pollen-food allergy syndrome.

26.9.4 Types of OFC

There are various types of OFC that may be clinically indicated, including open, single-blind, or double-blind, placebo-controlled trials. The choice of the type of OFC is based on a clinical assessment of the potential for bias in the interpretation of the results (Figure 26.1) (Nowak-Wegrzyn *et al.*, 2009).

Open OFC

Open OFC is the unmasked, unblinded administration of a food in its natural form (Nowak-Wegrzyn *et al.*, 2009). Open OFC is recommended for IFA due to its clear clinical manifestation. Moreover, in NFA, open OFC is effective in the clinical setting. NFA can be performed in out-patient base.

It is simple to plan and reproduce the natural exposure in terms of quantity and method of preparation. However, this type of test has the highest potential for bias because the results may depend on age, personality, and type of symptoms. Whereas a clearly negative open OFC rules out reactions to the food, a positive result with only subjective symptoms may need to be confirmed by a blinded challenge.

Open OFC is a cost-efficient procedure that saves substantial time and resources. It is thus considered a reasonable first choice to evaluate an adverse reaction to a food when the need for an OFC has been established.

Blinded OFC

Blinding and masking by mixing the challenge food with a masking vehicle or placing the food in vehicles including opaque capsules reduces bias (Nowak-Wegrzyn *et al.*, 2009). In the single-blind OFC, the observer, but not the patient, knows the food being tested. In the double-blind

OFC, the challenge material is provided by a third party, such as a dietician, whereas the patient, the patient's family, and the observer are unaware of when the test food is given. Thus, bias is minimized. Placebo-controlled challenges may be administered in either a single-blind or double-blind fashion.

Open or blinded OFC

Although DBPCFC is conducted in many studies, open OFC is adequate for IFA because of the clinical severity and the clarity the clinical manifestations of this disease. DBPCFC was tried for NFA, but open OFC is optimal in the clinical setting because of nutritional problems, inconveniences to daily living and the resulting loss of compliance among patients who are being tested for multiple foods (Kim *et al.*, 2002; Noh *et al.*, 2007).

26.10 Treatment of food allergy in AD

26.10.1 Avoidance

The current standard of care for patients with food allergies is based on avoidance of the trigger foods with the goal of eventually gaining tolerance (Kurihara, 2010). In those with suspected food allergies that are thought to be aggravating eczema, food avoidance is therefore important. However, it is also vital that such diets are carefully supervised because of the risk of nutritional deficiencies, particularly if calcium-rich foods are avoided (Fox *et al.*, 2004).

26.10.2 Tolerance induction for food allergies: immunotherapy for food allergy

The development of novel therapeutic modalities targeting the mucosal immune response including oral immunotherapy, has shown great promise in providing a definitive therapy for food allergies (Kurihara, 2010). Trials of TIFA have been increasing in recent years, and there is great hope for positive and radical treatments for food allergies in the future.

TIFA for IFA is a food desensitization process that starts with extremely low doses and gradually builds up to high doses over a period of several months or years. Another innovative approach is oral tolerance induction using IFN-γ within six months. However, for NFA, there is no way to induce tolerance in the clinical field except for oral tolerance induction using IFN-γ.

Oral immunotherapy using IFN-γ for food allergies was described as completely successful in human NFA cases by Noh and Lee, who named this therapy "Specific oral tolerance induction (SOTI)" in 2003. Thereafter, SOTI was attempted for IFA in humans by several investigators without the use of IFN-γ (Scurlock and Jones, 2010). However, its success rate was not high, and several limitations were observed, such as tolerable dosage. The complete SOTI protocol for IFA was accomplished using IFN-γ in 2009 by Noh and Lee. Further beneficial effects of IFN-γ in SOTI using IFN-γ as described below (Lee *et al.*, 2010).

Conventionally, subcutaneous immunotherapy was tried as an allergen immunotherapy via the injection route, but it is known to be ineffective for food allergies. Additionally, other therapeutic modalities are being investigated, including traditional Chinese medicine, humanized monoclonal anti-IgE therapy, peptide immunotherapy, epicutaneous immunotherapy, and sublingual immunotherapy.

26.11 Specific oral tolerance induction

26.11.1 Classical SOTI

There have been several trials of this method, and their relevant protocols showed slight differences with the same basic concepts (Kurihara, 2010). Moreover, the results of these treatments showed differences according to the groups. Generally, in this type of protocol, food allergens are increased gradually from extremely small doses to high doses over a long period of time, up to a year. Several limitations were observed, and relatively low success rates were found compared to protocols that used IFN-γ. This treatment is used only in IFA.

26.11.2 SOTI using IFN-g

Oral immunotherapy for food allergies with SOTI using IFN-γ was well described by Lee *et al.* (2010). Although the immunologic concept is same in IFA and NFA, the basic practical concepts show important differences. IFA showed low dose allergy and tolerance to low dose is very important. Without overcoming allergic reactions to low doses, tolerance to higher doses cannot be achieved for IFA (Figure 26.4). To treat IFA with SOTI using IFN-γ, the initial and incremental doses are extremely low, and the acquisition of tolerance is gradual, involving small, incremental doses. This is in contrast to NFA treatment protocols, in which tolerance to high-dose allergens is the goal. Regardless of the allergy provocation caused by the prior dose, tolerance is achieved with the next dose. Accordingly, the therapeutic period in IFA is relatively longer (generally within 6 months) than the period of NFA (generally 7-10 days). The immunologic mechanisms for the differences in these clinical characteristics have not yet been elucidated.

Currently, the only effective treatment for NFA is SOTI using IFN-γ because this is the only protocol that effectively induces immune tolerance. It is surprising to see tolerance acquisition for periods as short as a week in NFA with the use of IFN-γ. Moreover, IFN-γ has the additional advantages of improving Th1/Th2 imbalance in allergic state, creating tolerogenic effects for other allergens, and enhancing Th1 effects, which increase immunity to infections according to the natures of IFN-γ.

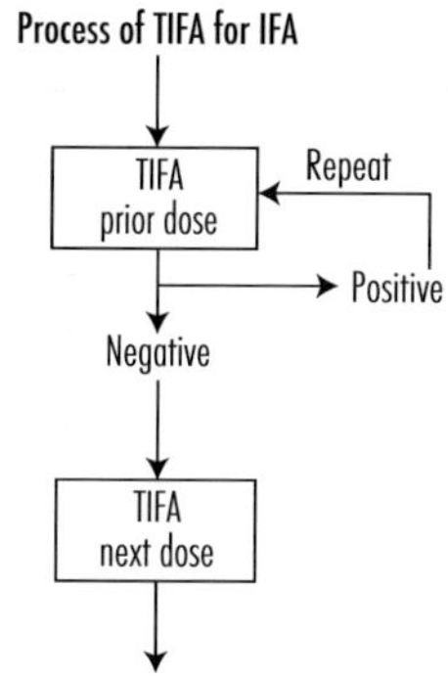

Figure 26.4. Principle of overcome of prior dose. In tolerance induction of food allergy in IFA, tolerance cannot be acquired without the acquisition of tolerance to the prior lower dose. Thus, during tolerance induction of food allergy for IFA, if patients show any clinical manifestations to the prior dose, the treatment should be repeated until the patients do not react to the dose.

26.12 Concluding remarks

Food allergy is strongly associated with AD. The proper diagnosis and management of food allergy is essential for the treatment of AD. It is of paramount importance to distinguish between IFA and NFA in order to manage food allergy in AD. Thus, IgE-based tests should be applied limited to IgE-mediated allergy. After the precise diagnosis of IFA or NFA, the relevant tolerance induction treatment should be given. Through these processes, food allergies may be effectively managed from diagnosis to treatment. Integrated guidelines for both IFA and NFA are suggested in Section 26.13.

26.13 Guidelines for the management of food allergy in AD

There is no available guideline for the management of food allergy in AD because IFA and NFA are not strictly distinguished in terms of their immunologic foundations and clinical presentations. Therefore, in this manual, integrated guidelines for IFA and NFA are suggested.

26.13.1 Diagnosis of food allergy in AD

Diagnosis of AD

When the patients are suspected or confirmed to have AD, food allergies should be investigated as a possible cause. First of all, an exact discrimination between IFA and NFA should be considered in all processes, including the diagnosis and management of AD (Figure 26.5).

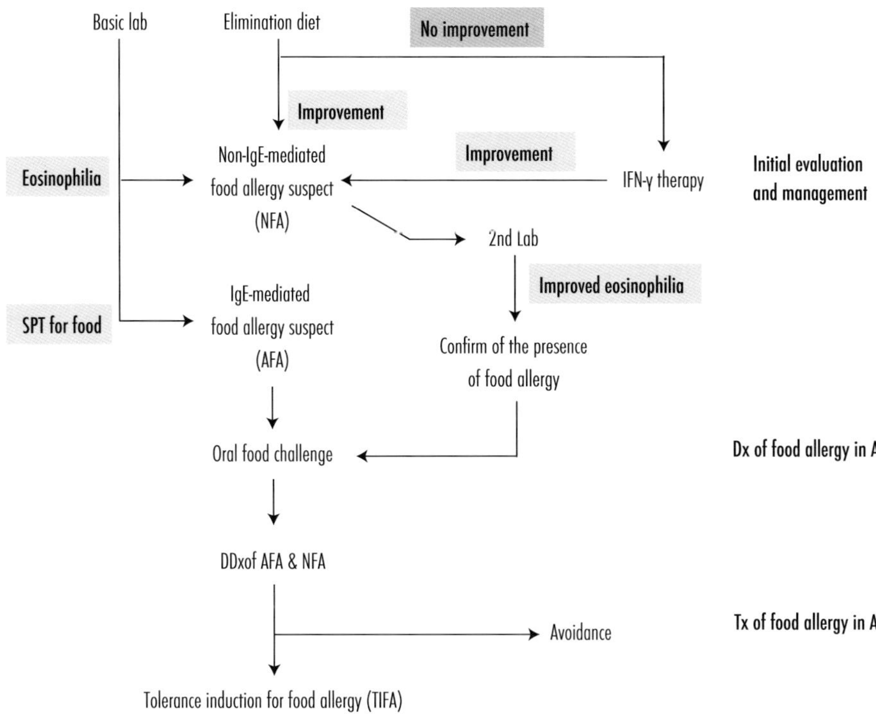

Figure 26.5. Basic concept for food allergy management in AD. Initially, an elimination diet is conducted with basic laboratory tests, including blood eosinophils, serum total IgE, allergen-specific IgE and skin prick tests for allergens. After an initial evaluation, OFC is performed for NFA and IFA. After listing allergenic foods, avoidance or tolerance induction of food allergies may be tried.

Initial management of AD

- History taking, especially concerning food allergies:
  - possible foods causing AD should be evaluated completely;
  - exercise-induced food allergies should be considered and excluded.
- Clinical severity scoring should be done at:
  - initial evaluation;
  - every visit;
  - before and after in every procedure.
- Dietary record:
  In the whole processes of management for food allergies in AD, daily diet diaries should be used to record complete basic recipes for each meal. All food should be recorded for the supervision of dietary management, including elimination diets, open OFCs, and tolerance induction. This is especially important in breast-feeding infants. In such cases, the diet of the mother should be recorded along with the diet of the infant, including times of ingestion, and the time for the appearance of new AD lesions or aggravation of old AD lesions.

- Basic laboratory tests:
  - CBC with differential counts, esp. for eosinophils;
  - serum total IgE;
  - serum eosinophil cationic protein levels;
  - specific IgE for necessary allergens (basic statistics & history);
  - SPT for basic inhalants & foods, as many as can be screened;
  - allergy patch test;
  - other special tests if necessary (i.e. exercise-induced food allergies).
- Elimination diet:
  - initial elimination diet according to the basic statistics, history, SPT;
  - second elimination diet according to the results of allergen-specific IgEs.
- Basic general management:
  - without basic care and the control of allergens other than foods, the precise diagnosis of food allergies may be confusing sometimes;
  - skin care;
  - house dust mite care and other inhalant allergen care (dander, pollen, fungi and other);
  - other causes (infection control, other).
- Prerequisites for AD patients for the diagnosis of food allergies:
  - discontinue all medications that are able to affect allergy provocation (steroids, antihistamines, other immunosuppressants, drugs of unknown effects (herbal medications), etc.);
  - should expect allergy provocation by inhalants or environmental allergens.

Oral food challenge

The diagnosis of food allergies should follow Goldman's principle. The principle of diagnosis for food allergies is different between infants/children and adults. In infants or children who are affected by elimination diets even for a short period, due to nutritional requirements, the principle of the diagnosis of food allergies is to find a safe food that does not provoke allergic reactions; in adults, who are relatively stable with respect to nutritional problems in the short term, the diagnostic principle is to find the allergenic foods that are the cause of AD.

Once patients show improvement through an initial elimination diet, oral food challenges are subsequently conducted. Other foods that are not contained in the initial elimination diet should be completely considered. IFN-γ therapy can be considered when patients do not improve by the initial elimination diet, because IFN-γ therapy itself does not affect allergy provocation during the next OFC.

- The suspected foods should be listed, and a relevant elimination diet should result in an improvement of AD symptoms.
- The order for the OFC should prioritize NFA before IFA. Additionally, the importance of certain foods should be considered according to the patient's wishes.

- OFC for NFA:
  - OFC for NFA should generally precede OFC for IFA. OFC for NFA is conducted according to the relevant protocols (Figure 26.6, Tables 26.1, 26.2, 26.3 and 26.4).
  - OFC for IFA is conducted according to the relevant protocols.

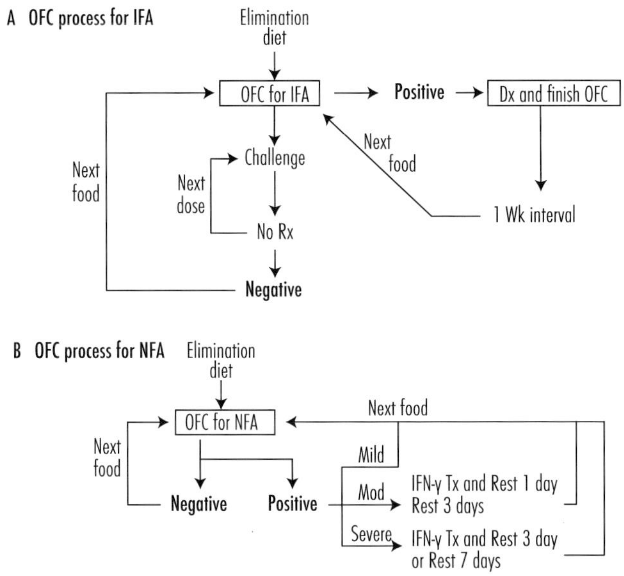

Figure 26.6. OFC process for IFA and NFA. (A) OFC for IFA is conducted by increasing the dose of test foods gradually according to the protocol. After positive reactions, the test is delayed at least for a week. (B) OFC for NFA is performed simply by eating foods as much as possible according to the protocol. The interval for the next test is modified according to the degree of clinical manifestations. IFN-γ can be used for the effective progress of OFC if it is indicated.

Table 26.1. Initial dose for OFC in IFA. In OFC for IFA, the dosage and clinical severity of the allergy provocation is important to determine the initial and incremental doses for tolerance induction. (Unit: milk in ml and soybeans, wheat, eggs in g).

IFA	OFC dose		Wet Wt		Initial/1 portion	
Initial	severe	mod to mild	adult full dose	target dose	severe	mod to mild
Milk	0.1	1	200	300	1/2,000	1/200
Egg	0.025	0.25	50	75	1/2,000	1/200
Wheat	0.1	1	200	300	1/2,000	1/200
Soybean	0.1	1	200	300	1/2,000	1/200

Table 26.2. Progress of OFC for IFA (milk, soybeans, wheat). (Unit: milk in ml and soybeans, wheat in g).

IFA	OFC progression					
Dose	0.1[a]	0.2	0.3	0.5	1[b]	2
Observation time	15 min.	15 min.	15 min.	15 min.	30 min.	30 min.
Dose	5	10	20	50	100	200
Observation time	30 min.	30 min.	1 hr	1 hr	1 hr	1 hr

[a] Initial dose for severe form.
[b] Initial dose for mod to mild form.

Table 26.3. Progress of OFC for IFA (eggs). (Unit: eggs in g).

IFA	OFC progression					
Dose	0.025[a]	0.05	0.75	0.125	0.25[b]	0.5
Observation time	15 min.	15 min.	15 min.	15 min.	30 min.	30 min.
Dose	0.75	1.25	5	12.5	25	50
Observation time	30 min.	30 min.	1 hr	1 hr	1 hr	1 hr

[a] Initial dose for severe form.
[b] Initial dose for mod to mild form.

Table 26.4. Protocols for NFA. The presence of NFA is determined only through OFC. (Unit: milk in ml and soybeans, wheat, eggs in g).

	Milk	Egg	Wheat	Soybeans	Beef	Chicken	Pork
Dose	200	50	100	100	200	120	120

26.13.2 Management of food allergy in AD

Positive foods should be listed with a classification of IFA, NFA, or mixed-form of IFA and NFA. Avoidance should always be attempted as an easy way to control food allergies. However, when indicated, tolerance induction for food allergies is recommended. The basic management of food allergies is modified according to the patients' circumstances, including age and feeding status:

- Before 1 year:
  - waiting & avoidance;
  - tolerance induction may be tried if it is absolutely necessary.
- Before 3 years:
  - waiting & avoidance;
  - tolerance induction may be tried if it is absolutely necessary.
- After 3 years:
  - waiting & avoidance;
  - tolerance induction when it is indicated.

Avoidance

The general principle of management for food allergies is avoidance. However, therapeutic modalities for tolerance induction of food allergies have recently become remarkably advanced.

Tolerance induction of food allergy in AD

Tolerance induction of food allergies is recommended when indicated.
- nutritional deficiency;
- inevitable conditions: patient is unable to avoid allergen (e.g. a wheat allergy in a baker).
- no available formula for infant;
- quality of life for patients or parents.

The immunologic concept and relevant procedures are significantly different between IFA and NFA, whereas basic immunologic concepts for tolerance induction for IFA and NFA are the same. IFA is an allergy at low dose ranges and NFA is an allergy at high dose ranges. The initial and incremental doses are extremely low in IFA, whereas these parameters are larger in NFA. Moreover, the principle of overcome of prior dose is a very important point in the tolerance induction for IFA but is not applied in NFA. Thus, an exact differential diagnosis between IFA and NFA is the most important step in the treatment of food allergies in AD:
- NFA

  In most cases, tolerance induction for NFA should precede tolerance induction for IFA. Tolerance induction for NFA is easy and simple, so the therapeutic term is short. However, foods that are absolutely necessary (if, for example, substitute foods for formula are not affordable) or cannot be avoided in daily livings, should precede other foods regardless of IFA or NFA status. For tolerance induction of NFA, food is challenged gradually from low doses to high doses simultaneously with the administration of IFN-γ.
- IFA

  For tolerance induction in IFA, two methods are present according to the use of IFN-γ. Without IFN-γ, the initial dose and incremental doses are extremely low. However, with IFN-γ, the initial dose is also low, but is much more than the dose given without IFN-γ. The treatment term is up to 6 months with IFN-γ, while without IFN-γ, it may last up to a year.

The protocols for tolerance induction show slight differences with different investigators, although the basic dosing concepts are the same.

- TIFA
  - TIFA for NFA should generally precede TIFA for IFA. TIFA for NFA is conducted according to the relevant protocols (Table 26.5 and 26.6).
  - TIFA for IFA is conducted according to the relevant protocols.

Table 26.5. Dosage protocol for tolerance induction of food allergy in IFA. The initial and incremental doses are determined by the dose and severity that cause minimal allergy provocation. Tolerance induction is conduced according to the schedule.

IFA therapeutic dose		Increment	10% rule									
-10 ml	Severe	0.1	0.1	0.2	0.3	0.4	0.5	0.6	0.7	0.8	0.9	1
		0.2	1.2	1.4	1.6	1.8	2	2.2	2.4	2.6	2.8	3
		0.3	3.3	3.6	3.9	4.2	4.5					
		0.4	4.6	5.0								
		0.5	5.5	6.0	6.5	7.0	7.5	8.0	8.5	9.0	9.5	10
	Mod to mild	1	1	2	3	4	5	6	7	8	9	10
10-300 ml		1	11	12	13	14	15	16	17	18	19	20
		2	22	24	26	28	30					
		3	33	36	39	42	45	48	51	54	57	60
		5	65	70	75	80	85	90	95	100		
		10	110	120	130	140	150	160	170	180	190	200
		20	220	240	260	280	300					

Table 26.6. Dosage protocol for tolerance induction of food allergy in NFA (unit: milk in ml and soybeans, wheat, eggs, beef, pork, chicken in g).

30 kg dose	Milk	Egg	Wheat	Soybeans	Beef	Chicken	Pork
1st	25	5	10	10	25	15	15
2nd	50	10	20	20	50	30	30
3rd	75	15	40	40	75	45	45
4th	100	20	80	80	100	60	60
5th	150	30	120	120	150	90	90
6th	200	50	180	180	200	120	120
7th	300	75	240	240	300	180	180

26.13.3 Special conditions for food allergies in AD

Breast-fed or formula-fed infants

a. Breast-feeding infants
   The diagnosis of food allergies in infants is made through the mother. When the mother eats, the baby is fed with breast milk between 1-2 hrs. of the mother's intake. The reactions of the baby are then observed. The primary purpose of the diagnosis of food allergy is to set up safe recipes for the infant and the mother. Subsequently, the test results will be used during the phase of solid food introduction. Safe (negative) foods will be introduced with priority. OFC is conducted on the mother using normal protocols; the sole difference is that foods given to the mother and reactions are observed in the infant.
b. Formula-feeding infants
   The primary purpose of treatment is to set up safe formula feeding, whether with casein hydrolysate, soy formula or others:
   - When the infant is suffering from AD with formula feeding of cow's milk, casein hydrolysate is tried. If the infant improves with casein hydrolysate, he or she is not allergic to casein and casein hydrolysate may be used.
   - When the infant does not improve with casein hydrolysate, soy formula is tried. If the infant improves with soy formula, soy formula is used as a safe formula.
   - If the infant does not improve with either soy formula or casein hydrolysate, the infant should find another formula. If another formula is not available, it is possible to induce tolerance for cow's milk or soy formulas.
c. Mixed-feeding infant (breast feeding with formula feeding)
   - An initial elimination diet for the mother is conducted; simultaneously, a safe formula feeding is set up as described in (b).
   - After the initiation of safe formula feeding, OFC of mother for the diagnosis of food allergy in the infant should be performed as described in (a).

AD in infants who are being introduced to solid foods (~5-12 months)

a. Breast-feeding baby
   - The primary purpose of diagnosis is to set up a safe recipe for the infant to prevent nutritional deficiencies due to elimination diets or food allergies.
   - Elimination diet should be conducted in the mother; all solid foods should be stopped in the infant (elimination diet of infant).
   - Sequential introduction of solid foods may be performed according to the schedule of weaning. However, the sequence of foods can be changed according to the circumstances of the infant and mother.
   - Mother and infant should then be simultaneously fed with challenge foods. The reactions are observed in the infant.
   - Nutritional care for the infant is essential.

b. Formula-feeding baby
   - The primary purpose of the diagnosis is to set up safe recipes for the infant to prevent nutritional deficiencies due to elimination diets or food allergies as well as to set up safe formula feeding.
   - Stop all solid foods to the infant (elimination diet in the infant) and simultaneously set up a safe formula feeding during the elimination diet in the infant as described in the Section on breast-fed or formula-fed infants (b).
   - Sequential introduction of solid foods should be performed according to the schedule of weaning. However, the sequence of food kinds can be changed according to the circumstances of the infant.
   - Nutritional care for the infant is essential.

c. Mixed-feeding baby
   - Initially, an elimination diet should be initiated in the mother and infant (all solid foods must be stopped); simultaneously, a safe formula feeding should be set up as described in (b).
   - After the initiation of safe formula feeding, OFC of the mother and infant for the diagnosis of food allergy in the infant should be conducted as described in (a).

Children older than 1 year

- Protocols for the management of food allergy in AD are the same as those of adults.
- However, infants or children should be monitored for nutritional deficiency.
- Tolerance induction for food allergy before the age of 3 years may be tried if indicated.
- Tolerance induction for food allergy is indicated after the age of 3 years.

References

Anonymous, 2008. Official report of 'Atopy Safe School Project', Health Guidance Division of the Seodaemun-Gu Health Center, Seoul, Korea.

Bieber, T., 2008 Atopic dermatitis. New England Journal of Medicine 358, 1483-1494.

Breuer, K., Heratizadeh, A., Wulf, A., Baumann, U., Constien, A., Tetau, D., Kapp, A.and Werfel, T., 2004. Late eczematous reactions to food in children with atopic dermatitis. Clinical and Experimental Allergy 34, 817-824.

Burks, A.W., Mallory, S.B., Williams, L.W. and Shirrell, M.A., 1988. Atopic dermatitis: Clinical relevance of food hypersensitivity reactions. The Journal of Pediatrics 113, 447-451.

Burks, A.W., James, J.M., Hiegel, A., Wilson, G., Wheeler, J.G., Jones, S.M. and Zuerlein, N., 1998. Atopic dermatitis and food hypersensitivity reactions. The Journal of Pediatrics 132, 132-136.

Fox, A.T., Du Toit, G., Lang, A. and Lack, G., 2004. Food allergy as a risk factor for nutritional rickets. Pediatric Allergy and Immunology 15, 566-569.

Hahn, E.L. and Bacharier, L.B., 2005. The atopic march: The pattern of allergic disease development in childhood. Immunology and Allergy Clinics of North America 25, 231-246.

Hamid, Q., Boguniewicz, M. and Leung, D.Y., 1994. Differential in situ cytokine gene expression in acute versus chronic atopic dermatitis. Journal of Clinical Investigation 94, 870-876.

Hanifin, J.M. and Rajka, G., 1980. Diagnostic features of atopic dermatitis. Acta Dermato-venereologica (Stockh) 92, 44-47.

Jyonouchi, H., 2008. Non-IgE mediated food allergy, Inflammation and Allergy Drug Targets 3, 173-180.

Kim, T.E., Park, S.W., Noh, G. and Lee, S., 2002. Comparison of skin prick test results between crude allergen extracts from foods and commercial allergen extracts in atopic dermatitis by double-blind placebo-controlled food challenge for milk, egg, and soybean. Yonsei Medical Journal 43, 613-620.

Kurihara, K., 2010. Immunotherapy for Food Allergy. Allergology International. 59, 9-14.

Lee, S.S., Lee, K.Y. and Noh, G., 2001a. The necessity of diet therapy for successful interferon-γ therapy in atopic dermatitis. Yonsei Medical Journal 42, 161-171.

Lee, S.S., Noh, G. and Lee, K.Y., 2001b. Clinical application of histamine prick test for food challenge in atopic dermatitis. Journal of Korean Medical Science 16, 276-282.

Lee, J.H., Noh, G., Noh, J., Lee, S., Choi, W.S., Kim, H.S., Lee, K., Choi, S., Jin, H., Cho, S. and Lee, S., 2010. Clinical characteristics of oral tolerance induction of IgE-mediated and non-IgE-mediated food allergy using interferon gamma. Allergy and Asthma Proceedings 31, e39-e47.

Noh, G., Ji, E.J., Park, J.N., Kim, K.H., Do, M.H., Lee, E.K. and Lee, S.S., 1999. The importance of food open challenge test in atopic dermatitis: The comparison of allergy history, skin-prick test, and specific IgE detection. Nutritional Science 2, 119-124.

Noh, G. and Lee, S.S., 2003. Effects of IFN-γ on milk-specific oral tolerance induction for milk allergy in atopic dermatitis: Food specific oral tolerance induction using IFN-γ as adjuvant. In: Marone, G. (Ed.) Clinical Immunology and Allergy in Medicine. Naples, Italy: JGC Editions, pp. 475-496.

Noh, G., Ahn, H.S., Cho, N.Y., Lee, S. and Oh, J.W., 2007. The clinical significance of food specific IgE/IgG4 in food specific atopic dermatitis. Pediatric Allergy and Immunology 18, 63-70.

Noh, G. and Lee, S.S., 2009. A pilot study of IFN-γ-induced specific oral tolerance induction for IgE-mediated anaphylactic food allergy. Journal of Interferon and Cytokine Research 29, 667-676.

Noh, G., Jin, H., Lee, J., Noh, J., Lee, W.M. and Lee, S.S., 2010 Eosinophilia as a predictor of food allergy in atopic Dermatitis Allergy and Asthma Proceedings 31, e18-e24.

Nowak-Wegrzyn, A., Assa'ad, A.H., Bahna, S.L., Bock, S.A., Sicherer, S.H. and Teuber, S.S., 2009. Work Group report: Oral food challenge testing. Journal Allergy and Clinical Immunology 123, s365-s383.

Rance, F., 2008. Food allergy in children suffering from atopic eczema. Pediatric Allergy and Immunology 19, 279-284.

Ricci, G., Patrizi, A., Giannetti, A., Dondi, A., Bendandi, B. and Masi, M., 2010. Does improvement management of atopic dermatitis influence the appearance of respiratory allergic diseases? A follow-up study. Clinical and Molecular Allergy 8, 8-19.

Schultz Larsen, F.V. and Holm, N.V., 1985. Atopic dermatitis in a population based twin series. Concordance rates and heritability estimation. Acta Dermato-venereologica, Supplementum (Stockh) 114, 159.

Scurlock, A.M. and Jones, S.M., 2010. An update on immunotherapy for food allergy. Current Opinions in Allergy Clininical Immunololgy 10, 587-593.

Suh, K.Y., 2010. Food allergy and atopic dermatitis: separating fact from fiction. Seminars in Cutaneous Medicine and Surgery 29, 72-78.

Vieluf, D., Wieben, A. and Ring, J., 1999. Oral provocation tests with food additives in atopic eczema. International Achieves of Allergy and immunology 118, 232-233.

Worm, M., Ehlers, I., Sterry, W. and Zuberbier, T., 2000. Clinical relevance of food additives in adult patients with atopic dermatitis. Clinical and Experimental Allergy 30, 407-414.

Key facts of gut microbiota

- The collection of microorganisms (e.g. bacteria, archaea, fungi, and viruses) that live in the gastrointestinal tract is referred to as the gut microbiota.
- The human gut microbiota contains 500-1000 bacterial species that belong to over 50 bacterial phyla dominated by the *Bacteroidetes* and the *Firmicutes*.
- The bacterial density in the human gut is rich in the distal region and reaches 10^{12} cells/g contents in the colon.
- The intestinal tract is sterile *in utero*, and the colonization by microorganisms begins immediately at birth.
- In addition to mother's microbiota, many other factors including diet, environmental exposures, and antibiotic therapies contribute to shape the gut microbiota.
- The gut microbiota contributes to host physiology including nutrition, immunology, and behavior.
- The composition of gut microbiota is analyzed by using not only conventional culture techniques, but also culture-independent molecular biological methods based on 16S rRNA gene sequence.

Summary points

- The indigenous gut microbiota present early in life is thought to play an important role in the pathogenesis of allergic diseases including atopic dermatitis.
- Breast milk is protective against atopic dermatitis, and gut microbiota in breast-fed infants is dominated by bifidobacteria, due to the fermentation of non-digestible oligosaccharides in human milk.
- Infant formula supplemented with non-digestible oligosaccharides including fructo-oligosaccharides increases bifidobacteria in the gut.
- Supplementation of the galacto-oligosaccharides/fructo-oligosaccharides mixture to infant formula is protective against atopic dermatitis.
- Supplementation of kestose reduces disease severity in infants with atopic dermatitis.
- Dietary fructo-oligosaccharides modulate the composition of gut microbiota and reduce the allergic skin inflammation in several mouse models.
- Additional well-designed prospective clinical trials, as well as mechanistic studies using appropriate animal models, are required to advance knowledge in this field.

27. Fructo-oligosaccharides and skin inflammation

K. Sonoyama
Division of Applied Bioscience, Research Faculty of Agriculture, Hokkaido University, Kita-9, Nishi-9, Kita-ku, Sapporo 060-8589, Japan; ksnym@chem.agr.hokudai.ac.jp

Abstract

The indigenous gut microbiota present early in life is thought to play an important role in the pathogenesis of allergic diseases including atopic dermatitis, and strategies to manipulate the gut microbiota in infancy have been considered in preventing the development of such diseases. Gut microbiota in breast-fed infants is dominated by bifidobacteria, and breast milk is protective against atopic dermatitis. Considering that fermentation of breast milk oligosaccharides in the gut results in the proliferation of bifidobacteria, supplementation of non-digestible oligosaccharides, such as fructo-oligosaccharides, to infant formula could increase bifidobacteria in the gut, which may in turn protect against atopic dermatitis. Indeed, recent human interventions have demonstrated that supplementation of oligosaccharide mixtures containing high-molecular mass fructo-oligosaccharides to infant formula reduced the incidence of atopic dermatitis. In addition, supplementation of kestose, a low-molecular mass fructo-oligosaccharide, reportedly reduced the disease severity in infants with atopic dermatitis. These human studies suggest that fructo-oligosaccharides exert preventive and therapeutic effects on atopic dermatitis in infants. Animal studies have also shown that fructo-oligosaccharide supplementation reduced the severity of atopic dermatitis-like skin inflammation and allergic contact dermatitis. Although the mechanism by which fructo-oligosaccharides affect the development of atopic dermatitis remains to be elucidated, the development of animal models would enable mechanistic studies. Previous animal studies showed that dietary fructo-oligosaccharides increased production of secretory immunoglobulin A and interleukin-10 in the gut, which may be advantageous in the prevention of food allergy. One possible mechanism by which fructo-oligosaccharides modulate the immune response is linked to gut microbiota. Recognition of microbe-associated molecular patterns and short-chain fatty acids (i.e. fermentation products of non-digestible oligosaccharides in the gut) by specific receptors in the gut may modulate immune responses. Another possible mechanism is that fructo-oligosaccharides may be absorbed intact in the gut and interact directly with immune cells outside the intestinal tract. Because the clinical data are currently insufficient to recommend prebiotics as part of a standard protocol in the prevention and treatment of atopic dermatitis, additional well-designed prospective clinical trials and mechanistic studies using appropriate animal models are needed to advance knowledge in this field.

Keywords: allergy, atopic dermatitis, gut microbiota, hygiene hypothesis, prebiotics

Abbreviations

AD	Atopic dermatitis
AOS	Acidic oligosaccharide
CHS	Contact hypersensitivity
DNFB	2,4-dinitrofluorobenzene
DP	Degree of polymerization
FOS	Fructo-oligosaccharide
GOS	Galacto-oligosaccharide
Ig	Immunoglobulin
IL	Interleukin
RT-qPCR	Real time-quantitative PCR
SCFA	Short-chain fatty acid
SIgA	Secretory IgA

27.1 Introduction

Several hypotheses have been proposed to explain the rise in allergic diseases, including AD, over the last decades in developed countries. Among them, the most conspicuous is the hygiene hypothesis, which suggests that improved hygiene and decreased exposure of the immune system to microbes in early life could be an important environmental aetiology of the allergy (Strachan, 1989). Among the microbes that influence host immunity, indigenous gut microbiota early in life is thought to play an important role in the pathogenesis of allergic diseases (Van der Aa *et al.*, 2010). This idea is supported by epidemiological data demonstrating that differences in the composition of gut microbiota in infancy precede the development of AD (reviewed in Van der Aa *et al.*, 2010). Babies who developed allergy were less often colonized with enterococci during the first month of life, and with bifidobacteria during the first year of life, when compared with healthy babies. In addition, allergic infants had higher clostridial counts and increased prevalence of *Staphylococcus aureus* colonization, whereas the counts of *Bacteroides* were lower. Furthermore, *Escherichia coli* and *Clostridium difficile* were reportedly associated with infants who developed AD. Therefore, strategies to manipulate the gut microbiota in infancy have been considered in preventing the development of allergic diseases (Van der Aa *et al.*, 2010). Indeed, clinical trials showed that maternal administration of *Lactobacillus rhamnosus* GG (i.e. probiotics) during pregnancy and lactation was beneficial in preventing the development of AD in at-risk children during the first 7 years of life (Kalliomäki *et al.*, 2007).

Breast milk is protective against AD (Gdalevich *et al.*, 2001). Breast-fed infants have a relatively simple composition of gut microbiota dominated by bifidobacteria, whereas formula-fed infants have a more adult-type diverse microbiota in which *Enterobacteriaceae*, *Clostridium*, and *Bacteroides* predominate (Harmsen *et al.*, 2000). This difference in the composition of gut microbiota is possibly involved in the lower prevalence of AD in breast-fed infants. The predominance of bifidobacteria in the gut microbiota of breast-fed infants is thought to result from the fermentation of [in human milk. Therefore, the supplementation of non-digestible oligosaccharides (i.e. prebiotics) to infant formula could increase bifidobacteria in the gut, which may in turn protect against AD. In addition, enrichment of bifidobacteria indigenous to the host by non-digestible oligosaccharides might be more effective than administration of probiotics exotic to the host, because exotic bacteria cannot colonize or live long in the host gut. This article describes the prevention and treatment of AD by supplementation of a non-digestible oligosaccharide, FOS.

27.2 Fructo-oligosaccharides

FOS (Figure 27.1) can be produced by enzymatic hydrolysis of inulin, a polymer of D-fructose linked by β(2,1) bonds with a terminal α(1,2) linked D-glucose (Roberfroid, 2007). Inulin occurs naturally in several edible fruits and vegetables, and chicory is most commonly used for the extraction of inulin. The DP of native chicory inulin varies from 2 to 60, with an average of 12. The enzymatic hydrolysis of inulin using an endoinulinase (EC 3.2.1.7) produces a mixture of FOS in which the DP varies from 2 to 8, with an average of 4. Alternatively, FOS can be obtained by transfructosylation action of the fungal enzyme β-fructosidase (EC 3.2.1.7) from *Aspergillus niger* on sucrose. The DP of synthesized FOS varies from 2 to 4, with an average of 3.6.

Because inulin-type fructans, including FOS, resist hydrolysis by small intestine digestive enzymes, they are classified as non-digestible oligosaccharides (Roberfroid, 2007). Upon reaching the colon, they are broken down specifically by bifidobacteria. Thus, bifidobacteria have a competitive advantage for the inulin-type fructans in a mixed culture environment such as the human gut (Roberfroid, 2007). Indeed, human studies have demonstrated that consumption of

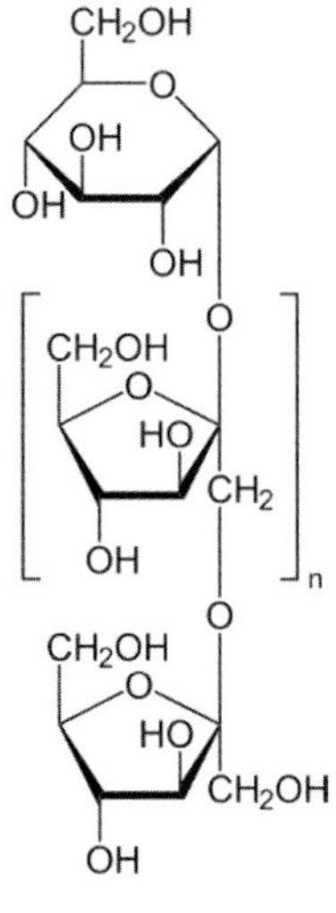

Figure 27.1. Structure of inulin-type fructans, n = 1-60.

inulin-type fructans, including FOS, increased faecal bifidobacteria (reviewed in Roberfroid, 2007). Therefore, FOS is a candidate formula milk supplement that can reproduce the bifidogenic effect of human milk oligosaccharides.

Based on the analysis of human milk oligosaccharides, Boehm *et al.* (2004) developed an oligosaccharide mixture consisting of 90% low-molecular mass GOS with DP<4 and 10% high-molecular mass FOS with DP>10. Several studies compared the effect of a cow's milk formula supplemented with the GOS/FOS mixture, unsupplemented formula, and human breast milk on the composition of the faecal bacterial flora in infants. The results showed that the proportion of bifidobacteria in the GOS/FOS group was significantly higher than that in the unsupplemented control group and was comparable to that in the breast milk group (reviewed in Scholtens *et al.* 2008). As described below, human feeding studies have been performed to examine the effect of supplementation with the GOS/FOS mixture on the development of AD.

27.3 Effect of FOS on AD – human studies

Moro *et al.* (2006) conducted a prospective, double-blind, randomized, placebo controlled trial to test whether supplementation with the GOS/FOS mixture affected the incidence of AD in at-risk infants. Subjects were 259 term infants with a maternal history of AD, allergic rhinitis, or asthma. They received a hypoallergenic formula with extensively hydrolysed cows' milk whey protein supplemented either with the GOS/FOS mixture (8 g/l) or maltodextrin (8 g/l) as placebo for 6 months. A total of 102 infants in the GOS/FOS group and 104 infants in the placebo group completed the study. The cumulative incidence of AD during the feeding period was significantly lower in the GOS/FOS group (9.8%) than in the placebo group (23.1%) (Figure 27.2). In addition, faecal bifidobacteria counts at 3 and 6 months of age were significantly higher in the GOS/FOS group than in the placebo group. This study suggests that supplementation with the GOS/FOS mixture in the hypoallergenic formula is protective against AD in at-risk infants, probably through expansion of intestinal bifidobacteria.

Arslanoglu *et al.* (2008) reported the results of a follow up of the study by Moro *et al.* (2006). Of 152 participants, 134 infants (68 in placebo, 66 in GOS/FOS group) completed the follow-up. At age 2, the cumulative incidence of AD, recurrent wheezing, and allergic urticaria was still significantly lower in the GOS/FOS group (13.6%, 7.6%, and 1.5%, respectively) than in the placebo group (27.9%, 20.6%, and 10.3%, respectively) (Figure 27.3), suggesting that the protective effect of the GOS/FOS mixture continues beyond the intervention period in at-risk infants (Arslanoglu *et al.*, 2008).

Although a positive family history for atopic diseases is the strongest predictor for the development of AD, the majority of children who develop AD come from families with a negative family history (Bergmann *et al.*, 1998). Therefore, Grüber *et al.* (2010) conducted a prospective, double-blind, randomized, placebo controlled trial to test whether infants with a negative family history, as did infants with a positive family history, respond to the prebiotic supplementation. Healthy

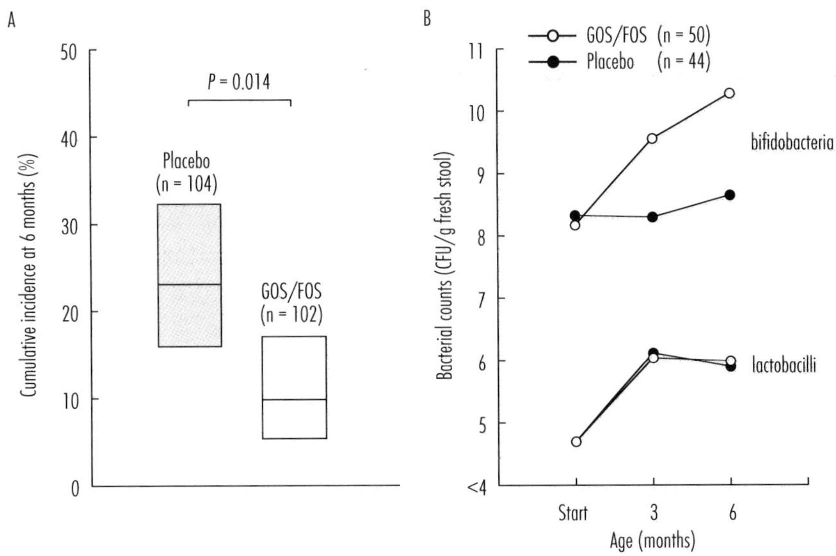

Figure 27.2 Cumulative incidence of atopic dermatitis at 6 months of age (A) and changes in faecal bifidobacteria and lactobacilli counts (B) in at-risk infants fed a hypoallergenic cow's milk formula supplemented with the galacto-oligosaccharides/fructo-oligosaccharides mixture or placebo (maltodextrins). Data are presented as means with 95% confidence intervals and medians in charts A and B, respectively.
Adapted from Moro *et al.* (2006).

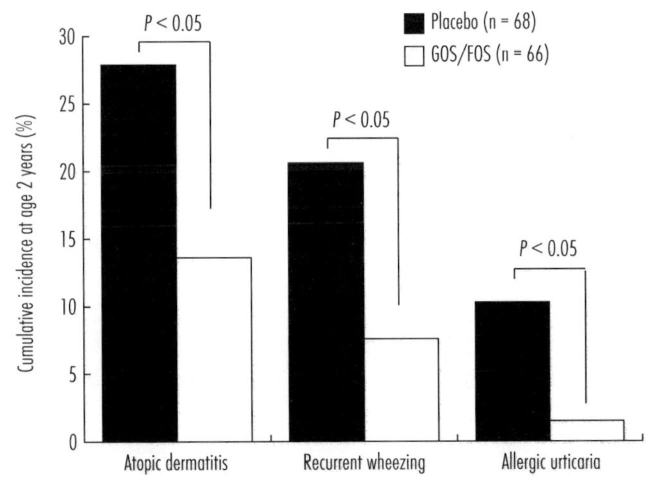

Figure 27.3. Cumulative incidence of allergic diseases at the end of 2-year follow up period in the galacto-oligosaccharides/fructo-oligosaccharides mixture and placebo groups of at-risk infants. Data are presented as means.
Adapted from Arslanoglu *et al.* (2008).

term infants with a negative family history from Austria, Germany, Italy, The Netherlands, and Switzerland were recruited before the age of 8 weeks. They received a regular cow's milk formula supplemented either with (414 infants) or without (416 infants) a combination of the GOS/FOS mixture (6.8 g/l) and pectin-derived acidic oligosaccharides (AOS, 1.2 g/l). In addition, a total of 300 infants exclusively breast-fed were included in the nonrandomized reference group. A total of 361, 374, and 266 infants in the GOS/FOS/AOS, control, and breast-fed groups, respectively, completed the observation up to the first birthday. The cumulative incidence of AD during the first year of life was significantly lower in the GOS/FOS/AOS group (5.7%) than in the control group (9.7%)(Figure 27.4). The incidence of AD in the GOS/FOS/AOS group was in the low range of the breast-fed group (7.3%). This study suggests that supplementation of the GOS/FOS/AOS mixture is protective against AD in low-risk infants. It remains unclear to what extent the addition of AOS to the GOS/FOS mixture contributes to a clinical advantage in the primary prevention of AD (Grüber *et al.*, 2010).

Grüber *et al.* (2010) reported that there were no differences in serum total and hen's egg- or cow's milk-specific immunoglobulin (Ig)E levels between the GOS/FOS/AOS, control, and breast-fed groups at age 6 and 12 months. However, Van Hoffen *et al.* (2008) reported conflicting results. They analysed the immune response in infants enrolled in the study conducted by Moro *et al.* (2006). At 6 months of age, plasma total IgE, IgG1, IgG2, and IgG3, and cow's milk-specific IgG1 levels were significantly lower in the GOS/FOS group (41 infants) than in the placebo group (43 infants) (Van Hoffen *et al.*, 2008). Interestingly, no differences between the groups were detected with respect to the vaccination response towards diphtheria, tetanus, and polio. Van Hoffen *et al.*

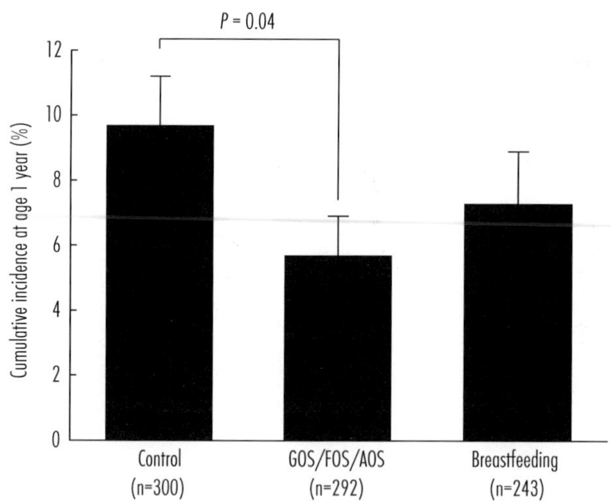

Figure 27.4. Cumulative incidence of atopic dermatitis at age 1 year in low-risk infants fed breast milk or standard cow milk formula supplemented with or without the galacto-oligosaccharides/fructo-oligosaccharides/acidic oligosaccharides mixture. Data are presented as means with standard errors.
Adapted from Grüber *et al.* (2010).

(2008) mentioned that GOS/FOS supplementation might be specifically able to alter the response towards antigens in the intestinal tract. Further investigations are needed to clarify the effect of GOS/FOS supplementation on the immune response in infants.

Shibata *et al.* (2009) reported the therapeutic effect of kestose (β-D-fructofuranosyl-(2,1)-β-D -fructofuranosyl-(2,1)-α-D-glucopyranoside, DP2) on established AD in Japan. They conducted a randomized, double-blind, placebo controlled trial. Infants with AD were randomized to kestose (15 infants) and placebo (14 infants) groups, and received 1~2 g of kestose and maltose, respectively, every day for 12 weeks. Disease severity (SCORAD score) was significantly lower in the kestose group than in the placebo group on week 6 and week 12 (Figure 27.5). This is the first study showing the therapeutic effect of FOS on AD. In contrast, the severity of AD was not affected by the supplementation of GOS/FOS or GOS/FOS/AOS in the feeding trials described above (Arslanoglu *et al.*, 2008; Grüber *et al.*, 2010; Moro *et al.*, 2006). With regard to these data, Grüber *et al.* (2010) indicated that prebiotics might be less effective once the disease has been established.

In a trial conducted by Shibata *et al.* (2009), kestose administration failed to increase the faecal bifidobacteria count estimated by genus-specific real time-quantitative PCR (RT-qPCR) based on the 16S rRNA gene sequence, suggesting that a mechanism other than an alteration in the bifidobacteria population as a whole played a major role in the improvement of clinical symptoms in AD patients treated with kestose. They indicated that an increase of particular *Bifidobacterium* species, rather than the entire genus, might explain why the increase of total bifidobacteria count

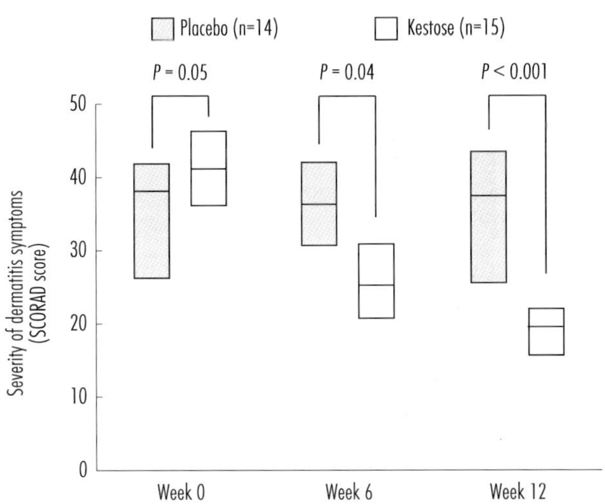

Figure 27.5. Changes in the severity of atopic dermatitis in infants fed kestose or placebo (maltose). Data are presented as medians with interquartile ranges.
Adapted from Shibata *et al.* (2009).

was rather minimal in comparison with the significant improvement of the SCORAD score in the kestose-treated subjects (Shibata *et al.*, 2009).

27.4 Effect of FOS on AD: animal studies

Several human interventions demonstrated the primary prevention and severity improvement of AD in infants by administration of FOS or oligosaccharide mixtures containing FOS as described above. However, the mechanism by which FOS affects the development of AD remains to be elucidated. Therefore, the development of animal models would enable mechanistic studies.

Fujiwara *et al.* (2010b) examined the effects of maternal dietary supplementation with FOS on the development of AD-like skin inflammation in the offspring of NC/Nga mice. NC/Nga mice provide an excellent animal model for human AD (Kawamoto and Matsuda, 2003). When NC/Nga mice are kept in air-uncontrolled conventional surroundings, they develop skin lesions. In contrast, when kept in a specific pathogen-free room, they exhibit no clinical signs. Therefore, environmental allergens such as mite antigens are thought to contribute to the development of skin lesions. Clinical signs begin with scratching behavior and, starting at the age of 8 weeks, IgE elevation, followed by the onset of eczematous conditions along with the infiltration of various inflammatory cells in the skin lesions (Kawamoto and Matsuda, 2003). Elevated expression of Th2 cytokines and chemokines is observed in lesional skin areas (Kawamoto and Matsuda, 2003). Previously, Fujiwara *et al.* (2008) demonstrated that supplementation with FOS (50 g/kg diet) in female BALB/c mice during pregnancy and lactation altered the composition of gut microbiota in their sucklings (Figure 27.6). This finding provided the impetus for the study examining whether modulation of gut microbiota by FOS in infancy influences the development of AD-like skin inflammation later in life. In order to test this, Fujiwara *et al.* (2010b) fed female NC/Nga mice diets either with or without FOS supplementation (50 g/kg diet) during pregnancy and lactation. The FOS was composed of D-glucose and D-fructose (1.3%), sucrose (2.5%), kestose (DP2, 37.3%), nystose (DP3, 49.1%), and fructosylnystose (DP4, 9.8%). After weaning, offspring were fed the diets supplemented with or without FOS for 11 weeks in an air-uncontrolled conventional room. Although maternal supplementation with FOS suppressed the increase in skin severity score and scratching behavior in offspring, dietary FOS after weaning was less effective (Figure 27.7). The diminution of skin lesions was accompanied by lower serum concentrations of total IgG1 and lower expression levels of tumor necrosis factor-α in the lesional tissue. These data indicate that modulation of the gut microbiota in infancy by maternal consumption of FOS diminished the severity of AD-like skin inflammation in the offspring of NC/Nga mice. Analysis of 16S rRNA gene profiles in the faeces of offspring suggested that maternal supplementation with FOS modulated the gut microbiota in sucklings, and that the gut microbiota in offspring after weaning was controlled by the diet consumed. Therefore, this study supports the idea that the time point for modulation of gut microbiota is important in preventing the development of AD.

Allergic contact dermatitis is one of the most prevalent human skin diseases. This pathological condition arises after CHS. CHS is a T cell-mediated, antigen-specific type of skin inflammation

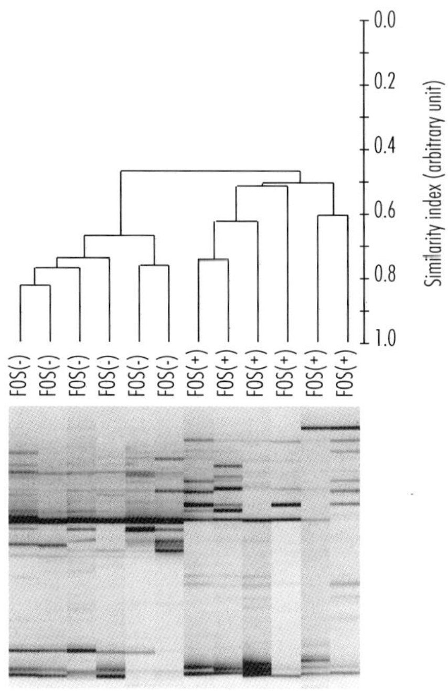

Figure 27.6. Similarity of the composition of caecal microbiota based on 16S rRNA gene sequences in Ng/Nga mice fed a semi-purified diet supplemented with (FOS(+)) or without (FOS(-)) fructo-oligosaccharides. Gel image of PCR-denaturing gradient gel electrophoresis and the dendrogram of band profiles are shown. Each lane in the gel and each line in the dendrogram represent an individual mouse.
Adapted from Fujiwara *et al.* (2010a).

that is induced by topical skin contact with haptens in a previously sensitized host. Watanabe *et al.* (2008) tested whether FOS supplementation affects the CHS in mice. Female BALB/c mice were fed a semi-purified diet with or without FOS supplementation (50 g/kg diet) for 3 weeks and were then epicutaneously immunized with DNFB. Afterwards, mice continued to receive their respective diets. At 5 days after immunization, the mice were ear challenged with the hapten. Ear swelling after the challenge was significantly reduced in the mice fed the diet supplemented with FOS compared to mice fed control diet. Dietary FOS altered the composition of gut microbiota analysed by PCR-denaturing gradient gel electrophoresis based on 16S rRNA gene sequence. The numbers of bifidobacteria were significantly higher in mice fed the FOS-supplemented diet than in mice fed the control diet, and ear swelling was negatively correlated with the numbers of bifidobacteria in the faeces.

Bifidobacterium pseudolongum was revealed to be the most predominant bifidobacteria in the intestine of mice fed the FOS-supplemented diet (Watanabe *et al.*, 2008); therefore, Sasajima *et al.* (2010) examined whether oral administration of *B. pseudolongum* affects CHS response. They isolated viable *B. pseudolongum* from mouse faeces and administered daily the bacterium (2×10^7 cells) to female BALB/c mice fed a semi-purified diet with or without FOS supplementation

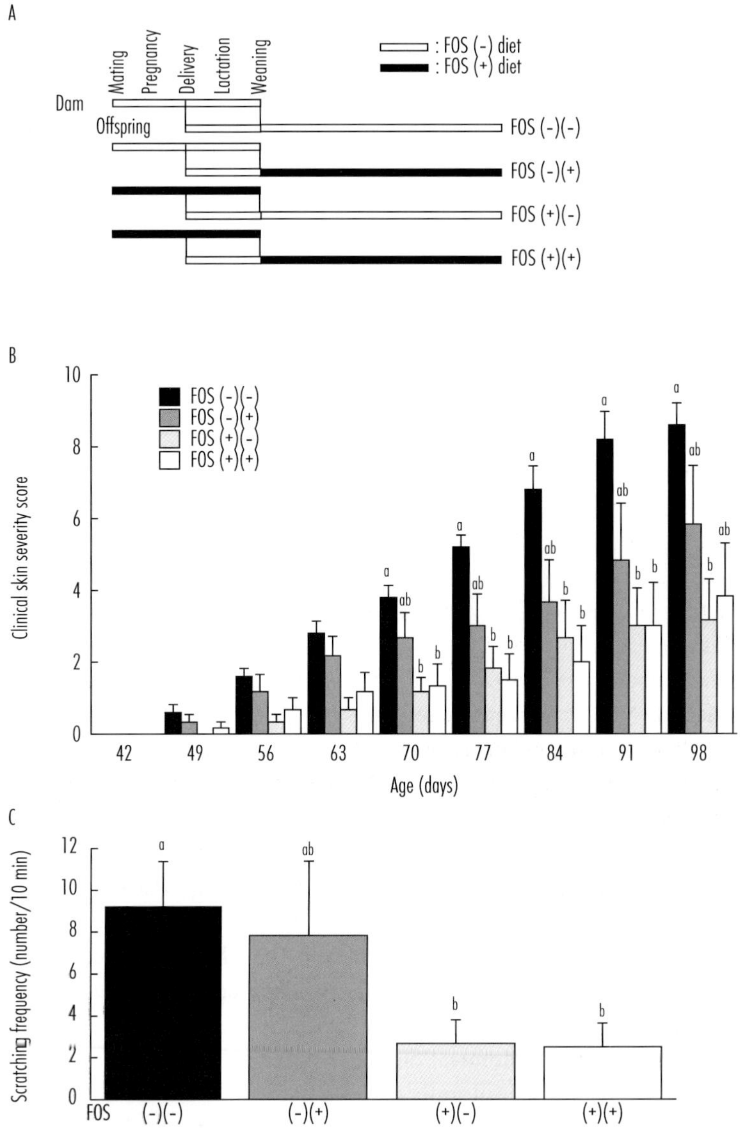

Figure 27.7. Effect of maternal consumption of fructo-oligosaccharides on the severity of atopic dermatitis-like skin lesions in the offspring of NC/Nga mice. (A) Schematic presentation of the experimental design. Female NC/Nga mice were fed a semi-purified diet supplemented with (FOS(+)) or without (FOS(-)) fructo-oligosaccharides during pregnancy and lactation. After weaning at age 21 days, offspring were fed either FOS(+) or FOS(-) for 11 weeks. FOS(+)-fed offspring whose dam was fed FOS(+) were referred to as FOS(+)(+). Thus, offspring were divided into four groups: FOS(+)(+), FOS(+)(-), FOS(-)(+) and FOS(-)(-). (B) Temporal changes in the skin severity score in the offspring. (C) Frequency of scratching behavior for 10 min at age 91 days in the offspring. Data are presented as means with standard errors. Mean values with different letters were significantly different ($P<0.05$). Adapted from Fujiwara *et al.* (2010b).

(50 g/kg diet). Two weeks after starting the test diets, mice received DNFB on the ear auricle twice at 7-day intervals. Although dietary FOS reduced the CHS response as demonstrated by ear swelling, *B. pseudolongum* administration resulted in a reduction in the initial phase only of the CHS response. *B. pseudolongum* administration increased hapten-specific IgG1, while dietary FOS decreased serum IgG2a. Administration of FOS and *B. pseudolongum* decreased interferon-γ production and increased IL-10 production in cervical lymph node cells restimulated with hapten *in vitro*. From these results, Sasajima *et al.* (2010) concluded that *B. pseudolongum* proliferation in the intestinal tract is partially responsible for the reduction in DNFB-induced CHS response by dietary supplementation with FOS in mice, which may be mediated by the modulation of antigen-induced cytokine production.

Fujiwara *et al.* (2010a) tested whether maternal dietary supplementation with FOS affects CHS in the offspring of NC/Nga mice. Initially, they showed that FOS supplementation (50 g/kg diet) reduced DNFB-induced CHS response as demonstrated by ear swelling in adult NC/Nga mice (Figure 27.8), which was consistent with the previous results in BALB/c mice (Watanabe *et al.*, 2008). Additionally, myeloperoxidase activity in the ear tissue was significantly lower in mice fed the FOS-supplemented diet, suggesting reduced neutrophil infiltration in the inflamed ear auricles by dietary FOS. RT-qPCR analysis showed that mRNA levels for IL-10, IL-12p40, and IL-17 in the lesional ear skin were significantly lower in mice fed FOS. Thus, dietary FOS might reduce local inflammatory responses of DNFB-induced CHS in NC/Nga mice (Fujiwara *et al.*, 2010a). Next, female NC/Nga mice were fed diets either with or without FOS during pregnancy and lactation. After weaning, offspring were fed the diets supplemented with or without FOS. Three weeks after weaning, offspring received DNFB on the ear auricle 4 times at 7-day intervals. Although FOS supplementation after weaning reduced ear swelling, maternal FOS consumption was ineffective in offspring (Figure 27.9). Thus, the findings are in contrast to another observation that the severity of AD-like skin inflammation in the offspring of NC/Nga mice was reduced by maternal FOS consumption, and to a lesser extent, by post-weaning FOS consumption (Fujiwara *et al.*, 2010b). Although further investigations are needed to resolve this discrepancy, it is assumed that maternal FOS consumption and, perhaps modulation of gut microbiota in infancy by maternal FOS consumption, might have different impacts on the development of symptoms in different types of allergic reaction in the offspring.

Together, the animal studies described here successfully demonstrated that FOS supplementation affects the development of allergic skin inflammation. The results reproduce the effects observed in human studies in which FOS supplementation prevented or improved the AD (Arslanoglu *et al.*, 2008, Gruber *et al.*, 2010, Moro *et al.*, 2006, Shibata *et al.*, 2009). Therefore, these animal models would contribute to the elucidation of the mechanism for prevention and improvement of AD by FOS supplementation in humans.

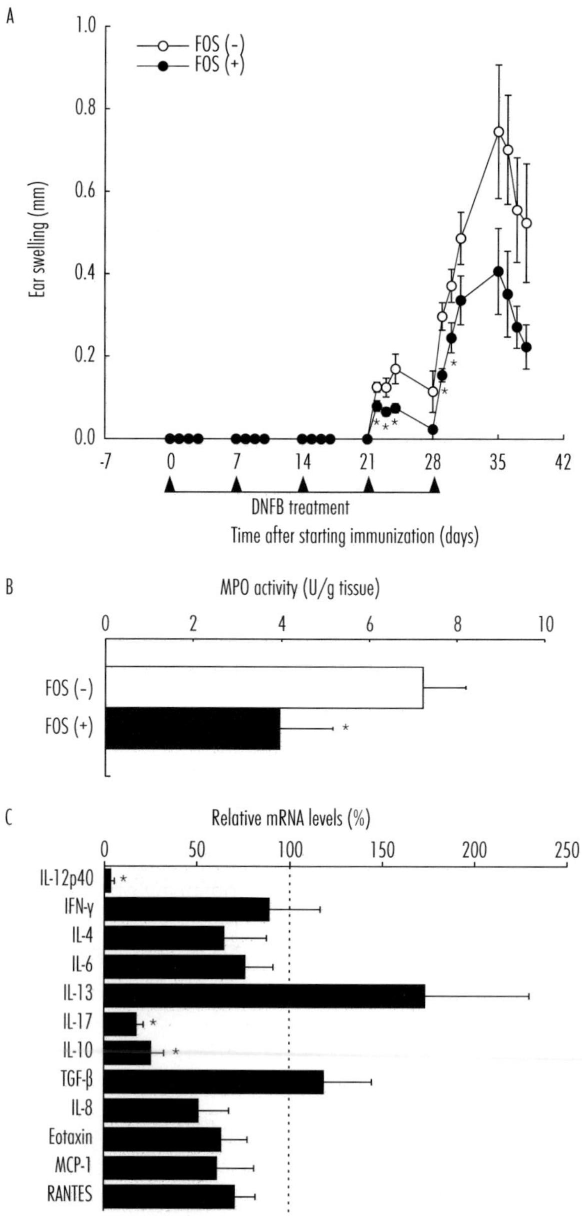

Figure 27.8. Contact hypersensitivity response in adult NC/Nga mice fed a semi-purified diet supplemented with (FOS(+)) or without (FOS(-)) fructo-oligosaccharides. (A) Temporal changes in ear swelling after starting DNFB application. (B) Myeloperoxidase activity in the ear auricles. (C) Relative mRNA levels for cytokines and chemokines in the inflamed ear tissue, and values of mice fed FOS(+) are shown relative to the levels in mice fed FOS(-). Data are presented as means with standard errors. Mean values with asterisks were significantly different ($P<0.05$).

Adapted from Fujiwara *et al.* (2010a).

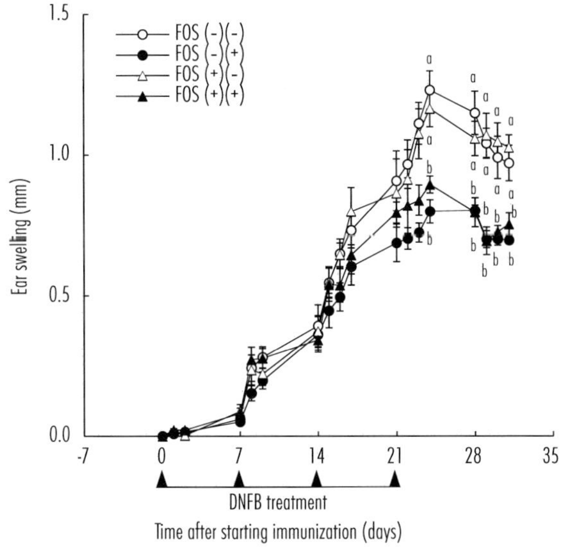

Figure 27.9. Effect of maternal consumption of fructo-oligosaccharides on the contact hypersensitivity response in the offspring of NC/Nga mice. Temporal changes in ear swelling after starting 2,4-dinitrofluorobenzene application are shown. Data are presented as means with standard errors. Mean values with different letters were significantly different ($P<0.05$).
Adapted from Fujiwara *et al.* (2010a).

27.5 Mechanisms

Several studies have reported on the modulation of immune response by dietary FOS in animals. Hosono *et al.* (2003) demonstrated that dietary supplementation with FOS (DP2~4, 25 g/kg diet) increased the SIgA excretion in faeces and IgA production in cultured Peyer's patch lymphocytes in response to stimulation with a bifidobacterial homogenate in BALB/c mice. In keeping with this, Nakamura *et al.* (2004) showed that feeding a purified diet supplemented with FOS (DP2~4, 50 g/kg diet) increased IgA concentrations and polymeric Ig receptor expression in the small and large intestinal tissues, as well as ileal SIgA secretion, and IgA^+ cells in Peyer's patches in BALB/c mice. In addition, Roller *et al.* (2004) reported that supplementation with Synergy (100 g/kg diet), a mixture of FOS and inulin, increased the IgA concentrations in the caecal contents of F344 rats. Furthermore, IL-10 production was higher in cultured Peyer's patch cells of FOS-fed animals (Hosono *et al.*, 2003; Roller *et al.*, 2004). SIgA plays a role in maintaining a noninflammatory local homeostasis as well as mucosal barrier function in the gut (Corthësy, 2009), and IL-10 is one of the key players in the oral immune tolerance induction (Kunisawa and Kiyono, 2010). Although further studies are required to clarify if the modulated immune responses described here are involved in the prevention and improvement of AD, elevated production of SIgA and IL-10 in the gut may be advantageous in the prevention of food allergy; AD is often associated with food allergy in children.

As previously mentioned, one possible mechanism by which FOS modulates immune response is linked to gut microbiota. FOS-induced changes in the composition of gut microbiota may alter the presence of microbe-associated molecular patterns in the gut and, through pattern recognition receptors such as Toll-like receptors, local immune cells may respond to these molecular motifs (Akira *et al.*, 2001). In addition, SCFAs, fermentation products of non-digestible oligosaccharides such as FOS in the gut, may mediate the modulation of immune response. As G-protein-coupled receptors (GPR41 and GPR43) in immune and non-immune cells respond to SCFA (Brown *et al.*, 2005), increases in SCFA by consumption of FOS may activate these cells via GPR41 and GPR43, which may in turn modulate immune responses.

Another possible mechanism is that FOS may directly interact with the immune cells outside the intestinal tract. Obermeier *et al.* (1999) demonstrated that oligosaccharides in breast milk are partially absorbed intact in the gut of breast-fed infants. In addition, Watanabe *et al.* (2004) reported that 60 and 40 μmol/l raffinose (β-D-fructofuranosyl-*O*-α-D-galactopyranosyl-(1,6)- α -D-glucopyranoside) were detected in portal venous and abdominal arterial plasma, respectively, after the intragastric administration of this non-digestible oligosaccharide in rats. Furthermore, intraperitoneal administration of raffinose, as well as dietary supplementation with raffinose, reportedly reduced allergic airway inflammation in ovalbumin-sensitized Brown Norway rats (Watanabe *et al.*, 2004). From these findings, it is speculated that non-digestible oligosaccharides may be partially absorbed intact in the gut and then modulate the immune responses through direct interaction with the immune cells. However, it is unclear whether FOS are absorbed intact in the gut.

27.6 Conclusion

Prebiotics such as FOS have been increasingly gaining attention for counteracting the current rise of allergic diseases including AD. Although human trials so far demonstrated the primary prevention and improvement of AD by supplementation with a FOS-containing oligosaccharide mixture and kestose in infants, the data are currently insufficient to recommend prebiotics as part of a standard protocol in the prevention and treatment of AD. Clearly, additional well-designed prospective clinical trials, as well as mechanistic studies using appropriate animal models, are required to advance knowledge in this field.

References

Akira, S., Takeda, K. and Kaisho, T., 2001. Toll-like receptors: critical proteins linking innate and acquired immunity. Nature Immunology 2, 675-680.

Arslanoglu, S., Moro, G.E., Schmitt, J., Tandoi, L., Rizzardi, S. and Boehm, G., 2008. Early dietary intervention with a mixture of prebiotic oligosaccharides reduces the incidence of allergic manifestations and infections during the first two years of life. The Journal of Nutrition 138, 1091-1095.

Bergmann, R.L., Bergmann, K.E. and Wahn, U., 1998. Can we predict atopic disease using perinatal risk factors? Clinical and Experimental Allergy 28, 905-907.

Boehm, G., Jelinek, J., Stahl, B., van Laere, K., Knol, J., Fanaro, S., Moro, G. and Vigi, V., 2004. Prebiotics in infant formulas. Journal of Clinical Gastroenterology 38 (Suppl), S76-S79.

Brown, A.J., Jupe, S. and Briscoe, C.P., 2005. A family of fatty acid binding receptors. DNA and Cell Biology 24, 54-61.

Corthësy, B., 2009. Secretory immunoglobulin A: well beyond immune exclusion at mucosal surfaces. Immunopharmacology and Immunotoxicology 31, 174-179.

Fujiwara, R., Sasajima, N., Takemura, N., Ozawa, K., Nagasaka, Y., Okubo, T., Sahasakul, Y., Watanabe, J. and Sonoyama, K., 2010a. 2,4-Dinitrofluorobenzene-induced contact hypersensitivity response in NC/Nga mice fed fructo-oligosaccharide. Journal of Nutritional Science and Vitaminology (Tokyo) 56, 260-265.

Fujiwara, R., Takemura, N., Watanabe, J. and Sonoyama, K., 2010b. Maternal consumption of fructo-oligosaccharide diminishes the severity of skin inflammation in offspring of NC/Nga mice. British Journal of Nutrition 103, 530-538.

Fujiwara, R., Watanabe, J. and Sonoyama, K., 2008. Assessing changes in composition of intestinal microbiota in neonatal BALB/c mice through cluster analysis of molecular markers. British Journal of Nutrition 99, 1174-1177.

Gdalevich, M., Mimouni, D., David, M. and Mimouni, M., 2001. Breast-feeding and the onset of atopic dermatitis in childhood: a systematic review and meta-analysis of prospective studies. Journal of the American Academy of Dermatology 45, 520-527.

Grüber, C., Van Stuijvenberg, M., Mosca, F., Moro, G., Chirico, G., Braegger, C.P., Riedler, J., Boehm, G. and Wahn, U.; MIPS 1 Working Group., 2010. Reduced occurrence of early atopic dermatitis because of immunoactive prebiotics among low-atopy-risk infants. Journal of Allergy and Clinical Immunology 126, 791-797.

Harmsen, H.J., Wildeboer-Veloo, A.C., Raangs, G.C., Wagendorp, A.A., Klijn, N., Bindels, J.G. and Welling, G.W., 2000. Analysis of intestinal flora development in breast-fed and formula-fed infants by using molecular identification and detection methods. Journal of Pediatric Gastroenterology and Nutrition 30, 61-67.

Hosono, A., Ozawa, A., Kato, R., Ohnishi, Y., Nakanishi, Y., Kimura, T. and Nakamura, R., 2003. Dietary fructooligosaccharides induce immunoregulation of intestinal IgA secretion by murine Peyer's patch cells. Bioscience, Biotechnology, and Biochemistry 67, 758-764.

Kalliomäki, M., Salminen, S., Poussa, T. and Isolauri, E., 2007. Probiotics during the first 7 years of life: a cumulative risk reduction of eczema in a randomized, placebo-controlled trial. Journal of Allergy and Clinical Immunology 119, 1019-1021.

Kawamoto, K. and Matsuda, H., 2003. Spontaneous mouse model of atopic dermatitis in NC/Nga mice. In: Chan, L.S. (ed.), Animal Models of Human Inflammatory Skin Diseases, CRC Press, San Diego, CA, USA, pp. 372-384.

Kunisawa, J. and Kiyono, H., 2010. Aberrant interaction of the gut immune system with environmental factors in the development of food allergies. Current Allergy and Asthma Reports 10, 215-221.

Moro, G., Arslanoglu, S., Stahl, B., Jelinek, J., Wahn, U. and Boehm, G., 2006. A mixture of prebiotic oligosaccharides reduces the incidence of atopic dermatitis during the first six months of age. Archives of Disease in Childhood 91, 814-819.

Nakamura, Y., Nosaka, S., Suzuki, M., Nagafuchi, S., Takahashi, T., Yajima, T., Takenouchi-Ohkubo, N., Iwase, T. and Moro, I., 2004. Dietary fructooligosaccharides up-regulate immunoglobulin A response and polymeric immunoglobulin receptor expression in intestines of infant mice. Clinical and Experimental Immunology 137, 52-58.

Obermeier, S., Rudloff, S., Pohlentz, G., Lentze, M.J. and Kunz, C., 1999. Secretion of ^{13}C-labelled oligosaccharides into human milk and infant's urine after an oral [^{13}C]galactose load. Isotopes in Environmental and Health Studies 35, 119-125.

Roberfroid, M.B., 2007. Inulin-type fructans: functional food ingredients. The Journal of Nutrition 137 (Suppl), 2493S-2502S.

Roller, M., Rechkemmer, G. and Watzl, B., 2004. Prebiotic inulin enriched with oligofructose in combination with the probiotics *Lactobacillus rhamnosus* and *Bifidobacterium lactis* modulates intestinal immune functions in rats. The Journal of Nutrition 134, 153-156.

Sasajima, N., Ogasawara, T., Takemura, N., Fujiwara, R., Watanabe, J. and Sonoyama, K., 2010. Role of intestinal *Bifidobacterium pseudolongum* in dietary fructo-oligosaccharide inhibition of 2,4-dinitrofluorobenzene-induced contact hypersensitivity in mice. British Journal of Nutrition 103, 539-548.

Scholtens, P.A., Alliet, P., Raes, M., Alles, M.S., Kroes, H., Boehm, G., Knippels, L.M., Knol, J. and Vandenplas, Y., 2008. Fecal secretory immunoglobulin A is increased in healthy infants who receive a formula with short-chain galacto-oligosaccharides and long-chain fructo-oligosaccharides. The Journal of Nutrition 138, 1141-1147.

Shibata, R., Kimura, M., Takahashi, H., Mikami, K., Aiba, Y., Takeda, H. and Koga, Y., 2009. Clinical effects of kestose, a prebiotic oligosaccharide, on the treatment of atopic dermatitis in infants. Clinical and Experimental Allergy 39, 1397-1403.

Strachan, D.P., 1989. Hay fever, hygiene, and household size. British Medical Journal 299, 1259-1260.

Van der Aa, L.B., Heymans, H.S., van Aalderen, W.M. and Sprikkelman, A.B., 2010. Probiotics and prebiotics in atopic dermatitis: review of the theoretical background and clinical evidence. Pediatric Allergy and Immunology 21, e355-e367.

Van Hoffen, E., Ruiter, B., Faber, J., M'Rabet, L., Knol, E.F., Stahl, B., Arslanoglu, S., Moro, G., Boehm, G. and Garssen, J., 2009. A specific mixture of short-chain galacto-oligosaccharides and long-chain fructo-oligosaccharides induces a beneficial immunoglobulin profile in infants at high risk for allergy. Allergy 64, 484-487.

Watanabe, H., Sonoyama, K., Watanabe, J., Yamaguchi, N., Kikuchi, H., Nagura, T., Aritsuka, T., Fukumoto, K. and Kasai, T., 2004. Reduction of allergic airway eosinophilia by dietary raffinose in Brown Norway rats. British Journal of Nutrition 92, 247-255.

Watanabe, J., Sasajima, N., Aramaki, A. and Sonoyama, K., 2008. Consumption of fructo-oligosaccharide reduces 2,4-dinitrofluorobenzene-induced contact hypersensitivity in mice. British Journal of Nutrition 100, 339-346.

Index

Index

W

X

Y

Z

About the editor

Victor R. Preedy BSc, PhD, DSc, FSB, FRCPath, FRSPH is Professor of Nutritional Biochemistry, King's College London, Professor of Clinical Biochemistry, Kings College Hospital (Honorary) and Director of the Genomics Centre, King's College London. Presently he is a member of the Kings College London School of Medicine. Professor Preedy graduated in 1974 with an Honours Degree in Biology and Physiology with Pharmacology. He gained his University of London PhD in 1981 when he was based at the Hospital for Tropical Disease and The London School of Hygiene and Tropical Medicine. In 1992, he received his Membership of the Royal College of Pathologists and in 1993 he gained his second doctoral degree, i.e. DSc, for his outstanding contribution to protein metabolism in health and disease. Professor Preedy was elected as a Fellow to the Institute of Biology (FIBiol) in 1995 and to the Royal College of Pathologists in 2000. Since then he has been elected as a Fellow to the Royal Society for the Promotion of Health (FRSH; 2004) and The Royal Institute of Public Health (FRIPHH; 2004). In 2009, Professor Preedy became a Fellow of the Royal Society for Public Health. In his career Professor Preedy has carried out research at the National Heart Hospital (part of Imperial College London) and at the MRC Centre at Northwick Park Hospital. He has collaborated with research groups in Finland, Japan, Australia, USA and Germany. He is a leading expert on the mechanisms of disease and has lectured nationally and internationally. He has published over 570 articles, which includes over 165 peer-reviewed manuscripts based on original research, 90 reviews and over 20 books.